Operator Theory
Advances and Applications
Vol. 73

Editor
I. Gohberg

Nonselfadjoint Operators and Related Topics

Workshop on Operator Theory and Its Applications, Beersheva, February 24–28, 1992

Edited by

A. Feintuch
I. Gohberg

Springer Basel AG

Volume Editorial Office:

Raymond and Beverly Sackler Faculty of Exact Sciences
School of Mathematical Sciences
Tel Aviv University
IL-69978 Tel Aviv
Israel

A CIP catalogue record for this book is available from the Library of Congress, Washington D.C., USA

Deutsche Bibliothek Cataloging-in-Publication Data
Nonselfadjoint operators and related topics / Workshop on Operator Theory and Its Applications, Beersheva, February 24 – 28, 1992. Ed. by A. Feintuch ; I. Gohberg. [Vol. ed. office: Raymond and Beverly Sackler Faculty of Exact Sciences, School of Mathematical Sciences, Tel Aviv University]. – Basel ; Boston ; Berlin : Birkhäuser, 1994
(Operator theory ; Vol. 73)
ISBN 978-3-0348-9663-4 ISBN 978-3-0348-8522-5 (eBook)
DOI 10.1007/978-3-0348-8522-5

NE: Feintuch, Avraham [Hrsg.]; Bêt has-Sēfer le-Maddā'ê ham-Mātēmāṭ¬¬24qā <Rāmat-Āv¬¬24v> / haf- Fāqûlṭā le-Maddā'¬¬24m Medûyyāq¬¬24m'al Šēm Raymônd û-Beverly Saqler; Workshop on Operator Theory and Its Applications <1992, Be'ēr Ševa'>; GT

Originally published by Birkhäuser Verlag in 1994
Camera-ready copy prepared by the editors
Printed on acid-free paper produced from chlorine-free pulp
Cover design: Heinz Hiltbrunner, Basel

ISBN 978-3-0348-9663-4

9 8 7 6 5 4 3 2 1

Table of Contents

EDITORIAL INTRODUCTION

This volume presents the Proceedings of the joint U.S. Israel Workshop on Operator Theory and its Applications, held February 24 to 28, 1992, at the Ben Gurion University of the Negev, Beersheva. This event was sponsored by the United States-Israel Binational Science Foundation and the Ben-Gurion University of the Negev, and many outstanding experts in operator theory participated in it. The workshop honoured Professor Emeritus Moshe Livsic on the occasion of his retirement.

The volume contains a selection of papers covering a wide range of topics in modern operator theory and its applications, from abstract operator theory to system theory and computers in operator models. The papers treat linear and nonlinear problems, and study operators from different abstract and concrete classes. Many of the topics are from the area in which the contributions of Moshe Livsic were extremely important.

Moshe Livsic, together with his family spent the past fifteen years in Israel. His presence in the Ben Gurion University of the Negev played an important role in turning this university into an international centre of operator theory and its applications. During this period he was mostly interested in the extension of the theory of single nonselfadjoint operators to the case of two or more commuting operators. This subject was the major focus of the weekly seminar in analysis of operator theory which was established by Moshe. Over the years this seminar has hosted many leading operator theorists, and it is well known internationally for its open atmosphere and lively discussions, instituted by Moshe.

A short time after settling in Israel Moshe began lecturing undergraduate courses in Hebrew. More than that, he was put in charge of the programme for talented high school students, a programme with which he remains involved to this day. A number of excellent young students were trained within this programme. One of these students was V. Vinnikov, who, at an early age, wrote his dissertation under Moshe's direction. Even after his retirment Moshe remains an active member of the department. He continues to take part in the weekly seminar and is writing a book with N. Kravitsky and A. Markus on the theory of commuting operators. (For a more detailed biography of Moshe Livsic see the book "Topics in Operator Theory and Interpolation," Birkhauser Verlag, OT 29, 1988.)

On behalf of the contributors to this volume we wish Moshe Livsic many years of good health and continuous activity, which is so important to all of us.

A. Feintuch I. Gohberg
February, 1994

Operator Theory:
Advances and Applications, Vol. 73

JOINT SPECTRUM AND DISCRIMINANT VARIETIES OF COMMUTING NONSELFADJOINT OPERATORS

M.S.Livšic and A.S. Markus

1 Introduction

The theory of commuting operators with finite-dimensional imaginary parts which has been developed during the last decade yielded fruitful connections with algebraic geometry. Let $A = (A_1, \ldots, A_n)$ be an n-tuple of commuting bounded operators with finite non-Hermitian rank in a Hilbert space H and let $A^* = (A_1^*, \ldots, A_n^*)$ be the adjoint n-tuple. We construct two sets Δ^{in} and Δ^{out} of polynomials of n complex variables $(z_1, \ldots, z_n)$ such that

$$p(A_1^*, \ldots, A_n^*)|\mathcal{H} = 0 \quad (p \in \Delta^{in}) \tag{1.1}$$

and

$$q(A_1, \ldots, A_n)|\mathcal{H} = 0 \quad (q \in \Delta^{out}) \tag{1.2}$$

where $\mathcal{H}$ is the so called "principal" subspace. The subspaces $\mathcal{H}$ and $H \ominus \mathcal{H}$ are invariant subspaces of operators $A_1, \ldots, A_n$ and $A_1^*, \ldots, A_n^*$. All the operators $A_1, \ldots A_n$ are selfadjoint on the subspace $H \ominus \mathcal{H}$. In many cases of interest this subspace is not important for the theory of nonselfadjoint operators and, in such cases, it can be omitted. We define two algebraic varieties D^{in} and D^{out} taking the intersections of all the possible varieties in $\mathbb{C}^n$ satisfying the equations

$$p(z_1, \ldots, z_n) = 0 \quad (p \in \Delta^{in}),$$

$$q(z_1, \ldots, z_n) = 0 \quad (q \in \Delta^{out})$$

respectively. These varieties are said to be the *discriminant* varieties and they play an important role in the problem of classification of commuting n-tuples $(A_1, \dots, A_n)$. It is known [6] that $D^{in} = D^{out}$ in the case $n = 2$. One of the main purposes of this work is to investigate the general case $n \geq 2$. *It turns out that in the case $n \geq 2$ these two discriminant varieties D^{in} and D^{out} coincide up to a finite number of isolated points.* This result can not be improved even in the case of finite-dimensional space H and $n = 3$.

In connection with the above problem it became necessary to investigate the joint spectrum of commuting operators with finite-dimensional imaginary parts. We consider the slightly more general case of compact imaginary parts.

The complement in $\mathbb{C}^n$ to the set of all regular points of the n-tuple $A = (A_1, \dots, A_n)$ is called the *joint spectrum* of this n-tuple and is denoted by $\sigma(A)$. Here we call a point $\lambda = (\lambda_1, \dots, \lambda_n)$ the *regular point* of the n-tuple A if there exist operators $B_1, \dots, B_n$ and $C_1, \dots, C_n$ such that

$$\sum_{k=1}^{n} B_k(A_k - \lambda_k I) = I, \ \sum_{k=1}^{n} (A_k - \lambda_k I) C_k = I. \tag{1.3}$$

These natural definitions belong to Harte [2]. There exist other definitions of joint spectrum but in the case we consider, all of them are equivalent. Note that in the case $dim H < \infty$ the joint spectrum of commuting operators $A_1, \dots, A_n$ coincides with the set of their joint eigenvalues, i.e. points $\lambda = (\lambda_1, \dots, \lambda_n)$ such that $A_k f = \lambda_k f$ $(k = 1, \dots, n)$ for some nonzero vector f (joint eigenvector).

The joint spectrum of commuting operators has a number of important properties. The main one is the spectral mapping theorem [2]. This theorem and the equalities (1.1), (1.2) imply that for the n-tuple $A = (A_1, \dots, A_n)$ of commuting operators with finite-dimensional imaginary parts the following inclusions are valid:

$$\sigma(A^*|\mathcal{H}) \subset D^{in}, \ \sigma(A|\mathcal{H}) \subset D^{out}.$$

In Section 2 some special properties of the joint spectrum of commuting operators with compact imaginary parts are established. We prove that if one of the equalities (1.3) holds, then another one holds too. In this case we can set $C_k = B_k$ and assume that these operators commute with all A_k and with each other. The main point in Section 2 is the problem of the connection between the joint spectrum of the n-tuple A and the spectra of linear combinations of the operators A_k. We establish that in a number of cases (for example, if $n = 2$ or $\lambda \notin \mathbb{R}^n$) the point $\lambda = (\lambda_1, \dots, \lambda_n)$ belongs to $\sigma(A)$ if and only if

$$\sum_{k=1}^{n} \xi_k \lambda_k \in \sigma(\sum_{k=1}^{n} \xi_k A_k)$$

for any numbers ξ_k. We also prove that if a point $\lambda(\notin \mathbb{R}^n)$ is a regular point of both n-tuples A and A^*, then there exist numbers $\xi_1, \ldots, \xi_n$ such that

$$\sum_{k=1}^{n} \xi_k \lambda_k \notin \sigma(\sum_{k=1}^{n} \xi_k A_k), \ \sum_{k=1}^{n} \xi_k \lambda_k \notin \sigma(\sum_{k=1}^{n} \xi_k A_k^*).$$

These statements on the connection between the joint spectrum and the spectra of linear combinations are used in the proof of the principal result about the realtion between the varieties D^{in} and D^{out}.

2 Joint Spectra of Commuting Operators with Compact Imaginary Parts.

2.1. We start with the definition of joint spectra.

Definition 2.1 *Let $A = (A_1, \ldots A_n)$ be an n-tuple of bounded linear operators in a Hilbert space H. A point $z = (z_1, \ldots, z_n) \in \mathbb{C}^n$ is said to be a point of the left joint spectrum $\sigma_l(A)$ of the n-tuple A, if an n-tuple $B = (B_1, \ldots, B_n)$ of linear bounded operators in H, such that*

$$\sum_{k=l}^{n} B_k(A_k - z_k I) = I$$

does not exist.

The *right joint spectrum* $\sigma_r(A)$ of an n-tuple A is defined similarly. Finally, the *joint spectrum* of an n-tuple A is defined by

$$\sigma(A) = \sigma_l(A) \cup \sigma_r(A). \tag{2.1}$$

It is readily seen that for $n = 1$ the definition (2.1) is equivalent to the conventional definition of the spectrum of an operator. The spectrum of an operator A is also denoted by $\sigma(A)$.

The following properties of the joint spectra are deduced directly from the definitions:

2.1°. The sets $\sigma_l(A), \sigma_r(A)$ and $\sigma(A)$ are compacts.

2.2°. The sets $\sigma_l(A)$ and $\sigma_r(A^*)$ ($\sigma(A)$ and $\sigma(A^*)$) are symmetrical with respect to $\mathbb{R}^n$.

2.3°.

$$\sigma_l(A) \subset \sigma_l(A_1) \times \ldots \times \sigma_l(A_n), \sigma_r(A) \subset \sigma_r(A_1) \times \ldots \times \sigma_l(A_n), \sigma(A) \subset \sigma(A_1) \times \ldots \times \sigma(A_n).$$

2.4°. If the space H is decomposed into the direct sum of subspaces H_1 and H_2 which are invariant with respect to all the operators A_k ($k = 1, \dots, n$), then

$$\sigma_l(A) = \sigma_l(A|H_1) \cup \sigma_l(A|H_2), \quad \sigma_r(A) = \sigma_r(A|H_1) \cup \sigma_r(A|H_2),$$

$$\sigma(A) = \sigma(A|H_1) \cup \sigma(A|H_2).$$

All these properties of the joint spectra differ in no way from the case of a single operator; however, it turns out that for $n > 1$ the joint spectrum may happen to be empty. But in the case of *commuting* operators A_k such a situation is impossible, and in this case we have (for $H \neq \{0\}$)

$$\sigma_l(A_1, \dots, A_n) \neq \emptyset, \ \sigma_r(A_1, \dots, A_n) \neq \emptyset, \ \sigma(A_1, \dots, A_n) \neq \emptyset. \tag{2.2}$$

This fact is a corollary of the following central result concerning the joint spectra which is called the *spectral mapping theorem.*

Theorem 2.2 (Harte [2]) *Consider an n-tuple $A = (A_1, \dots, A_n)$ of commuting operators in the space H and an m-tuple of polynomials in n variables*

$$p(z) = (p_1(z_1, \dots, z_n), \dots, (p_m(z_1, \dots, z_n)).$$

Then

$$\sigma_l(p(A)) = p(\sigma_l(A)), \ \sigma_r(p(A)) = p(\sigma_r(A)), \ \sigma(p(A)) = p(\sigma(A)).$$

2.2. Let us begin this subsection with the following well-known result (see, e.g., [1], Chapter 1, §5).

Lemma 2.3 *Let A be a linear bounded operator with a compact imaginary part (i.e., $A - A^*$ is a compact operator). The non-real spectrum of A is at most countable and may have only real limit points. Each point of the non-real spectrum is a normal eigenvalue.*

Recall (see [1], Chapter 1, §2) that the eigenvalue λ_0 of the operator A is called *normal,* if the space H is decomposable into a direct sum $H = L \dotplus M$ of two invariant subspaces of the operator A and

$$\dim L < \infty, \ \sigma(A|L) = \{\lambda_0\}, \quad \lambda_0 \notin \sigma(A|M).$$

Subspaces L and M are given by

$$L = \bigcup_{k=1}^{\infty} Ker(A - \lambda_0 I)^k, \quad M = \bigcap_{k=1}^{\infty} (A - \lambda_0 I)^k H,$$

and therefore they are invariant with respect to any operator commuting with A.

Lemma 2.4 *Let $A_1, \ldots, A_n$ be commuting operators with compact imaginary parts and $A = (A_1, \ldots, A_n)$. If $\lambda \in \sigma(A)$ and $\lambda \notin \mathbb{R}^n$, then λ is an isolated point of $\sigma(A)$.*

Proof: There exists a number k $(1 \leq k \leq n)$ such that $\lambda_k \notin \mathbb{R}$. By the property 2.3°, $\lambda_k \in \sigma(A_k)$ and, by virtue of Lemma 2.3, λ_k is a normal eigenvalue of the operator A_k. Hence, $H = L\dot{+}M$, where subspaces L and M have the above-mentioned properties. In particular, these subspaces are invariant for all the operators A_j. Since $\lambda_k \notin \sigma(A_k|M)$, $\lambda \notin \sigma(A|M)$ (the property 2.3°), and, by 2.1°, there exists neighbourhood of the point λ not intersecting with $\sigma(A|M)$.

On the other hand, since $\dim L < \infty$, $\sigma(A|L)$ is a finite set. Therefore, there exists a punctured neighbourhood of the point λ, not intersecting with $\sigma(A|L)$. Now the statement of the lemma follows from the said above by virtue of the property 2.4°.

Let us prove now two simple lemmas concerning sparse sets in $\mathbb{C}^n$ and $\mathbb{C}^2$.

Lemma 2.5 *Let Γ be at most countable set in $\mathbb{C}^n$, not containing the point 0. Then there exists a vector $\xi = (\xi^k)_1^n \in \mathbb{R}^n$ such that*

$$\sum_1^n \xi^k \lambda_k \neq 0$$

for all $\lambda \in \Gamma$. The set of such vectors is a set of the second category in $\mathbb{R}^n$.

Proof: Let Ξ be a set of vectors $\xi \in \mathbb{R}^n$ with the property formulated in the statement of the lemma. The complementary set $\mathbb{R}^n \backslash \Xi$ consists of vectors $\eta = (\eta^k)_1^n \in \mathbb{R}^n$ such that for each of them the equalities

$$\sum_1^n \eta^k Re\lambda_k = 0, \quad \sum_1^n \eta^k Im\lambda_k = 0 \tag{2.3}$$

hold for some $\lambda \in \Gamma$.

Since $\lambda \neq 0$, the set of vectors η, satisfying the conditions (2.3), is nowhere dense in $\mathbb{R}^n$, and since Γ is at most countable, $\mathbb{R}^n \backslash \Xi$ is a set of the first category in $\mathbb{R}^n$. The lemma is proved.

Lemma 2.6 *Let F be a set of complex numbers, all elements of which, excluding at most a countable subset, are real. Then there exist numbers $\xi^1, \xi^2 \in \mathbb{C}$ such that*

$$\xi^1\lambda_1 + \xi^2\lambda_2 \neq 0 \quad (\lambda_1, \lambda_2 \in F, |\lambda_1| + |\lambda_2| > 0).$$

The set of such points (ξ^1, ξ^2) is a set of the second category in $\mathbb{C}^2$.

Proof: The set of complex numbers

$$G = \{\lambda_1/\lambda_2 : \lambda_1, \lambda_2 \in F, \ \lambda_2 \neq 0\}$$

lies on a countable number of rays originating from the zero point. Let ξ_0 be an arbitrary complex number, not belonging to any of these rays. Then

$$\lambda_1 - \lambda_2\xi_0 \neq 0 \ \ \{\lambda_1, \lambda_2 \in F, \ \lambda_2 \neq 0\} \tag{2.4}$$

Obviously, the inequality (2.4) is true also when $\lambda_2 = 0, \lambda_1 \neq 0$. Therefore, the numbers $\xi^1 = \xi$, $\xi^2 = -\xi_0\xi$ possess the required property for any $\xi \neq 0$ ($\xi \in \mathbb{C}$). The lemma is proved.

2.3. Here we consider the connection between the joint spectra and the spectra of linear combinations of operators

Lemma 2.7 *Let $A_1, \ldots, A_n$ be commuting operators with compact imaginary parts, $A = (A_1, \ldots, A_n)$ and let $\lambda = (\lambda_1, \ldots, \lambda_n) \in \mathbb{C}^n$, at least one of these λ_k being non-real. The point $\lambda \in \sigma_l(A)$ if and only if for any numbers $\xi^1, \ldots, \xi^n \in \mathbb{R}$,*

$$\sum_1^n \xi^k \lambda_k \in \sigma_l\left(\sum_1^n \xi^k A_k\right). \tag{2.5}$$

Proof: If $\lambda \in \sigma_l(A)$, then (2.5) follows from Theorem 2.2 (for $m = 1$ and $p_1(z) = \xi z$).

Let now $\lambda \notin \sigma_l(A)$. Consider the coordinate λ_k which is non-real. If $\lambda_k \notin \sigma(A_k)$, then $\lambda_k \notin \sigma_l(A_k)$. This means that we have chosen a point $\xi = (\delta^{jk})_{j=1}^n$, for which (2.5) does not hold. Thus it remains only to consider the case $\lambda_k \in \sigma(A_k)$. Since $\lambda_k \notin \mathbb{R}$, by Lemma 2.3 λ_k is a normal eigenvalue of the operator A_k. Therefore, $H = L \dot{+} M$, where the subspaces L ($\dim L < \infty$) and M are invariant for all the operators A_j ($j = 1, \ldots, n$). For the restrictions of the operator A_k onto the subspaces L and M the conditions

$$\sigma(A_k|L) = \{\lambda_k\}, \ \lambda_k \notin \sigma(A_k|M)$$

are valid.

By virtue of 2.3° and the finiteness of $dimL$, the set $\sigma_l(A|L)$ is finite. Since $\lambda \notin \sigma_l(A)$, $\lambda \notin \sigma_l(A|L)$ (the property 2.4°). By applying Lemma 2.5 to the set

$$\Gamma = \sigma_l(A|L) - \lambda$$

we obtain that for a set of vectors ξ dense in $\mathbb{R}^n$ the condition

$$\sum_{k=1}^n \xi^k (z_k - \lambda_k) \neq 0 \tag{2.6}$$

is valid for all $z = (z_k)_1^n \in \sigma_l(A|L)$. Since $\lambda_k \notin \sigma(A_k|M)$, according to $2.3°\lambda \notin \sigma(A|M)$. Consequently, if $\xi = (\delta^{jk})_{j=1}^n$, condition (2.6) holds for all $z \in \sigma(A|M)$. Obviously, this is true for all vectors $\xi \in \mathbb{R}^n$ sufficiently close to the vector $(\delta^{jk})_{j=1}^n$.

All this implies that a vector $\xi \in \mathbb{R}^n$ may be chosen so that the condition (2.2) is true for all $z \in \sigma_l(A|L) \cup \sigma(A|M)$. According to (2.1) and the property 2.4°, the condition (2.6) will hold for all $z \in \sigma_l(A)$. But then from Theorem 2.2 (for $m = 1$ and $p_1(z) = \xi(\lambda - z)$) we obtain

$$0 \notin \sigma_l\left(\sum_1^n \xi^k(A_k - \lambda_k I)\right).$$

The lemma is proved.

Remark 2.1 The property 2.2° implies that Lemma 2.7 is valid for the right joint spectrum as well.

By the definition of the joint spectrum (2.1) we obtain from Lemma 2.7 and Remark 2.1 the following result.

Theorem 2.8 *Let $A_1, \ldots, A_n$ be commuting operators with compact imaginary parts , $A = (A_1, \ldots, A_n)$ and let $\lambda = (\lambda_1, \ldots, \lambda_n)$, at least one of these λ_k being non-real. The point λ is a point of the joint spectrum $\sigma(A)$ if and only if for any numbers $\xi^1, \ldots, \xi^n \in \mathbb{R}$,*

$$\sum_{k=1}^n \xi^k \lambda_k \in \sigma\left(\sum_{k=1}^n \xi^k A_k\right).$$

We shall see below, that for $\lambda \in \mathbb{R}^n$ Lemma 2.7 and Theorem 2.8 become invalid (examples 2.1 and 2.2). In the following lemma the point $\lambda = 0$ is considered (we can go to this point by shift from any point $\lambda \in \mathbb{R}^n$). We shall show that in this case one may obtain the description of the joint spectrum of an n-tuple A in terms of spectra of linear combinations of the operators A_k^2.

Lemma 2.9 *Let $A_1, \ldots, A_n$ be commuting operators with compact imaginary parts. The point $0 \in \sigma_l(A)$ if and only if*

$$0 \in \sigma_l\left(\sum_{k=1}^n \xi^k A_k^2\right) \tag{2.7}$$

for any numbers $\xi^k \in \mathbb{R}$.

Proof: If $0 \in \sigma_l(A)$, then (2.7) follows from Theorem 2.2 (for $m = 1$ and $p_1(z) = \sum \xi^k z_k^2$).

Let now $0 \notin \sigma_l(A)$. Take $B_k = A_k^2$. By Theorem 2.2 (for $m = n$ and $p_k(z) = z_k^2$) we have $0 \notin \sigma_l(B)$. Take

$$\rho = \min_{\lambda \in \sigma_l(B)} |\lambda| \ (> 0), \quad \varepsilon = \rho(2n)^{-1}.$$

Let

$$M_0 = \{\lambda : Re\lambda > 0, \ |\mathrm{Im}\lambda| < \varepsilon\} \bigcup \{\lambda : Re\lambda \leq 0, \ |\lambda| < \varepsilon\},$$

$$M_1 \in \mathbb{C} \setminus M_0.$$

Denote by P_k $(k = 1, \ldots, n)$ the Riesz projection of the operator B_k, corresponding to the part of its spectrum located in M_1. Since $B_k = G_k + T_k$, where $G_k \geq 0$ and T_k is a compact operator, $dim P_k < \infty$ (see [1], Chapter 1, Theorems 2.2 and 5.2.).

Take

$$Q_k = 1 - P_k, \ Q = Q_1 Q_2 \ldots Q_n, \ P = I - Q, \ R_1 = PH, \ R_0 = QH.$$

It can be readily verified, that the subspaces R_1 and R_0 are invariant for all the operators B_k, $dim R_1 < \infty$ and $\sigma(B_k|R_0) \subset M_0$ $(k = 1, \ldots, n)$.

Moreover, since $0 \notin \sigma_l(B)$, by the property 2.4°,

$$0 \notin \sigma_l(B|R_0), \ \ 0 \notin \sigma_l(B|R_1) \tag{2.8}$$

Let us prove now that

$$Re \sum_1^n \lambda_k > 0, \quad (\lambda = (\lambda_k)_1^n \in \sigma_l(B|R_0)). \tag{2.9}$$

Denote by l $(0 \leq l \leq n)$ the quantity of the numbers λ_k lying in the right half-plane $Re\lambda > 0$. Assume for convenience $Re\lambda_k > 0$ $(k \leq l)$, $Re\lambda_k \leq 0$ $(k > l)$. Since $\lambda_k \in M_0$ $(k = 1, \ldots, n)$,

$$\sum_{j>l} |\lambda_j|^2 < \varepsilon^2 (n - l), \tag{2.10}$$

and since $|\lambda| \geq \rho$ $(\lambda \in \sigma_l(B))$,

$$\sum_1^l |\lambda_j|^2 = |\lambda|^2 - \sum_{j>l} |\lambda_j|^2 \geq \rho^2 - \varepsilon^2(n - l).$$

Since $Re\lambda_k > 0$ $(k \leq l)$ and $\varepsilon = \rho(2n)^{-1}$,

$$\sum_1^l Re\lambda_k \geq \Big(\sum_1^l (Re\lambda_k)^2\Big)^{\frac{1}{2}} = \Big(\sum_{j=1}^l |\lambda_j|^2 - \sum_{j=1}^l (\mathrm{Im}\lambda_j)^2\Big)^{\frac{1}{2}} >$$

$$> (\rho^2 - \varepsilon^2(n - l) - \varepsilon^2 l)^{\frac{1}{2}} = (\rho^2 - n\varepsilon^2)^{\frac{1}{2}} > n\varepsilon\sqrt{3}.$$

This, together with (2.10), implies

$$\sum_1^n Re\lambda_k \geq \sum_1^l Re\lambda_k - \sum_{j>l} |\lambda_j| > n\varepsilon\sqrt{3} - (n - l)^{\frac{1}{2}} \Big(\sum_{j>l} |\lambda_j|^2\Big)^{\frac{1}{2}} >$$

$$> n\varepsilon\sqrt{3} - \varepsilon(n-l) > 0.$$

Thus, the inequality (2.9) is proved. This inequality entails

$$\sum_{1}^{n} \lambda_k \neq 0 \quad (\lambda \in \sigma_l(B|R_0))$$

and by Theorem 2.2 (for $m = 1$ and $p(\lambda) = \sum \lambda_k$) we obtain that

$$0 \notin \sigma_l(\sum_{1}^{n} B_k|R_o). \tag{2.11}$$

Apply now Lemma 2.5 to the finite set $\Gamma = \sigma_l(B|R_1)$ (the condition $0 \notin \Gamma$ holds by virtue of (2.8)). By this lemma, there exists a set of vectors ξ dense in $\mathbb{R}^n$ such that

$$\sum_{k=1}^{n} \xi^k \lambda_k \neq 0$$

for all $\lambda \in \sigma_l(B|R_1)$. By Theorem 2.2, this implies

$$0 \notin \sigma_l(\sum_{1}^{n} \xi^k B_k|R_1). \tag{2.12}$$

If the numbers $|\xi^k - 1|$ $(k = 1, \dots, n)$ are sufficiently small, then from (2.11) it follows that

$$0 \notin \sigma_l(\sum^{n} \xi^k B_k|R_0). \tag{2.13}$$

Let us select the numbers $\xi^k \in \mathbb{R}$ such that both conditions (2.12) and (2.13) are satisfied. Then by the property 2.4°,

$$0 \notin \sigma_l(\sum^{n} \xi^k B_k) = \sigma_l(\sum^{n} \xi^k A_k^2).$$

The lemma is proved.

Remark 2.2 Lemma 2.9 holds for both right joint spectrum and joint spectrum as well. This follows from 2.2° and (2.1).

2.4. Here we prove the coincidence of the left and the right joint spectra for the operators with compact imaginary parts. Let us begin with the simple case $n = 1$.

Lemma 2.10 *For any operator A with a compact imaginary part*

$$\sigma_l(A) = \sigma_r(A) = \sigma(A). \tag{2.14}$$

Proof: It is well known that $\partial\sigma(A) \subset \sigma_l(A)$. Since $\sigma_r(A)$ and $\sigma_l(A^*)$ are symmetrical with respect to $\mathbb{R}$ (see 5.2°), it follows that $\partial\sigma(A) \subset \sigma_r(A)$. But, by Lemma 2.3, $\partial\sigma(A) = \sigma(A)$ and we obtain (2.14).

Theorem 2.11 *If $A_1, \dots, A_n$ are commuting operators with compact imaginary parts, then*

$$\sigma_l(A_1, \dots, A_n) = \sigma_r(A_1, \dots, A_n) = \sigma(A_1, \dots, A_n).$$

Proof: Let first $\lambda \notin \mathbb{R}^n$. If $\lambda \notin \sigma_l(A)$, then, by Lemma 2.7, there exist numbers $\xi^k \in \mathbb{R}$, such that

$$\sum_1^n \xi^k \lambda_k \notin \sigma_l(\sum_1^n \xi^k A_k).$$

Then Lemma 2.10 entails

$$\sum_1^n \xi^k \lambda_k \notin \sigma_r(\sum_1^n \xi^k A_k),$$

and, by Remark 2.1, $\lambda \notin \sigma_r(A)$.

If $\lambda \in \mathbb{R}^n$ and $\lambda \notin \sigma_l(A)$, then, by Lemma 2.9, there exist numbers $\xi^k \in \mathbb{R}$ such that

$$0 \notin \sigma_l(\sum_1^n \xi^k (A_k - \lambda_k I)^2).$$

Since the operator $\sum_1^n \xi^k (A_k - \lambda_k I)^2$ has a compact imaginary part, by Lemma 2.10,

$$0 \notin \sigma_r(\sum_1^n \xi^k (A_k - \lambda_k I)^2).$$

Therefore, it follows from Remark 2.2 that $\lambda \notin \sigma_r(A)$. Thus we have proved that $\sigma_r(A) \subset \sigma_l(A)$.

Applying property 2.2°, we obtain the inverse inclusion. Hence, $\sigma_r(A) = \sigma_l(A)$. Recalling (2.1), we obtain the assertion of the theorem.

Remark 2.3 The results of Subsections 2.3, 2.4 are valid for an arbitrary "joint spectrum" which has the spectral mapping property and which coincides with the usual spectrum in the case $n = 1$. In particular, *the Harte joint spectrum* $\sigma(A)$ (see (2.1)) *coincides with the Taylor joint spectrum* (see [8]) *for any n-tuple* $A = (A_1, \dots, A_n)$ *of commuting operators with compact imaginary parts.*

2.5. We show that for $\lambda \in \mathbb{R}^n$ Theorem 2.8 ceases to be valid.

Example 2.1 Let A_1 and A_2 be commuting self-adjoint operators such that

$$\sigma(A_1, A_2) = \{ (x_1, x_2) \in \mathbb{R}^2 \; : \; x_1^2 + x_2^2 = 1 \}. \tag{2.15}$$

To construct such a pair of operators it suffices, for example, to select a sequence of points $\{(\alpha_k, \beta_k)\}_1^\infty$, dense onto the circle $x_1^2 + x_2^2 = 1$ and to define for some orthonormal basis $\{e_k\}_1^\infty$ of the space H:

$$A_1 e_k = \alpha_k e_k, \;\; A_2 e_k = \beta_k e_k \;\; (k = 1, 2, \dots).$$

Let us show that $0 \in \sigma(\xi^1 A_1 + \xi^2 A_2)$ for any $\xi^1, \xi^2 \in \mathbb{R}$. For $\xi^1 = \xi^2 = 0$ it is obvious. If $(\xi^1)^2 + (\xi^2)^2 \neq 0$, then assume

$$\lambda_1 = \frac{\xi^2}{((\xi^1)^2 + (\xi^2)^2)^{\frac{1}{2}}}, \quad \lambda_2 = \frac{-\xi^1}{((\xi^1)^2 + (\xi^2)^2)^{\frac{1}{2}}}.$$

Since $(\lambda_1)^2 + (\lambda_2)^2 = 1$, $(\lambda_1, \lambda_2) \in \sigma(A_1, A_2)$ and, by Theorem 2.2, $\xi^1 \lambda_1 + \xi^2 \lambda_2 \in \sigma(\xi^1 A_1 + \xi^2 A_2)$, i.e. $0 \in \sigma(\xi^1 A_1 + \xi^2 A_2)$. On the other hand, by (2.15), $(0,0) \notin \sigma(A_1, A_2)$.

We shall indicate here certain conditions ensuring the validity of the assertion of Theorem 2.8 also for $\lambda \in \mathbb{R}^n$.

Theorem 2.12 *Let $A_1, \ldots, A_n$ be commuting operators with at most countable spectra and $A = (A_1, \ldots, A_n)$. A point $\lambda \in \mathbb{C}^n$ belongs to $\sigma(A)$ if and only if*

$$\xi\lambda \in \sigma(\xi A) \tag{2.16}$$

for any point $\xi \in \mathbb{R}^n$.

Proof: If $\lambda \in \sigma(A)$, then (2.16) follows from Theorem 2.2.

If, on the contrary, $\lambda \notin \sigma(A)$, then the set $\Gamma = \sigma(A) - \lambda$ is at most countable (this follows from the property 2.3°) and does not contain the point 0. By Lemma 2.5, there exists a point $\xi \in \mathbb{R}^n$ such that

$$\xi\lambda \neq \xi z$$

for any point $z \in \sigma(A)$. By Theorem 2.2, this entails

$$\xi\lambda \notin \sigma(\xi A).$$

The theorem is proved.

Note that the conditions of Theorem 2.12 obviously hold if $dim H < \infty$ or if all the operators A_k are compact.

In the following theorem we consider linear combinations with complex coefficients (by virtue of Example 2.1, we cannot confine ourselves to real coefficients here).

Theorem 2.13 *Let A_1 and A_2 be commuting operators with compact imaginary parts. A point $(\lambda_1, \lambda_2) \in \mathbb{C}^2$ belongs to $\sigma(A_1, A_2)$ if and only if*

$$\xi^1 \lambda_1 + \xi^2 \lambda_2 \in \sigma(\xi^1 A_1 + \xi^2 A_2) \tag{2.17}$$

for any $\xi^1, \xi^2 \in \mathbb{C}$.

Proof: If $(\lambda_1, \lambda_2) \in \sigma(A_1, A_2)$, then (2.17) follows from Theorem 2.2.

Let now $(\lambda_1, \lambda_2) \notin \sigma(A_1, A_2)$. If at least one of the numbers λ_1, λ_2 is non-real, then Theorem 2.8 entails the existence of numbers ξ^1, ξ^2 (even real), such that

$$\xi^1\lambda_1 + \xi^2\lambda_2 \notin \sigma(\xi^1 A_1 + \xi^2 A_2). \tag{2.18}$$

If $\lambda_1, \lambda_2 \in \mathbb{R}$, then, by Lemma 2.3, the set

$$F = (\sigma(A_1) - \lambda_1) \cup (\sigma(A_2) - \lambda_2)$$

contains at most a countable number of non-real numbers. By Lemma 2.6, numbers $\xi^1, \xi^2 \in \mathbb{C}$ can be found such that $\xi^1 z_1 + \xi^2 z_2 \neq 0$ for all $z_1, z_2 \in F$, $|z_1| + |z_2| > 0$ (the set of such points (ξ^1, ξ^2) is a set of the second category in $\mathbb{C}^2$).

Since $\sigma(A_1, A_2) \subset \sigma(A_1) \times \sigma(A_2)$ (the property 2.3°), and since $(\lambda_1, \lambda_2) \notin \sigma(A_1, A_2)$, it follows that

$$\xi^1\lambda_1 + \xi^2\lambda_2 \neq \xi^1\mu_1 + \xi^2\mu_2$$

for any point $(\mu_1, \mu_2) \in \sigma(A_1, A_2)$. By Theorem 2.2, we obtain (2.18). The theorem is proved.

Let us show that Theorem 2.13 does not allow generalization to the case $n > 2$.

Example 2.2. Let A_1, A_2, A_3 be commuting self-adjoint operators such that

$$\sigma(A_1, A_2, A_3) = \{\,(x_1, x_2, x_3) \in \mathbb{R}^3 \;:\; x_1^2 + x_2^2 + x_3^2 = 1\} \tag{2.19}$$

Let us show that

$$0 \in \sigma(\xi^1 A_1 + \xi^2 A_2 + \xi^3 A_3) \tag{2.20}$$

for any numbers $\xi^1, \xi^2, \xi^3 \in \mathbb{C}$. Let $\xi^k = \alpha^k + i\beta^k$ $(\alpha^k, \beta^k \in \mathbb{R})$. The system of two equations in three unknowns

$$\begin{cases} \alpha^1 x_1 + \alpha^2 x_2 + \alpha^3 x_3 = 0 \\ \beta^1 x_1 + \beta^2 x_2 + \beta^3 x_3 = 0 \end{cases}$$

has a non-trivial solution $(x_1, x_2, x_3) \in \mathbb{R}^3$. Consequently, it has also solution (x_1, x_2, x_3), for which $x_1^2 + x_2^2 + x_3^2 = 1$. By (2.19), this means that

$$\xi^1 x_1 + \xi^2 x_2 + \xi^3 x_3 = 0$$

for some point $(x_1, x_2, x_3) \in \sigma(A_1, A_2, A_3)$, and we obtain (2.20). On the other hand, by (2.19), $(0, 0, 0) \notin \sigma(A_1, A_2, A_3)$.

Let us give here one more statement about the joint spectrum of operators with compact imaginary parts.

Theorem 2.14 *Let $A = (A_1, \ldots, A_n)$ be an n-tuple of commuting operators with compact imaginary parts and let $\lambda \in \mathbb{C}^n$. If $\lambda \notin \sigma(A)$, then there exist operators $B^1, \ldots, B^n$, commuting with A_k $(k = 1, \ldots, n)$ and with each other such that*

$$\sum_{k=1}^{n} B^k(A_k - \lambda_k I) = \sum_{k=1}^{n}(A_k - \lambda_k I)B^k = I.$$

Proof: If $\lambda \notin \mathbb{R}^n$, then, by Theorem 2.8, numbers $\xi^1, \ldots \xi^n \in \mathbb{R}$ exist such that

$$\sum_1^n \xi^k \lambda_k \notin \sigma(\sum_1^n \xi^k A_k).$$

Therefore it suffices to put

$$B^j = \xi^j (\sum_{k=1}^{n} \xi^k (A_k - \lambda_k I))^{-1} \quad (j = 1, 2, \ldots, n).$$

Let now $\lambda \in \mathbb{R}^n$. By Remark 2.2, numbers $\xi^1, \ldots, \xi^n \in \mathbb{R}$ can be found such that

$$0 \notin \sigma(\sum_{k=1}^{n} \xi^k (A_k - \lambda_k I)^2),$$

and now it suffices to put

$$B^j = \xi^j (A_j - \lambda_j I)(\sum_{k=1}^{n} \xi^k (A_k - \lambda_k I)^2)^{-1}.$$

The theorem is proved.

2.6. In the following two theorems we deal with spectra of similar linear combinations of the operators A_k and the operators A_k^*.

Theorem 2.15 *Let $A = (A_1, \ldots, A_n)$ be n-tuple of commuting operators with compact imaginary parts and let $\lambda \in \mathbb{C}^n$ with $\lambda \notin \mathbb{R}^n$. If $\lambda \notin \sigma(A)$ and $\lambda \notin \sigma(A^*)$, then a point $\xi \in \mathbb{R}^n$ can be found such that*

$$\xi\lambda \notin \sigma(\xi A), \; \xi\lambda \notin \sigma(\xi A^*). \tag{2.21}$$

Proof: Repeating the arguments used in the proof of Lemma 2.7, we obtain that, given $\lambda \notin \sigma(A)$, the first of the relations (2.21) holds for any point ξ of a set $M (\subset \mathbb{R}^n)$, which is a set of the second category in a certain neighborhood U of the point $e^k = (\delta^{jk})_{j=1}^n$ (the number k is chosen so that $\lambda_k \notin R$). Since $\lambda \notin \sigma(A^*)$, the second of the relations (2.21) holds for any point ξ of a set M_*, which is a set of the second category in U. Now, it suffices to take any point $\xi \in M \cap M_*$. The theorem is proved.

Theorem 2.16 *Let A_1 and A_2 be commuting operators with compact imaginary parts and let $\lambda_1, \lambda_2 \in \mathbb{C}$. If*

$$(\lambda_1, \lambda_2) \notin \sigma(A_1, A_2), \quad (\lambda_1, \lambda_2) \notin \sigma(A_1^*, A_2^*),$$

then numbers $\xi^1, \xi^2 \in \mathbb{C}$ can be found such that

$$\xi^1\lambda_1 + \xi^2\lambda_2 \notin \sigma(\xi^1 A_1 + \xi^2 A_2), \quad \xi^1\lambda_1 + \xi^2\lambda_2 \notin \sigma(\xi^1 A_1^* + \xi^2 A_2^*). \tag{2.22}$$

Proof: As was shown in the proof of Theorem 2.13, provided $(\lambda_1, \lambda_2) \notin \sigma(A_1, A_2)$, there exists a set M of the second category in $\mathbb{C}^2$ such that the first of the relations (2.22) holds for any point $(\xi^1, \xi^2) \in M$. Since $(\lambda_1, \lambda_2) \notin \sigma(A_1^*, A_2^*)$, there exists also a set M_* of the second category in $\mathbb{C}^2$ such that the second of the relations (2.22) holds for any point $(\xi^1, \xi^2) \in M_*$. Taking $(\xi^1, \xi^2) \in M \cap M_*$, we obtain the assertion of the theorem.

Remark 2.4 If the operators $A_1, \dots, A_n$ satisfy the conditions of Theorem 2.12, then the assertion of Theorem 2.15 is true for any $\lambda \in \mathbb{C}^n$. The proof of this fact is similar to that of Theorem 2.16.

3 Colligations and Vessels

3.1. Let H be a Hilbert space (finite or infinite dimensional) and let E be a finite dimensional space, equipped with a scalar product. Consider a linear mapping $\Phi : H \to E$ and the adjoint mapping $\Phi^* : E \to H$.

Definition 3.1 *Let A_k $(k = 1, 2, \dots, n)$ be a set of linear bounded operators in H and let σ_k be a set of selfadjoint operators in E. The collection*

$$X = (A_1, \dots, A_n; H, \Phi, E; \sigma_1, \dots, \sigma_n)$$

is said to be a colligation if the relations

$$\tfrac{1}{i}(A_k - A_k{}^*) = \Phi^* \sigma_k \Phi \tag{3.1}$$

hold. The space H is said to be the inner space, and the space E is said to be the outer space (or the coupling space) of the colligation.

The operators σ_k are said to be the rates, and the mapping Φ is said to be the window of X.

A colligation is said to be *strict* if the following conditions are fulfilled:
1) $range\Phi = E$. 2) $\bigcap Ker\sigma_k = 0$.

A colligation is called *commutative* if $A_k A_j = A_j A_k$, $k, j = 1, 2, \dots, n$.

Every given set of operators $A_1, \dots, A_n$ with finite dimensional imaginary parts can be embedded in a strict colligation. To construct such an embedding one can use the non-Hermitian subspace

$$G = \sum_{k=1}^{n} (A_k - A_k{}^*)H.$$

Taking $E = G$ and putting

$$\sigma_k = \tfrac{1}{i}(A_k - A_k{}^*)|_G$$

we obtain the conditions (3.1) with $\Phi = P_G$ where P_G is the orthoprojector of H onto G.

Proposition 3.2 *The window Φ of a strict colligation has the following properties: 1) $\Phi^*(E) = G$. 2) $\Phi(G) = E$. 3) $\Phi(H \ominus G) = 0$.*

Proof: The relations (3.1) imply

$$G = \sum_{k=1}^{n} (A_k - A_k{}^*)H = \Phi^* \sum_{k=1}^{n} \sigma_k(E).$$

For strict colligations the relation

$$\sum_{k=1}^{n} \sigma_k(E) = E$$

holds, and, therefore, $G = \Phi^*(E)$. To prove 2) assume that $(\Phi G, u_0) = 0$ $(u_0 \in E)$. Then $(G, \Phi^* u_0) = 0$ and, using $G = \Phi^*(E)$, we obtain $\Phi^* u_0 = 0$. Hence,

$$(\Phi H, u_0) = (H, \Phi^* u_0) = 0.$$

Using $\Phi H = E$ we obtain $u_0 = 0$.

From $\Phi(G) = E$ and $\Phi^*(E) = G$ we conclude that $dim G = dim E$ and that Φ^* and Φ are one-to-one mappings of E onto G and of G onto E respectively.

It is clear that selfadjoint operators $\frac{1}{i}(A_k - A_k{}^*)$ annihilate the orthogonal complement $G^\perp = H \ominus G$. Using the conditions (3.1) we obtain

$$\Phi^* \sigma_k \Phi(H \ominus G) = 0$$

and

$$\sigma_k \Phi(H \ominus G) = 0$$

which, for strict colligations, is equivalent to $\Phi(H \ominus G) = 0$.

Let now X be a commutative colligation. The relations $A_k A_j = A_j A_k$ imply that the ranges of self-adjoint operators

$$\tfrac{1}{i}(A_k A_j^* - A_j A_k^*) = \tfrac{1}{i}[(A_k - A_k{}^*)A_j^* - (A_j - A_j{}^*)A_k^*] \tag{3.2}$$

and

$$\tfrac{1}{i}(A_j^*A_k - A_k^*A_j) = \tfrac{1}{i}[(A_j^* - A_j)A_k - (A_k^* - A_k)A_j] \tag{3.3}$$

belong to the non-Hermitian subspace G.

Proposition 3.3 *If X is a commutative and strict colligation then there exist two sets of selfadjoint operators $\gamma_{kj} = -\gamma_{jk}$ and $\tilde{\gamma}_{kj} = -\tilde{\gamma}_{jk}$ in E such that*

$$\tfrac{1}{i}(A_kA_j^* - A_jA_k^*) = \Phi^*\gamma_{kj}\Phi \tag{3.4}$$

and

$$\tfrac{1}{i}(A_j^*A_k - A_k^*A_j) = \Phi^*\tilde{\gamma}_{kj}\Phi, \tag{3.5}$$

where Φ is the window of X.

Indeed, the operators γ_{kj} and $\tilde{\gamma}_{kj}$ can be defined by the formulas

$$\gamma_{kj} = (\Phi_0^*)^{-1}\left[\tfrac{1}{i}(A_kA_j^* - A_jA_k^*)\right](\Phi_0)^{-1}$$

and

$$\tilde{\gamma}_{kj} = (\Phi_0^*)^{-1}\left[\tfrac{1}{i}(A_j^*A_k - A_k^*A_j)\right](\Phi_0)^{-1},$$

where $\Phi_0 = \Phi|_G$ maps G onto E.

Relations (3.1), (3.2), (3.4) imply

$$\Phi^*(\sigma_k\Phi A_j^* - \sigma_j\Phi A_k^*) = \Phi^*\gamma_{kj}\Phi. \tag{3.6}$$

In the case of strict colligations the operator Φ^* can be cancelled and we obtain the following important relation:

$$\sigma_k\Phi A_j^* - \sigma_j\Phi A_k^* = \gamma_{kj}\Phi \tag{3.7}$$

Similarly, one can obtain the relations

$$\sigma_k\Phi A_j - \sigma_j\Phi A_k = \tilde{\gamma}_{kj}\Phi. \tag{3.8}$$

Subtracting these equalitites we will obtain for strict commutative colligations

$$\tilde{\gamma}_{kj} = \gamma_{kj} + i(\sigma_k\Phi\Phi^*\sigma_j - \sigma_j\Phi\Phi^*\sigma_k). \tag{3.9}$$

The condition $\Phi H = E$ of a strict colligation is too restrictive: the projection of a strict colligation on an invariant subspace is not in general strict. Moreover, a colligation with a one-dimensional inner space can not be strict, unless $dim E = 1$. To overcome these difficulties we shall use the notion of a regular colligation which was introduced by N. Kravitsky [3]. A colligation, by definition, is said to be *regular* if there exist selfadjoint operators $\gamma_{kj} = -\gamma_{jk}$, satisfying the conditions (3.7). Then we *define* the operators $\tilde{\gamma}_{kj}$ with the help of (3.9). It is easy to check that the conditions (3.7), (3.9), (3.1) imply the relation (3.8).

3.2. Investigations of the last decade have shown that the basic notion reflecting the properties of several nonselfadjoint operators is the notion of an operator *vessel* which can be defined as folows:

Definition 3.4 *The collection*

$$V = (A_k; H, \Phi, E; \sigma_k, \gamma_{kj}, \tilde{\gamma}_{kj}) \tag{3.10}$$

where $\sigma_k = \sigma_k^*$ and $\gamma_{kj} = -\gamma_{jk}, \tilde{\gamma}_{kj} = -\tilde{\gamma}_{jk}$ are operators in E is said to be a vessel if the following conditions are fulfilled:

$$\begin{cases} \frac{1}{i}(A_k - A_k{}^*) = \Phi^*\sigma_k\Phi & (3.11)\\ \sigma_k\Phi A_j^* - \sigma_j\Phi A_k^* = \gamma_{kj}\Phi & (3.12)\\ \tilde{\gamma}_{kj} = \gamma_{kj} + i(\sigma_k\Phi\Phi^*\sigma_j - \sigma_j\Phi\Phi^*\sigma_k) & (3.13)\\ \sigma_k\Phi A_j - \sigma_j\Phi A_k = \tilde{\gamma}_{kj}\Phi. & (3.14) \end{cases}$$

Taking the difference between the left-hand sides of (3.12), (3.14) and using (3.11), (3.13) we obtain

$$(\sigma_k\Phi A_j - \sigma_j\Phi A_k) - (\sigma_k\Phi A_k^* - \sigma_j\Phi A_k^*) = \tilde{\gamma}_{kj}\Phi - \gamma_{kj}\Phi. \tag{3.15}$$

Hence, the conditions (3.14) follow from (3.11)-(3.13) and, vice versa, the condition (3.12) is the consequence of (3.11), (3.13), (3.14). The equalities (3.13) are said to be *linkage* conditions.

The equalities (3.7)-(3.9) imply that *every commutative strict colligation can be embedded in a vessel with self-adjoint operators γ_{kj} and $\tilde{\gamma}_{kj}$.*

Such a vessel (with commuting A_k and self-adjoint γ_{kj}, $\tilde{\gamma}_{kj}$) is called a *commutative* vessel. In this paper we consider only commutative vessels.

We will use the following notations:

$$\gamma_{kj} = \gamma_{kj}^{in}, \ \tilde{\gamma}_{kj} = \gamma_{kj}^{out}.$$

3.3. Let X be the colligation

$$X = (A_1, \dots, A_n; H, \Phi, E; \sigma_1, \dots, \sigma_n)$$

and let $z \notin \sigma(\sum_{k=1}^n \xi_k A_k)$ $(z \in \mathbb{C}, \xi = (\xi_1, \dots, \xi_n) \in \mathbb{C}^m)$.

The operator function

$$S(\xi, z) = I - i\Phi(\sum_{k=1}^n \xi_k A_k - zI)^{-1}\Phi^* \sum_{j=1}^n \xi_j\sigma_j$$

is said to be the *complete characteristic function* of the colligation X.

Consider the colligation

$$X' = (A'_1, \ldots, A'_n; H', \Phi', E; \sigma_1, \ldots, \sigma_n)$$

such that

$$A'_k = UA_kU^{-1}, \; \Phi' = \Phi U^{-1},$$

where U is an isometric mapping of H onto H'. The colligations X and X' in this case are said to be *unitary equivalent.* The complete characteristic functions (CCF's) of the unitary equivalent colligations coincide:

$$S'(\xi, z) = I - i\Phi'(\textstyle\sum_{k=1}^n \xi_k A'_k - zI')^{-1}\Phi'^* \sum_{j=1}^n \xi_j \sigma_j =$$
$$= I - i\Phi' U^{-1}(U \textstyle\sum_{k=1}^n \xi_k A_k U^{-1} - zI')^{-1} U\Phi'^* \sum_{j=1}^n \xi_j \sigma_j = S(\xi, z).$$

For commutative colligations the CCF defines the corresponding colligation up to the unitary equivalence on the so called "principal" subspace. To introduce this notion we need the following

Lemma 3.5 *If X is a commutative colligation then the linear closed envelope*

$$\mathcal{H} = span\{A_1^{k_1} \ldots A_n^{k_n} \Phi^* E\} \; (0 \leq k_1, \ldots, k_n < \infty),$$

coincides with the linear closed envelope

$$\mathcal{H}^* = span\{A_1^{*k_1} \ldots A_n^{*k_n} \Phi^* E\}.$$

Proof: From the colligation conditions

$$(A_k - A_k{}^*)f = i\Phi^* \sigma_k \Phi f \in \Phi^* E$$

we conclude that $A_k f - A_k^* f \in \mathcal{H}^*$ $(f \in H)$ and the sum of subspaces

$$\Phi^* E + A_k \Phi^* E = \Phi^* E + A_k^* \Phi^* E \subset \mathcal{H}^*.$$

Assume that

$$A_1^{k_1} \ldots A_n^{k_n} \Phi^* E \subset \mathcal{H}^* \;\; (0 \leq k_1 \leq N_1, \ldots, 0 \leq k_n \leq N_n).$$

Taking into account that the subspace $\mathcal{H}^*$ is an invariant subspace of the operators A_k^* we conclude that the subspace

$$A_1^{N_1} \ldots A_k^{N_k+1} \ldots A_n^{N_n} \Phi^* E = (A_k - A_k{}^*) A_1^{N_1} \ldots A_n^{N_n} \Phi^* E + A_k^* A_1^{N_1} \ldots A_n^{N_n} \Phi^* E$$

lies in $\mathcal{H}^*$ and, by induction, $\mathcal{H} \subseteq \mathcal{H}^*$. Analogously, we can obtain the inclusion $\mathcal{H}^* \subseteq \mathcal{H}$, which implies $\mathcal{H} = \mathcal{H}^*$.

The subspace $\mathcal{H}$ is said to be the *principal* subspace of the colligation. It is evident that $\mathcal{H}$ is an invariant subspace of A_k and A_k^*. Hence, the orthogonal complement $H_0 = H \ominus \mathcal{H}$ is also an invariant subspace of A_k, A_k^*. If $f_0 \in H_0$ then $(f_0, \Phi^* E) = 0$ and therefore $(\Phi f_0, E) = (f_0, \Phi^* E) = 0$. The element Φf_0 belongs to E and Φf_0 is orthogonal to E. Hence $\Phi f_0 = 0$ and

$$(A_k - A_k^*) f_0 = i\Phi^* \sigma_k \Phi f_0 = 0,$$

which implies that the restriction of A_k to H_0 is selfadjoint. Thus we have proved the following:

Theorem 3.6 *If X is a commutative colligation then the inner space H admits the orthogonal decomposition $H = \mathcal{H} \oplus H_0$, where $\mathcal{H}$ and H_0 are invariant subspaces of $A_1, \ldots, A_n$ and of $A_1^*, \ldots A_n^*$. All the operators A_k are selfadjoint on H_0 and the operator Φ vanishes identically on H_0.*

In many cases of interest the subspace H_0 is not important for the theory of non-selfadjoint operators and, in such cases, it can be dropped. If the principal subspace $\mathcal{H}$ coincides with the space H then the colligation is said to be *irreducible.*

If a colligation is embedded in a vessel we shall use the terms "principal subspace" and "irreducible" for the vessel as well.

Denote by $S_*(\xi, z)$ the CCF of the *adjoint* colligation

$$X^* = (A_1^*, \ldots, A_n^*; H, -\Phi, E; -\sigma_1, \ldots, -\sigma_n),$$

i.e.

$$S_*(\xi, z) = I + i\Phi(\sum_{k=1}^{n} \xi_k A_k^* - zI)^{-1} \Phi^* \sum_{j=1}^{n} \xi_j \sigma_j$$

for $z \notin \sigma(\sum_{k=1}^{n} \xi_k A_k^*)$.

Theorem 3.7 *Between CCF's of adjoint colligations X and X^* there exists the following connection:*

$$S(\xi, z) S_*(\xi, z) = I. \tag{3.16}$$

Proof: If $z \notin \sigma(\sum_{k=1}^{n} \xi_k A_k) \bigcup \sigma(\sum_{k=1}^{n} \xi_k A_k^*)$, then

$$S(\xi, z) S_*(\xi, z) - I = -i\Phi(\sum_{k=1}^{n} \xi_k A_k - zI)^{-1} \Phi^* \sum_{j=1}^{n} \xi_j \sigma_j +$$

$$+ i\Phi(\sum_{k=1}^{n} \xi_k A_k^* - zI)^{-1} \Phi^* \sum_{j=1}^{n} \xi_j \sigma_j +$$

$$+\Phi(\sum_{k=1}^{n}\xi_k A_k - zI)^{-1}\Phi^*\sum_{j=1}^{n}\xi_j\sigma_j\Phi(\sum_{k=1}^{n}\xi_k A_k^* - zI)^{-1}\Phi^*\sum_{j=1}^{n}\xi_j\sigma_j =$$

$$= \Phi(\sum_{k=1}^{n}\xi_k A_k - zI)^{-1}[i\sum_{k=1}^{n}\xi_k(A_k - A_k^*) + \Phi^*\sum_{j=1}^{n}\xi_j\sigma_j\Phi](\sum_{k=1}^{n}\xi_k A_k^* - zI)^{-1}\Phi^*\sum_{j=1}^{n}\xi_j\sigma_j.$$

Using the condition (3.11) we obtain (3.16).

4 The Discriminant Varieties

4.1. With the given commutative vessel

$$V = (A_k; H, \Phi, E; \sigma_k, \gamma_{kj}^{in}, \gamma_{kj}^{out})$$

one can associate the following two sets of polynomials in the complex variables $z_1, \ldots, z_n$:

$$\Delta_\Gamma^{in}(z) = det\sum_{j,k}\Gamma^{jk}(z_j\sigma_k - z_k\sigma_j + \gamma_{jk}^{in}) \tag{4.1}$$

and

$$\Delta_\Gamma^{out}(z) = det\sum_{j,k}\Gamma^{jk}(z_j\sigma_k - z_k\sigma_j + \gamma_{jk}^{out}) \tag{4.2}$$

where $\Gamma^{jk} = -\Gamma^{kj}$ are arbitrary operators in the coupling space E.

Definition 4.1 *Polynomials of the form (4.1) or (4.2) are said to be input discriminant polynomials or output disciminant polynomials respectively.*

The intersection D^{in} of the varieties in $\mathbb{C}^n$ satisfying the equations

$$\Delta_\Gamma^{in}(z) = 0$$

for all possible sets $\Gamma = (\Gamma^{jk})$ is said to be the *input discriminant variety* of the vessel. Analogously can be defined D^{out}-*output discriminant variety*: it is the intersection of all the varieties satisfying the equations

$$\Delta_\Gamma^{out}(z) = 0.$$

The following *generalized Cayley-Hamilton theorem* plays an important role in this theory.

Theorem 4.2 (Livšic [5]) *Let*

$$V = (A_k; H, \Phi, E; \sigma_k, \gamma_{kj}^{in}, \gamma_{kj}^{out})$$

be a commutative irreducible vessel and $\Delta_\Gamma^{in}(z_1, \ldots, z_n)$, $\Delta_\Gamma^{out}(z_1, \ldots, z_n)$ be arbitrary given discriminant polynomials of this vessel. Then the operators $A_1, \ldots, A_n$ satisfy the algebraic equations

$$\Delta_\Gamma^{in}(A_1^*, \ldots, A_n^*) = 0$$

and

$$\Delta_\Gamma^{out}(A_1,\ldots,A_n)=0.$$

We will not assume that the vessel under consideration is irreducible but it is supposed throughout here that $\Phi\neq 0$. This assumption is quite natural since all the operators A_k are self-adjoint for $\Phi=0$.

Since $\Phi\neq 0$, the principal subspace of the vessel V is non-trivial: $\mathcal{H}\neq\{0\}$.

Consider a new vessel

$$\tilde V=(A_k|\mathcal{H};\mathcal{H},\Phi|\mathcal{H},E;\sigma_{kj},\gamma_{kj}^{in},\gamma_{kj}^{out}).$$

Since the space E and the operators $\sigma_k,\gamma_{kj}^{in},\gamma_{kj}^{out}$ remain unchanged, so do the discriminant varieties D^{in} and D^{out} upon transition from V to $\tilde V$. Thus, Theorem 4.2 implies

$$\Delta_\Gamma^{in}(A^*|\mathcal{H})=0,\ \Delta_\Gamma^{out}(A|\mathcal{H})=0$$

for $A=(A_1,\ldots,A_n)$ and for any input discriminant polynomial $\Delta_\Gamma^{in}(z)$ and any output discriminant polynomial $\Delta_\Gamma^{out}(z)$.

By virtue of Theorem 2.2 this entails $\Delta_\Gamma^{in}(z)=0$ for any point $z\in\sigma(A^*|\mathcal{H})$ and $\Delta_\Gamma^{out}(z)=0$ for any point $z\in\sigma(A|\mathcal{H})$. In other words, the following statement is established.

Theorem 4.3 $\sigma(A^*|\mathcal{H})\subset D^{in},\ \sigma(A|\mathcal{H})\subset D^{out}$.

Since $\mathcal{H}\neq\{0\}$, then $\sigma(A^*|\mathcal{H})\neq\emptyset,\ \sigma(A|\mathcal{H})\neq\emptyset$, and we have

Corollary 4.4 *The discriminant varieties D^{in} and D^{out} are non-empty.*

4.2. Let us consider the following subspaces of E:

$$E^{in}(z)=\bigcap Ker(z_j\sigma_k-z_k\sigma_j+\gamma_{jk}^{in}) \tag{4.3}$$

and

$$E^{out}(z)=\bigcap Ker(z_j\sigma_k-z_k\sigma_j+\gamma_{jk}^{out}). \tag{4.4}$$

Here $z=(z_1,\ldots,z_n)\in\mathbb{C}^n$.

Lemma 4.5 *The discriminant varieties D^{in} and D^{out} are connected with the subspaces $E^{in}(z)$ and $E^{out}(z)$ in the following way:*

$$D^{in}=\{\,z\in\mathbb{C}^n\mid dimE^{in}(z)>0\},\quad D^{out}=\{\,z\in\mathbb{C}^n\mid dimE^{out}(z)>0\}.$$

Proof: If $u \neq 0$ belongs to $E^{in}(z)$ then

$$(z_j\sigma_k - z_k\sigma_j + \gamma_{jk}^{in})u = 0 \quad (j, k = 1, \ldots, n) \tag{4.5}$$

and u is a nontrivial solution of the equations

$$\sum_{j,k} \Gamma^{jk}(z_j\sigma_k - z_k\sigma_j + \gamma_{jk}^{in})u = 0.$$

Hence, $\Delta_\Gamma^{in}(z) = 0$, and $z \in D^{in}$. Suppose now that $z \in D^{in}$. Then all the determinants vanish:

$$det \sum_{j,k} \Gamma^{jk}(z_j\sigma_k - z_k\sigma_j + \gamma_{jk}^{in}) = 0. \tag{4.6}$$

The equations (4.5) form a system of linear equations in the r-dimensional space. Choosing appropriate matrices in (4.6) we conclude that all the possible $r \times r$ determinants of the system (4.5) vanish and, therefore, there exists a common nontrivial solution of these equations. The lemma is proved.

Suppose that $A = (A_1, \ldots, A_n)$ and that for a given point $z \in \mathbb{C}^n$ there exist a point $\xi \in \mathbb{C}^n$ such that

$$\sum_{k=1}^{n} \xi_k z_k \notin \sigma(\sum_{k=1}^{n} \xi_k A_k). \tag{4.7}$$

By Theorem 2.2, for the existence of the point ξ it is necessary that $z \notin \sigma(A)$. In Section 2 (Theorems 2.8, 2.12, 2.13) several cases have been indicated for which this condition is also sufficient (but for $n > 2$ it is not always true, see Example 2.2).

If the condition (4.7) is satisfied, consider the operator

$$S(\xi, \xi z) = I - i\Phi(\sum_{k=1}^{n} \xi_k(A_k - z_k I))^{-1}\Phi^* \sum_{j=1}^{n} \xi_j\sigma_j$$

acting in the space E, where $S(\xi, z)$ is the complete characteristic function introduced in Section 3.

Lemma 4.6 *The operator $S(\xi, \xi z)$ transforms the subspace $E^{in}(z)$ into the subspace $E^{out}(z)$.*

Proof: Consider an arbitrary vector $u_0 \in E^{in}(z)$. By the definition of $E^{in}(z)$ (see (4.3)),

$$(z_j\sigma_k - z_k\sigma_j + \gamma_{jk}^{in})u_0 = 0 \quad (j, k = 1, \ldots, n). \tag{4.8}$$

Take the operator equality adjoint to (3.12) and apply its both sides to the vector u_0:

$$(A_j\Phi^*\sigma_k - A_k\Phi^*\sigma_j)u_0 = \Phi^*\gamma_{jk}^{in}u_0. \tag{4.9}$$

Substract equality (4.8) multiplied from the left by Φ^*:

$$(A_j - z_j I)\Phi^* \sigma_k u_0 = (A_k - z_k I)\Phi^* \sigma_j u_0. \tag{4.10}$$

Take

$$f_0 = (\sum_{k=1}^{n} \xi_k (A_k - z_k I))^{-1} \Phi^* \sum_{j=1}^{n} \xi_j \sigma_j u_0.$$

Using equality (4.10), it is easily verified that

$$(A_m - z_m I) f_0 = \Phi^* \sigma_m u_0 \quad (m = 1, \dots, n). \tag{4.11}$$

Denote

$$v_0 = S(\xi, \xi z) u_0 = u_0 - i\Phi f_0$$

and let us verify that $v_0 \in E^{out}(z)$ (this is just the assertion of the lemma).

By (3.13), (4.8) and (4.11),

$$\begin{aligned} (z_j\sigma_k - z_k\sigma_j + \gamma_{jk}^{out})v_0 &= (z_j\sigma_k - z_k\sigma_j + \gamma_{jk}^{out})u_0 - \\ &-i(z_j\sigma_k - z_k\sigma_j + \gamma_{jk}^{out})\Phi f_0 = \\ &= i(\sigma_j\Phi\Phi^*\sigma_k - \sigma_k\Phi\Phi^*\sigma_j)u_0 - i(z_j\sigma_k - z_k\sigma_j + \gamma_{jk}^{out})\Phi f_0 = \\ &= i\sigma_j\Phi(A_k - z_k I)f_0 - i\sigma_k\Phi(A_j - z_j I)f_0 - i(z_j\sigma_k - z_k\sigma_j + \gamma_{jk}^{out})\Phi f_0 = \\ &= -i(\sigma_k\Phi A_j - \sigma_j\Phi A_k + \gamma_{jk}^{out}\Phi)f_0. \end{aligned}$$

But this expression equals 0 by (3.14). By virtue of (4.4) this means that $v_0 \in E^{out}(z)$. The lemma is proved.

If the condition

$$\sum_{k=1}^{n} \xi_k z_k \notin \sigma(\sum_{k=1}^{n} \xi_k A_k^*) \tag{4.12}$$

obtained by substituting A_k^* for A_k in (4.7) is satisfied, then the following operator acting in E can be defined:

$$S_*(\xi, \xi z) = I + i\Phi(\sum_{k=1}^{n} \xi_k (A_k^* - z_k I))^{-1} \Phi^* \sum_{j=1}^{n} \xi_j \sigma_j.$$

Lemma 4.7 *If the conditions (4.7) and (4.12) are satisfied then the operator $S_*(\xi, \xi z)$ is the inverse of the operator $S(\xi, \xi z)$.*

This lemma follows from Theorem 3.7.

Lemma 4.8 *If there exists a point $\xi \in \mathbb{C}^n$ such that both conditions (4.7) and (4.12) are satisfied, then*

$$dim E^{in}(z) = dim E^{out}(z). \tag{4.13}$$

Therefore, $z \in D^{in}$ if and only if $z \in D^{out}$.

Proof: By Lemma 4.6,

$$S(\xi,\xi z)E^{in}(z)\subset E^{out}(z),$$

and since the operator $S(\xi,\xi z)$ is invertible by Lemma 4.7,

$$dim E^{in}(z)\le dim E^{out}(z). \tag{4.14}$$

Along with the vessel V, consider the *adjoint vessel*

$$V^*=(A_k^*;H,-\Phi,E;-\sigma_k,-\gamma_{kj}^{out},-\gamma_{kj}^{in}).$$

It is easily seen that upon passing from the vessel V to the adjoint vessel V^* the subspaces $E^{in}(z)$ and $E^{out}(z)$ exchange the roles. Hence, the inequality sign in (4.14) can be replaced by the opposite one, and we obtain the equalities (4.13).

4.3. In the papers [4,6] it has been proved that for $n=2$ the equality $D^{in}=D^{out}$ is always true. It turns out that this is not true for $n>2$.

Example 4.1 Let $H=E=\mathbb{C}^3$ (with the standard scalar product) $\Phi=I$,

$$A_1=\begin{bmatrix}0&1&1\\0&0&0\\0&0&0\end{bmatrix}, A_2=\begin{bmatrix}0&1&0\\0&0&0\\0&0&0\end{bmatrix}, A_3=iI,\ \sigma_k=(1/i)(A_k-A_k{}^*)(k=1,2,3),$$

$$\gamma_{jk}^{in}=(1/i)(A_kA_j^*-A_jA_k^*),\ \gamma_{jk}^{out}=(1/i)(A_j^*A_k-A_k^*A_j)\ (j,k=1,2,3).$$

Obviously

$$V=(A_k;H,\Phi,E;\sigma_k,\gamma_{kj}^{in},\gamma_{kj}^{out})$$

is a commutative irreducible vessel.

Direct calculations show that in this example D^{in} consists of two points, and D^{out} consists of one point:

$$D^{in}=\{(0,0,\pm i)\}, D^{out}=\{(0,0,i)\}.$$

We intend to show in this section, that the varieties D^{in} and D^{out} differ at most by a finite number of points. At first, we consider the case $z\in\mathbb{R}^n$. This case is more complicated in a certain sense, since we cannot confine ourselves to linear combinations of the operators A_k (see Example 2.2). However, in this case the obtained result is more complete - a point of $\mathbb{R}^n$ cannot belong to only one of the varieties D^{in} and D^{out}.

Due to the impossibilibilty in general of constructing the operator $S(\xi,\xi z)$ for a point $z\in\mathbb{R}^n$, $z\notin\sigma(A)$, we are forced to look for some substitute for this operator.

We can pass to the point 0 by a shift transformation, and so we introduce the following definition for the point $z=0$ only.

As before, we consider a commutative vessel V with $\Phi\ne0$. As has been pointed out, in this case we may assume that the vessel V is irreducible.

Suppose that $0 \notin \sigma(A)$. Then, by Remark 2.2, there exist numbers $\xi^k \in \mathbb{R}$ such that the operator

$$\sum_{k=1}^{n} \xi^k A_k^2$$

is invertible. Take

$$S^{(0)} = I - i\Phi(\sum_{k=1}^{n} \xi^k A_k^2)^{-1} \sum_{j=1}^{n} \xi^j A_j \Phi^* \sigma_j. \tag{4.15}$$

Lemma 4.9 $S^{(0)} E^{in}(0) \subset E^{out}(0)$.

Proof: Let $u_0 \in E^{in}(0)$, i.e.

$$\gamma_{jk}^{in} u_0 = 0 \quad (j, k =, 1, \dots, n). \tag{4.16}$$

It follows from (4.9) and (4.16) that

$$A_j \Phi^* \sigma_k u_0 = A_k \Phi^* \sigma_j u_0 \quad (j, k =, 1, \dots, n). \tag{4.17}$$

Take

$$f_0 = (\sum_{k=1}^{n} \xi^k A_k^2)^{-1} \sum_{j=1}^{n} \xi^j A_j \Phi^* \sigma_j u_0.$$

By (4.17) we have, for any $m = 1, \dots, n$,

$$A_m f_0 = (\sum_{k=1}^{n} \xi^k A_k^2)^{-1} \sum_{j=1}^{n} \xi^j A_j A_m \Phi^* \sigma_j u_0 =$$

$$= (\sum_{k=1}^{n} \xi^k A_k^2)^{-1} \sum_{j=1}^{n} \xi^j A_j^2 \Phi^* \sigma_m u_0 = \Phi^* \sigma_m u_0. \tag{4.18}$$

The equalities (3.13), (4.16), (4.18) and (3.14) give

$$\gamma_{jk}^{out} S^{(0)} u_0 = \gamma_{jk}^{out}(u_0 - i\Phi f_0) =$$

$$\gamma_{jk}^{out} u_0 - i(\sigma_k \Phi \Phi^* \sigma_j - \sigma_j \Phi \Phi^* \sigma_k) u_0 - i\gamma_{jk}^{out} \Phi f_0 =$$

$$= -i(\sigma_k \Phi A_j - \sigma_j \Phi A_k) f_0 - i\gamma_{jk}^{out} \Phi f_0 = 0 \quad (j, k = 1, \dots, n).$$

But this means that $S^{(0)} u_0 \in E^{out}(0)$. The lemma is proved.

Theorem 4.10 $D^{in} \cap \mathbb{R}^n = D^{out} \cap \mathbb{R}^n$.

Proof: Let $\lambda \in \mathbb{R}^n$. If $\lambda \in \sigma(A)$, then, by 2.2° $\lambda \in \sigma(A^*)$, and Theorem 4.3 entails $\lambda \in D^{in} \cap D^{out}$.

We consider now the case $\lambda \notin \sigma(A)$ and prove that $\lambda \in D^{in}$ iff $\lambda \in D^{out}$. Since $\lambda_k \in \mathbb{R}$ ($k = 1, \ldots, n$), we can consider the operators $A_k - \lambda_k I$ instead of A_k, and thus it suffices to consider the case $\lambda = 0$.

If $0 \notin \sigma(A)$, then, by virtue of Remark 2.2, there exist numbers $\xi^k \in R$ such that the operator $\sum_{k=1}^{n} \xi^k A_k^2$ is invertible. In this case we may assume that the number

$$\sum_{k=1}^{n} |\xi^k - 1| \tag{4.19}$$

is arbitrarily small.

Note now that the operator

$$\sum_{1}^{n} A_k A_k^* \tag{4.20}$$

is invertible. Indeed, for any $f \in H$,

$$\| \sum_{1}^{n} A_k A_k^* f \| \|f\| \geq (\sum_{1}^{n} A_k A_k^* f, f) = \sum_{1}^{n} \|A_k^* f\|^2. \tag{4.21}$$

Since $0 \notin \sigma(A)$, by 2.2° $0 \notin \sigma(A^*)$ and, hence, $0 \notin \sigma_l(A^*)$. Therefore, there exists a number $\delta > 0$ such that

$$\sum_{1}^{n} \|A_k^* f\|^2 \geq \delta \|f\|^2. \tag{4.22}$$

It follows from (4.21) and (4.22), that the operator (4.20) is invertible. But then the operator $\sum_{k=1}^{n} \xi^k A_k A_k^*$ is also invertible provided that the number (4.19) is sufficiently small. Thus, there exist numbers $\xi^k \in \mathbb{R}$ such that both operators

$$\sum_{1}^{n} \xi^k A_k^2, \quad \sum_{1}^{n} \xi^k A_k A_k^*$$

are invertible. We define the operator

$$S_*^{(0)} = I + i\Phi(\sum_{k=1}^{n} \xi^k A_k A_k^*)^{-1} \sum_{j=1}^{n} \xi^j A_j \Phi^* \sigma_j$$

acting in E, and we shall show that it is the inverse of the operator $S^{(0)}$ (see (4.15)).

Indeed,

$$S^{(0)} S_*^{(0)} = I + \Phi(\sum_{1}^{n} \xi^k A_k^2)^{-1} \sum_{1}^{n} \xi^j A_j \Phi^* \sigma_j \Phi(\sum_{1}^{n} \xi^k A_k A_k^*)^{-1} \sum_{1}^{n} \xi^j A_j \Phi^* \sigma_j -$$

$$-i\Phi(\sum_{1}^{n} \xi^k A_k^2)^{-1} \sum_{1}^{n} \xi^j A_j \Phi^* \sigma_j + i\Phi(\sum_{1}^{n} \xi^k A_k A_k^*)^{-1} \sum_{1}^{n} \xi^j A_j \Phi^* \sigma_j. \tag{4.23}$$

Let us transform the second term using the equality (3.11):

$$\Phi(\sum_1^n \xi^k A_k^2)^{-1} \sum_1^n \xi^j A_j \Phi^* \sigma_j \Phi(\sum_1^n \xi^k A_k A_k^*)^{-1} \sum_1^n \xi^j A_j \Phi^* \sigma_j =$$

$$= -i\Phi(\sum_1^n \xi^k A_k^2)^{-1} \sum_1^n \xi^j A_j (A_j - A_j^*)(\sum_1^n \xi^k A_k A_k^*)^{-1} \sum_1^n \xi^j A_j \Phi^* \sigma_j =$$

$$= -i\Phi(\sum_1^n \xi^k A_k A_k^*)^{-1} \sum_1^n \xi^j A_j \Phi^* \sigma_j + i\Phi(\sum_1^n \xi^k A_k^2)^{-1} \sum_1^n \xi^j A_j \Phi^* \sigma_j.$$

Replacing the second term of the right-hand side of (4.23) by the last expression, we obtain the equality $S^{(0)} S_*^{(0)} = I$.

Thus, the operator $S^{(0)}$ is invertible. Therefore, Lemma 4.9 implies

$$dim E^{in}(0) \leq dim E^{out}(0).$$

Applying the last equality to the adjoint vessel V^*, we obtain that $dim E^{in}(0) \geq dim E^{out}(0)$ (see the proof of Lemma 4.8). Hence, $dim E^{in}(0) = dim E^{out}(0)$ and therefore,

$$0 \in D^{in} \Leftrightarrow 0 \in D^{out}.$$

The theorem is proved.

We proceed now to the consideration of points not belonging to $\mathbb{R}^n$.

Lemma 4.11 *Let $\lambda \in \mathbb{C}^n \backslash \mathbb{R}^n$. If the point λ belongs to neither $\sigma(A)$ nor $\sigma(A^*)$, then*

$$\lambda \in D^{in} \Leftrightarrow \lambda \in D^{out}.$$

Proof: By Theorem 2.15, there exists a point $\xi \in \mathbb{R}^n$ such that both conditions (4.7) and (4.12) are satisfied. Therefore, the assertion of the lemma follows from Lemma 4.8.

Lemma 4.12 *If a point $\lambda (\in \mathbb{C}^n)$ belongs to D^{in} and does not belong to D^{out}, then it belongs to $\sigma(A^*)$ and does not belong to $\sigma(A)$.*

Proof: Since $\lambda \notin D^{out}$, by Theorem 4.3 $\lambda \notin \sigma(A)$, and it remains only to prove that $\lambda \in \sigma(A^*)$.

Assume that this is not true. Since, by Theorem 4.10, $\lambda \notin \mathbb{R}^n$, it follows from this assumption and from Theorem 2.15, that there exists a vector $\xi \in \mathbb{R}^n$ such that

$$\sum_{k=1}^n \xi_k \lambda_k \notin \sigma(\sum_{k=1}^n \xi_k A_k), \ \sum_{k=1}^n \xi_k \lambda_k \notin \sigma(\sum_{k=1}^n \xi_k A_k^*).$$

But then, by Lemma 4.8,

$$\lambda \in D^{in} \Leftrightarrow \lambda \in D^{out}.$$

which contradicts the condition of the lemma. Thus, the lemma is proved.

4.4. It follows from Lemmas 4.12 and 2.4 that the varieties D^{in} and D^{out} differ by at most a countable number of points. In fact, the number of these points is finite.

Theorem 4.13 *The sets D^{in} and D^{out} differ by at most a finite number of points. If some point belongs to one of these sets and does not belong to the other one, then it is an isolated point of the corresponding set.*

Proof: Consider any point λ belonging to D^{in} and not belonging to D^{out}. By virtue of Theorem 4.10 $\lambda \notin \mathbb{R}^n$, and by virtue of Lemma 4.12 $\lambda \in \sigma(A^*)$. By Lemma 2.4, λ is an isolated point of $\sigma(A^*)$, and thus a neighbourhood U of the point λ can be found such that

$$\lambda' \notin \sigma(A^*) \quad (\lambda' \in U,\ \lambda' \neq \lambda). \tag{4.24}$$

By virtue of Lemma 4.12 $\lambda \notin \sigma(A)$, and, since the joint spectrum is closed (see 2.1°), it may also be assumed that

$$\lambda' \notin \sigma(A)\ (\lambda' \in U). \tag{4.25}$$

Since $\lambda \notin \mathbb{R}^n$, we can also suppose that

$$U \cap \mathbb{R}^n = \emptyset. \tag{4.26}$$

Let the point λ be not isolated in the set D^{in}. Then there exists a sequence of points $\lambda^{(m)}$ converging to λ and such that

$$\lambda^{(m)} \in D^{in},\ \lambda^{(m)} \in U,\ \lambda^{(m)} \neq \lambda. \tag{4.27}$$

The relations (4.24)-(4.27) and Lemma 4.11 entail $\lambda^{(m)} \in D^{out}$. But then, since D^{out} is closed, we obtain $\lambda \in D^{out}$, which contradicts the choice of λ.

Thus, the set of points belonging to D^{in} and not belonging to D^{out} is isolated in D^{in}. But the algebraic set D^{in} may contain at most a finite number of isolated points (see, e.g., [7], Chapter 1, §3 and Chapter VII, §2) and, therefore, the set $D^{in} \setminus (D^{in} \cap D^{out})$ is finite. Considering the adjoint vessel V^*, we obtain the assertion of the theorem for the set $D^{out} \setminus (D^{out} \cap D^{in})$. The theorem is proved.

In the case $n = 2$ the input discriminant polynomials have the following form:

$$\begin{aligned}\Delta^{in}_{\Gamma}(z_1, z_2) &= det[\Gamma^{12}(z_1\sigma_2 - z_2\sigma_1 + \gamma^{in}_{12}) + \Gamma^{21}(z_2\sigma_1 - z_1\sigma_2 + \gamma^{in}_{21})] = \\ &= det(2\Gamma^{12})det(z_1\sigma_2 - z_2\sigma_1 + \gamma^{in}_{12}).\end{aligned}$$

Therefore, there is only one input discriminant polynomial (up to a constant factor):

$$\Delta^{in}(z_1, z_2) = det(z_1\sigma_2 - z_2\sigma_1 + \gamma^{in}_{12}).$$

Analogously, the output discriminant polynomial has the form

$$\Delta^{out}(z_1, z_2) = det(z_1\sigma_2 - z_2\sigma_1 + \gamma^{out}_{12}).$$

Since

$$D^{in} = \{ (z_1, z_2) \in \mathbb{C}^2 \mid \Delta^{in}(z_1, z_2) = 0\},$$

there are two cases:

1) $\Delta^{in}(z_1, z_2) \not\equiv 0$, D^{in} is an algebraic curve.

2) $\Delta^{in}(z_1, z_2) \equiv 0$, $D^{in} = \mathbb{C}^2$.

In both cases the set D^{in} does not contain isolated points. This assertion holds for the set D^{out} as well. Hence Theorem 4.13 implies that *the discriminant varieties D^{in} and D^{out} coincide in the case $n = 2$.*

This result was pointed out in [6] as a corrolary of the following remarkable equality [4]:

$$\Delta^{in}(z_1, z_2) = \Delta^{out}(z_1, z_2).$$

References

1. Gohberg I.C. and Krein M.G., *Introduction to the Theory of Linear Nonselfadjoint Operators*, Amer. Math. Soc., Providence, 1969.
2. Harte R.E., *Spectral mapping theorems.* Proc. Roy. Irish Acad. Sect. A, **72** (1972), 89-107.
3. Kravitsky N., *Regular colligations for several operators in Banach space*, Integral Equations Operator Theory, **6** (1983), 224-249.
4. Livšic M.S., *A method for constructing triangular canonical models of commuting operators based on connections with aglebraic curves*, Integral Equations Operator Theory, **3** (1980), 489-507.
5. Livšic M.S., *Commuting operators and fields of systems distributed in Euclidean space*, in *Toeplitz Centennial*, Operator Theory: Advances and Applciations, Vol 4, Birkhäuser Verlag, Basel, 1982, pp. 377-413.
6. Livšic M.S., *Cayley-Hamilton theorem, vector bundles and divisors of commtuing operators*, Integral Equations Operator Theory, **6** (1983), 250-273.
7. Shafarevich I.R., *Basic Algebraic Geometry*, Springer-Verlag, 1974.
8. Taylor J.L. *A joint spectrum for commuting families of operators*, J. Funct. Anal., **6** (1970), 172-191.

Department of Mathematics and Computer Sciences
Ben-Gurion University of the Negev
Beer-Sheva, Israel

MSC 1991: Primary 47A13, Secondary 47A48

Operator Theory:
Advances and Applications, Vol. 73

On the differential structure of matrix–valued rational inner functions

Daniel Alpay, Laurent Baratchart, and Andrea Gombani *

Dedicated to Professor M.S. Livsic, with admiration

Contents

---

*This research took partly place while this author visited the Center for Mathematics and Computer Science, Amsterdam, The Netherlands, with the support of a NATO fellowship

1 Introduction and preliminaries

1.1 Introduction

The present work is devoted to a differential-geometric study of rather classical objects in analysis namely inner matrix-valued functions. More precisely, we prove that the collection of such matrices with prescribed size and McMillan degree is an embedded submanifold in the Hardy space, two parametrizations of which are derived. We do this with an eye on linear control systems not only because several methods from system-theory are relevant to such a study but also because the geometry of inner matrix-valued functions in itself impinges on the parametrization of systems *via* the Douglas-Shapiro-Shields factorization of transfer-functions introduced in [29].

First recall what Schur and inner functions of the disk are. A complex-valued function is called a Schur function if it is analytic in the open unit disk $\mathbb{D}$ and bounded by one in modulus there. Now, it is a theorem of Fatou that such a function has non-tangential boundary values almost everywhere on the unit circle $\mathbb{T}$; a Schur function is then called inner if these boundary values are of modulus one almost everywhere on $\mathbb{T}$. The multiplicative structure of Schur and inner functions is well-known (see e.g. [30] [25], [40]); in particular, rational inner functions, also called finite Blaschke products, are of the form:

$$b(z) = c\prod_{i=1}^{n} \frac{z - w_i}{1 - z\bar{w}_i} \tag{1.1}$$

where the w_i are in $\mathbb{D}$ and c is a complex number of modulus one. The number n is the degree of b, and we will denote by I_n^1 the set of Blaschke products of degree n. Considering I_n^1 as a subset of some Hardy space H_q where $1 \leq q \leq \infty$ (the actual value of q turns out to be irrelevant), we want to analyse the smoothness of this object. Expression (1.1), however explicit it may be, is not adequate for this purpose because the functions $b \to w_i$ are not differentiable at branching points. Of course, it is obvious how to remedy this: we simply set

$$p(z) = \prod_{i=1}^{n}(z - w_i) \tag{1.2}$$

and define the reciprocal polynomial of p by

$$\tilde{p}(z) = \prod_{i=1}^{n}(1 - z\bar{w}_i) = z^n\bar{p}(\frac{1}{z}) \tag{1.3}$$

so that

$$b(z) = c\frac{p(z)}{\tilde{p}(z)}. \tag{1.4}$$

Since p and $\tilde{p}$ are coprime, (1.4) defines a one-to-one correspondance between I_n^1 and pairs (c, p), so we can choose c (ranging over $\mathbb{T}$) and the coefficients of p except for the leading one (ranging over some open subset of $\mathbb{C}^n$) as coordinates.

There is, however, a more subtle way of describing I_n^1 which was introduced by Schur in his celebrated paper [49]: starting from a Schur function f, one defines recursively a sequence (f_k) of Schur functions by setting $f_0 = f$ and

$$f_{k+1}(z) = \frac{f_k(z) - f_k(0)}{z(1 - \overline{f_k(0)}\, f_k(z))}. \tag{1.5}$$

The process stops if, at some stage, $f_k(0)$ has modulus one. The sequence of numbers $\rho_k \overset{\text{def}}{=} f_k(0)$ are called the Schur coefficients of f, and they completely characterize the function. Moreover, f belongs to I_n^1 if and only if $|\rho_k| < 1$ for $k < n$ and $|\rho_n| = 1$. This leads to another proof of the smoothness of I_n^1 [12] by showing that it is diffeomorphic to the product of n copies of the open unit disk and of a copy of the unit circle. It should also be clear that everything in what precedes can be specialized to the subset RI_n^1 of I_n^1 consisting of real Blaschke products of degree n, namely those satisfying $b(\bar{z}) = \overline{b(z)}$. In this case, parameters c and p as above are real and so are the Schur coefficients. For more information on the Schur algorithm and some of its applications to signal processing, we refer to [32].

In this paper, our interest lies in matrix-valued functions rather than just scalar ones. A $\mathbb{C}^{p\times p}$-valued rational function Q is called inner if it is analytic in $\mathbb{D}$ and takes unitary values on the unit circle $\mathbb{T}$. The set of $\mathbb{C}^{p\times p}$-valued rational inner functions of degree n will be denoted by I_n^p, where the degree is now meant to be the McMillan degree. Our objective is to study the differential structure of I_n^p and also of RI_n^p which is the subset of I_n^p consisting of real functions (i.e. satisfying $Q(\bar{z}) = \overline{Q(z)}$). As we shall see, it is again true that I_n^p and RI_n^p are smooth, but it is more demanding to obtain effective parametrizations. A multiplicative decomposition into elementary factors still exists as was shown by Potapov [47], but again runs short of smoothness at branching points. Also, a direct analysis based on some explicit description of the matrix analogous to (1.4) is still possible when $p = 2$ [21] but runs into difficulties for $p > 2$. In contrast, the seemingly more involved approach using Schur parameters does carry over to the matrix case.

To state this more precisely, let us introduce the set $\mathcal{U}_p$ (resp. $\mathcal{O}_p$) of unitary (resp. orthogonal) $p \times p$ matrices and recall that $\mathcal{U}_p$ (resp. $\mathcal{O}_p$) is a manifold of dimension p^2 (resp. $p(p-1)/2$) (see e.g. [34]). Now, it will follow from the matrix version of the Schur algorithm that I_n^p, considered as a subset of some H_q with $1 \le q \le \infty$, is locally diffeomorphic to the product of n copies of the open unit ball in $\mathbb{C}^p$ and a copy of $\mathcal{U}_p$. Similarly, RI_n^p is locally diffeomorphic to the product of n copies of the open unit ball in $\mathbb{R}^p$ and a copy of $\mathcal{O}_p$. Hence the spaces I_n^p and RI_n^p are smooth (and even real analytic) manifolds of dimension $2np+p^2$ and $np+p(p-1)/2$ respectively. We shall say further that two members Q_1 and Q_2 of I_n^p are equivalent if there exists a unitary matrix U such that $Q_1 = Q_2U$. A similar equivalence is defined in RI_n^p, U now being orthogonal. We will denote by $I_n^p/\mathcal{U}_p$ and $RI_n^p/\mathcal{O}_p$ respectively the associated quotient spaces, and we shall see that these spaces of "normalized" inner functions are also smooth manifolds.

Yet another way to proceed is to consider inverses of inner matrices rather than I_n^p itself. Indeed, the inverse of $Q \in I_n^p$ is a proper transfer function of McMillan degree n and we may resort to classical tools from system-theory like realizations and coprime factorizations. This time, however, charts will be obtained in terms of realizations showing in particular that $I_n^p/\mathcal{U}_p$ (resp. $RI_n^p/\mathcal{O}_p$) is diffeomorphic to the manifold of observable pairs (C, A) where $C \in \mathbb{C}^{p\times n}$ (resp. $\mathbb{R}^{p\times n}$) and $A \in \mathbb{C}^{n\times n}$ (resp. $\mathbb{R}^{n\times n}$). This, in some sense, can be expected from the Beurling-Lax theorem because members of $I_n^p/\mathcal{U}_p$ are in one-to-one correspondance with shift invariant subspaces of H_2^p of codimension n and the orthogonal complement of such a subspace, being n-dimensional and invariant under the left shift, is therefore the span of the columns of some $C(I_n - zA)^{-1}$. This system theoretic approach will be taken in section 2 and the Schur algorithm applied in section 3. We discuss in the final section a specific link to rational approximation in Hardy spaces and identification which is stressed in [12], [21] and [13].

1.2 Some preliminaries and notations

In this subsection, we fix notations and review a number of facts on matrix-valued functions which will be needed throughout the paper. Recall that the open unit disk and the unit circle are denoted by $\mathbb{D}$ and $\mathbb{T}$ respectively. The symbol $\mathbb{C}^{n\times m}$ stands for the space of $n \times m$ matrices with complex entries. When m is equal to 1, we will write $\mathbb{C}^n$ for short. We let $GL(n)$ denote the group of square $n \times n$ complex matrices with non zero determinant. We set $\mathcal{H}_p$ (resp. $\mathcal{S}_p$) to be the set of hermitian (resp. skew-hermitian) matrices of size p. The identity matrix of $\mathbb{C}^{n\times n}$ will be denoted by I_n, or I if n is understood from the context. The symbol A^* will designate the transposed conjugate of the matrix A as well as the adjoint of an operator between Hilbert spaces. In particular, when $z \in \mathbb{C}$, z^* is the conjugate of z. The transpose of the matrix A will be denoted by A^t. If H is a Hilbert space and M is a closed subspace of H, the orthogonal complement of M in H will be denoted by $H \ominus M$.

The abbreviation $l.s.\{v_i\}$ is used to mean the linear span of the vectors v_i.

A complex scalar or matrix-valued function F, defined over a subset of $\mathbb{C}$ which is stable under conjugation, is said to be real if

$$F(z^*) = F(z)^*.$$

We let $L_q(\mathbb{T})$ stand for the usual Lebesgue space of the circle and $\|f\|_q$ means the norm of f in $L_q(\mathbb{T})$. The Hardy space with exponent q of the open unit disk will be denoted by H_q; recall that H_q is the space of functions f holomorphic in $\mathbb{D}$ and such that

$$\sup_{r<1} \|f(re^{i\theta})\|_{L_q(\mathbb{T})} < \infty. \tag{1.6}$$

Such a f has a nontangential limit at almost every point of $\mathbb{T}$ and defines in this way a function belonging to $L_q(\mathbb{T})$ whose Fourier coefficients of negative index do vanish. Conversely, any such function in $L_q(\mathbb{T})$ is the nontangential limit of some uniquely defined $f \in H_q$. It is then customary to consider H_q as a closed subspace of $L_q(\mathbb{T})$ by identifying f and its boundary function. The norm of the boundary function in $L_q(\mathbb{T})$ is, by definition, the norm of f in H_q and turns out to be equal to the *sup* in (1.6) (see e.g. [30], [25] or [40]). In particular, it follows from Parseval's theorem that

$$H_2 = \{f(z) = \sum_0^\infty a_k z^k \ , \ a_k \in \mathbb{C} \text{ and } \|f\|_2^2 = \sum_0^\infty |a_k|^2 < \infty\}.$$

Also, H_∞ is the space of bounded holomorphic functions on $\mathbb{D}$ endowed with the *sup* norm. We designate by $\mathcal{A}$ the disk algebra which is the closed subalgebra of H_∞ comprising functions that are continuous on $\overline{\mathbb{D}}$. The symbol RH_q will denote the real Hardy space of functions in H_q which are real or, equivalently, whose Fourier coefficients in $L_q(\mathbb{T})$ are real. The symbol $\bar{H}_q$ stands for the conjugate Hardy space of functions f analytic outside $\bar{\mathbb{D}}$ (including at infinity) and such that $f(1/z) \in H_q$. We define $R\overline{H}_q$ accordingly. Just like before, $\bar{H}_q$ identifies to a subspace of $L_q(\mathbb{T})$ but the Fourier coefficients vanish this time on positive indices.

When $1 \le q < \infty$, the space $L_q^{p\times m}(\mathbb{T})$ of $p \times m$ matrices with entries in $L_q(\mathbb{T})$ will be endowed with the following norm: if M is such a matrix with entries $m_{i,j}$, we set

$$\|M\|_q = \left(\sum_{i,j} \|m_{i,j}\|_q^q\right)^{\frac{1}{q}}.$$

When $q = \infty$, we define

$$\|M\|_\infty = ess.\ \sup_{\mathbb{T}} \|M(e^{i\theta})\|,$$

where $\|A\|$ denotes the operator norm $\mathbb{C}^m \to \mathbb{C}^p$ of the complex $p \times m$ matrix A. Of course, we may work with many other equivalent norms in $L_q^{p\times m}(\mathbb{T})$ and the above choice is mainly for definiteness. When $q = \infty$, for instance, we may also take the *sup* of the $\|m_{i,j}\|_\infty$'s. The present definition, however, has the advantage of making $L_\infty^{p\times p}(\mathbb{T})$ into a Banach algebra and is the one usually adopted in control theory. The subspaces $H_q^{p\times m}$ and $\bar{H}_q^{p\times m}$ of $L_q^{p\times m}(\mathbb{T})$ are equipped with the induced norm and so are the real subspaces $RH_q^{p\times m}$ and $R\bar{H}_q^{p\times m}$.

For the convenience of the reader, we recall some basic facts from matrix-valued rational functions. If such a function W is analytic at infinity, it can be written as:

$$W(z) = D + C(zI_n - A)^{-1}B \tag{1.7}$$

where (A, B, C, D) are matrices of adequate sizes, D being merely the value at infinity. The expression (1.7) is called a realization of W. If the size n of A is minimal, the realization is said to be minimal. Two minimal realizations of a given function W are always similar, namely if

$$W(z) = D + C_i(zI_n - A_i)^{-1}B_i, \text{ for } i = 1, 2$$

are two minimal realizations, then there exists a unique invertible matrix S such that:

$$\begin{pmatrix} A_2 & B_2 \\ C_2 & D \end{pmatrix} = \begin{pmatrix} S & 0 \\ 0 & I \end{pmatrix} \begin{pmatrix} A_1 & B_1 \\ C_1 & D \end{pmatrix} \begin{pmatrix} S^{-1} & 0 \\ 0 & I \end{pmatrix}.$$

Recall a pair (C, A) of matrices in $\mathbb{C}^{p\times n} \times \mathbb{C}^{n\times n}$ is said to be observable if

$$\bigcap_{k=0}^{k=\infty} \operatorname{Ker}\{CA^k\} = \{0\}$$

and that a pair (A, B) is said to be reachable is (B^t, A^t) is observable, It is actually well-known that a realization (A, B, C, D) is minimal if and only if (C, A) is observable and (A, B) is reachable [42]. In this case, the poles of the rational function $W(z)$ are the eigenvalues of A.

Essentially equivalent to the above is the fact that any complex vector space $\mathcal{M}$ of germs of $\mathbb{C}^p$-valued analytic functions at 0 which is both finite-dimensional and invariant under R_0, the left shift at 0 defined by

$$R_0 f(z) = \frac{f(z) - f(0)}{z},$$

is in fact made of rational functions and can be described as the span of the columns of some matrix

$$C(I_n - zA)^{-1} \tag{1.8}$$

where (C, A) is observable and n is the dimension of $\mathcal{M}$. Suppose now that W is real or equivalently that the rational matrix $W(z)$ has real coefficients. Since realization theory is valid over any field, the matrices A, B, C and D in (1.7) may be chosen so as to be real. Accordingly, we say that a space $\mathcal{M}$ of analytic germs as above is real if

$$\sum a_k z^k \in \mathcal{M} \Longrightarrow \sum \bar{a}_k z^k \in \mathcal{M}.$$

In this case, the matrices A and C in (1.8) may also be chosen so as to be real.

The minimal feasible n in (1.7) is called the McMillan degree of W. It is invariant under Moebius tansfomations of the argument and, since any W can be construed to be analytic at infinity by performing such a transformation, this provides a definition of the McMillan degree for general rational matrix functions. Also, if W is square and $\det W$ does not vanish identically, then W and W^{-1} have same McMillan degree. We refer to the monographs [42], [16], [7], [29] and [33] for more details.

In the sequel, $\Sigma_{p,m}(n)$ denotes the set of $p \times m$ rational matrices of McMillan degree n that are analytic at infinity. We single out the subset $\Sigma_{p,m}^{-}(n)$ of $\Sigma_{p,m}(n)$ made of matrices having no poles outside $\mathbb{D}$. It is obvious that for any $q \geq 1$

$$\Sigma_{p,m}^{-}(n) = \Sigma_{p,m}(n) \cap \bar{H}_q^{p \times m}.$$

The subset $R\Sigma_{p,m}(n)$ of real elements in $\Sigma_{p,m}(n)$ is just the collection of transfer-functions of causal discrete-time linear dynamical systems with m inputs, p outputs and n-dimensional minimal state. The corresponding subset $R\Sigma_{p,m}^{-}(n)$ of $\Sigma_{p,m}^{-}(n)$ consists of stable transfer-functions.

Any $W \in \Sigma_{p,m}^{-}(n)$ can be factorized as

$$W = Q^{-1} R \tag{1.9}$$

where Q belongs to I_n^p while R is a rational matrix which is analytic in the closed disk. Moreover, such a pair (Q, R) is unique up to left multiplication by some unitary matrix. If, furthermore, W happens to be real, the pair (Q, R) can also be chosen so as to be real. Expression (1.9) will be referred to as the left Douglas–Shapiro–Shields factorization of W and holds more generally for strictly noncyclic functions [24] [29]. An elementary account of the rational version above, which is all we shall need, and of the real case can be found in [13]. The fact that Q is, up to a unitary factor, uniquely determined by the property of having the same McMillan degree as W can be viewed as a consequence of Fuhrmann's realization theory. Working with the transpose allows one to define similarly a right Douglas–Shapiro–Shields factorization $W = R_1 Q_1^{-1}$, where this time $Q_1 \in I_n^m$.

For z and w two complex numbers, we set

$$\rho_w(z) = 1 - zw^* \quad (1.10)$$

$$b_w(z) = (z-w)/(1-zw^*). \quad (1.11)$$

Given a matrix valued function $A(z)$, the function $A^\sharp(z)$ is defined to be $(A(\frac{1}{z^*}))^*$.

Throughout, the terms smooth and C^∞ are used interchangeably. If M is a manifold, modelled on some Banach space, the tangent space to M at x will be denoted by $\mathcal{T}_x(M)$ or simply by $\mathcal{T}_x$ if M is understood from the context. If $f : M_1 \to M_2$ is a smooth map between two manifolds, the symbol $\mathbf{D}f(x)$ is intended to mean the derivative of f at the point $x \in M_1$ which is a linear map $\mathcal{T}_x(M_1) \to \mathcal{T}_{f(x)}(M_2)$. The effect of $\mathbf{D}f(x)$ on the vector v will be denoted by $\mathbf{D}f(x).v$. If $f(x_1, x_2, ..., x_k)$ is a function of k arguments, $\mathbf{D}_j f(x)$ designates partial derivative with respect to x_j. For these and other basic notions in differential geometry (such as charts, submanifolds, embeddings and the like), we refer to [44], [1] and [34].

The symbol ■ will mark the end of a proof.

2 The differential structure of I_n^p

Let us define

$$(I_n^p)^\sharp = \{Q^\sharp;\ Q \in I_n^p\},$$

and observe that $Q^\sharp$ is then also Q^{-1} by definition of I_n^p. It is clear that $(I_n^p)^\sharp$ is the subset of $\Sigma_{p,p}^-(n)$ consisting of those matrices M satisfying $MM^\sharp = I$. In this section, we study I_n^p by applying to $(I_n^p)^\sharp$ some standard devices from system–theory. We first proceed with some preliminaries on the geometry of $\Sigma_{p,m}^-(n)$.

2.1 Embedding $\Sigma_{p,m}^-(n)$ in $\bar{H}_q^{p\times m}$

We denote by Ω the open subset of $\mathbb{C}^{n\times n} \times \mathbb{C}^{n\times m} \times \mathbb{C}^{p\times n} \times \mathbb{C}^{p\times m}$ consisting of all minimal 4-tuples (A, B, C, D). Let

$$\Pi : \Omega \to \Sigma_{p,m}(n)$$

be given by

$$(A, B, C, D) \longrightarrow D + C(zI_n - A)^{-1}B. \quad (2.1)$$

We obtain a topology on $\Sigma_{p,m}(n)$ by requiring Π to be a quotient map for the similarity relation. It is by now a standard procedure, originally due to Hazewinkel and Kalman, to make $\Sigma_{p,m}(n)$ into an abstract smooth manifold by constructing local sections of the map Π. More precisely, $\Sigma_{p,m}(n)$ can be covered by a finite collection of open sets $(\mathcal{V}_k)$ such that each $\mathcal{V}_k$ is the domain of a map $\phi_k : \mathcal{V}_k \to \Omega$ with the property that

$$\Pi \circ \phi_k = id$$

where id denotes the identity map. For $W \in \mathcal{V}_k$, a certain subcollection of the entries of $\phi_k(W)$ will serve as coordinates on $\Sigma_{p,m}(n)$ which is endowed in that way with a complex structure of dimension $mp + n(m+p)$, see for instance [41], [37], [22], [36]. With this definition Π becomes a smooth map. This complex structure in itself plays here no role since it does not carry over to $(I_n^p)^\sharp$. For this reason, we shall deal only with the underlying real structure of dimension $2mp + 2n(m+p)$. However, many subsequent computations look more natural over $\mathbb{C}$ and we shall often find it convenient to use complex dimensions rather than real ones. In the sequel, dimension means real dimension unless the contrary is explicitely stated. Restricting ourselves to real-valued coordinates in the above process, we get a parametrization of $R\Sigma_{p,m}(n)$ which is thus a submanifold of $\Sigma_{p,m}(n)$ of dimension $mp + n(m+p)$.

Now, let Ω^- denote the open subset of Ω consisting of 4-tuples (A, B, C, D) such that the eigenvalues of A all belong to $\mathbb{D}$. It is easy to check that Π is an open map (see e.g. [10]), so that $\Sigma^-_{p,m}(n) = \Pi(\Omega^-)$ is an open subset of $\Sigma_{p,m}(n)$ hence a submanifold of the same dimension, charts being obtained by restriction. Let

$$j_q : \Sigma^-_{p,m}(n) \longrightarrow \bar{H}_q^{p\times m}$$

be the natural inclusion. It is proved in [11], using results from [22], that j_2 restricted to $R\Sigma^-_{p,m}(n)$ is an embedding, that is by definition a smooth immersion which is a homeomorphism onto its image. This we strengthen as follows.

Theorem 2.1 *For $1 \leq q \leq \infty$, the map j_q is an embedding. By restriction, j_q also induces an embedding $R\Sigma^-_{p,m}(n) \to R\bar{H}_q^{p\times m}$.*

Proof: We show smoothness first. Since the natural inclusion $\bar{H}_\infty \to \bar{H}_q$ is continuous, we may restrict ourselves to $q = \infty$. Because j_q can be locally expressed as $j_q \circ \Pi \circ \phi_k$ and ϕ_k is smooth by definition, we need only show that $j_\infty \circ \Pi : \Omega^- \to \bar{H}_\infty$, which is given by (2.1) is smooth. Being linear and continuous, the natural inclusions

$$\mathbb{C}^{p\times m} \to \bar{H}_\infty^{p\times m} \quad \mathbb{C}^{p\times n} \to \bar{H}_\infty^{p\times n} \quad \text{and} \quad \mathbb{C}^{n\times m} \to \bar{H}_\infty^{n\times m}$$

are smooth and so is the affine map

$$A \to (zI - A)$$

from $\mathbb{C}^{n\times n}$ to $L_{\infty}^{n\times n}(\mathbb{T})$. Taking the inverse being a smooth operation on the set of invertible elements in any Banach algebra, we have that

$$A \to (zI - A)^{-1}$$

is also smooth and it assumes values in $\bar{H}_{\infty}^{n\times n}$. Finally, since multiplication is a continuous 3-linear map

$$\bar{H}_{\infty}^{p\times n} \times \bar{H}_{\infty}^{n\times n} \times \bar{H}_{\infty}^{n\times m} \longrightarrow \bar{H}_{\infty}^{p\times m},$$

we get smoothness of (2.1) as desired.

For j_q to be an immersion, we still have to show that the derivative $\mathbf{D}j_q(W)$ is injective at every $W \in \Sigma_{p,m}^{-}(n)$ and that its image splits. This last condition arises because we are embedding $\Sigma_{p,m}^{-}(n)$ in an infinite–dimensional space [44] but is automatically satisfied here since $\operatorname{Im}\mathbf{D}j_q(W)$ is finite dimensional hence splits in the Banach space $\bar{H}_{\infty}^{p\times m}$ (see e.g. [48], [51]). Letting (A, B, C, D) be a minimal realization of W, the image of $\mathbf{D}j_q(W)$ obviously contains the image of the derivative of $j_q \circ \Pi$ at (A, B, C, D). If we differentiate this map, which is formally given by (2.1), with respect to the arguments, we find that

$$\mathbf{D}_4(j_q \circ \Pi) \ : \mathbb{C}^{p\times m} \to \bar{H}_{\infty}^{p\times m} \quad \text{is given by} \quad U \to U \tag{2.2}$$

$$\mathbf{D}_3(j_q \circ \Pi) \ : \mathbb{C}^{p\times n} \to \bar{H}_{\infty}^{p\times m} \quad \text{is given by} \quad V \to V(zI - A)^{-1}B \tag{2.3}$$

$$\mathbf{D}_2(j_q \circ \Pi) \ : \mathbb{C}^{n\times p} \to \bar{H}_{\infty}^{p\times m} \quad \text{is given by} \quad W \to C(zI - A)^{-1}W \tag{2.4}$$

and, using the standard rule to compute the derivative of the inverse in a Banach algebra, we see that

$$\mathbf{D}_1(j_q \circ \Pi) : \mathbb{C}^{n\times n} \to \bar{H}_{\infty}^{p\times m} \tag{2.5}$$

is given by

$$X \to -C(zI - A)^{-1}X(zI - A)^{-1}B.$$

Define $\mathcal{M}_{A,B,C}$ to be the subspace of $\bar{H}_{\infty}^{p\times m}$ spanned by the images of the derivatives (2.3), (2.4) and (2.5) so that

$$\operatorname{Im}\mathbf{D}[j_q \circ \Pi](A, B, C, D) = \operatorname{Im}\mathbf{D}_4(j_q \circ \Pi) + \mathcal{M}_{A,B,C}.$$

Observe the sum is direct since the first term is the space of constant matrix functions while each member of $\mathcal{M}_{A,B,C}$ vanishes at infinity. Therefore, counting complex dimensions leads to

$$\begin{aligned}\dim_{\mathbb{C}}(\operatorname{Im}\mathbf{D}[j_q\circ\Pi](A,B,C,D)) &= mp+\dim_{\mathbb{C}}\mathcal{M}_{A,B,C}\\ &\leq \dim_{\mathbb{C}}(\operatorname{Im}\mathbf{D}j_q(W))\\ &\leq \dim_{\mathbb{C}}\Sigma^-_{p,m}(n)=mp+n(m+p).\end{aligned}$$

To establish injectivity, we will prove that

$$\dim_{\mathbb{C}}\mathcal{M}_{A,B,C}\geq n(m+p) \tag{2.6}$$

by using nice selections, a tool originally due to Brunovsky and Kalman which is also instrumental in defining the manifold structure of $\Sigma_{p,m}(n)$. Specifically, we choose k column vectors $b_{i_1},b_{i_2},...,b_{i_k}$ of B such that

$$b_{i_1},Ab_{i_1},...,A^{\kappa_1}b_{i_1},b_{i_2},Ab_{i_2},...,A^{\kappa_2}b_{i_2},...,b_{i_k},Ab_{i_k},...,A^{\kappa_k}b_{i_k}$$

is a basis of $\mathbb{C}^n$ and it is easy to see that controllability implies the existence of such vectors. This entails of course that $\sum_j\kappa_j=n-k$. Define $\mathcal{W}$ (resp. $\mathcal{X}$) to be the subspace of $\mathbb{C}^{n\times m}$ (resp. $\mathbb{C}^{n\times n}$) consisting of complex matrices whose kernel contains the elements e_{i_j}'s of the canonical basis of $\mathbb{C}^m$ (resp. contains the $A^\kappa b_{i_j}$'s whenever $\kappa<\kappa_j$). The complex dimensions of $\mathcal{W}$ and $\mathcal{X}$ are thus $n(m-k)$ and nk respectively.
Observe now that $\mathcal{M}_{A,B,C}$ is the collection of rational matrix functions of the form

$$V(zI-A)^{-1}B-C(zI-A)^{-1}X(zI-A)^{-1}B+C(zI-A)^{-1}W \tag{2.7}$$

where V, W and X range over $\mathbb{C}^{p\times n}$, $\mathbb{C}^{n\times m}$ and $\mathbb{C}^{n\times n}$ respectively. If we restrict W to $\mathcal{W}$ and X to $\mathcal{X}$, we claim that (2.7) cannot be zero unless V, W and X are all zero. This will show that $\mathcal{M}_{A,B,C}$ contains at least

$$\dim_{\mathbb{C}}\mathbb{C}^{p\times n}+\dim_{\mathbb{C}}\mathcal{W}+\dim_{\mathbb{C}}\mathcal{X}=n(m+p)$$

independent vectors over $\mathbb{C}$ so that (2.6) will hold.
To establish the claim, we assume that (2.7) is identically zero and we evaluate at e_{i_j} for some $j\in\{1,...,k\}$. We get from the definition of $\mathcal{W}$

$$V(zI-A)^{-1}b_{i_j}=C(zI-A)^{-1}X(zI-A)^{-1}b_{i_j}. \tag{2.8}$$

The Taylor expansion of $X(zI-A)^{-1}b_{i_j}$ at infinity is

$$\sum_{l=1}^{\infty}XA^{l-1}b_{i_j}z^{-l}$$

and has a zero of order at least $\kappa_j + 1$ by definition of $\mathcal{X}$.
Because $C(zI - A)^{-1}$ vanishes at infinity, the right hand-side of (2.8) has a zero at infinity of order $\kappa_j + 2$ at least. But then, this must hold for the left hand-side as well. Computing the Taylor expansion yields

$$VA^l b_{i_j} = 0 \text{ for } 0 \leq l \leq \kappa_j.$$

Since $j \in \{1, ..., k\}$ was arbitrary, we get from the definition of the b_{i_j}'s that $V = 0$ and (2.8) implies

$$C(zI - A)^{-1}X(zI - A)^{-1}b_{i_j} = 0 \quad \text{for} \quad 1 \leq j \leq k. \tag{2.9}$$

Now, the identity

$$X(zI - A)^{-1}A^l b_{i_j} = -XA^{l-1}b_{i_j} + zX(zI - A)^{-1}A^{l-1}b_{i_j}$$

and the definition of $\mathcal{X}$ shows that

$$C(zI - A)^{-1}X(zI - A)^{-1}A^l b_{i_j} = zC(zI - A)^{-1}X(zI - A)^{-1}A^{l-1}b_{i_j},$$

provided $1 \leq l \leq \kappa_{i_j}$. From (2.9), we therefore get by induction

$$C(zI - A)^{-1}X(zI - A)^{-1}A^l b_{i_j} = 0 \quad \text{for } 1 \leq j \leq k \text{ and } 0 \leq l \leq \kappa_{i_j}.$$

The fact that the $A^l b_{i_j}$'s form a basis of $\mathbb{C}^n$ implies now

$$C(zI - A)^{-1}X(zI - A)^{-1} = 0 \quad \text{hence} \quad C(zI - A)^{-1}X = 0.$$

Computing the Taylor expansion at infinity and using observability yields $X = 0$. Since we assumed that (2.7) is zero, we finally conclude in the same manner that $W = 0$ thereby achieving the proof of the claim.

To see that j_q is an embedding, there remains for us to show that it is a homeomorphism onto its image, in other words that it has a continuous inverse. Let W_k be a sequence of rational matrix functions of McMillan degree n converging in $\bar{H}_q^{p\times m}$ to some W which is also of degree n. The Cauchy formula implies that the convergence is uniform over any compact subset of $\{|z| > 1\}$, so that $W_k(\infty) \to W(\infty)$ and, say, the first $2n - 1$ derivatives at infinity of W_k also converge to those of W. From this collection of derivatives, one can construct the associated $np \times nm$ block-Hankel matrices Ha_k and Ha of W_k and W and from them, knowing that degree of all the rational matrices involved is n, one can derive realizations $(A_k, B_k, C_k, W_k(\infty))$ of W_k and $(A, B, C, W(\infty))$ of W using for instance Ho's algorithm [42] which yields rational formulae for the coefficients of the realization in terms of the entries of Ha_k and Ha. The only arbitrary choice in the algorithm is that of a nonsingular submatrix

of the Hankel matrix but, if such a choice is made on Ha, the corresponding submatrix of Ha_k will be nonsingular too when k is large enough since $\mathrm{Ha}_k \to \mathrm{Ha}$. In this manner, we may arrange things so that $(A_k, B_k, C_k, W_k(\infty))$ converges to $(A, B, C, W(\infty))$ in Ω^-. Since Π is continuous, we get that $W_k = \Pi(A_k, B_k, C_k, W_k(\infty))$ converges to $W = \Pi(A, B, C, W(\infty))$ in $\Sigma^-_{p,m}(n)$ as desired.
Since $R\Sigma^-_{p,m}(n)$ is a submanifold of $\Sigma^-_{p,m}(n)$, the assertion on the restriction of j_q to real matrix functions is obvious. ∎

Theorem 2.1 makes it possible to identify the tangent space $\mathcal{T}_W$ of $\Sigma^-_{p,m}(n)$ at the point W with

$$\operatorname{Im} \mathbf{D}j_q(W) = \operatorname{Im} \mathbf{D}\Pi(A, B, C, D))$$

which is a subspace of $\bar{H}^{p\times m}_q$ of complex dimension $mp + n(m+p)$. This subspace we shall now describe in more details. To this effect, we need some pieces of notation.

If $Q \in I^p_n$ and $Q_1 \in I^m_n$, we denote by $\mathcal{L}_Q$ (resp. $\mathcal{R}_{Q_1}$) the space of rational $p \times m$ matrix functions M such that M has no pole in the closed unit disk and $Q^{-1}M$ (resp. MQ_1^{-1}) has no pole in $\{|z| \geq 1\}$ and vanishes at infinity (resp. has no pole in $\{|z| \geq 1\} \cup \infty$). Note that $\mathcal{L}_Q$ may alternatively be defined as the space of matrices whose columns belong to $H^p_2 \ominus QH^p_2$ which is the prototype of the left shift–invariant subspace of complex dimension n that will appear in section 3. Consequently, $\mathcal{L}_Q$ has complex dimension nm. Similarly, we observe that M belongs to $\mathcal{R}_{Q_1}$ if and only if M^t belongs to $\mathcal{L}_{zQ_1^t}$. Since the McMillan degree of zQ_1^t is $m + n$, the complex dimension of $\mathcal{R}_{Q_1}$ is $p(m + n)$.

We now define $\mathcal{W}(Q, Q_1)$ to be the space of rational matrix functions M analytic in $\{|z| \leq 1\}$ such that $Q^{-1}MQ_1^{-1}$ has no poles in $\{|z| \geq 1\}$ including at infinity. It is plain that

$$\mathcal{L}_Q \subset \mathcal{W}(Q, Q_1) \quad \text{and} \quad Q\mathcal{R}_{Q_1} \subset \mathcal{W}(Q, Q_1).$$

Therefore,

$$\mathcal{L}_Q + Q\mathcal{R}_{Q_1} \subset \mathcal{W}(Q, Q_1) \tag{2.10}$$

and the sum is direct since any member M in the intersection must satisfy

$$Q^{-1}M \in z^{-1}\bar{H}^{p\times m}_2 \cap H^{p\times m}_2 = 0.$$

Conversely, let M be an element of $\mathcal{W}(Q, Q_1)$. Then $Q^{-1}M$ belongs to $L^{p\times m}_2(\mathbb{T})$ and can be written as

$$Q^{-1}M = h_+ + h_- \quad \text{where } h_+ \in H^{p\times m}_2 \text{ and } h_- \in \bar{H}^{p\times m}_2. \tag{2.11}$$

Multiplying (2.11) on the right by Q_1^{-1}, we find that

$$h_+Q_1^{-1} = Q^{-1}MQ_1^{-1} - h_-Q_1^{-1} \in \bar{H}_2^{p\times m},$$

so that $h_+ \in \mathcal{R}_{Q_1}$. Multiplying now (2.11) to the left by Q, we obtain

$$Qh_- = M - Qh_+ \in H_2^{p\times m}.$$

Hence $Qh_- \in \mathcal{L}_Q$ and (2.10) is in fact an equality:

$$\mathcal{L}_Q \oplus Q\mathcal{R}_{Q_1} = \mathcal{W}(Q, Q_1). \tag{2.12}$$

In particular, we see that

$$dim_{\mathbb{C}}\mathcal{W}(Q, Q_1) = dim_{\mathbb{C}}\mathcal{L}_Q + dim_{\mathbb{C}}\mathcal{R}_{Q_1} = n(m+p) + pm. \tag{2.13}$$

We may now identify our tangent space as follows.

Proposition 2.1 *Let q satisfy $1 \le q \le \infty$. Let W belong to $\Sigma_{p,m}^-(n)$ and $W = Q^{-1}R = R_1Q_1^{-1}$ be left and right Douglas–Shapiro–Shields factorizations. Then, at W, the tangent space to $\Sigma_{p,m}^-(n)$ viewed as an embedded submanifold in $\bar{H}_q^{p\times m}$ is given by*

$$\mathcal{T}_W = Q^{-1}\mathcal{W}(Q, Q_1)Q_1^{-1}. \tag{2.14}$$

At $W \in R\Sigma_{p,m}^-(n)$, the tangent space to $R\Sigma_{p,m}^-(n)$ is still given by (2.14) provided Q and Q_1 are chosen real and we limit ourselves to real members of $\mathcal{W}(Q, Q_1)$.

Proof: Let (A, B, C, D) be a minimal realization of W. From Theorem 2.1, the space $\mathcal{T}_W$ is made of matrix–valued functions of the form

$$\begin{aligned} &U + V(zI - A)^{-1}B \\ &\quad -C(zI-A)^{-1}X(zI-A)^{-1}B + C(zI-A)^{-1}W \end{aligned} \tag{2.15}$$

where U, V, W and X range over $\mathbb{C}^{p\times m}$, $\mathbb{C}^{p\times n}$, $\mathbb{C}^{n\times m}$ and $\mathbb{C}^{n\times n}$ respectively. Let

$$C(zI - A)^{-1} = \Xi^{-1}\Delta \quad \text{and} \quad (zI - A)^{-1}B = \Delta_1\Xi_1^{-1}$$

be Douglas–Shapiro–Shields factorizations. It is obvious that (2.15) can be written as

$$\Xi^{-1}\Lambda\Xi_1^{-1}, \tag{2.16}$$

for some Λ which is rational and analytic in the closed disk. We claim that $\Xi = UQ$ where U is a unitary matrix. Indeed, since $C(zI - A)^{-1}$ has McMillan degree n,

$$W = \Xi^{-1}(\Delta B + \Xi D) \quad \text{and} \quad W = Q^{-1}R$$

are two left Douglas–Shapiro–Shields factorizations of W. Since such a factorization is unique up to left multiplication by a unitary factor, we are done. A similar argument on the right shows that $\Xi_1 = Q_1 U_1$ where U_1 is again unitary.
Therefore, functions of the form (2.16) are in $Q^{-1}\mathcal{W}(Q, Q_1)Q_1^{-1}$ so that $\mathcal{T}_W$ is included in the right hand-side of (2.14). But we must then have equality since complex dimensions coincide as follows from 2.13. The proof of the real case is *mutatis mutandis* the same. ∎

2.2 The differential structure of I_n^p

In this section, we prove that I_n^p, $I_n^p/\mathcal{U}_p$ and their real analogues are smooth manifolds. Since we now deal with square matrices, m is to be set equal to p in the preceding results.

Proposition 2.2 *The set $(I_n^p)^\sharp$ is an embedded submanifold in $\Sigma_{p,p}^-(n)$ of dimension p^2+2np. When embedded in $\bar{H}_q^{p\times p}$ with $1 \le q \le \infty$, the tangent space to $(I_n^p)^\sharp$ at $Q^\sharp$ is given by*

$$\mathcal{T}_{Q^\sharp}(I_n^p)^\sharp = \{(S + F - F^\sharp)Q^{-1}\}, \tag{2.17}$$

where $S \in \mathcal{S}_p$ and $F \in Q^{-1}\mathcal{L}_Q$. The subset $(RI_n^p)^\sharp$ is also an embedded submanifold of dimension $p(p-1)/2 + np$. Its tangent space is still given by (2.17) provided S is restricted to range over skew-symmetric matrices and F over real elements of $\mathcal{L}_Q$.

Proof: Embed $\Sigma_{p,p}^-(n)$ into $\bar{H}_\infty^{p\times p}$ and define a map

$$\Upsilon : \Sigma_{p,p}^-(n) \longrightarrow L_\infty^{p\times p}(\mathbb{T})$$

by the formula

$$\Upsilon(W) = WW^\sharp.$$

Since $W \to W^\sharp$ is a linear continuous map $H_\infty^{p\times p} \to \bar{H}_\infty^{p\times p}$ and multiplication is bilinear and continuous in $L_\infty^{p\times p}(\mathbb{T})$, it is obvious that Υ is smooth, and its derivative at W

$$\mathbf{D}\Upsilon(W) : \mathcal{T}_W \longrightarrow L_\infty^{p\times p}(\mathbb{T})$$

is given by

$$M \longrightarrow MW^\sharp + WM^\sharp. \tag{2.18}$$

Let $W = Q^{-1}R = R_1 Q_1^{-1}$ be right and left Douglas–Shapiro–Shields factorizations of W. By Proposition 2.1, we have that

$$\mathcal{T}_W = Q^{-1}\mathcal{W}(Q, Q_1)Q_1^{-1}$$

and from (2.12) we can write every $M \in \mathcal{T}_W$ as

$$M = Q^{-1}(\ell_Q + Qr_{Q_1})Q_1^{-1} \tag{2.19}$$

for some $\ell_Q \in \mathcal{L}_Q$ and some $r_{Q_1} \in \mathcal{R}_{Q_1}$. Plugging this into (2.18) and taking into account that $Q^{-1} = Q^\sharp$ and $Q_1^{-1} = Q_1^\sharp$, we get

$$\mathbf{D}\Upsilon(W).M = Q^\sharp \ell_Q R_1^\sharp + r_{Q_1} R_1^\sharp + R_1 \ell_Q^\sharp Q + R_1 r_{Q_1}^\sharp. \tag{2.20}$$

Note that $Q^\sharp \ell_Q R_1^\sharp$ belongs to $z^{-1}\bar{H}_2^{p\times p}$, so the Fourier coefficient of index 0 in (2.20) is $F_0 + F_0^*$ where F_0 is the zero–th Fourier coefficient of $r_{Q_1} R_1^\sharp$. If we choose $r_{Q_1} = KQ_1$ where K is a constant matrix, we get $F_0 = KW^\sharp(0) = KW(\infty)^*$. Let us first assume that $W(\infty)$ is invertible so that the linear map $\alpha : \mathcal{R}_{Q_1} \to \mathbb{C}^{p\times p}$ defined by $r_{Q_1} \to F_0$ is surjective. Then, the subspace $\mathcal{R}'_{Q_1} = \alpha^{-1}(\mathcal{S}_p)$ has codimension p^2 in $\mathcal{R}_{Q_1}$ hence dimension $p^2 + 2np$.

Suppose now that $M \in \ker \mathbf{D}\Upsilon(W)$ so that (2.20) is zero. Then $r_{Q_1} \in \mathcal{R}'_{Q_1}$ and ℓ_Q is completely determined by r_{Q_1}. Indeed, R_1 is invertible in $L_\infty^{p\times p}(\mathbb{T})$ by our assumption on $W(\infty)$, and if we denote by P_+ the orthogonal projection onto $H_2^{p\times p}$, we have

$$\ell_Q^\sharp = -R_1^{-1}\left[P_+(r_{Q_1} R_1^\sharp + R_1 r_{Q_1}^\sharp)\right] Q^\sharp.$$

Therefore,

$$\dim\ker \mathbf{D}\Upsilon(W) \le \dim \mathcal{R}'_{Q_1} = p^2 + 2np. \tag{2.21}$$

For $a \in \mathbb{D}$, the map $\varphi_a : H \to H \circ b_a$ (where b_a is defined by 1.11) is a linear isometry of $L_\infty^{p\times p}$ preserving $\Sigma_{p,p}^-(n)$ and it is immediately checked that φ_a commutes with Υ. Applying the chain rule, we get

$$\mathbf{D}\Upsilon(\varphi_a(W)) \circ \varphi_a = \varphi_a \circ \mathbf{D}\Upsilon(W)$$

so that

$$\dim\ker \mathbf{D}\Upsilon(W) = \dim\ker \mathbf{D}\Upsilon(\varphi_a(W)). \tag{2.22}$$

Now, if we merely assume that W is invertible in $L_\infty^{p\times p}(\mathbb{T})$, it is possible to adjust a so that $W_a(\infty)$ is invertible. In view of (2.22), we conclude that (2.21) still holds in this case.

We claim that (2.21) is an equality whenever $W \in (I_n^p)^\sharp$. Since the rank cannot decrease locally, we will then conclude that the kernel of $\mathbf{D}\Upsilon$ has dimension $p^2 + 2np$ on a neighborhood of $(I_n^p)^\sharp$ and the infinite–dimensional version of the constant rank theorem (see e.g. [17] 5.10.5-6) will imply that $(I_n^p)^\sharp = \Upsilon^{-1}(I_p)$ is a smooth manifold whose tangent space at W is $\ker \mathbf{D}\Upsilon(W)$.

To prove the claim, set $W = Q^\sharp$. Plugging $R_1 = I_p$ and $Q_1 = Q$ in (2.20) yields

$$\mathbf{D}\Upsilon(W).M = [r_Q + \ell_Q^\sharp Q] + [r_Q + \ell_Q^\sharp Q]^\sharp. \tag{2.23}$$

The first bracket in the right hand-side of (2.23) is analytic in the closed unit disk by the definition of ℓ_Q and r_Q so that (2.23) is zero if and only if

$$r_Q + \ell_Q^\sharp Q = S \tag{2.24}$$

where S is a skew–Hermitian constant. Now, for any $\ell_Q \in \mathcal{L}_Q$ and any $S \in \mathcal{S}_p$, (2.24) determines a unique r_Q which is readily checked to belong to $\mathcal{R}_Q$. Moreover, r_Q can be zero only if ℓ_Q and S are both zero since $\ell_Q^\sharp Q$ vanishes at zero. The kernel of $\mathbf{D}\Upsilon(W)$ is thus of dimension $p^2 + 2np$ and this was the claim.

To establish (2.17), observe that $\ker \mathbf{D}\Upsilon(W)$ at $W = Q^\sharp$ is the set of matrix-valued functions of the form (2.19) where $Q_1 = Q$ and such that (2.24) holds. A simple computation shows that this set is just the collection

$$(S + (Q^{-1}\ell_Q) - (Q^{-1}\ell_Q)^\sharp)Q^{-1},$$

where $S \in \mathcal{S}_p$ and $\ell_Q \in \mathcal{L}_Q$. But it is easily checked from the definitions that $\mathcal{L}_Q = QP_-(Q^{-1}H_2^{p\times p})$ and this implies (2.17).

In the real case, the proof is identical provided a is chosen real. Note also that skew–hermitian has to be replaced by skew–symmetric so that $R\mathcal{R}'_{Q_1}$ has codimension $p^2 - p(p-1)/2$ in $R\mathcal{R}_{Q_1}$ hence dimension $p(p-1)/2 + np$. ∎

We now introduce the set $NI_n^p \subset I_n^p$ of functions Q normalized by the condition $Q(1) = I_p$ and its real counterpart RNI_n^p. These functions will serve as representatives in $I_n^p/\mathcal{U}_p$ and we will study $(NI_n^p)^\sharp$.

Proposition 2.3 *The set $(NI_n^p)^\sharp$ is a smooth submanifold of $(I_n^p)^\sharp$ of dimension $2np$ and the product map*

$$(NI_n^p)^\sharp \times \mathcal{U}_p \longrightarrow (I_n^p)^\sharp \tag{2.25}$$

is a diffeomorphism. A similar statement holds in the real case replacing $\mathcal{U}_p$ by $\mathcal{O}_p$, the dimension of $(RNI_n^p)^\sharp$ being np.

Proof: Let E be the evaluation map at 1. It is linear and continuous $\mathcal{A}^{p\times p} \to \mathbb{C}^{p\times p}$ (recall $\mathcal{A}$ is the disk algebra). Since $(I_n^p)^\sharp \subset \mathcal{A}^{p\times p}$, the restriction

$$E : (I_n^p)^\sharp \to \mathcal{U}_p$$

is smooth and its derivative at any $Q^\sharp$, which is defined from $\mathcal{T}_{Q^\sharp}(I_n^p)^\sharp$ into $\mathcal{T}_{Q^\sharp(1)}\mathcal{U}_p$ is again evaluation at 1. Since the tangent space to $\mathcal{U}_p$ at I_p is $\mathcal{S}_p$, it follows from (2.17) that E is submersive at every point of $E^{-1}(I_p) = (NI_n^p)^\sharp$ which is thus a submanifold of codimension p^2 hence of dimension $2np$. Consider now the map (2.25). It is obviously smooth together with its inverse

$$Q^\sharp \to (Q^\sharp Q(1), Q^\sharp(1)).$$

The proof in the real case is similar. ∎

Recall that two members of $(I_n^p)^\sharp$ (resp. $(RI_n^p)^\sharp$) are equivalent if they are equal up to right multiplication by a unitary (resp. orthogonal) factor and that that the sets of equivalence classes are denoted by $(I_n^p)^\sharp/\mathcal{U}_p$ and $(RI_n^p)^\sharp/\mathcal{O}_p$ respectively. We endow them with the quotient topology and it is an immediate consequence of proposition 2.3 that the natural map $(NI_n^p)^\sharp \to (I_n^p)^\sharp/\mathcal{U}_p$ is a homeomorphism allowing us to carry over to $(I_n^p)^\sharp/\mathcal{U}_p$ the manifold structure of $(NI_n^p)^\sharp$. We proceed similarly in the real case.

Changing z into $1/z$, we are now able to restate the results of this section in terms of I_n^p:

Theorem 2.2 *The set I_n^p is a smooth manifold of dimension $p^2 + 2np$ embedded in $H_q^{p\times p}$ for $1 \leq q \leq \infty$. The set NI_n^p is a smooth submanifold of dimension $2np$ homeomorphic to $I_n^p/\mathcal{U}_p$. As well, the subset RI_n^p is a smooth manifold of dimension $p(p-1)/2 + np$ embedded in $RH_q^{p\times p}$, and the set RNI_n^p is a smooth submanifold of dimension np homeomorphic to $RI_n^p/\mathcal{U}_p$.*

We were able so far to establish the smoothness of I_n^p and to identify its tangent space in $H_q^{p\times p}$. However, the above approach does not provide us with explicit charts since we did rely on the constant rank theorem. It is the purpose of the next subsection to fill this gap by making contact with the classical manifold of observable pairs.

2.3 Constructing charts from realizations

We need to introduce the set of observable pairs. Let Ω_o denote the open subset of $\mathbb{C}^{p\times n} \times \mathbb{C}^{n\times n}$ consisting of all observable pairs (C, A). Two pairs (C_1, A_1) and (C_2, A_2) will be said to be equivalent if there exists $T \in GL(n)$ such that

$$\begin{aligned} C_1 &= C_2T \\ A_1 &= T^{-1}A_2T. \end{aligned}$$

Let $Obs_p(n)$ be the set of equivalence classes for the above relation and

$$\Pi_o : \Omega_o \to Obs_p(n)$$

be the natural map. When $Obs_p(n)$ is equipped with the quotient topology, the Hazewinkel–Kalman method already mentioned in section 2.1 (see [37], [38], [10]) allows one to construct local sections of Π_o endowing $Obs_p(n)$ with a complex structure of complex dimension np. With this definition, Π_o is smooth. Again, we shall only work with the underlying real structure of dimension $2np$. The set $RObs_p(n)$ of equivalence classes of real pairs is in turn a submanifold of dimension np. We single out the subset Ω_o^- of Ω_o consisting of pairs (C, A) for which the spectrum of A is in $\mathbb{D}$. Being open in $Obs_p(n)$, the set $Obs_p^-(n) = \Pi_o(\Omega_o^-)$ is naturally a smooth manifold.

We first characterize realizations of members of $(I_n^p)^\sharp$. Representations similar to the one below can be found in [31], [5], [7] or [28] in the case of the half-plane. We give a simple proof for sake of completeness.

Proposition 2.4 *Let W be a rational $\mathbb{C}^{p\times p}$–valued function analytic at infinity and let $W(z) = D + C(zI_n - A)^{-1}B$ be a minimal realization of W. Then, the following statements are equivalent:*

(i) *The function W belongs to $(I_n^p)^\sharp$.*

(ii) *There exists a positive definite matrix P such that:*

$$A^*PA + C^*C = P \tag{2.26}$$

$$A^*PB + C^*D = 0 \tag{2.27}$$

$$B^*PB + D^*D = I \tag{2.28}$$

Proof: Let us prove that (i) implies (ii). Since the given realization is minimal the spectrum of A is contained in a closed disk

$$E_\alpha = \{z \in \mathbb{C}; |z| \leq \alpha < 1\}.$$

Therefore the function $W(z)W(z)^*$ is in $L_2^{p\times p}(\mathbb{T})$ and, for $|z|=1$ we can write the power series expansions:

$$\begin{aligned}W^*(z)W(z) &= (D^*+B^*(z^{-1}I_n-A^*)^{-1}C^*)(D+C(zI_n-A)^{-1}B)\\ &= D^*D+D^*C\left(z^{-1}\sum_{j=0}^{\infty}(z^{-1}A)^j\right)B+B^*\left(z\sum_{j=0}^{\infty}(zA^*)^j\right)C^*D\\ &+B^*\left(z\sum_{j=0}^{\infty}(zA^*)^j\right)C^*C\left(z^{-1}\sum_{j=0}^{\infty}(z^{-1}A)^j\right)B.\end{aligned}$$

Equating Fourier coefficients in the equation $W^*W=I_p$, we get

$$\frac{1}{2i\pi}\int_{\mathbb{T}}W(z)^*W(z)\frac{dz}{z}=D^*D+B^*\sum_{j=0}^{\infty}\left(A^{*j}C^*CA^j\right)B=I_p \tag{2.29}$$

and for $k>0$

$$\begin{aligned}&\frac{1}{2i\pi}\int_{\mathbb{T}}z^{-k}W(z)^*W(z)\frac{dz}{z}=\\ &B^*(A^*)^{k-1}C^*D+B^*A^{*k}\sum_{j=0}^{\infty}\left(A^{*j}C^*CA^j\right)B=0.\end{aligned} \tag{2.30}$$

Define

$$P=\sum_{j=0}^{\infty}\left((A^*)^jC^*CA^j\right). \tag{2.31}$$

Observability of the pair (A,C) implies that P is positive definite. From the definition of P, (2.26) holds and (2.28) follows from (2.29). Rewrite (2.30) as

$$B^*A^{*(k-1)}(C^*D+A^*PB)=0$$

to conclude that (2.27) holds, thanks to the controllability of the pair (A,B). To prove that (ii) implies (i), observe that the unique solution to (2.26) is given by (2.31) and reverse the above arguments. ∎

Note that equations (2.26)–(2.28) may be capsulized as

$$\begin{pmatrix}A^*P & C\\ B^*P & D^*\end{pmatrix}\begin{pmatrix}AP^{-1} & B\\ CP^{-1} & D\end{pmatrix}=\begin{pmatrix}I_n & 0\\ 0 & I_p\end{pmatrix}. \tag{2.32}$$

We now state the main result of this section.

Proposition 2.5 *Let $\nu:\Omega^-\longrightarrow\Omega_o^-$ be the natural projection*

$$(A,B,C,D)\longrightarrow(C,A).$$

The map $\Pi_o\circ\nu\circ\Pi^{-1}$ is a diffeomorphism from $(NI_n^p)^\sharp$ onto $Obs_p^-(n)$. By restriction, ν induces a diffeomorphism between $(RNI_n^p)^\sharp$ and $RObs_p^-(n)$.

Proof: Set $\xi = \Pi_o \circ \nu \circ \Pi^{-1}$. This application is clearly well–defined and continuous thanks to the smoothness of the local sections ϕ_k of Π. To construct a set–theoretic inverse, we first need to show that any pair $(C, A) \in \Omega_o^-$ can be completed into a minimal 4–tuple (A, B, C, D) satisfying (2.26)–(2.28) fo some strictly positive matrix P (which is then given by (2.31). If A is invertible, this is done in [5, Theorem 3.10]. When A is not invertible, we approximate A by a sequence of invertible A_k such that $(C, A_k) \in \Omega_o^-$, and we let B_k and D_k denote completions of (C, A_k) satisfying (2.26)–(2.28). The corresponding matrices P_k are bounded thanks to equation (2.31) and have a convergent subsequence with limit, say, P. By a classical inertia theorem (see e.g. [5, Theorem 3.15]), P is invertible. Therefore, formula (2.28) implies that B_k and D_k are uniformly bounded and thus have convergent subsequences, with limit, say, B and D respectively, which realize the desired completion. Next, we claim that such a completion is unique up to right multiplication of B and D by the same unitary constant. Assume indeed that B_1 and D_1 is another one. Since P is uniquely determined by A and C, (2.32) shows that (B^*P, D^*) and (B_1^*P, D_1^*) have the same span of their rows. Thus there exists $P_1 \in GL(n)$ such that $B_1^* = P_1 B^*$ and $D_1^* = P_1 D^*$. Again from (2.32), P_1 is unitary.

Now, let W be defined by $W(z) = D + C(zI_n - A)^{-1}B$. The set–theoretic inverse to ξ is given by $WW(1)^\sharp$.

To complete the proof, it is enough to show that the above completion process $\Omega_o^- \to \Omega^-$ may be performed locally by a smooth function.

For $(C, A) \in \Omega_o^-$, let $P = P(A, C)$ be the smooth function of (C, A) defined by (2.31). Consider the map

$$\chi : \Omega^- \mapsto \mathbb{C}^{n \times p} \times \mathcal{H}_p$$

as

$$\chi(A, B, C, D) = \left\{ \begin{array}{c} A^*P(A, C)B + C^*D \\ B^*P(A, C)B + D^*D - I \end{array} \right\}$$

it follows from Proposition 2.4 that $\Pi^{-1}(I_n^p)^\sharp = \chi^{-1}(0)$.

Now χ is obviously smooth and we claim that the partial derivative with respect to (B, D) is surjective at every point of $\chi^{-1}(0)$. To this effect we compute

$$\mathbf{D}_2\chi(A, B, C, D) : V \mapsto \left[\begin{array}{c} A^*PV \\ V^*PB + B^*PV \end{array} \right] \tag{2.33}$$

$$\mathbf{D}_4\chi(A, B, C, D) : X \mapsto \left[\begin{array}{c} C^*X \\ X^*D + D^*X \end{array} \right] \tag{2.34}$$

Let $\left[\begin{array}{c} S_1 \\ S_2 \end{array} \right]$ belong to $\mathbb{C}^{n \times p} \times \mathcal{H}_p$. Since P is invertible, (2.26) implies that

$$A^*CV_1 + C^*X_1 = S_1$$

is solvable. Define $V_2 = BS_2/2$ and $X_2 = DS_2/2$. Then we get:

$$\begin{aligned}
[\mathbf{D}_2\chi.V_2 + \mathbf{D}_4\chi.X_2] &= \begin{bmatrix} A^*PV_2 + C^*X_2 \\ V_2^*PB + B^*PV_2 + X_2^*D + D^*X_2 \end{bmatrix} \\
&= \begin{bmatrix} (A^*PB + C^*D)S_2/2 \\ S_2/2(B^*PB + D^*D) + (B^*PB + D^*D)S_2/2 \end{bmatrix} \\
&= \begin{bmatrix} 0 \\ S_2 \end{bmatrix}.
\end{aligned}$$

Setting $X = X_1 + X_2$ and $Y = Y_1 + Y_2$, we obtain

$$[\mathbf{D}_2\chi.V_2 + \mathbf{D}_4\chi.X_2] = \begin{bmatrix} S_1 \\ S_2 \end{bmatrix}, \tag{2.35}$$

thereby proving our claim.

It is now easy to construct a smooth local section of ν: start from (C, A) and take a completion (A, B, C, D). Extract from (B, D) a set of variables with respect to which the partial derivative is an isomorphism and apply the implicit function theorem. The proof in the real case is similar. ∎

Changing z into z^{-1}, we obtain at once

Corollary 2.1 *The space I_n^p is diffeomorphic to $I_n^p/\mathcal{U}_p \times \mathcal{U}_p$, and $I_n^p/\mathcal{U}_p$ is diffeomorphic to $Obs_p^-(n)$. Similar statements hold for RI_n^p, $\mathcal{O}_p$ and $RObs_p^-(n)$.*

Beside the fact that it expresses our spaces of inner matrix-valued functions in terms of familiar objects, this corollary has some topological consequences. For E a topological space, let us denote by $\mathbf{H}_k(E, G)$ the n-th homology group of E with coefficients in G. We further set $\mathcal{G}_{k,l}$ and $R\mathcal{G}_{k,l}$ to be the complex and real Grassmann manifolds of k–subspaces of an l–space.

Corollary 2.2 *$I_n^p/\mathcal{U}_p$, $RI_n^p/\mathcal{O}_p$, and I_n^p are connected while RI_n^p has two connected components. For $k \geq 0$, there are isomorphisms*

$$\mathbf{H}_k(I_n^p/\mathcal{U}_p, \mathbb{Z}) = \mathbf{H}_k(\mathcal{G}_{n,n+p-1}, \mathbb{Z}), \tag{2.36}$$

$$\mathbf{H}_k(I_n^p, \mathbb{Q}) = \sum_{l_1+l_2=k} \mathbf{H}_{l_1}(\mathcal{G}_{n,n+p-1}, \mathbb{Q}) \otimes \mathbf{H}_{l_2}(\mathcal{U}_p, \mathbb{Q}), \tag{2.37}$$

$$\mathbf{H}_k(RI_n^p/\mathcal{O}_p, \mathbb{Z}/2\mathbb{Z}) = \mathbf{H}_k(R\mathcal{G}_{n,n+p-1}, \mathbb{Z}/2\mathbb{Z}), \tag{2.38}$$

$$\mathbf{H}_k(RI_n^p, \mathbb{Z}/2\mathbb{Z}) = \sum_{l_1+l_2=k} \mathbf{H}_{l_1}(R\mathcal{G}_{n,n+p-1}, \mathbb{Z}/2\mathbb{Z}) \otimes \mathbf{H}_{l_2}(\mathcal{O}_p, \mathbb{Z}/2\mathbb{Z}). \tag{2.39}$$

Proof: We may replace $Obs_p^-(n)$ by $Obs_p(n)$ and $RObs_p^-(n)$ by $RObs_p(n)$ in corollary 2.1 because it is shown in [35] that these two spaces are diffeomorphic (the theorem in [35] is for $\Sigma_{p,m}(n)$ and $\Sigma_{p,m}^-(n)$ but its proof would yield the same conclusion for observable pairs). Then (2.36) and (2.37) are implied by the fact that $Obs_p(n)$ and $\mathcal{G}_{n,n+p-1}$ (resp. $RObs_p(n)$ and $R\mathcal{G}_{n,n+p-1}$) have the same integral (resp. mod 2) homology by the results of [38] also reported in [39]. Now, the Künneth theorem (see e.g. [46]) implies (2.37) and (2.39). Since $\mathbf{H}_0(E, G)$ is a direct sum on the number of connected components, the first statement follows from the connectivity properties of $\mathcal{U}_p$, $\mathcal{O}_p$, and the Grassmann manifolds. ∎

The connectivity statement can of course be proved without resorting to homology from known properties of minimal pairs. For the computation of the homology of $\mathcal{G}_{k,l}$, $\mathcal{U}_p$, and $\mathcal{O}_p$ we refer the reader to [43].

3 Charts using the Schur algorithm

In this section, we study I_n^p and related manifolds using Schur analysis. We begin with a review of finite–dimensional reproducing kernel spaces, focusing on two special cases, namely $H(Q)$ and $H(\Theta)$ spaces. Relationships between these spaces lead to the tangential Schur algorithm (theorem 3.4). Spaces $H(Q)$ are the orthogonal complements of Beurling-Lax invariant spaces (see e.g. [29]) while $H(\Theta)$ spaces were first introduced and studied by L. de Branges in [18]. Analogues of $H(Q)$ spaces for general Schur functions B were defined and studied by L. de Branges and J. Rovnyak in [19], [20]. Some of the material appears in a more general context in [4].

3.1 Preliminaries and H(Q) spaces

A Hilbert space H of $\mathbb{C}^p$-valued functions defined on some set $\mathcal{E}$ is called a reproducing kernel Hilbert space if there exists a $\mathbb{C}^{p\times p}$–valued function $K(z, w)$ such that:

(1) For every choice of $c \in \mathbb{C}^p$ and $w \in \mathcal{E}$, the function $z \to K(z, w)c$ belongs to H

(2) For every $f \in H$ and w, c as above,

$$< f \, , \, K(. \, , \, w)c \; >= c^* f(w) \tag{3.1}$$

where $< \, , \, >$ denotes the inner product in H.

Note that, by the Riesz representation theorem, it is equivalent to assume that the functionals

$$f \longrightarrow c^* f(w)$$

are all continuous. The function $K(z,w)$ is uniquely defined and called the reproducing kernel of H. It is positive in the sense that for every $r \in \mathbb{N}$, every $w_1, ..., w_r \in \mathcal{E}$ and $c_1, ..., c_r \in \mathbb{C}^p$, the $r \times r$ matrix with ij entry $c_j^* K(w_j, w_i) c_i$ is nonnegative. Moreover, there is a one-to-one correspondence between positive functions and reproducing kernel Hilbert spaces: whenever $K(z,w)$ is positive for z, w in $\mathcal{E}$, the completion of the linear space generated by the functions $z \to K(z,w)c$ with $w \in \mathcal{E}$ and $c \in \mathbb{C}^p$ endowed with the scalar product

$$< K(.,w)c, K(.,v)d >= d^* K(v,w)c \tag{3.2}$$

is the reproducing kernel space with kernel $K(z,w)$; see [6], [50].

A classical example is the Hardy space H_2^p. Its reproducing kernel is $I_p/\rho_w(z)$ with z and w in $\mathbb{D}$ and ρ_w is defined in (1.10). In this case, equation (3.1) is just the Cauchy formula.

When H is finite dimensional, the kernel can be expressed explicitly as follows: let $f_1, ..., f_N$ be a basis of H over $\mathbb{C}$ and P be the matrix with $i-j$ entry $< f_j \, , \, f_i >$; then

$$K(z,w) = [f_1(z), ..., f_N(z)] P^{-1} [f_1(w), ..., f_N(w)]^*. \tag{3.3}$$

Let now Q be an element of I_n^p. Left multiplication by Q is an isometry M_Q from H_2^p into itself and for $w \in \mathbb{D}$ and $c \in \mathbb{C}^p$

$$M_Q^*(c/\rho_w) = Q(w)^* c/\rho_w.$$

It follows that the function

$$K_Q(z,w) = (I_p - Q(z)Q(w)^*)/\rho_w(z) \tag{3.4}$$

is positive for $z, w \in \mathbb{D}$. The main properties of the associated reproducing kernel space, denoted by $H(Q)$, are gathered in the next theorem.

Theorem 3.1 *Let Q be in I_n^p. Then $H(Q) = H_2^p \ominus Q H_2^p$ and its complex dimension is equal to the McMillan degree n of Q. Furthermore, $H(Q)$ is R_0-invariant so that there exists an observable pair of matrices $C \in \mathbb{C}^{p \times n}$ and $A \in \mathbb{C}^{n \times n}$ with its spectrum in $\mathbb{D}$ such that $H(Q)$ is spanned by the columns of*

$$C(I_n - zA)^{-1}. \tag{3.5}$$

The space $H(Q)$ determines Q uniquely up to right multiplication by some unitary factor. A possible choice for Q is

$$Q(z) = I_p - (1-z)C(I_n - zA)^{-1} P^{-1} (I_n - A)^{-*} C^*, \tag{3.6}$$

where P is the solution to the Stein equation:

$$P - A^*PA = C^*C. \tag{3.7}$$

The space $H(Q)$ is real if and only if there exists a unitary matrix U such that QU is real. In this case, C and A in (3.5) may also be chosen real.

All the assertions in the theorem are widely known, at least in the complex case, and we refer the reader to [4] or [26] for a proof. Formula (3.6) deserves perhaps a word of explanation: if we take as a basis of $H(Q)$ the columns of (3.5), the corresponding Gram matrix in the H_2^p inner product is precisely the solution P of (3.7). Equating (3.4) and (3.3) yields then

$$(I_p - Q(z)Q(w)^*)/\rho_w(z) = C(I_n - zA)^{-1}P^{-1}(I_n - wA)^{-*}C^*.$$

Specializing to $w = 1$ and normalizing Q so that $Q \in NI_n^p$, we obtain (3.6). From this, the real case is easy, because A and C can be chosen real as mentionned in section 1.2 and P is also real.

3.2 $H(\Theta)$ spaces and the tangential Schur algorithm

We shall deal in the sequel with a special instance of what is called the class of J-contractive functions which were first studied by Potapov in [47]. We content ourselves with defining rational J-inner functions as follows. Put

$$J = \begin{pmatrix} I_p & 0 \\ 0 & -I_p \end{pmatrix} \quad J_+ = \begin{pmatrix} I_p & 0 \\ 0 & 0 \end{pmatrix} \quad J_- = \begin{pmatrix} 0 & 0 \\ 0 & I_p \end{pmatrix}.$$

A $\mathbb{C}^{2p\times 2p}$-valued rational function Θ will be called J-inner if at every point of analyticity of Θ in $\mathbb{D}$,

$$\Theta(z)J\Theta(z)^* \leq J$$

and equality holds for z point of analyticity on $\mathbb{T}$.
If Θ is a rational J-inner function, the function

$$\Sigma(z) = (J_+ - \Theta(z)J_-)^{-1}(\Theta(z)J_+ - J_-)$$

is inner and the function

$$K_\Theta(z,w) = (J - \Theta(z)J\Theta(w)^*)/\rho_w(z) \tag{3.8}$$

is positive for z and w varying in the domain of analyticity of Θ, as follows from the equality

$$\frac{I_{2p} - \Sigma(z)\Sigma(w)^*}{\rho_w(z)} = (J_+ - \Theta(z)J_-)^{-1} K_\Theta(z,w) (J_+ - \Theta(w)J_-)^{-*}.$$

We will denote by $H(\Theta)$ the associated reproducing kernel Hilbert space of $\mathbb{C}^{2p}$ valued functions. The space $H(\Theta)$ has properties similar to that of $H(Q)$ (see [4] and [26]). In particular, it is finite–dimensional, and the complex dimension is still the McMillan degree of Θ. A noteworthy difference with $H(Q)$ is that functions in $H(\Theta)$ need not lie in H_2^p, but here we shall only be concerned with the case where $H(\Theta) \subset H_2^p$. Then, 0 belongs to the domain of analyticity of Θ, and $H(\Theta)$ is again R_0-invariant. It can thus be represented as the span of the columns of a matrix of the form (3.5). Conversely we have:

Theorem 3.2 *Let (C, A) be an observable pair in $C \in \mathbb{C}^{2p \times n} \times \mathbb{C}^{n \times n}$ such that the spectrum of A lies in $\mathbb{D}$. Let $f_1, ..., f_n$ be the vector-valued functions defined by*

$$[f_1, ..., f_n](z) = C(I_n - zA)^{-1},$$

and $\mathcal{M} = l.s.\{f_i\}$ be endowed with the inner product

$$< f_j, f_i >_{\mathcal{M}} = < f_j, Jf_i >_{H_2^{2p}} .$$

If this inner product is positive definite, then $\mathcal{M}$ is a reproducing kernel Hilbert space with reproducing kernel of the form (3.8). In this case the function Θ is unique up to a J-unitary right multiplicative factor and, defining the matrix P by $P_{ij} = < f_j, f_i >_{\mathcal{M}}$, a possible choice for Θ is given by

$$\Theta(z) = I_{2p} - (1 - z\zeta_0^*)C(I_n - zA)^{-1}P^{-1}(I_n - \zeta_0 A)^{-*}C^*J \tag{3.9}$$

where ζ_0 is arbitrary in $\mathbb{T}$.

This result is essentially the finite–dimensional version of the Beurling-Lax theorem in H_2^p endowed with the above metric, see [8]. For a proof in the setting of reproducing kernel spaces see e.g. [26] and [5]. A direct way of seeing it is to start from (3.9) and to plug it in (3.8). Then it is easy to check that this expression for K_Θ satisfies (3.3) using the fact that P satisfies the equation:

$$P - A^*PA = C^*JC. \tag{3.10}$$

■

Corollary 3.1 *Let $w_0, ...w_{m-1}$ be m points in $\mathbb{D}$ and $x_0, ..., x_{m-1}$ be m elements of $\mathbb{C}^{2p}$. Suppose that the $m \times m$ matrix P with ij entry $P_{ij} = \frac{x_i^* J x_j}{\rho_{w_j}(w_i)}$ is strictly positive. Then the space*

$$\mathcal{M} = l.s.\{x_0/\rho_{w_0}, ..., x_{m-1}/\rho_{w_{m-1}}\}$$

endowed with the inner product defined by P is a finite dimensional $H(\Theta)$ space.

The function Θ can be chosen real when the w_i's are real and the x_i's are in $\mathbb{R}^{2p}$.

The next theorem describes the deep link between the spaces $H(\Theta)$ and $H(Q)$ on which the tangential Schur algorithm ultimately rests. It is a consequence of a general result due to de Branges and Rovnyak (see [19] and [3]).

For $f \in H_2^{2p}$ and $Q \in I_n^p$ define $\tau f \in H_2^p$ as

$$\tau f = [I_p, \ -Q]f. \tag{3.11}$$

Theorem 3.3 *Let Q be in I_n^p and let Θ be a J-inner $2p \times 2p$ rational function of McMillan degree d.*

Let $\begin{pmatrix} \Theta_{11} & \Theta_{12} \\ \Theta_{21} & \Theta_{22} \end{pmatrix}$ be the decomposition of Θ into four $\mathbb{C}^{p\times p}$ valued blocks. Then τ defines an isometry from $H(\Theta)$ into $H(Q)$ if and only if there exists an inner function $Q^{(1)} \in I_{n-d}^p$ such that

$$Q = T_\Theta(Q^{(1)}) = (\Theta_{11}Q^{(1)} + \Theta_{12})(\Theta_{21}Q^{(1)} + \Theta_{22})^{-1}. \tag{3.12}$$

Finally, if Q and Θ are real, so is $Q^{(1)}$.

Proof: Let us suppose that τ is an isometry; in particular, it is a contraction and, by the result of de Branges and Rovnyak, we can write Q as in (3.12) for some function $Q^{(1)}$ analytic and contractive in $\mathbb{D}$. The inverse of the linear fractional transformation T_Θ is $T_{\Theta^{-1}}$, so that $Q^{(1)}$ is rational. By the maximum modulus principle, $Q^{(1)}$ is in fact unitary on $\mathbb{T}$. We now show that $Q^{(1)} \in I_{(n-d)}^p$.

The operators $\tau\tau^*$ and $I - \tau\tau^*$ are orthogonal projections and thus

$$H(Q) = \mathrm{Im}(I - \tau\tau^*) \oplus \mathrm{Im}(\tau\tau^*).$$

For $u \in \mathbb{C}^p$ and $w \in \mathbb{D}$ it is easily verified (see e.g. [19] or [3]) that

$$\tau^*(K_Q(.,w)u) = K_\Theta(.,w)\begin{pmatrix} u \\ -Q(w)^*u \end{pmatrix}. \tag{3.13}$$

Thus,

$$(I - \tau\tau^*)(K_Q(.,w)u)(z) = (\Theta_{11} - Q\Theta_{21})(z)K_{Q^{(1)}}(z,w)((\Theta_{11} - Q\Theta_{21})(w))^*u, \tag{3.14}$$

and since the reproducing kernel of the projection is the projection of the reproducing kernel we find that

$$(\Theta_{11} - Q\Theta_{12})(z)K_{Q^{(1)}}(z,w)((\Theta_{11} - Q\Theta_{12})(w))^*$$

is the reproducing kernel of $\mathrm{Im}(I - \tau\tau^*)$. Because τ is an isometry, this space has complex dimension $n - d$. Moreover, the function $\Theta_{11} - Q\Theta_{12}$ has nonzero determinant, so that the above complex dimension is also equal to the complex dimension of $H(Q^{(1)})$. Thus the McMillan degree of $Q^{(1)}$ is $n - d$.

Conversely, suppose that $Q = T_\Theta(Q^{(1)})$ where $Q^{(1)}$ has degree $n-d$; using again the theorem of de Branges and Rovnyak, we can assert that the map τ is a contraction. Now, a general theorem [27] states that, if Γ_1 and Γ_2 are two positive operators in a Hilbert space, then

$$\mathbf{Im}(\Gamma_1 + \Gamma_2)^{1/2} = \mathbf{Im}\Gamma_1^{1/2} + \mathbf{Im}\Gamma_2^{1/2} \tag{3.15}$$

and this decomposition is orthogonal if and only if it is direct. Since our spaces are finite–dimensional, we can dispense with the square roots:

$$H(Q) = \mathbf{Im}(I - \tau\tau^*) + \mathbf{Im}(\tau\tau^*). \tag{3.16}$$

The space $\mathbf{Im}(\tau)$ has complex dimension at most d and by (3.14) $\mathbf{Im}(I - \tau\tau^*)$ has complex dimension $n - d$. It follows that the decomposition (3.16) is direct hence orthogonal by the above mentioned theorem, and that τ is injective. Consequently, τ is an isometry.

We shall draw two consequences of Theorem 3.3 and Corollary 3.1. The first one describes the solution to the Nevanlinna–Pick problem.

Proposition 3.1 *Let n and m be integers with $n > 0$ and $0 \leq m \leq n$. For $Q \in I_n^p$, there exists unit vectors $u_0, \ldots, u_{m-1}$ in $\mathbb{C}^p$ and points $w_0, \ldots, w_{m-1}$ in $\mathbb{D}$ such that the $m \times m$ matrix P with ij entry*

$$P_{ij} = u_i^* K_Q(w_i, w_j) u_j \tag{3.17}$$

is strictly positive. The space $\mathcal{M}$ defined in Corollary 3.1 with

$$x_i = \begin{pmatrix} u_i \\ Q(w_i)^* u_i \end{pmatrix},$$

endowed with the inner product P, is then a $H(\Theta)$ space for some J-inner rational function Θ of degree m. There exists an element $\Sigma \in I_{n-m}^p$ such that $Q = T_\Theta(\Sigma)$. Finally, when Q is real, the w_i can be chosen real and the u_i in $\mathbb{R}^p$.

Proof: We first prove by induction on m that one can find vectors u_i as prescribed. The claim is true for $m = 1$; in fact, even more is true in this case since w_0 can be assigned arbitrarily in $\mathbb{D}$. Suppose indeed that $Q(w)^*u$ is of unit norm for all unit vectors $u \in \mathbb{C}^p$. Then,

$$u^* K_Q(w,w)u = < K_Q(.,w)u, K_Q(.,w)u >_{H(Q)}$$

is zero for all $u \in \mathbb{C}^p$, so that $K(.,w)$ is zero. But this is impossible for (3.4) would imply that Q is a constant.
Suppose now that the claim holds for some $m < n$ and let $u_0, \ldots, u_m$ and $w_0, \ldots, w_m$ satisfy our requirements. If the claim does not hold for $m+1$, then every unit vector $u \in \mathbb{C}^p$ and every $w \in \mathbb{D}$, is such that the space spanned by the $K_Q(.,w_i)u_i$'s and $K_Q(.,w)u$ is degenerate in $H(Q)$. Since the $K_Q(.,w_i)u_i$'s are linearly independent, this forces $\dim_{\mathbb{C}} H(Q) = m < n$ contradicting the assumption.

By corollary 3.1, $\mathcal{M}$ is a $H(\Theta)$ space and the map τ defined in (3.11) an isometry from $H(\Theta)$ onto $H(Q)$. By theorem 3.3, $Q = T_\Theta(\Sigma)$ for some $\Sigma \in I^p_{n-m}$. The case where Q is real is similar. ∎

The special case where $m = p$, all the w_i are equal, and the u_i form a basis of $\mathbb{C}^p$ correspond to the Schur algorithm of [23]. The case where $m = 1$ will be of special interest to us and is singled out in the next corollary.

Corollary 3.2 *Let Q be in I^p_n ,$n > 0$, and let $w \in \mathbb{D}$. Then, there exists $u \in \mathbb{C}^p$ of unit norm such that $\|Q(w)^*u\| < 1$. The complex one-dimensional space spanned by*

$$f = \begin{pmatrix} u \\ Q(w)^*u \end{pmatrix} / \rho_w, \tag{3.18}$$

endowed with the inner product

$$< f, f > = u^*(I_p - Q(w)Q(w)^*)u/(1 - |w|^2),$$

is a $H(\Theta)$ space where Θ is given by formula (3.9) with $\zeta_0 = 1$, $A = w$, and

$$C = \begin{pmatrix} u \\ Q(w)^*u \end{pmatrix}.$$

When Q is in RI^p_n, w and u may be chosen real. The function Θ is then real.

Combining Theorem 3.3 and Corollary 3.2, we obtain the tangential version of the Schur algorithm. The latter will be a tool to obtain charts of I^p_n, as explained in the next section. We first need to mention some relationships between $H(\Theta)$ spaces and interpolation (see [26],[2]).

Proposition 3.2 *Let $w_i, i = 0, \ldots, m-1$ be m points in $\mathbb{D}$, $u_i, i = 0, \ldots, m-1$ and $v_i, i = 0, \ldots, m-1$ be vectors in $\mathbb{C}^p$. Let*

$$f_i(z) = \frac{\begin{pmatrix} u_i \\ v_i \end{pmatrix}}{1 - w_i^* z} \tag{3.19}$$

and let $\mathcal{M}$ denote the linear span of the functions f_i. Let us suppose that

$$< f_i, f_j >_{\mathcal{M}} = \frac{u_i^* u_j - v_i^* v_j}{1 - w_i w_j^*} \tag{3.20}$$

defines a positive quadratic form. Then, there exists a rational J-inner function Θ such that $\mathcal{M} = H(\Theta)$ and there is a one-to-one correspondance between elements Q in I_n^p such that

$$Q(w_i)^* u_i = v_i \qquad i = 1, \ldots, m \tag{3.21}$$

and the set of $T_\Theta(\Sigma)$, where $\Sigma \in I_{n-m}^p$.

Proof: Let Σ range over the set of $p \times p$ Schur functions. From [26], it follows that the set of $Q = T_\Theta(\Sigma)$ describes all Schur functions which satisfy the interpolation condition (i). On another hand, the interpolation conditions (3.21) are to the effect that the map τ (defined in (3.11) is an isometry from $H(\Theta)$ onto $H(Q)$.The conclusion now follows from theorem 3.3. ∎

Theorem 3.4 *(The tangential Schur algorithm) Let $Q \in I_n^p$ and let $w_0, \ldots, w_{n-1}$ be n (possibly non distinct points) in the open unit disk. Then, for $0 \le i \le n-1$, there exist $Q^{(i)} \in I_{n-i}^p$, unit vectors u_i, and J-inner rational functions of degree one Θ_i given by formula (3.9) where $\xi_0 = 1$, $A = w_i^*$, and*

$$C = \begin{pmatrix} u_i \\ Q(w_i)^{(i)*} u_i \end{pmatrix},$$

such that $Q^{(0)} = Q$ and

$$Q^{(i)} = T_{\Theta_{i+1}}(Q^{(i+1)}) \quad i = 0, \ldots n-1. \tag{3.22}$$

In particular, setting $\Theta = \Theta_0 \ldots . \Theta_{n-1}$, we get $Q = T_\Theta(U)$ where $U = Q^{(n)} = Q(1)$ is a constant unitary matrix.

Finally, when Q is real, the u_i's and the w_i's can also be chosen real, and so are the Θ_i's and the $Q^{(i)}$'s.

Proof: The theorem is a recursive application of Theorem 3.3 and Corollary 3.2. Let w_0 be in $\mathbb{D}$. Applying Corollary 3.2, we build Θ_0 which takes the value I_{2p} at $z = 1$ and is such that the map τ is an isometry from $H(\Theta_0)$ into $H(Q)$. From Theorem 3.3, we obtain a linear fractional transformation $Q = T_{\Theta_0}(Q^{(1)})$ where $Q^{(1)}$ is in I^p_{n-1}. Iterating this procedure $n-2$ times, we obtain a constant $Q^{(n)} \in \mathcal{U}_p$. Since all the functions Θ_i take the value I_{2p} at $z = 1$, we have that $Q^{(n)} = Q(1)$. One proceeds similarly in the real case. ∎

As explained in [4], Theorem 3.4 reduces to the classical Schur algorithm when $p = 1$ and at each stage the point w is taken to be the origin. It is recursive: at each step, one chooses a point w and a direction u, in order to compute the Blaschke factor. Alternatively, one may proceed in one shot from n points and n directions using Proposition 3.2.

3.3 Constructing the charts

In this section, we construct new charts on I^p_n in terms of transfer functions rather than realizations. We develop two (equivalent) atlases, one based on Proposition 3.2 and another one on the tangential Schur algorithm.

For $u_0, ..., , u_{n-1} \in \mathbb{C}$ of unit length, $w_0, ..., w_{n-1}$ in the open unit disk and $(\mathcal{V}, \vartheta)$ a chart on $\mathcal{U}_p$, we define a chart (W, ψ) by its domain:

$$\begin{aligned} &W(u_0, ..., u_{n-1}, w_0, ..., w_{n-1}, \mathcal{V}) \\ &= \{Q \in I^p_n \mid P > 0\ , Q(1) \in \mathcal{V}\} \end{aligned} \tag{3.23}$$

where P is the matrix defined in proposition 3.1, and its coordinate map

$$\psi(Q) = (Q(w_0)^* u_0, ..., Q(w_{n-1})^* u_{n-1}, \vartheta(Q(1))). \tag{3.24}$$

Theorem 3.5 *The family (W, ψ) defines a C^∞ atlas on I^p_n which is compatible with its natural structure of embedded submanifold of $H^{p\times p}_q$ fo any $1 \le q \le \infty$. If we choose real w_i's and and u_i's and if we restict ourselves to real coordinates and orthogonal matrices, we obtain an atlas for RI^p_n.*

Proof: That $W(u_0, ..., u_{n-1}, w_0, ..., w_{n-1}, \mathcal{V})$ is open in I^p_n is easily checked from the definition. It is equally clear that ψ is defined and smooth on some open subset of $H^{p\times p}_q$. Thanks to Proposition 3.2, the range of ψ is $\mathcal{B} \times \vartheta(\mathcal{V})$ where $\mathcal{B}$ denotes the set of $(v_0, \ldots, v_{n-1})$ such that the matrix defined in (3.20) is strictly positive, and is therefore open in $\mathbb{R}^{2np+p^2}$. Finally, ψ^{-1} is given by $Q = T_\Theta(Q(1))$, where Θ is rationally computed from $Q(w_i)^* u_i$ and $Q(1)$, hence is smooth. The real case is obvious. ∎

The preceding theorem gives explicit charts, one drawback being that the ranges of the charts are rather involved. One may alternatively use the tangential Schur algorithm to obtain charts whose range is the product of n copies of the unit ball in $\mathbb{C}^p$ with an open subset of $\mathcal{U}_p$–but the coordinate functions, this time, are more involved. This is our last result.

We define this new atlas as follows: the chart (V,ϕ) will have domain

$$(3.25)\qquad \begin{aligned} &V(u_0,\dots,u_{n-1},w_0,\dots,w_{n-1},\mathcal{V}) \\ &= \{Q\in I_n^p \mid \|Q(w_i)^{(i)*}u_i\|<1,\ i=0,\dots,n-1, Q(1)\in\mathcal{V}\}\end{aligned}$$

and the coordinate map will be given by

$$(3.26)\qquad \phi(Q)=(Q^{(0)*}(w_0)u_0,\dots,Q^{(n-1)*}(w_{n-1})u_{n-1},Q(1)),$$

where the functions $Q^{(i)}$ are defined recursively as in theorem 3.4. Namely, we first construct Θ_0 by setting $\xi_0=1$, $A=w_0^*$, and $C=\begin{pmatrix} u_0 \\ Q(w_0)^*u_0\end{pmatrix}$ in formula (3.9). Then, we define $Q^{(1)}$ by inverting the formula $Q=T_{\Theta_0}(Q^{(1)})$ and iterate the same procedure on $Q^{(1)}$. The function Q is uniquely determined by $\phi(Q)$ and if

$$\xi=(\xi_0,\dots\xi_{n-1},U)$$

where $U\in\mathcal{V}$ and each $\xi_i\in\mathbb{C}^p$ is of norm strictly less than one, we have from Proposition 3.2 that

$$(3.27)\qquad \phi^{-1}(\xi)=T_{\Theta_0\dots\Theta_{n-1}}(U)$$

where Θ_i is given by (3.9) with $A=w_i^*$, $\zeta_0=1$, and $C=\begin{pmatrix} u_i \\ \xi_i\end{pmatrix}$.

Hence, the range of ϕ consists of the announced product.

Theorem 3.6 *The family (V,ϕ) defines a C^∞ atlas on I_n^p which is compatible with that of Theorem 3.5. Restricting to real parameters in the charts as in the cited theorem, we get an atlas on RI_n^p.*

The proof of theorem 3.6 is analogous to the proof of theorem 3.5 and will be omitted.

4 Conclusion

Having shown that the set of $p\times p$ inner functions of degree n is a submanifold of $H_q^{p\times p}$ in the real and complex case, we produced two different parametrization for it, one based on the set of observable pairs, and the other on Schur coefficients. Both are well-known tools in system-theory and interpolation theory respectively, stressing here a link of a topological nature between two domains which are already known to interfere strongly from the analytic viewpoint. Along the same lines, further extensions to $J-inner$ and $J-unitary$ functions are to be expected.

Perhaps the main practical contribution of the paper is to provide a mean of applying differential calculus to the set I_n^p. Such a need arises, for instance, in rational approximation and this was part of the authors' motivation for studying these questions: in fact, it is easily shown (see [13], [21]) that obtaining the best L^2 approximant of degree n of a function in $H_2^{p\times m}$ is equivalent to minimize a nonlinear function on the set I_n^p. To justify a differential approach and to use gradient algorithms for the minimization, a differential structure together with an explicit parametrization are needed.

In the scalar case where $m = p = 1$, a numerical algorithm has been derived in [14] to generate local minima of the criterion, and a uniqueness theorem has been obtained in [15] for sufficiently stable Stieltjes functions (i.e. transfer-functions of relaxation systems). Both references use in an essential way the topological structure of the closure of $RI_n^1/\mathcal{O}_1$ in H_2 (which is completely different from its closure in H_∞ so that here the exponent does matter). This closure turns out to be a projective space and the global step of the uniqueness proof in [15] drops out from the corresponding Morse inequalities [12].

In the case where $p = 2$, this problem was considered in [21] and our results allow for such a study when p is arbitrary. A full generalization, however, would again require a detailed knowledge of the closure of $RI_n^p/\mathcal{O}_p$ in $H_2^{p\times p}$. For $p > 1$, this is by no mean well-understood.

References

[1] R. Abraham and J. Robbin. *Transversal mappings and flows.* Benjamin, New-York, 1967.

[2] D. Alpay, P. Bruinsma, A. Dijksma, and H. de Snoo. *Interpolation problems, extensions of symmetric operators and reproducing kernel spaces I*, pages 35–82. Operator theory: advances and applications OT50. Birkhäuser Verlag, Basel, 1991.

[3] D. Alpay and H. Dym. Hilbert spaces of analytic functions, inverse scattering and operator models, I. *Integral equations and operator theory*, 7:589–641, 1984.

[4] D. Alpay and H. Dym. *On applications of reproducing kernel spaces to the Schur algorithm and rational J-unitary factorization*, pages 89–159. Operator theory: Advances and applications OT18. Birkhäuser Verlag, Basel, 1986.

[5] D. Alpay and I. Gohberg. *Unitary rational matrix functions*, pages 175–222. Operator theory: advances and applications OT33. Birkhäuser Verlag,Basel, 1988.

[6] N. Aronszjan. Theory of reproducing kernels. *Transactions of the American mathematical society*, 68:337–404, 1950.

[7] J. Ball, I. Gohberg, and L. Rodman. *Interpolation of rational matrix functions.* Birkhäuser Verlag, Basel, 1990.

[8] J. Ball and J.W. Helton. A Beurling–Lax theoem for the Lie group $U(m,n)$ which contains most classical interpolation. *Journal of operator theory*, 8:107–142, 1983.

[9] L. Baratchart. *Sur l'approximation rationnelle L_2 pour les systèmes dynamiques linéaires.* Theše de doctorat, Université de Nice , 1987.

[10] L. Baratchart. On the parametrization of linear constant systems. *SIAM J. Control and optimization*, pages 752–773, 1985.

[11] L. Baratchart. Existence and generic properties for l^2 approximants of linear systems. *I.M.A. Journal of Math. Control & Information*, 3:89–101, 1986.

[12] L. Baratchart. On the topological structure of inner functions and its use in identification of linear systems. In B. Bonnard, B. Bride, J. P. Gauthier, and I. Kupka, editors, *Analysis of controlled dynamical systems.* Birkhäuser, Boston, 1991.

[13] L. Baratchart and M. Olivi. Inner-unstable factorization of stable rational transfer functions. In G.B. Di Masi, A. Gombani, and A. B. Kurzhansky, editors, *Modeling, estimation and control of systems with uncertainty.* Birkhäuser, Boston, 1991.

[14] L. Baratchart, M. Cardelli and M. Olivi. Identification and rational L_2 approximation: a gradient algorithm. *Automatica*, vol. 27, n.2, pp 413–418, 1991.

[15] L. Baratchart and F. Wielonsky. Rational approximation in H^2 and Stieltjes integrals : a uniqueness theorem. *Constructive Approximation*, to appear.

[16] H. Bart, I. Gohberg, and M. Kaashoek. *Minimal factorization of matrix and operator functions.* Birkhäuser Verlag, Basel, 1979.

[17] N. Bourbaki. *Variétés diffe'rentielles et analytiques* fascicule de résultats 1-7. Hermann, Paris, 1971.

[18] L. de Branges. Some Hilbert spaces of analytic functions I. *Transactions of the American mathematical society*, 106:445–468, 1963.

[19] L. de Branges and J. Rovnyak. Canonical models in quantum scattering theory. In C. Wilcox, editor, *Perturbation theory and its applications in quantum mechanics.* Holt, Rinehart and Winston, New–York, 1966.

[20] L. de Branges and J. Rovnyak. *Square summable power series.* Holt, Rinehart and Winston, New–york, 1966.

[21] M. Cardelli. *Contributions à l'approximation rationnelle L_2 des fonctions de transfert.* PhD thesis, Université de Nice Sophia -Antipolis, 1990.

[22] J.M.C. Clark. The consistent selection of local coordinates in linear system identification. In *Proceedings of the joint automatic control conference*, pages 576–580, 1976.

[23] P. Delsarte, Y. Genin, and Y. Kamp. Schur parametrization of positive definite block-Toeplitz systems. *SIAM journal in applied mathematics*, pages 34–46, 1979.

[24] R.G. Douglas, H.S. Shapiro, and A.L. Shields. Cyclic vectors and invariant subspaces for the backward shift operator. *Annales de l' Institut Fourier (Grenoble)*, 20:37–76, 1970.

[25] P.L. Duren. *Theory of H^p spaces.* Academic press, New–York, 1970.

[26] H. Dym. *J-contractive matrix functions, reproducing kernel spaces and interpolation*, volume 71 of *CBMS lecture notes.* American mathematical society, Rhodes island, 1989.

[27] P.A. Fillmore and J.P. Williams. On operator ranges. *Advances in Mathematics*, 7:254–281, 1971.

[28] L. Finesso and G. Picci. A characterization of minimal square spectral factors. *IEEE transactions on automatic control*, 27:122–127, 1982.

[29] P.A. Fuhrmann. *Linear systems and operators in Hilbert space.* McGraw-Hill international book company, 1981.

[30] J.B. Garnett. *Bounded analytic functions.* Academic press, San Diego, 1981.

[31] Y. Genin. P. van Dooren, T. Kailath, J.M. Delosme, and M. Morf. On Σ–lossless transfer functions and related questions. *Linear algebra and its applications*, 50:251–275, 1983.

[32] I. Gohberg, editor. *I. Schur methods in operator theory and signal processing.* Operator theory: advances and applications. Birkhäuser Verlag, Basel, 1986.

[33] I. Gohberg, P. Lancaster, and L. Rodman. *Invariant subspaces of matrices and applications.* Wiley, New–York, 1986.

[34] V. Guillemin and A. Pollack. *Differential topology.* Prentice–Hall, Englewood Cliffs, N.J., 1974

[35] B. Hanzon. On the differential manifold of fixed order stable linear systems. In *Systems and Control Letters*, vol. 13, pp 345-352, 1989.

[36] M. Hazewinkel. Moduli and canonical forms for linear dynamical linear systems II: the topological case. *Math. system theory*, 10:363–385, 1977.

[37] M. Hazewinkel and R.E. Kalman. Moduli and canonical forms for linear systems. Technical report, Economic Institute, Erasmus University, Rotterdam, 1974.

[38] U Helmke. *Zur Topologie des Raumes linearer Kontrollsysteme.* PhD thesis, Uni. Bremen, 1982.

[39] U. Helmke. The topology of the space of linear systems. In *Proceedings of the 21-th conferenceon decision and control*, pages 948–949, 1982.

[40] K Hoffman. *Banach spaces of analytic functions.* Prentice–Hall, Englewood Cliffs, N J., 1962

[41] R.E. Kalman. Algebraic geometric description of the class of linear systems of constant dimension. In *8 th annual Princeton conference on information sciences and systems*, 1974.

[42] R.E. Kalman, P.L. Falb, and M.A.K. Arbib. *Topics in mathematical system theory.* Mc Graw–Hill, New-York, 1969.

[43] A.W. Knapp. *Lie groups, Lie algebras, and cohomology.* Math. Notes Princeton Univ. Press, 1988

[44] S. Lang. *Differential manifolds.* Addison-Wesley, 1972.

[45] L. Ljung. *System identification. Theory for the user.* Prentice-Hall, 1987.

[46] J.R. Munkres. *Elements of algebraic topology.* Addison-Wesley, 1984.

[47] V.P. Potapov. The multiplicative structure of J-contractive matrix functions. *American mathematical society translations*, 15, 1960.

[48] H.H. Schaefer. *Topological vector spaces.* Springer, New-York, 1986.

[49] I. Schur. Über die potenzreihen, die im "Innern des Einheitkreises Beschrankt sind. *Journal fur die reine und angewandte Mathematik*, 147, 1917. English translation in: I. Schur methods in operator theory and signal processing. (Operator theory: advances and applications OT 18 (1986), Birkhäuser Verlag), Basel.

[50] L. Schwartz. Sous espaces hilbertiens d'espaces vectoriels topologiques et noyaux associés (noyaux reproduisants). *Journal d'analyse mathématique*, pages 115–256, 1964.

[51] L. Schwartz. *Topologie et analyse fonctionnelle.* Hermann, 1970.

Daniel Alpay
Department of Mathematics
Ben–Gurion University of the Negev
POB 653. 84105 Beer-Sheva
Israel

Laurent Baratchart
Institut National de Recherche en Informatique et Automatique
Route des Lucioles
Sophia–Antipolis
06560 Valbonne
France

Andrea Gombani
Ladseb–CNR
Corso Stati Uniti 4
35020 Padova
Italy

MSC classification: 93B29, 46E22

Operator Theory:
Advances and Applications, Vol. 73

CONSERVATIVE DYNAMICAL SYSTEMS AND NONLINEAR LIVSIC-BRODSKII NODES

Dedicated to Moshe Livsic

Joseph A. Ball[1]

We review some ideas on conservative dynamical systems and show how this leads to the notion of a nonlinear Livsic-Brodskii node, where the vector field $A(x)$ defining the internal dynamics for the system replaces the linear operator A appearing in the linear theory. We show that under certain hypotheses two vector fields leading to the same Livsic-Brodskii characteristic input-outmap map are equivalent via an energy-preserving diffeomorphic change of variable, a nonlinear analogue of one of the main theorems in the linear theory. We also survey related issues concerning lossless and dissipative systems for both the linear and nonlinear case, including an application to the standard problem of H^∞-control theory.

INTRODUCTION

In the late 1940's Livsic introduced an object called an operator node (sometimes also translated as operator colligation) which not only began a whole new approach to the study of nonselfadjoint operators on a Hilbert space but also anticipated many later developments in the field now known as mathematical system theory. The original work was aimed at building a spectral theory for a single nonselfadjoint operator (see [Br]) but motivation also came from engineering and physics and a quest for applications of operator theory to these areas (see [L, LY]). In more recent work since moving to Israel, Prof. Livsic has been very active with his Israeli colleagues in extending the theory to handle an n-tuple of commuting nonselfadjoint operators (see e.g. [LW, V]); this

[1]The author was partially supported by NSF Grant DMS-9101400.

work calls up techniques from algebraic geometry and Riemann surface theory. Our goal here is to discuss a different direction in which to extend the theory, namely, to nonlinear nodes where the Hilbert space H is replaced by a differentiable manifold X with an energy function replacing the norm and the linear operator A is replaced by a vector field $x \to A(x)$ on X.

It is now well understood (see [H, BC]) how the Livsic model theory and the related model theories of Sz.-Nagy-Foias and de Branges-Rovnyak relate to the type of system theory associated with the name of Kalman (see [An] for a nice survey of developments in the past thirty years). In particular the operator model theory is a theory up to unitary equivalence whereas system theory is an affine theory where equivalence is up to similarity (or more generally quasisimilarty). Most advances in the nonlinear theory have been with respect to this affine theory (see the recent texts [I, NvS]). However there is a nonlinear analogue of the Hilbert space equivalence up to unitary transformations, namely, conservative systems; these together with the related class of dissipative, dynamical systems have been studied in some detail by Willems [W1] with further elaboration by Hill and Moylan [HM1-3].

Physical motivation comes from the theory of energy conserving systems in classical mechanics (where no inputs or outputs are present and one's goal is simply to describe rather than to influence the evolution of the system) and from lossless networks in electrical circuit theory (see [AV, WCGGG2]). My main purpose here is to describe how this is a natural setting in which to define a nonlinear version of a Livsic-Brodskii node and characteristic operator function. Many questions remain but one result which is set down here is that under appropriate conditions two vector fields producting the same characteristic input-output map necessarily are the same after a diffeomorphic energy-preserving change of variable on the state space.

The paper is organized as follows. Section 1 summarizes what is needed concerning lossless dissipative systems, especially for the nonlinear case. Section 2 discusses nonlinear Livsic-Brodskii nodes and characteristic input-output maps and establishes the characteristic input-output map as an invariant of a vector field. Construction of a nonlinear Livsic-Brodskii node can be viewed as embedding a given vector field as the internal dynamics of a lossless, dissipative dynamical system. The last Section 3, expository in nature, surveys some recent research results, all of which can be viewed as solving for some piece of a lossless or dissipative systems in terms of the other pieces.

Included are interpolation problems, Darlington synthesis and the H^∞-control problem.

Finally, it is a pleasure to thank D.Z. Arov for useful comments on the manuscript.

1. CONSERVATIVE SYSTEMS

A fundamental notion in physics is that of a *conservative system*, i.e. a dynamical system in which a physical state $x(t)$ evolves in such a way that some energy function of the state $e(x(t))$ is constant along paths of the system (see e.g. [LP]). In the simplest linear case the evolution of the state is described by a differential equation

$$\dot{x} = Ax \tag{1.1}$$

where A is a bounded operator on a Hilbert space H and the state vector $x(t)$ assumes values in H and the energy function $e(x) = \|x\|^2$ is simply the norm-squared. (A more realistic scenario is the case where A is an unbounded operator with some dense domain, but for technical simplicity here we avoid the details of this more elaborate theory). In this case the energy conservation property is equivalent to

$$\begin{aligned} 0 &= \frac{d}{dt}\{\|x(t)\|^2\} \\ &= 2 < \dot{x}(t), x(t) > \\ &=< (A + A^*)x(t), x(t) > \end{aligned}$$

for all $x(t)$, i.e.

$$A = -A^* \text{ if } e(x) = \|x\|^2. \tag{1.2}$$

More generally the energy function may be given by a quadratic form $e(x) =< Px, x >$ associated with a selfadjoint operator P. Then a similar computation yields that (1.1) is energy conserving if and only if

$$PA + A^*P = 0. \tag{1.3}$$

Our interest here is the extension of these ideas to nonlinear settings. To concentrate on a technically simple but nevertheless illustrative case, we assume that the state space X is equal to a subset of $\mathbb{R}^n$, or more generally, is an n-dimensional manifold. Then we assume that the evolution of the system is described by a differential equation

$$\dot{x} = A(x) \tag{1.4}$$

where $A(x)$ is a smooth function from $\mathbb{R}^n$ to $\mathbb{R}^n$, or more generally, a vector field on the manifold X. We assume that the energy function $e: X \to \mathbb{R}$ is some given smooth function. Then the energy conservation property becomes $0 = \frac{d}{dt}\{e(x(t))\} = \nabla e(x(t)) \cdot A(x(t))$ for all $x(t)$, and hence energy conservation means

$$\nabla e(x) \cdot A(x) = 0 \text{ for all } x \text{ in } X. \tag{1.5}$$

(Here the dot product is the standard inner product on $\mathbb{R}^n$; if X is a manifold, we assume that there is defined an inner product on the tangent space, although this is not essential.) Note that if e is given by a quadratic form $e(x) =< Px, x >$ then $\nabla e(x) = 2Px$ and (1.5) becomes

$$2x^T PA(x) = 0$$

for all $x \in X$. If $A(x) = Ax$ is linear, this in turn collapses to (1.3).

However in practice, particularly in circuit and control theory, many systems are not energy conserving due to interaction with the outside environment. If one expands the energy book-keeping to include the energy absorbed from and lost to the outside environment, then the total enlarged system is conservative; this is essentially the idea of a lossless system in circuit theory (see [AV,WCGGG2]). More formally, suppose first that the dynamics is linear as in (1.11). Introduce a linear output map $y = Cx$ such that

$$A + A^T = -C^T JC \tag{1.6}$$

where J is an appropriate signature matrix. If $x(t)$ is any solution of (1.1) we have

$$\frac{d}{dt}\{\|x(t)\|^2 + \int_{t_0}^{t} < JCx(s), CXx(s) > ds\}$$
$$=< [A + A^T + C^T JC]x(t), x(t) >= 0$$

and hence

$$\|x(t)\|^2 - \|x(t_0)\|^2 = -\int_{t_0}^{t} < JCx(s), Cx(s) > ds. \tag{1.7}$$

If we interpret $< JCx(t), Cx(t) >$ as the power imparted to the outside environment and $\|x\|^2$ as the energy of the state x, then (1.7) can be interpreted as saying that the loss of energy of the state vector x in the period from time t_0 to time t is exactly accounted for by the total amount of energy dissipated to the outside environment by the output $y = Cx$. More generally, the energy of the state x may be given by the quadratic form $e(x) =< Px, x >$ in which case (1.6) should be adjusted to the Lyapunov equation

$$AP + A^T P + C^T JC = 0. \tag{1.8}$$

Then the same calculation as above gives us the energy balance equation

$$< Px(t_2), x(t_2) > - < Px(t_1), x(t_1) >= - \int_{t_1}^{t_2} < JCx(s), Cx(s) > ds. \tag{1.9}$$

The next layer is to consider a linear system with both inputs and outputs such as

$$\begin{aligned} \dot{x} &= Ax + Bu \\ y &= Cx + Du \end{aligned} \tag{1.10}$$

where the net power absorbed by the system (equal to the power absorbed from the inputs minus the power disssipated by the outputs) is taken to be $< Ju(t), u(t) > - < Jy(t), y(t) >$. If we assume that the energy function e is given by the quadratic form $e(x) =< Px, x >$, then differentiation of the desired energy balance equation

$$< Px(t), x(t) > - < Px(t_0), x(t_0) >= \int_{t_0}^{t} \{< Ju(s), u(s) > - < Jy(s), y(s) >\} ds \tag{1.11}$$

yields the infinitesimal form of the energy balance equation

$$2 < P(Ax(t) + Bu(t)), x(t) >=< Ju(t), u(t) > - < J(Cx(t) + Du(t)), Cx(t) + Du(t) > \tag{1.12}$$

for all t, or

$$[x^T \quad u^T] \begin{bmatrix} PA + A^T P + C^T JC & PB + C^T JD \\ B^T P + D^T JC & -J + D^T JD \end{bmatrix} \begin{bmatrix} x \\ u \end{bmatrix} = 0$$

for all x, u. This leads to the conditions

$$PA + A^T P + C^T JC = 0 \tag{1.13a}$$

$$PB = -C^T JD \tag{1.13b}$$

$$D^T JD = J \tag{1.13c}$$

A linear system (1.10) which satisfies the energy balance equation (1.11), or equivalently (1.12) or (1.13), is said to be lossless (with respect to the *energy* or *storage function* $e(x) =< Px, x >$

and *supply rate* $S(u,y) = u^Tu - y^Ty$); see [AV] for more discussion and examples from linear circuit theory. The nonlinear version of this setup has been studied by Willems [W1] with further elaboration by Hill and Moylan [HM1-3] and Wyatt *et al* [WCGGG2]. Rather than considering the most general nonlinear systems, we consider systems of the form

$$\begin{aligned} \dot{x} &= A(x) + B(x)u \\ y &= C(x) + D(x)u \end{aligned} \tag{1.14}$$

which are affine in the input variable u. Let us assume that a smooth energy or storage function $e : X \to \mathbb{R}$ is given and that the supply rate $S(u,y)$ on input-output pairs is again taken to be

$$S(u,y) = < Ju, u > - < Jy, y > .$$

Then the defining property for the system (1.14) to be *conservative* is that the energy balance equation

$$e(x(t)) - e(x(t_0)) = \int_{t_0}^{t} \{< Ju(s), u(s) > - < Jy(s), y(s) >\} ds \tag{1.15}$$

hold. Assuming that the energy function is smooth, we may differentiate with respcet to t to obtain the infinitesimal version of the energy balance equation

$$\nabla e(x) \cdot [A(x) + B(x)u] + < J(C(x) + D(x)u), C(x) + D(x)u > - < Ju, u > = 0. \tag{1.16}$$

To simplify (1.16), we assume

$$D(x) = I. \tag{1.17}$$

Restricting to the case $u = 0$ in (1.16) gives the Hamilton-Jacobi equation

$$\nabla e(x) \cdot A(x) + C(x)^T JC(x) = 0, \tag{1.18}$$

the nonlinear analogue of the Lyapunov equation (1.13a). Using (1.17) and (1.18), we can then simplify (1.16) to

$$\nabla e(x) \cdot B(x)u + 2C(x)^T Ju = 0$$

for all u and x, or

$$\nabla e(x) \cdot B(x) + 2C(x)^T J = 0, \tag{1.19}$$

the nonlinear analogue of (1.13b). Conversely, if (1.17) - (1.19) hold, then we get (1.16). Hence, if we assume (1.17), then (1.18) and (1.19) are necessary and sufficient for the nonlinear input-affine system (1.14) to be conservative with respect to the smooth storage function e and supply rate $< Ju, u > - < Jy, y >$. Less explicit equations are derived in [W1] and [HM1] for conservativeness (as well as dissipativity to be defined below) with respect to a more general supply rate $S(u, y)$. We next give a general class of conservative systems motivated from circuit theory.

Example 1.1 We take the state space X and the input-output space U both to be $\mathbb{R}^n$. We suppose that $x \to f(x)$ is an $n \times n$ matrix valued function on $\mathbb{R}^n$ such that $f(x)$ is invertible for all x and moreover the $\mathbb{R}^n$ -vector valued function $2(f(x)^{-1})^T x$ (matrix multiplication with the second x considered as a column vector) is the gradient of a real valued function e defined on $\mathbb{R}^n$ (thus $2(f(x)^{-1})^T x = \nabla e(x)$). An example to have in mind is

$$f(x) = \begin{bmatrix} f_1(x_1) & & \\ & \ddots & \\ & & f_n(x_n) \end{bmatrix}$$

(where $x = (x_1, \dots, x_n)$) where each f_j is a function of one variable satifying $f_j(t) \geq \epsilon > 0$ for all t in $\mathbb{R}$ and where $e(x_1, \dots, x_n) = 2\sum_{j=1}^n \int_0^{x_j} f_j(t)^{-1} t dt$. Then consider the input affine system (1.14) with

$$A(x) = -f(x)x, \quad B(x) = \sqrt{2} f(x)$$
$$C(x) = -\sqrt{2}x, \quad D(x) = I$$

Then it is easily checked that

$$\nabla e(x) \cdot A(x) + C(x)^T C(x) = -2x \cdot x + 2x \cdot x = 0$$

and

$$\nabla e(x) \cdot B(x) u = 2^{3/2} x \cdot u = -2C(x)^T$$

Hence (1.18) and (1.19) are verified and we conclude that this system is lossless with energy function e and supply rate $u^T u - y^T y$. This example amounts to the equations coming from a LC-circuit expressed in the scattering formalism. In a hybrid impedance-admittance description, the system has the simpler representation

$$\dot{x} = f(x)u$$

$$y = x$$

(i.e. $A(x) = 0$, $B(x) = f(x)$, $C(x) = x$, $D(x) = 0$) and we have the identity

$$\nabla e(x) \cdot f(x)u = 2y^T x$$

if $f(x)^T \nabla e(x) = 2x$. For an LC-circuit with k capacitors and $n - k$ inductors the state vector is $(x_1, \ldots, x_n) = (V_{C_1}, \ldots, V_{C_k}, I_{L_1}, \ldots, I_{L_{n-k}})$ where V_{C_i} is the current through the ith capacitor and V_{L_i} is the voltage across the jth inductor. The output vector is simply the readout of the state. Finally, if the characteristic of the kth capacitor is given by

$$I_{C_k} = C_k(V_{C_k}) \frac{dV_{C_k}}{dt}$$

and the characteristic of the jth inductor is

$$V_{L_j} = L_j(I_{L_j}) \frac{dI_{L_j}}{dt}$$

then we take $f(x)$ to be the diagonal matrix function

$$f(x) = diag.\{C_1(x_1)^{-1}, \ldots, C_k(x_k)^{-1}, L_1(x_{k+1})^{-1}, \ldots, L_{n-k}(x_n)^{-1}\}$$

(where $x = (x_1, \ldots, x_n)$). For a fuller discussion see [AV, W2] for the linear case and [W1] for the nonlinear case. Here we are also implicitly assuming that the capacitor voltages and inductor currents uniquely determine the capacitor currents and inductor voltages; this happens generically as a consequence of Kirchoff's laws. For a fuller discussion of this last point, see [HS, Sm]. Here we are suppressing the details of the connection laws for the various simple components of the circuit.

In many applications the related notion of a dissipative system is key; in fact most of the work in [W1] and [HM] is concerned with dissipative systems. For a dissipative system one requires that the energy or storage function assume only nonnegative values ($e : X \to \mathbb{R}^+$) and the energy balance equation (1.15) is required to hold only with inequality

$$e(x(t)) - e(x(t_0)) \leq \int_{t_0}^{t} \{< Ju(s), u(s) > - < Jy(s), y(s) >\} ds \qquad (1.20)$$

An important class of systems consists of those which are simultaneously conservative and dissipative (with the same energy function e), i.e. those for which (1.15) is satisfied by an e with values in $\mathbb{R}^+$. Note that Livsic-Brodskii nodes (where $e(x) = \|x\|^2$) are dissipative.

The distinction between lossless and dissipative has to do with work done in the state space being independent of path ([WCGGG1-2]). (The reader should be aware that some authors define dissipativeness to be part of the definition of losslessness; we prefer to keep these concepts separate.)

Infinitesimal conditions analogous to (1.18) - (1.19) characterizing dissipativity (for the case of a smooth energy function e) can be derived as follows. We will make the simplifying assumption for the dissipative case that $D(x) = 0$ (which occurs often in applications), so $y(t) = C(x(t))$. Differentation of (1.20) with respect to t gives

$$H(x,u) := \nabla e(x) \cdot (A(x) + B(x)u) + C(x)^T JC(x) - u^T Ju \leq 0.$$

or

$$\max_u \quad H(x,u) \leq 0. \tag{1.21}$$

Since $H(x,u)$ is quadratic in u we can compute its maximum explicitly by setting the derivative equal to zero (or equivalently by "completing the square"). Differentiating with respect to u in the direction h gives

$$D_u H(x,u)[h] = \nabla e(x) \cdot B(x)h - 2u^T Jh.$$

Demanding that $D_u H(x,u)[h] = 0$ for all directions h and solving for u yields

$$u_{crit} = \frac{1}{2} JB(x)^T \nabla e(x).$$

Plugging u_{crit} back into $H(x,u)$ gives us the condition

$$\begin{aligned} \max_u \quad H(x,u) &= H(x, u_{crit}) \\ &= \nabla e(x)^T A(x) - \frac{1}{4} \nabla e(x)^T B(x) JB(x)^T \nabla e(x) + C(x)^T JC(X) \leq 0. \end{aligned} \tag{1.22}$$

A sufficient way for (1.22) to hold is that it hold with equality. In the linear case where $e(x) = \langle Hx, x \rangle$ is taken to be quadratic, this leads to an algebraic Riccati equation for the unknown H. In the nonlinear case this is a Hamilton-Jacobi equation of a fairly general type as studied in classical mechanics; a solution procedure is described in [vS] (see also Section 3.4).

For a linear input-output system of the form (1.10) an often used tool is the associated transfer function of the complex variable s

$$W(s) = D + C(sI - A)^{-1}B.$$

There are at least two ways to motivate the transfer function. One way is to apply the Laplace transform to the equations in (1.10) where we also impose the initial condition $x(0) = 0$. Eliminating $\hat{x}(s)$ and solving for $\hat{y}(s)$ in terms of $\hat{u}(s)$ (where $\hat{f}(s) = \int_0^\infty e^{-st} f(t)dt$ is the Laplace transform of $f(t)$) yields

$$\hat{y}(s) = W(s)\hat{u}(s).$$

For the case that A is *stable* (i.e. all eigenvalues of A are in the left half plane), another way is via considering the ouput $y(t)$ associated with a periodic input $u(t) = e^{st}\xi$ $(s + \bar{s} = 0,\ \xi \in \mathbb{R}^m)$ where again we take $x(0) = 0$. The calculation gives

$$\begin{aligned} x(t) &= \int_0^t e^{(t-\tau)A} B e^{s\tau} d\tau \xi \\ &= \int_0^t e^{tA+\tau(sI-A)} d\tau B\xi \\ &= e^{tA}(sI - A)^{-1} \quad \{e^{\tau(sI-A)}\}|_{\tau=0}^{\tau=t} \quad B\xi \\ &= e^{tA}(sI - A)^{-1}[e^{ts}e^{-tA} - I]B\xi \\ &= (sI - A)^{-1}B(e^{ts}\xi) - e^{tA}(sI - A)^{-1}B\xi \end{aligned}$$

Thus, up to the transient term $Ce^{tA}(sI - A){-}1B\xi$ which we discard, we have

$$y(t) = Cx(t) + Du(t) \approx e^{ts}(W(s)\xi).$$

Thus, with neglect of the transient term, the output $y(t)$ caused by the input $u(t) = e^{st}\xi$ of frequency s in directon ξ is again periodic with the same frequency s but in direction $W(s)\xi$ rather than ξ.

In the nonlinear case, we still eliminate the state vector and talk about an input-output map $u(t) \to y(t)$ (where we assume that $x(0)$ is equal to an eqilibrium point x_0 for the system) in the time domain but the transform of this map to the frequency domain via the Laplace transform is not so useful. An analogue of the second interpretation of the transfer function for the nonlinear case has been proposed by Isidori and Astolfi [IA].

A general consequence of the energy balance equation (1.15) is that

$$\int_0^t < Jy(s), y(s) > ds = \int_0^t < Ju(s), u(s) > ds \tag{1.23}$$

whenever $e(x(t)) = e(x(0))$. In particular, suppose we initialize the system at $x(t_0) = x_0$ where x_0 is an equilibrium state where $e(x_0) = 0$ and let us suppose that this equilibrium position is *stable* in the sense that $\lim_{t \to \infty} x(t) = x_0$ for any admissible input $u(t)$ in $L^2(0, \infty)$. Then a consequence of (1.23) is that

$$\int_0^\infty < Jy(s), y(s) > ds = \int_0^\infty < Ju(s), u(s) > ds \tag{1.24}$$

for any such input u, i.e. the IO map T_Σ leaves the J-quadratic form invariant on the space of admissible input signals u. A system having the input-output property (1.24) is usually called *lossless*. In the linear case this property has as a consequence that the transfer function $W(s)$ has J-isometric values on the imaginary line and this property is often used to define losslessness (with respect to J). In the nonlinear case the relation In the linear case, conservativeness and losslessness are closely related but not equivalent properties; in the nonlinear case the connections are even more subtle (see [WCGGG2]).

2. NONLINEAR LIVSIC-BRODSKII NODES: MODELS FOR A GIVEN DYNAMICS UP TO ENERGY PRESERVING DIFFEOMORPHIC CHANGE OF VARIABLE

We now present the original approach of Livsic to the study of nonselfadjoint operators on a Hilbert space (see e.g. [Br, L, LY]). To keep the notation consistent with section 1, we take as the fundamental object a nonskewadjoint opertor; thus the defect of the operator A is taken to be $A + A^*$, a measure of the departure from being skewadjoint ($A + A^* = 0$), rather than the defect equal to the measure $A - A^*$ of departure from selfadjointness as in the Livsic theory. To convert our notation to that of the Livsic school, simply replace our A by iA. If $A = -A^*$ then the spectral theory is complete via the spectral theorem. Given an operator which is not skewadjoint but whose defect operator is small in some appropriate sense (e.g. is finite rank or of trace class), the operator theoretic goal is to obtain a spectral theory or understand the invariant subspace structure of the operator A. One idea of Livsic was to use conservative systems as a tool for pursuing these operator theory objectives. Specifically, the idea is, given A, to construct B, C and $D = I$ so that the linear system

$$\dot{x} = Ax + Bu$$

$$y = Cx + Du \tag{2.1}$$

is conservative with respect to energy function $e(x) = \|x\|^2$ and an appropriate supply rate $S(u, y) = < Ju, u > - < Jy, y >$. To do this, we must construct B, C and $D = I$ so that (1.13a) - (1.13c) are satisfied with $P = I$ for an appropriate J. This entails as a first step the factorization

$$A + A^* = -C^*JC \tag{2.2}$$

of the defect operator, where $C: H \to U$, dim U =rank $(A + A^*)$ and J is an apppropriate signature operator, so (1.13a) is satisfied. The condition (1.13c) is automatic if we take $D = I$. Finally, use (1.13b) to define B as

$$B = -C^*J \tag{2.3}$$

The resulting system Σ given by

$$\dot{x} = Ax - C^*u$$

$$y = Cx + u$$

where

$$A + A^* = -C^*JC \tag{2.4}$$

is conservative in the sense discussed in Section 1 and is called the operator node or colligation associated with the nonskewadjoint operator A. The associated transfer function has the form

$$W_\Sigma(s) = I - C(sI - A)^{-1}C^*$$

and is called the *characteristic function* of the operator A. If A is completely nonskewadjoint, i.e. the restriction of A to any nonzero invariant subspace is not skewadjoint, then $W_\Sigma(s)$ is a complete, unitary invariant for A. More precisely, if A and $\tilde{A}$ are two completely nonskewadjoint operators on Hilbert spaces H and $\tilde{H}$ with associated nodes Σ and $\tilde{\Sigma}$ and characteristic operator functions W_Σ and $W_{\tilde{\Sigma}}$ such that $W_\Sigma(s) = W_{\tilde{\Sigma}}(s)$ for all s in a neighborhood of infinity, then there is a unitary operator from H onto $\widetilde{H}$ such that $\widetilde{A}U = UA$. Moreover, invariant subspaces of A match up with certain factorizations (called *regular factorizations*) of $W_\Sigma(s)$ and various operator theoretic properties of A can be matched up with corresponding function theoretic properties of $W_\Sigma(s)$. As was mentioned in the Introduction, more recently (see [LW, V]) there have been extensions of this theory to the setting where the single operator A is replaced by an n-tuple of commuting operators $(A_1, A_2, \ldots, A_n)$ which call upon techniques from algebraic geometry and Riemann surface theory; however these developments do not concern us here.

We now present a nonlinear analogue of the Livsic-Brodskii node for a nonskewadjoint operator A. We assume that the state space is R^n; a natural more general setting is to take the space X to be an n-dimensional manifold, but for notational and conceptual simplicity we will think of X as simply R^n. We are given a vector field $x \to A(x)$ on X and a smooth energy function $e: X \to R$. A nonlinear Livsic-Brodskii node associated with the vector field A and energy function e we take to mean a choice of vector fields $x \to B(x) = [\, B_1(x) \quad \ldots \quad B_m(x)]$ and mapping $C: X \to U$ (where U is an input-output space) such that the nonlinear input-affine system Σ given by

$$\begin{aligned} \dot{x} &= A(x) + B(x)u \\ y &= C(x) + u \end{aligned} \tag{2.5}$$

is conservative with respect to the energy function e and storage function $S(u, y) = < Ju, u > - < Jy, y >$ for an appropriate J. Thus construction of a Livsic-Brodskii node for a given vector field $A(x)$ and energy function e involves finding a vector field $B(x) = [\, B_1(x) \quad \ldots \quad B_m(x)\,]$, a mapping $C: X \to U$ and a signature operator J on U so that (1.18) and (1.19) are satisfied. As a first step, we must choose J so that (1.18) has a solution $C(x)$, i.e. we must solve

$$\nabla e(x) \cdot A(x) = C(x)^T J C(x). \tag{2.6}$$

Having picked $C(x)$ and J so that (2.6) is satisfied, we then must find a vector field $B(x)$ which satisfies (1.19), namely

$$\nabla e(x) \cdot B(x) + 2C(x)^T J = 0 \tag{2.7}$$

In the nonlinear case there is a lot of slack in the choice of solutions $C(x)$ and $B(x)$ of these equations, unlike in the linear case. We illustrate with a simple example where the equations can be easily solved.

Example 2.1 $A(x) = -x - x^3$, $e(x) = x^2/2$, $X = \mathbb{R}$.

Step 1. Solve (2.6) for $C(x)$:

$$-x^2 - x^4 = -C(x)^T J C(x).$$

Take $U = \mathbb{R}$, $J = 1$, $C(x) = x\sqrt{1+x^2}$.

Step 2. Find a solution $B(x)$ of (2.7):

$$xB(x) = -2x\sqrt{1+x^2}.$$

Take $B(x) = -2\sqrt{1+x^2}$. The resulting system

$$\begin{aligned} \dot{x} &= -x - x^3 - 2\sqrt{1+x^2}u \\ y &= x\sqrt{1+x^2} + u \end{aligned}$$

is conservative with dynamics given by $A(x) = -x - x^3$ as desired.

We now return to the general theory. Assume that we have constructed a conservative system Σ as in (2.5) whose main dynamics is equal to the prescribed vector field $A(x)$. We suppose also that 0 is an equilibrium point for the system (so $A(0) = 0, C(0) = 0$) and we take the initial state $x(0)$ to be the equilibrium point $x(0) = 0$. Then there is an induced input-output (IO) map $T_\Sigma: u(t) \to y(t)$ from an input function $u(t)$ to an ouput function $y(t)$. Here $u(t)$ is taken to be in some appropriate space of input functions, e.g. the space of piecewise continuous functions, such that we can solve the differential equation in (2.5) for $x(t)$ and then compute $y(t)$ from the second equation in (2.5). We take this IO map as the nonlinear analogue of the characteristic function in the linear theory. Now suppose that φ is a diffeomorhism on X taking 0 to 0 which leaves e invariant ($e \circ \varphi = e$). If we replace x with $x = \varphi(\tilde{x})$, then we can rewrite equations (2.5) in terms of the variable $\tilde{x}$ rather than x to get

$$\begin{aligned} \dot{\tilde{x}} &= D\varphi(\tilde{x})^{-1}[A \circ \varphi(\tilde{x})] + D\varphi(\tilde{x})^{-1} B \circ \varphi(\tilde{x})u \\ y &= C \circ \varphi(\tilde{x}) + u. \end{aligned}$$

This defines a new system $\tilde{\Sigma}$ where (A, B, C) has been transformed to $(\tilde{A}(x) = D\varphi(x)^{-1} A \circ \varphi$, $\tilde{B}(x) = D\varphi(x)^{-1} B \circ \varphi$, $\tilde{C}(x) = C \circ \varphi(x))$. Clearly, since $\tilde{\Sigma}$ was derived from Σ simply via a change of coordinates in the state space, the associated IO maps are the same ($T_{\tilde{\Sigma}} = T_\Sigma$). Since we assume that $e \circ \varphi = e$, it also follows from the assumed conservativeness of Σ that $\tilde{\Sigma}$ is also conservative. Thus $\tilde{\Sigma}$ is a nonlinear Livsic-Brodskii node associated with the new vector field $\tilde{A}(x) = D\varphi(x)^{-1} A \circ \varphi(x)$ such that $T_{\tilde{\Sigma}} = T_\Sigma$. Let us say that vector fields A and $\tilde{A}$ are *e-equivalent* if $\tilde{A}(x) = D\varphi(x)^{-1} A \circ \varphi(x)$ where φ is a diffeomorphism on X such that $e \circ \varphi = e$. Note that if A, $\tilde{A}$ and φ are all assumed to be linear and $e(x) = \frac{1}{2}\|x\|^2$, this notion collapses to unitary equivalence of the linear operators A and $\tilde{A}$. In the above dissussion we showed that if A has characteristic IO map T_Σ and $\tilde{A}$ is e-equivalent to A, then T_Σ is also a characteristic IO map for $\tilde{A}$. We would like to prove a converse statement, namely: if T_Σ is a characteristic IO map for both A and $\tilde{A}$, then A and $\tilde{A}$ are e-equivalent.

For such an assertion to be true we need some additional assumptions. Consider an input-affine system Σ as in (2.5). Following [J], we say that the system Σ is weakly reachable if, given any state $x_1 \in X$ there is a piecewise constant input u defined on some time interval $[0,T]$ so that $x(T) = x_1$, where $x(t)$ is determined by

$$\dot{x} = A(x) + B(x)u, \qquad x(0) = x_0.$$

Similarly, we say that the system (2.5) is observable if, given any two states $x_1, x_2 \in X$ with $x_1 \neq x_2$ there is a choice of piecewise constant input $u(t)$ defined on some time interval $[0,T]$ such that $y_1(T) \neq y_2(T)$, where $y_i(t)$ is determined by

$$\begin{aligned} \dot{x}_i &= A(x_i) + B(x_i)u, \qquad x_i(0) = x_i \\ y_i &= C(x_i) + u. \end{aligned}$$

The system Σ is said to *minimal* if it is both weakly reachable and observable, and to be *analytic* if A, B, C are analytic functions of x. We also assume that $A(x)$ and the various columns $B_1(x), \ldots, B_m(x)$ of $B(x)$ are complete vector fields so the differential equation defining the state vector has a solution $x(t)$ existing for all time t. We now state a global uniqueness theorem.

THEOREM 2.1. *Suppose that Σ_1 and Σ_2 given by*

$$\begin{aligned} \dot{x} &= A_i(x) + B_i(x)u \\ y &= C_i(x) + u \qquad (i = 1,2) \end{aligned}$$

are two minimal analytic systems which are conservative with respect to the energy function e and the storage rate $S(u,y) = \langle Ju, u\rangle - \langle Jy, y\rangle$ and induce the same IO maps $T_{\Sigma_1} = T_{\Sigma_2}$. Then there is a analytic diffeomorphism φ on the state space X such that

(i) $A_2(x) = D\varphi(x)^{-1} A_1 \circ \varphi(x)$

(ii) $B_2(x) = D\varphi(x)^{-1} B_1(x) \circ \varphi(x)$

(iii) $C_2(x) = C_1 \circ \varphi(x)$

and

(iv) $e \circ \varphi(x) = e(x)$

for all x in X. In particular the vector fields $A_1(x)$ and $A_2(x)$ are e-equivalent.

PROOF: From the main result in [J1] (see also [S]) it follows that there is an analytic diffeomorphism φ on X which satisfies (i), (ii), (iii). The map φ is constructed in the following way. Suppose a is a piecewise constant input function defined on the interval $[0,T]$

$$a(t) = \begin{cases} u_i \text{ for } T_{i-1} \le t < T_i \text{ and } 1 \le i \le N \\ u_0 \text{ for } t = T \end{cases}$$

where $0 = T_0 < T_1 < \ldots < T_N = T$ is a partition of $[0,T]$ and $u_i (0 \le i \le N)$ are elements of the input value space $\mathbb{R}^m$. Denote by $\psi_a^i(x_0)$ (for $i = 1,2$) the value $x_i(T)$ obtained by solving

$$\dot{x}_i = A_i(x_i) + B_i(x_i)a, \qquad x_i(0) = x_0 \tag{2.8}$$

for $0 \le t \le T$. Define a map φ from X into X by $\varphi(x) = \psi_a^2(x_0)$ whenever $x = \psi_a^1(x_0)$ for some piecewise constant input a defined on some interval $[0,T]$. Then by using the reachability and observability hypothesis and that $T_{\Sigma_1} = T_{\Sigma_2}$, one can show [J1] that φ is well-defined and bijective. By using that Σ_1 and Σ_2 are analytic, as also shown in [J1], one can show that φ is an analytic diffeomorphism. All that we add here is that, with the additional assumption that Σ_1 and Σ_2 are conservative with respect to the energy function e, it also follows that $e \circ \varphi = e$. To see this, note that if $x = \psi_a^1(x_0)$ and we set $y_a^i(x_0)(t) = C_1(x_i(t)) + a(t)$ for $0 \le t \le T$ where x_i is as in (2.7), then

$$e(x) = e(x_0) + \int_0^T \{\langle Jy_1(t), y_1(t)\rangle - \langle a(t), a(t)\rangle\} dt$$

by the energy balance equation for Σ_1

$$= e(x_0) \int_0^T \{< Jy_2(t), y_2(t) > - < a(t), a(t) >\}dt$$

since $T_{\Sigma_1} = T_{\Sigma_2}$

$$= e(\psi_a^2(x_0)) = e \circ \psi(x)$$

by the energy balance equation for the system Σ_2 and the definition of σ. Thus $e \circ \psi = e$ as asserted.

We remark that Theorem 2.1 is only one of many possible versions for a uniqueness theorem. Using other results from [J1], one can also do a global C^∞ or C^k version.

Using results from [J2] it is also possible to state results where the conclusion involves uniqueness only up to local diffeomorphisms preserving e defined on a neighborhood of a given point x in X and the hypotheses involve only local weak accessibility and local weak observability. For more on the various notions of observability and controllability in the nonlinear context, see [I, vS].

3. OTHER PARTIONINGS OF THE CAST OF CHARACTERS INTO KNOWNS AND UNKNOWNS

In Section 1 we introduced the notion of an input-affine conservative system. The needed ingredients, in addition to a state space X and an input-output space U, consisted of

(LS1) a vector field $A(x)$ on X

(LS2) m vector fields $B_1(x), \ldots, B_m(x)$ on X where m=dim U

(LS3) an output map $C: X \to U$

(LS4) an energy or storage function $e: X \to R$

(LS5) an input-output map T_Σ.

There is also the supply rate $S(u, y) = u^T Ju - y^T Jy$ but we shall consider this as fixed and given. For the discussion here we make the simplifying assumption that the feedthrough term $D(x)$ is the identity map I. If these objects are all associated with the same conservative system there are relations among them which we now recall. Specifically, with the assumption that $D(x) = I$, there are the algebraic relations

$$\nabla e(x) \cdot A(x) + C(x)^T JC(x) = 0 \tag{3.1}$$

$$\nabla e(x) \cdot B(x) + 2C(x)^T J = 0 \tag{3.2}$$

(where $B(x) = [\, B_1(x) \;\; \ldots \;\; B_m(x) \,]$ and $U = \mathbb{R}^m$) which together are equivalent to the energy balance equation. Once we are given $A(x), B(x)$ and $C(x)$ (i.e. (LS1) - (LS3)), then the input-output map T_Σ (LS5) is uniquely determined by

$$T_\Sigma(u)(t) = y(t)$$

where

$$\begin{aligned} \dot{x} &= A(x) + B(x)u, \qquad x(0) = x_0 \\ y &= C(x) + u. \end{aligned}$$

The construction of a nonlinear Luvsic-Brodskii node involves a compatible choice of (LS2) and (LS3) (as well as an appropriate J with which to define the supply rate) for a prespecified vector field $A(x)$ (LS1) and energy function $e(x) = \|x\|^2$ (LS4), i.e. (LS1) and (LS4) are given and one must solve for the remaining (LS2), (LS3) and (LS5). One can organize a large body of recent work in the system and control theory literature (especially in the linear case) as solving the problems associated with other possible partitionings of (LS1)-(LS5) into knowns and unknowns. We list several samples.

3.1. Given (LS1)-(LS3)

In this scenario we are given the complete internal description of the system (LS1) - (LS3) and the signature matrix giving the supply rate and seek to determine if the system is conservative. Note that (LS1) -(LS3) completely determine the IO map T_Σ; the only missing ingredient is the energy function e. To solve the problem one looks at the required indentities (3.1) and (3.2). A systematic treatment of the linear multivariate case is given in [AV]. Other more recent treatments are in [D, G, AG]. For the nonlinear case we refer to [W1], [HM1] and [WCGGG2].

3.2. Given (LS5)

Now we assume that we are given an input-output description (LS5) and seek the internal workings (LS1)-(LS3) of the system. In system theory this is known as the *realization problem.* In the Livsic linear theory, one assumes also that the state space is a Hilbert space and the energy function is the square of the norm ($e(x) = \|x\|^2$). The solution of this problem leads to an elegant triangular model for an operator having a given characteristic operator function (see[Br]). For the nonlinear case there does not appear to have been much work done; for the related class of nonlinear Hamiltonian control systems however, see [CI] and [CvS].

We mention a couple of variations (without going into much detail) where one wishes to construct (LS1) -(LS4) from some set of partial information concerning the input-output map (LS5).

3.2a. Given interpolation conditions

The problem (for the linear case) is to construct a realization for an all-pass matrix (i.e. the transfer function of a lossless system) which satisfies a prespecified collection of directional interpolation conditions. As the solution is highly nonunique it is of interest to construct such a matrix with the minimal McMillan degree. If one works with minimal, finite dimensional realizations, it is equivalent to construct a conservative realization of the given function. Recently there has appeared a series of papers dealing with the problem [AF1, AF2, BGR2, Sh]. Very preliminary formulation of such interpolation problems in the nonlinear case appears in [BH1].

3.2b. Darlington embedding

Mathematically the problem is as follows. One is given a Schur class matrix function S (i.e. its values on the closed right half plane are matrices of norm at most 1) and seeks to embed it as the (1,1) entry in a block 2×2 all-pass matrix

$$U = \begin{bmatrix} S & U_{12} \\ U_{21} & U_{22} \end{bmatrix}.$$

In circuit theory this has the interpretation of representing a given passive network as a lossless network cascade loaded to a resistor. Mathematical solutions in the frequency domain have been given in [Ar, DH]. A nice solution in terms of realization appears in [GR]. Patrick Dewilde has suggested that this would be of interest in the nonlinear case. We now give an analogue of the construction for nonlinear input-affine systems.

Consider the nonlinear input-affine system Σ (with feedthrough term equal to 0)

$$\begin{aligned} \dot{x} &= A(x) + B(x)u \\ y &= C(x). \end{aligned} \tag{3.3}$$

We assume that Σ is dissipative with storage rate given with the signature matrix J equal to the identity I and that e is a smooth energy function such that the energy dissipation inequality (1.22) holds with equality:

$$\nabla e(x) \cdot A(x) + -\frac{1}{4}\nabla e(x)^T B(x) B(x)^T \nabla e(x) + C(x)^T C(x) = 0 \tag{3.4}$$

The mathematical problem is to solve for vector fields $B_{21}(x), \ldots, B_{2m}(x)$ and output map $C_2(x)$ so that the enlarged system Σ_U given by

$$\begin{aligned} \dot{x} &= A(x) + B(x)u + B_2(x)u_2 \\ y &= C(x) + u_2 \\ y_2 &= C_2(x) + u \end{aligned} \tag{3.5}$$

(where $B(x) = [\, B_1(x) \quad \ldots \quad B_m(x)]$) is conservative with respect to the same energy function e. Physically the result has the following interpretation: the system Σ is the resulting closed loop system from cascade loading the conservative system Σ_U with a unit resistor in the scattering formalism. From (1.16) we see that the dilated system Σ_U is conservative if and only if the energy balance equation

$$\begin{aligned} &\nabla e(x) \cdot (A(x) + B(x)u + B_2(x)u_2) + \|C(x) + u_2\|^2 \\ &\|C_2(x) + u\|^2 - \|u\|^2 - \|u_2\|^2 = 0 \end{aligned} \tag{3.6}$$

holds. In particular, taking $u = 0$, $u_2 = 0$ gives

$$\nabla e(x) \cdot A(x) + C(x)^T C(x) + C_2(x)^T C_2(x) = 0. \tag{3.7}$$

From (3.3) we see that

$$C_2(x) = B(x)^T \frac{\nabla e(x)}{2} \tag{3.8}$$

leads to a solution of (3.7). Plug this back into (3.6) to get the condition

$$\nabla e(x) \cdot (B(x)u + B_2(x)u_2) + 2C(x)^T u_2 + 2C_2(x)^T u = 0. \tag{3.9}$$

Taking $u = 0$ leads to the condition on $B_2(x)$

$$\frac{\nabla e(x)}{2} B_2(x)u_2 = -C(x)^T u_2 \tag{3.10}$$

for all u_2. The following result summarizes the situation.

THEOREM 3.1. *Let Σ be a dissipative system as in (3.3) with smooth energy function e satisfying the dissipation equality (3.4). Set $C_2(x) = B(x)^T \frac{\nabla e(x)}{2}$ and let $B_2(x)$ be any smooth solution of*

$$\frac{\nabla e(x)}{2} \cdot B_2(x)u_2 = -C(x)^T u_2.$$

Then the enlarged system Σ_U given by

$$\dot{x} = A(x) + B(x)u + B_2(x)u_2$$
$$y = C(x) + u_2$$
$$y_2 = C_2(x) + u$$

is conservative with respect to the same energy function e.

We remark that this solution is consistent with the solution in [GR] for the linear case in the special case where the feedthrough term is 0. Indeed, in the linear case, the energy function e has the quadratic form $e(x) = < Hx, x >$ for an invertible Hermitian matrix H, $\nabla e(x) = 2Hx$ and one can solve (3.10) for $B_2(x) = B_2 x$ to get

$$B_2 = -H^{-1}C^T.$$

3.3 Given (LS1) together with (LS2) or (LS3)

We discuss the linear case first. We are given the dynamics operator A and the ouput operator C. From the point of view of the transfer function $W(s) = I + C(sI - A)^{-1}B$, if we assume that the realization is to be minimal, the given data consists of the poles and pole directions of the rational matrix function $W(s)$; for this reason, the pair (C, A) has been called a *pole pair* for $W(s)$ (see [BGR1]). Similar remarks apply to the pair (A, B); for reasons into which we do not go here, (C, A) is called a *right pole pair* and (A, B) is called a *left pole pair.* If, say, (C, A) is the given pair, then the problem is to find B so that $W(s) = I + C(sI - A)^{-1}B$ is a minimal realization for an all-pass matrix. Again the starting point is the set of relations (3.1). One first solves the Lyapunov equation (3.1) for $\nabla e(x) = Hx$ (H= Hermitian matrix). If H is invertible (as it must be for the square case if the realization is minimal), one can then use (3.2) to define B. This is essentially the procedure in [AG] (see also [BGR1]) where other issues and applications are addressed as well.

Another variation is to assume as given the input operator B together with the dynamics $A - BC$ for the inverse system; the given data consisting of the pair $(A^{\times}, B) = (A - BC, B)$ (called a *left null pair* for the system) can be interpreted as the zeros and zero directions of the transfer function. This version can be handled in a similar fashion and is also discussed in [AG].

The formulation of the problem and development of the same solution procedure in the nonlinear context appears in [BH2, BH3], but in the discrete time rather than in the continuous time setting. The nonlinear analogue of left null pair $(A^{\times}, B)$ is closely related to the notion of *zero dynamics* which has been introduced in connection with other nonlinear control problems [IM]. We discuss here a continuous time version of the problem for input-affine systems.

The problem is: we are given vector fields $A(x), B_1(x), \ldots, B_m(x)$ and a signature matrix J and seek an output map $C(x)$ so that the system Σ given by

$$\begin{aligned} \dot{x} &= A(x) + B(x)u \\ y &= C(x) + u, \end{aligned} \tag{3.11}$$

where $B(x) = [\, B_1(x) \quad \ldots \quad B_m(x)]$, is J-lossless and J-dissipative,(i.e. (1.20) and (1.24) hold). In the applications which we have in mind, the system Σ is actually the inverse of the lossless, dissipatve system of interest; hence we assume that the unknown energy function for Σ has the form $-e$ where e is nonnegative valued. Also in applications there often appears the additional constraint that

$$A(x) - B(x)C(x) \text{ is a stable vector field.} \tag{3.12}$$

To solve the problem, note that equations (3.1) and (3.2) with $-e$ in place of e become

$$-\nabla e(x) \cdot A(x) + C(x)^T JC(x) = 0 \tag{3.13}$$

$$-\nabla e(x) \cdot B(x) + 2C(x)^T J = 0. \tag{3.14}$$

Here $A(x)$ and $B(x)$ are known while $e(x)$ and $C(x)$ are the unknowns. From (3.14) we get

$$C(x)^T J = \frac{\nabla e(x)}{2} \cdot B(x).$$

Substitute this into (3.13) to get

$$-\nabla e(x) \cdot A(x) + \frac{1}{4} \nabla e(x) \cdot B(x) J B(x)^T \nabla e(x) = 0. \tag{3.15}$$

This is a Hamilton-Jacobi equation for the unknown energy function e, subject to the stability side condition (3.12). Once we have e, we use (3.14) to solve for C. A simple illustrative scalar example is the following.

Example 3.1 $A(x) = x + x^3$, $B(x) = 1$, $J = 1$.

We must solve

$$-\frac{de}{dx}(x)(x+x^3) + \frac{1}{4}\frac{de}{dx}(x)^2 = 0$$

subject to

$$x + x^3 - \frac{\frac{de}{dx}(x)}{2} \text{ is a stable vector field.}$$

The stability condition eliminates the possibility $\frac{de}{dx}(x) = 0$, so we seek a solution of

$$-(x+x^3) + \frac{1}{4}\frac{de}{dx}(x) = 0.$$

This leads to $e(x) = 2x^2 + x^4$. Then

$$\frac{dx}{dt} = x + x^3 + u$$
$$y = 2x + 2x^3 + u$$

is the desired system.

For more complicated examples, the Hamilton-Jacobi equation (with the stabilizing side condition) can be solved by finding the Lagrangian stable invariant manifold of a related Hamiltonian vector field, a canonical generalization of one of the standard methods for solving an algebraic Riccati equation. This point is discussed in some detail in [vS] in connection with the state feedback nonlinear H^∞-control problems.

In one approach to H^∞-control theory (see [BGR1, BR1, BR2]), the parametrization of the set of all solutions of the standard H^∞-control problem, at least for the simplest problems of "1-block" type, amounts to solving for a dissipative, conservative system having the same pole and zero structure as the given standard plant. In other words, in this approach, the standard problem of H^∞-control is reduced to a problem of the sort considered here. Implementation of this approach to the general linear time-invariant problem involves extra complications in the discussion of the zero structure of a rectangular rational matrix function. The same approach can be applied to nonlinear as well as to time-varying H^∞-control problems; see [BH1, BH2, BH3] and [BGK] for some preliminary results in this direction. The precise generality of this approach to handling nonlinear H^∞-control problems is a topic of work in progress.

3.4. Problems for dissipative systems

The same objects (LS1)-(LS5) discussed above for conservative systems are also attached to dissipative systems. For the class of nonlinear input-affine systems, if we assume that the feedthrough term $D(x)$ is 0 to keep the algebra simple, the system has the form

$$\begin{aligned} \dot{x} &= A(x) + B(x)u \\ y &= C(x) \end{aligned} \tag{3.16}$$

and the energy dissipation inequality from (1.22) becomes the Hamilton-Jacobi inequality

$$\nabla e(x) \cdot A(x) - \frac{1}{4}\nabla e(x)^T B(x) J B(x)^T \nabla e(x) + C(x)^T J C(x) \leq 0. \tag{3.17}$$

One can now propose to analyze the analogues of Problems (3.1)-(3.3) discussed above for the dissipative case. For example, if we consider the analogue of (3.4a), we are given the complete internal description ((LS1)-(LS3)) of the system together with the signature matrix J giving the supply rate, and we ask for the existence of a nonnegative valued energy function e on the state space which satisfies the energy dissipation inequality. If e is smooth, it necessarily must satisfy (3.17). In the scalar nonlinear case this criterion for dissipativity is the Kalman-Yakubovitch-Popov lemma. The multivariate linear case is worked out in [AV] and [GR]. For nonlinear dissipative systems there is a discussion about what e can be in [W1] and more precise results in [HM1].

Finally we mention another formulation of the H^∞-control problem which is directly related to dissipative systems. We discuss the problem in the context of input-affine nonlinear systems. We are given the so-called standard plant

$$\begin{aligned} \dot{x} &= A(x) + B_1(x)w + B_2(x)u \\ z &= C_1(x) + D_{12}(x)u \\ y &= C_2(x) + D_{21}(x)w \end{aligned} \tag{3.18}$$

and seek a compensator $K : y \to u$ assumed to have a state space represenatation of the form

$$\dot{\xi} = a(\xi) + b(\xi)y$$
$$u = c(\xi) \tag{3.19}$$

so that (1) *the closed loop system is internally stable* and (2) $\|z\|_2^2 \leq \|w\|_2^2$ *for all inputs* w *to the closed loop system*. One way to assure (2) is that the system be dissipative with respect to some smooth energy function e. The condition that e be dissipative for the closed loop system (3.18)-(3-19) amounts to finding nonlinear maps a, b, c so that

$$\begin{aligned}&\nabla_x e(x,\xi)\cdot(A(x) + B_1(x)w + B_2(x)c(\xi))\\ &+\nabla_\xi e(x,\xi)\cdot(a(\xi) + b(\xi)(C_2(x) + D_{21}(x)w))\\ &+\|C_1(x) + D_{12}(x)c(\xi)\|^2 - \|w\|^2 \leq 0\end{aligned}$$

for all w. Once such an e is found, under suitable hypotheses it can also be used as a Lyapunov function to prove that the system is stable in the sense of Lyapunov (i.e. condition (1) in the statement of the H^∞- problem). This idea of using e as a Lyapunov function to establish stability of a system was one of the main motivations for the study of dissipative systems in [W1] and was further developed in [HM2, HM3]. This strategy is the starting point for the analysis of the nonlinear H^∞-control problem in [IA] and [BHW]. The state feedback H^∞ control problem for the nonlinear case is handled in the same framework in [vS]. This approach is worked out in precise form for the linear, time-invariant case in [PAJ].

REFERENCES

[AG] D. Alpay and I. Gohberg, Unitary rational matrix functions, in *Topics in Interpolation Theory of Rational Matrix-valued Functions* (ed. I. Gohberg), Birkhäuser-Verlag, Basel-Berlin-Boston, 1988, pp. 175-222.

[AF1] T. Auba and Y. Funahashi, The structure of all-pass matrices that satisfy two-sided interpolation requirements, *IEEE Trans. Auto. Control* **AC-36** (1991), 1485-1489.

[AF2] T. Auba and Y. Funahashi, The structure of all-pass matrices that satisfy two-sided interpolation requirements, *IEEE Trans. Auto. Control* **AC-36** (1991), 1489-1493.

[AV] B.D.O. Anderson and S. Vongpanitlerd, *Network Analysis and Synthesis*, Prentice-Hall, Englewood Cliffs, N.J., 1973.

[An] A.C. Antoulas (ed.). *Mathematical System Theory: The Influence of R.E. Kalman*, Springer-Verlag, Berlin, 1991.

[Ar] D.Z. Arov, Darlington realization of matrix-valued functions, *Izv. Akad. Nauk SSSR Ser. Mat.*, **37** (1973), 1299-1331; English transl. in *Math. USSR Izv.*, **7** (1973).

[BC] J.A. Ball and N. Cohen, De Branges-Rovnyak operator models and systems theory: a survey, in *Topics in Matrix and Operator Theory* (ed. H. Bart, I. Gohberg and M.A. Kaashoek), OT50, Birkhäuser, Basel-Berlin-Boston, 1991, pp. 93-136.

[BGK] J.A. Ball, I. Gohberg and M.A. Kaashoek, Time-varying systems: Nevanlinna-Pick interpolation and sensitivity minimization, in *Recent Advances in Mathematical Theory of Systems, Control, Networks and Signal Processing I* (ed. H. Kimura and S. Kodama), Mita Press, Tokyo, 1992, pp. 53-58.

[BGR1] J.A. Ball, I. Gohberg and L. Rodman, *Interpolation of Rational Matrix Functions*, OT45 Birkhäuser-Verlag, Basel-Berlin-Boston, 1990.

[BGR2] J.A. Ball, I. Gohberg and L. Rodman, The structure of flat gain rational matrices that satisfy two-sided interpolation requirements, *Systems & Control Letters* 20 (1993), 401-412.

[BH1] J.A. Ball and J.W. Helton, Shift invariant manifolds and nonlinear analytic function theory, *Integral Equations and Operator Theory* 11 (1988), 615-725.

[BH2] J.A. Ball and J.W. Helton, Interpolation problems for null and pole structure of nonlinear systems, *Proc.* 27th *Conf. on Decision and Control*, Austin (1988), pp. 14-19.

[BH3] J.A. Ball and J.W. Helton, Inner-outer factorization of nonlinear operators, *J. Funct. Anal.*, 104 (1992), 363-413.

[BHW] J.A. Ball, J.W. Helton and M.L. Walker, H^∞ control for nonlinear systems with output feedback, *IEEE Trans. Auto. Control* **AC-38** (1993), 546-559.

[BR1] J.A. Ball and M. Rakowski, Interpolation by rational matrix functions and stability of feedback systems: the 2-block case, *J. Math. Systems, Estimation, and Control*, to appear.

[BR2] J.A. Ball and M. Rakowski, Interpolation by rational matrix functions and stability of feedback systems: the 4-block case, in *Operator Theory and Complex Analysis* (ed. T. Ando and I. Gohberg), Birkhäuser, Basel, 1992, pp. 96-142.

[BB] T. Basar and P. Bernhard, *H^∞-Optimal Control and Related Minimax Design Problems: A Dynamic Game Approach*, Birkhäuser-Verlag, Basel-Berlin-Boston, 1991.

[Br] M.S. Brodskii, *Triangular and Jordan Representations of Linear Operators*, Transl. of Mathematical Monographs, Amer. Math. Soc., Providence, 1971.

[CI] P.E. Crouch and M. Irving, On finite Volterra series which admit Hamiltonian realizations, *Math. Systems Theory*, 17(1984), 293-318.

[CvS] P.E. Crouch and A.J. van der Schaft, *Variational and Hamiltonian Control Systems*, Springer-Verlag, Berlin, 1987.

[DH] R.G. Douglas and J.W. Helton, Inner dilations of analytic matrix functions and Darlington systhesis, *Acta Sci. Math.* (Szeged), 34(1973), 61-67.

[D] J.C. Doyle, *Lecture notes in advances in multivariable control*, ONR/Honeywell Workshop, Minneapolis, 1984.

[Fr] B.C. Francis, *A Course in H_∞ Control Theory*, Springer-Verlag, Berlin, 1987.

[G] K. Glover, All optimal Hankel-norm approximations of linear multivariable systems and their L^∞ error bounds, *Int. J. Control*, 39(1984), 1115-1193.

[GR] I. Gohberg and S. Rubinstein, Proper contractions and their unitary minimal completions, in *Topics in Interpolation Theory of Rational Matrix-valued Functions* (ed I. Gohberg), Birkhäuser-Verlag, Basel-Berlin-Boston, 1988, pp. 223-247.

[H] J.W. Helton, Discrete time systems, operator models and scattering theory, *J. Funct. Anal.* 16(1974), 15-38.

[HM1] D.J. Hill and P.J. Moylan, Dissipative dynamical systems: basic input-output and state properties, *J. Franklin Inst.* 309(1980), 327-357.

[HM2] D.J. Hill and P.J. Moylan, The stability of nonlinear dissipative systems, *IEEE Trans. Auto. Control* **AC-21** (1976), 708-711.

[HM3] D.J. Hill and P.J. Moylan, Connections between finite-gain and asymptotic stability, *IEEE Tran. Auto. Control* **AC-25** (1980), 931-936.

[HS] M.W. Hirsch and S. Smale, *Differential Equations, Dynamical Systems, and Linear Algebra*, Associated Press, San Diego, 1974.

[I] A. Isidori, *Nonlinear Control Systems: An Introduction*, Springer-Verlag, Berlin, 1985.

[IA] A. Isidori and A. Astolfi, Disturbance attenuation and H_∞ control via measurement feedback in nonlinear systems, *IEEE Trans Auto. Control* **AC-37** (1992), 1283-1293.

[IM] A. Isidori and C.H. Moog, On the nonlinear equivalent of transmission zeros, in *Modelling and Adaptive Control* (ed. C.I. Byrnes and A. Kurzhanski), Springer-Verlag, Berlin, 1988, pp. 146-158.

[J1] B. Jakubczyk, Existence and uniqueness of realizations of nonlinear systems, *SIAM J. Control and Opt.* 18(1980), 445-471.

[J2] B. Jakubczyk, Local realizations of nonlinear causal operators, *SIAM J. Control and Opt.*, 24(1986), 230-242.

[J3] B. Jakubczyk, Existence of Hamiltonian realizations of nonlinear causal operators, *Bull. Pol. Acad. Sci., Ser. Math.* 34(1986), 737-747.

[LP] P.D. Lax and R.S. Phillips, *Scattering Theory*, Academic Press, New York, 1967.

[L] M.S. Livsic, *Operators, Oscillations, Waves: Open Systems*, Transl. Math. Mon., Amer. Math. Soc., Providence, 1973.

[LW] M.S. Livsic and L.L. Waksman, *Commuting Nonselfadjoint Operators in Hilbert Space*, Lecture Notes in Mathematics #1272, Springer-Verlag, Berlin, 1987.

[LY] M.S. Livshits(Livsic) and A.A. Yantsevich, *Operator Colligations in Hilbert Spaces*, Winston-Wiley, New York, 1979.

[NvS] H. Nijmeijer and A.J. van der Schaft, *Nonlinear Dynamical Control Systems*, Springer-Verlag, Berlin, 1990.

[PAJ] I.R. Petersen, B.D.O. Anderson and E.A. Jonckheere, A first principles solution to the non-singular H^∞ control problem, *Int. J. Robust and Nonlinear Control* 1, (1991), 171-185.

[Sh] U. Shaked, A two-sided interpolation approach to H^∞ optimization problems, *IEEE Trans. Auto. Control* **AC-34** (1989), 1293-1296.

[Sm] S. Smale, On the mathematical foundations of electrical circuit theory, *J. Differential Geom.* 7(1972), 193-210.

[Su] H.J. Sussmann, Existence and uniqueness of minimal realizations of nonlinear systems, *Math. Systems Theory* 10(1977), 263-284.

[vS] A.J. van der Schaft, L_2-gain analysis of nonlinear systems and nonlinear H_∞ control, *IEEE Trans. Auto. Control* **AC-37** (1992), 770-784.

[V] V. Vinnikov, Commuting nonselfadjoint operators and algebraic curves, in *Operator Theory and Complex Analysis* (ed. T. Ando and I. Gohberg), Birkhäuser-Verlag, Basel-Boston-Berlin, 1992, pp. 348-371.

[W1] J.C. Willems, Dissipative dynamical systems, Part I: General theory, *Arch. Rat. Mech. Anal.* 45(1972), 321-351.

[W2] J.C. Willems, Dissipative dynamical systems, Part II: Linear systems with quadratic supply rates, *Arch. Rational Mech. Anal.* 45(1972), 352-393.

[WCGGG1] J.L. Wyatt, L.O. Chua, J.W. Gannett, I.C. Goknar and D.N. Green, Energy concepts in the state-space theory of nonlinear n-ports: Part I—Passivity, *IEEE Trans. Circuits and Systems* **CAS-28** (1982), 48-60.

[WCGGG2] J.L. Wyatt, L.O. Chua, J.W. Gannett, I.C. Goknar and D.N. Green, Energy concepts in the state-space theory of nonlinear n-ports: Part II—Losslessness, *IEEE Trans. Circuits and Systems* **CAS-29** (1982), 417-430.

Department of Mathematics
Virginia Tech
Blacksburg, VA 24061
USA

MSC: 49L10, 47A48.

Operator Theory:
Advances and Applications, Vol. 73

ORTHOGONAL POLYNOMIALS OVER HILBERT MODULES

Asher Ben-Artzi and Israel Gohberg

Dedicated to M.S. Livšic with respect and admiration.

Orthogonalization with invertible squares is studied in the special case of modules over the C^*-algebra of block diagonal operators in ℓ_r^2. For orthogonalizations with invertible squares of the system $I, S, \ldots, S^m$, where S is a unilateral shift, are obtained results which describe their invertibility, Fredholm and index properties.

1. INTRODUCTION

Let B be a unital C^*-algebra. A right-B module M equipped with a B-valued inner product $\langle\cdot,\cdot\rangle : M \times M \to B$ satisfying $\langle \alpha_1 x_1 + \alpha_2 x_2, y\rangle = \alpha_1\langle x_1, y\rangle + \alpha_2\langle x_2, y\rangle$ $(\alpha_1, \alpha_2 \in \mathbf{C}\,; x_1, x_2, y \in M)$, and such that

$$\langle x, y\rangle = \langle y, x\rangle^* \qquad (x, y \in M)\,, \tag{1.1}$$

and

$$\langle xb, y\rangle = \langle x, y\rangle b \qquad (x, y \in M; b \in B)\,, \tag{1.2}$$

is called a *self-adjoint module* over B.

In the case when the inner product also satisfies the additional condition $\langle x, x\rangle > 0$ for each $0 \neq x \in M$, such a module is called in some sources a pre-Hilbert module (see for example [I] or [P]).

We are primarily interested in a concrete class of self-adjoint modules, which we now define. We begin by setting some notation. Denote by ℓ_r^2 the Hilbert space of all square summable sequences $(x_n)_{n=0}^{\infty}$ with $x_n \in \mathbf{C}^r$ $(n = 0, 1, \ldots)$. Let $\mathcal{L}(\ell_r^2)$ be the space of all linear bounded operators in ℓ_r^2. Each operator $T \in \mathcal{L}(\ell_r^2)$ admits a matrix representation $T = (t_{ij})_{ij=0}^{\infty}$ where t_{ij} are $r \times r$ matrices. We define B to be the C^*-algebra of all block diagonal operators, namely $T = (t_{ij})_{ij=0}^{\infty}$ belongs to B if and only if

$t_{ij} = 0$ for all $i \neq j$. Then the usual operator product Tb ($T \in \mathcal{L}(\ell_r^2)$; $b \in B$) turns $\mathcal{L}(\ell_r^2)$ into a right-B module. We will also use the projection diag: $\mathcal{L}(\ell_r^2) \to B$ defined in the following way. For $T = (t_{ij})_{i=0}^{\infty} \in \mathcal{L}(\ell_r^2)$ we define

$$\text{diag}\,(T) = (\delta_{ij} t_{ij})_{ij=0}^{\infty} \in B\ .$$

For each self-adjoint operator $R \in \mathcal{L}(\ell_r^2)$ we define a self-adjoint module M_R over B as follows. Let $M_R = \mathcal{L}(\ell_r^2)$ with its right-B module structure as above, and define the inner product $\langle\cdot,\cdot\rangle_R : M_R \times M_R \to B$ via

$$\langle T_1, T_2\rangle_R = \text{diag}\,(T_2^* R T_1)\ .$$

It is clear that M_R is a self-adjoint module.

We are interested in the generalization of the notion of orthogonalization in Euclidean spaces, to the module case.

DEFINITION. Let $T_0, \dots, T_m$ be elements of a self-adjoint module M over the unital C^*-algebra B. A system $V_0, \dots, V_m$ in M is called an *orthogonalization with invertible squares* of $T_0, \dots, T_m$, if the following conditions hold:

$$\langle V_i, V_j\rangle = 0 \quad (i,j = 0, \dots, m; \quad i \neq j)\ , \tag{1.3}$$

$$\text{span}_B(T_0, \dots, T_i) = \text{span}_B(V_0, \dots, V_i) \quad (i = 0, \dots, m)\ , \tag{1.4}$$

where $\text{span}_B(T_0, \dots, T_i) = \{T_0 b_0 + \dots + T_i b_i : b_0, \dots, b_i \in B\}$, and

$$\langle V_i, V_i\rangle \text{ is invertible} \quad (i = 0, \dots, m)\ . \tag{1.5}$$

If such a system $V_0, \dots, V_m$ exists then we say that the system $T_0, \dots, T_m$ *admits orthogonalization with invertible squares.* Condition (1.5) above is a type of normalization. Since the inner product in M is not necessarily positive definite, this condition replaces the usual condition $\langle V_i, V_i\rangle = 1$ of orthonormal systems in Euclidean spaces. We also define the Gramian $G(T_0, \dots, T_k)$ of a system $T_0, \dots, T_k$ via

$$G(T_0, \dots, T_k) = (\langle T_j, T_i\rangle)_{i,j=0}^{k}\ .$$

Thus, $G(T_0, \dots, T_k)$ belongs to $B_{k+1,k+1}$, which is the C^*-algebra of all $(k+1) \times (k+1)$ matrices with entries in B.

For the self-adjoint module M_R the following theorem holds.

THEOREM 1.1. *A system $T_0, \ldots, T_m$ in M_R admits an orthogonalization with invertible squares if and only if all the Gramians $G(T_0, \ldots, T_k)$ $(k = 0, \ldots, m)$ are invertible.*

Let us remark that this result is not true for general self-adjoint modules. For a counter-example consider $M' = \mathcal{L}(\ell_r^2)$ as a right module over the C^*-algebra $B' = \mathcal{L}(\ell_r^2)$ with inner product given by $\langle T_1, T_2 \rangle_{M'} = T_2^* T_1$, without the diagonal projection. Consider the system consisting of a single element $T_0 = S^*$ where $S = (\delta_{i,j+1} I_r)_{ij=0}^{\infty}$ is the unilateral block shift. Then $V_0 = I$, the identity operator, is an orthogonalization with invertible squares for T_0. In fact, condition (1.3) in the definition of orthogonalization with invertible squares is void because $m = 0$ in this case. Moreover, condition (1.4) holds because S^* is right invertible and therefore

$$\text{span}_{B'}(T_0) = \text{span}_{B'}(S^*) = \mathcal{L}(\ell_r^2) = \text{span}_{B'}(I) = \text{span}_{B'}(V_0) .$$

Finally, $\langle V_0, V_0 \rangle_{M'} = I$ is invertible and hence, condition (1.5) holds as well. On the other hand, the Gramian

$$G(T_0) = \langle S^*, S^* \rangle_{M'} = SS^*$$

is not invertible.

Let us now return to the example of the self-adjoint module M_R. Let $S = (\delta_{i,j+1} I_r)_{ij=0}^{\infty}$ be the forward block shift operator in ℓ_r^2. We are interested in the system $I, S, \ldots, S^m$, of elements of M_R. Note that if $R = I$, the identity operator in ℓ_r^2, then $\langle S^i, S^j \rangle_I = \text{diag}\,(S^{j*} S^i) = \delta_{ij} I$. Hence, if $R = I$ the system $I, S, \ldots, S^m$ is already orthogonal with invertible squares. Let us recall here that S^m is left invertible with

$$\text{codim Im } S^m = mr \quad (m = 0, 1 \ldots) . \tag{1.6}$$

We can now present the main result of this paper. We will use the following notation. For a finite self-adjoint matrix A, we denote by $\nu_+(A)$ the number of positive eigenvalues of A, counting multiplicities.

THEOREM 1.2. *Let $R = (R_{ij})_{ij=0}^{\infty}$ be a bounded self-adjoint operator in ℓ_r^2 such that the system $I, S, \ldots, S^m$ admits an orthogonalization with invertible squares $V_0, \ldots, V_m$ in the module M_R, where m is a positive integer. If*

$$\langle V_m, V_m \rangle_R \geq 0 , \tag{1.7}$$

then V_m is left invertible with

$$\text{codim Im } V_m = \nu_+\big((R_{ij})_{ij=n}^{n+m-1}\big) \tag{1.8}$$

for n sufficiently large.

A shown in [AG], without (1.7) the theorem does not hold even in the case when R is Toeplitz.

This paper is related to our earlier papers on nonstationary generalizations of orthogonal polynomials. In particular, the main theorem and Theorems 4.1 and 4.2 are a generalization and an improvement of the part of Theorem 1.1 of [BG1] concerning one side infinite matrices, and of Theorem 1.1 of [BG3].

This paper consists of four sections. In Section 2 we give some results about orthogonalization with invertible squares in the module M_R. Theorem 1.1 is contained in Theorem 2.1. Section 3 contains some preliminary results about inertia theorems for block weighted shifts which are used in the proof of Theorem 1.2. Section 4 contains the proof of Theorem 1.2 as well as a generalization.

We denote by I the identity operator, the space in which I is acting being clear from the context, and put $I_r = (\delta_{i,j})_{i,j=1}^r$ $(r = 1, 2, \ldots)$. For a Hilbert space H, we denote by $(u, v)_H$ the usual inner product of two vectors u and v in H.

2. ORTHOGONALIZATION WITH INVERTIBLE SQUARES

In this section we prove Theorem 2.1 below, which contains Theorem 1.1. Throughout this section, let R be a bounded self-adjoint operator in ℓ_r^2, and M_R be the self-adjoint module as defined in the introduction. Thus, $M_R = \mathcal{L}(\ell_r^2)$ is a right-B module over the C^*-algebra B of all block diagonal operators in ℓ_r^2, and the inner product $\langle\cdot,\cdot\rangle_R : M_R \times M_R \to B$ is given by $\langle T_1, T_2\rangle_R = \operatorname{diag}(T_2^* R T_1)$.

THEOREM 2.1. *Let $T_0, \ldots, T_m$ be a system in M_R. Then $T_0, \ldots, T_m$ admits an orthogonalization with invertible squares if and only if all the Gramians $G(T_0, \ldots, T_k)$ $(k = 0, \ldots, m)$ are invertible. Moreover, if $V_0, \ldots, V_m$ and $V_0', \ldots, V_m'$ are two orthogonalizations with invertible squares of $T_0, \ldots, T_m$, then there exist invertible elements $b_0, \ldots, b_m$ in B such that $V_i' = V_i b_i$ $(i = 0, \ldots, m)$. Finally, assume that all the Gramians $G(T_0, \ldots, T_k)$ $(k = 0, \ldots, m)$ are invertible. Then the system $W_0, \ldots, W_m$ obtained by the formula*

$$W_k = \sum_{i=0}^{k} T_i \Gamma_{ik}^k \quad (k = 0, \ldots, m) , \tag{2.1}$$

where

$$G(T_0, \ldots, T_k)^{-1} = (\Gamma_{ij}^k)_{ij=0}^k \quad (k = 0, \ldots, m) , \tag{2.2}$$

is an orthogonalization with invertible squares of $T_0, \ldots, T_m$. *For this system the equalities*

$$\langle W_k, W_k \rangle_R = \Gamma_{kk}^k \quad (k = 0, \ldots, m) \tag{2.3}$$

hold.

PROOF. Assume first that $V_0, \ldots, V_m$ is an orthogonalization with invertible squares of $T_0, \ldots, T_m$. Then

$$\operatorname{span}_B(V_0, \ldots, V_j) = \operatorname{span}_B(T_0, \ldots, T_j) \quad (j = 0, \ldots, m)$$

by (1.4). In particular, $V_j \in \operatorname{span}_B(T_0, \ldots, T_j)$ and hence there are elements α_{ij} $(j = 0, \ldots, m; i = 0, \ldots, j)$ in B such that

$$V_j = \sum_{i=0}^{j} T_i \alpha_{ij} \quad (j = 0, \ldots, m)\,. \tag{2.4}$$

Let us also note that by (1.4) again, $T_i \in \operatorname{span}_B(V_0, \ldots, V_i)$ $(i = 0, \ldots, m)$, and therefore, $T_i = \sum_{k=0}^{i} V_k \beta_{ki}$ for some $\beta_{ki} \in B$. For $k < j$ we have $\langle V_k, V_j \rangle_R = 0$ by (1.3). Hence, if $i < j$ we obtain from the preceding equalities that

$$\langle T_i, V_j \rangle_R = \left\langle \sum_{k=0}^{i} V_k \beta_{ki}, V_j \right\rangle_R = \sum_{k=0}^{i} \langle V_k, V_j \rangle_R \beta_{ki} = 0 \quad (i < j)\,. \tag{2.5}$$

We now take the inner product of (2.4) with V_j. Taking into account (2.5), it follows that

$$\langle V_j, V_j \rangle_R = \sum_{i=0}^{j} \langle T_i, V_j \rangle_R \alpha_{ij} = \langle T_j, V_j \rangle_R \, \alpha_{j,j} \quad (j = 0, \ldots, m)\,. \tag{2.6}$$

The invertible squares condition (1.5) implies that $\langle V_j, V_j \rangle_R$ is invertible, and therefore $\langle T_j, V_j \rangle_R$ is right invertible and $\alpha_{j,j}$ is left invertible. However, both $\langle T_j, V_j \rangle_R$ and $\alpha_{j,j}$ belong to B, where B is the algebra of all block diagonal operators in ℓ_r^2. Since $r < \infty$, B has the property that every right or left invertible element of B is invertible. Therefore, we conclude that

$$\langle T_j, V_j \rangle_R \quad \text{and} \quad \alpha_{j,j} \quad \text{are invertible } (j = 0, \ldots, m)\,. \tag{2.7}$$

Let us now remark that by taking the inner product of equality (2.4) for $j = 0, \ldots, m$, with T_h $(h = 0, \ldots, j)$, we obtain

$$\langle V_j, T_h \rangle_R = \sum_{i=0}^{j} \langle T_i, T_h \rangle_R \alpha_{ij} \quad (j = 0, \ldots, m; h = 0, \ldots, j) . \tag{2.8}$$

Let k be an integer with $0 \leq k \leq m$. We define two matrices in $B_{k+1,k+1}$ via $\Omega = \left(\langle V_j, T_i \rangle_R\right)_{ij=0}^{k}$ and $\Lambda = (\alpha_{ij})_{ij=0}^{k}$, where we set $\alpha_{ij} = 0$ for $i > j$. Then (2.8) implies

$$\Omega = G(T_0, \ldots, T_k)\Lambda . \tag{2.9}$$

Moreover, Ω and Λ are invertible. In fact, Λ is upper triangular by definition and Ω is lower triangular by (2.5) and (1.1). In addition, the diagonal elements of Ω are $\langle V_j, T_j \rangle_R$ and the diagonal elements of Λ are $\alpha_{j,j}$. These are invertible by (2.7) and consequently Ω and Λ are invertible. Equality (2.9) now implies that $G(T_0, \ldots, T_k)$ is invertible $(k = 0, \ldots, m)$. This proves the first part of the theorem in one direction. The reverse implication follows from the third part of the theorem which is proved below.

We now turn to the second part of the theorem. Let us note the following consequence of the previous considerations. Since Ω is lower triangular, the right column of equality (2.9) yields

$$\begin{pmatrix} 0 \\ \vdots \\ 0 \\ \langle V_k, T_k \rangle_R \end{pmatrix} = G(T_0, \ldots, T_k) \begin{pmatrix} \alpha_{1,k} \\ \vdots \\ \alpha_{k,k} \end{pmatrix} \quad (k = 0, \ldots, m) . \tag{2.10}$$

Moreover, $\langle V_k, T_k \rangle_R$ are invertible by (2.7) $(k = 0, \ldots, m)$. Now assume that $V_0', \ldots, V_m'$ is another orthogonalization of $T_0, \ldots, T_m$ with invertible squares. As above for $V_0, \ldots, V_m$, it follows that there exist elements α'_{ik} in B such that

$$V_k' = \sum_{i=0}^{k} T_i \alpha'_{ik} \quad (k = 0, \ldots, m) , \tag{2.11}$$

the elements $\langle V_k', T_k \rangle_R$ are invertible $(k = 0, \ldots, m)$, and the following equality which corresponds to (2.10) holds

$$\begin{pmatrix} 0 \\ \vdots \\ 0 \\ \langle V_k', T_k \rangle_R \end{pmatrix} = G(T_0, \ldots, T_k) \begin{pmatrix} \alpha'_{1,k} \\ \vdots \\ \alpha'_{k,k} \end{pmatrix} \quad (k = 0, \ldots, m) .$$

Since $G(T_0,\dots,T_k)$ is invertible, it follows from this equality and (2.10) that

$$\alpha'_{i,k} = \alpha_{i,k} b_k \quad (k = 0,\dots,m;\ i = 0,\dots,k) \tag{2.12}$$

where

$$b_k = \langle V_k, T_k\rangle^{-1}\langle V'_k, T_k\rangle \quad (k = 0,\dots,m) \tag{2.13}$$

are invertible. It follows immediately from (2.12) and the representations (2.4) and (2.11) that

$$V'_k = V_k b_k \quad (k = 0,\dots,m)\ .$$

This proves the uniqueness part of the theorem.

We now turn to the last part of the theorem. So assume that the Gramians $G_k = G(T_0,\dots,T_k)$ are invertible for each $k = 0,1,\dots,m$. Let Γ^k_{ij} be as in (2.2) $(k = 0,\dots,m; i,j = 0,\dots,k)$, and define W_k via (2.1) $(k = 0,\dots,m)$. We prove that $W_0,\dots,W_m$ is an orthogonalization with invertible squares of $T_0,\dots,T_m$.

First we show that $\Gamma^0_{0,0},\Gamma^1_{1,1},\dots,\Gamma^m_{m,m}$ are invertible. Let $k = 1,\dots,m$, and note that the following block decomposition holds

$$G_k = \begin{pmatrix} G_{k-1} & Z^* \\ Z & \langle T_k, T_k\rangle_R \end{pmatrix},$$

where $Z = (\langle T_1, T_k\rangle_R,\dots,\langle T_{k-1}, T_k\rangle_R)$. Since

$$G_k \begin{pmatrix} \Gamma^k_{1k} \\ \vdots \\ \Gamma^k_{kk} \end{pmatrix} = \begin{pmatrix} 0 \\ \vdots \\ 0 \\ I \end{pmatrix}, \tag{2.14}$$

we obtain

$$G_{k-1} \begin{pmatrix} \Gamma^k_{1k} \\ \vdots \\ \Gamma^k_{k-1,k} \end{pmatrix} + Z^*\Gamma^k_{k,k} = 0, \quad Z \begin{pmatrix} \Gamma^k_{1k} \\ \vdots \\ \Gamma^k_{k-1,k} \end{pmatrix} + \langle T_k, T_k\rangle_R \Gamma^k_{kk} = I\ .$$

This implies

$$(\langle T_k, T_k\rangle_R - Z G^{-1}_{k-1} Z^*)\Gamma^k_{kk} = I\ .$$

However Γ^k_{kk} is self-adjoint, and consequently Γ^k_{kk} is invertible $(k = 1,\dots,m)$. It is also clear from $(G_0)^{-1} = (\Gamma^0_{ij})^0_{ij=0}$ that Γ^0_{00} is invertible.

Let us now remark that (2.1) leads immediately to

$$\mathrm{span}_B(W_0,\ldots,W_k) \subset \mathrm{span}_B(T_0,\ldots,T_k) \quad (k=0,\ldots,m)\,. \tag{2.15}$$

We now prove that equality holds in (2.15) by induction. Since $W_0 = T_0\Gamma^0_{00}$ and Γ^0_{00} is invertible, it is clear that $\mathrm{span}_B(W_0) = \mathrm{span}_B(T_0)$. Let $k \in \{1,\ldots,m\}$ and suppose that we have shown that $\mathrm{span}_B(W_0,\ldots,W_{k-1}) = \mathrm{span}_B(T_0,\ldots,T_{k-1})$. Since Γ^k_{kk} is invertible we obtain from (2.1) that $T_k = \left(W_k - \sum_{i=0}^{k-1} T_i\Gamma^k_{ik}\right)(\Gamma^k_{kk})^{-1}$. Thus, $T_k \in \mathrm{span}_B(T_0,\ldots,T_{k-1},W_k)$, and therefore also

$$\mathrm{span}_B(T_0,\ldots,T_k) \subset \mathrm{span}_B(T_0,\ldots,T_{k-1},W_k)\,.$$

Since $\mathrm{span}_B(T_0,\ldots,T_{k-1}) = \mathrm{span}_B(W_0,\ldots,W_{k-1})$ by the induction hypothesis, we obtain

$$\mathrm{span}_B(T_0,\ldots,T_k) \subset \mathrm{span}_B(W_0,\ldots,W_k)\,.$$

Combining this with (2.15) it follows that

$$\mathrm{span}_B(T_0,\ldots,T_k) = \mathrm{span}_B(W_0,\ldots,W_k)\,. \tag{2.16}$$

Hence, (2.16) holds for all $k = 0,\ldots,m$.

We now prove that $W_0,\ldots,W_m$ is an orthogonalization of $T_0,\ldots,T_m$ with invertible squares by showing that

$$\langle W_h, W_k\rangle_R = 0 \quad (h<k)\,, \tag{2.17}$$

$$\langle W_k, W_k\rangle_R \text{ is invertible } (k=0,\ldots,m)\,. \tag{2.18}$$

Recall that by definition $G(T_0,\ldots,T_k) = \left(\langle T_j,T_i\rangle_R\right)^k_{ij=0}$. Hence, (2.14) leads to

$$\sum_{j=0}^{k}\langle T_j,T_i\rangle_R\Gamma^k_{jk} = \delta_{ik}I \quad (i=0,\ldots,k)\,.$$

This means that

$$\langle W_k,T_i\rangle_R = \left\langle \sum_{j=0}^{k} T_j\Gamma^k_{jk}, T_i\right\rangle_R = \sum_{j=0}^{k}\langle T_j,T_i\rangle_R\Gamma^k_{jk} = \delta_{ik}I \quad (i=0,\ldots,k)\,. \tag{2.19}$$

In particular, it follows from this equality that

$$\langle T_i, W_k\rangle_R = \langle W_k,T_i\rangle_R^* = 0 \quad (i=0,\ldots,k-1)\,. \tag{2.20}$$

Now let $h = 0, \ldots, k$. By (2.1) $W_h = \sum_{i=0}^{h} T_i \Gamma_{ih}^h$, whence

(2.21) $$\langle W_h, W_k \rangle_R = \sum_{i=0}^{h} \langle T_i, W_k \rangle_R \Gamma_{ih}^h = \sum_{i=0}^{h} \delta_{ik} \Gamma_{ih}^h = \delta_{hk} \Gamma_{kk}^k \quad (h = 0, \ldots, k) .$$

Since Γ_{kk}^k is invertible, this implies (2.17) and (2.18). By (2.16) - (2.18), the system $W_0, \ldots, W_m$ is an orthogonalization with invertible squares of $T_0, \ldots, T_m$. Finally, (2.3) follows from (2.21). □

For future reference we state here the following corollary which follows immediately from (2.3), the uniqueness of orthogonalization with invertible squares up to invertible factors given in Theorem 2.1, and condition (1.5).

COROLLARY 2.2. *Assume that the system $T_0, \ldots, T_m$ in M_R admits an orthogonalization with invertible squares $V_0, \ldots, V_m$, and denote $G(T_0, \ldots, T_m)^{-1} = (\Gamma_{ij})_{ij=0}^m$. Then the following four conditions are equivalent:* a) $\langle V_m, V_m \rangle_R \geq 0$, b) $\langle V_m, V_m \rangle_R \geq \varepsilon I$ *for some* $\varepsilon > 0$, c) $\Gamma_{m,m} \geq 0$, d) $\Gamma_{m,m} \geq \varepsilon_1 I$ *for some* $\varepsilon_1 > 0$.

We close this section with a lemma which will be used in the next section. Let k be a nonnegative integer. We consider the direct sum $(\ell_r^2)^{k+1}$ of $k+1$ copies of ℓ_r^2, and the space $\ell_{r(k+1)}^2$. We will use the unitary transformation $U : (\ell_r^2)^{k+1} \to \ell_{r(k+1)}^2$ defined by

(2.22) $$U\left(\left((x_{it})_{i=0}^{\infty}\right)_{t=0}^{k}\right) = \left((x_{it})_{t=0}^{k}\right)_{i=0}^{\infty} ,$$

where $x_{it} \in \mathbf{C}^r$, $(x_{it})_{i=0}^{\infty} \in \ell_r^2$, and $(x_{it})_{t=0}^{k} \in \mathbf{C}^{r(k+1)}$. The next lemma which we include for completeness, relates the matrix representing an operator in $(\ell_r^2)^{k+1}$ with the matrix of its U conjugate in $\ell_{r(k+1)}^2$.

LEMMA 2.3. *Let $Z = (Z_{st})_{st=0}^k$ be a bounded linear operator in $(\ell_r^2)^{k+1}$, where $Z_{st} \in \mathcal{L}(\ell_r^2)$ have the matrix representations $Z_{st} = (z_{ij,st})_{ij=0}^{\infty}$ and $z_{ij,st}$ are complex matrices of order $r \times r$. Then the matrix representation for the operator $UZU^* \in \mathcal{L}(\ell_{r(k+1)}^2)$ is given by*

(2.23) $$UZU^* = (M_{ij})_{ij=0}^{\infty}$$

where M_{ij} are $r(k+1) \times r(k+1)$ matrices given by

(2.24) $$M_{ij} = (z_{ij,st})_{st=0}^{k} .$$

PROOF. Let α and β be two nonnegative integers, u and v be two vectors in $\mathbf{C}^{r(k+1)}$, and consider the following vectors in $\ell_{r(k+1)}^2$

(2.25) $$x = (\delta_{i\alpha} u)_{i=0}^{\infty}, \qquad y = (\delta_{i\beta} v)_{i=0}^{\infty}.$$

In order to prove (2.23), it is enough to show that for such vectors x and y the following equality holds

$$(UZU^*x, y)_{\ell^2_{r(k+1)}} = (M_{\beta\alpha}u, v)_{\mathbf{C}^{r(k+1)}}, \tag{2.26}$$

where, here and in the sequel for a Hilbert space H, we denote by $(\cdot,\cdot)_H$ the inner product in H. We now prove (2.26). Denote the block entries of u and v by

$$u = (u_t)_{t=0}^k, \qquad v = (v_t)_{t=0}^k, \tag{2.27}$$

where u_t and v_t belong to $\mathbf{C}^r$ $(t = 0, \dots, k)$. By (2.24) we have

$$(M_{\beta\alpha}u, v)_{\mathbf{C}^{r(k+1)}} = \sum_{st=0}^{k} (z_{\beta\alpha,st}u_t, v_s)_{\mathbf{C}^r}. \tag{2.28}$$

In addition, we have by the definition (2.22) of U

$$U\left(\left((\delta_{i\alpha}u_t)_{i=0}^{\infty}\right)_{t=0}^{k}\right) = \left((\delta_{i\alpha}u_t)_{t=0}^{k}\right)_{i=0}^{\infty} = \left(\delta_{i\alpha}(u_t)_{t=0}^{k}\right)_{i=0}^{\infty}.$$

By (2.25) and (2.27), this implies

$$U\left(\left((\delta_{i\alpha}u_t)_{i=0}^{\infty}\right)_{t=0}^{k}\right) = (\delta_{i\alpha}u)_{i=0}^{\infty} = x.$$

Thus, $U^*x = \left((\delta_{i\alpha}u_t)_{i=0}^{\infty}\right)_{t=0}^{k}$. Similarly, $U^*y = \left((\delta_{i\beta}v_t)_{i=0}^{\infty}\right)_{t=0}^{k}$. Put

$$u_t' = (\delta_{i\alpha}u_t)_{i=0}^{\infty}, \quad v_t' = (\delta_{i\beta}v_t)_{i=0}^{\infty} \qquad (t = 0, \dots, k). \tag{2.29}$$

Then u_t' and v_t' belong to ℓ^2_r and we have $U^*x = (u_t')_{t=0}^k$ and $U^*y = (v_t')_{t=0}^k$. Therefore, by $Z = (Z_{st})_{st=0}^k$ we obtain

$$(UZU^*x, y)_{\ell^2_{r(k+1)}} = (ZU^*x, U^*y)_{(\ell^2_r)^{k+1}} = \sum_{st=0}^{k} (Z_{st}u_t', v_s')_{\ell^2_r}.$$

However, (2.29) and $Z_{st} = (z_{ij,st})_{ij=0}^{\infty}$ imply $(Z_{st}u_t', v_s')_{\ell^2_r} = (z_{\beta\alpha,st}u_t, v_s)_{\mathbf{C}^r}$. Hence,

$$(UZU^*x, y)_{\ell^2_{r(k+1)}} = \sum_{st=0}^{k} (z_{\beta\alpha,st}u_t, v_s)_{\mathbf{C}^r}.$$

This equality and (2.28) imply (2.26). □

We will use the following corollary in Section 4.

COROLLARY 2.4. *Let $R = (R_{ij})_{ij=0}^{\infty}$ be a bounded self-adjoint operator in ℓ_r^2, k a nonnegative integer, and consider the Gramian $G(I, S, \dots, S^k)$ in the module M_R, where $S = (\delta_{i,j+1} I_r)_{ij=0}^{\infty}$ is the block shift. Then*

$$UG(I,S,\dots,S^k)U^* = \left(\delta_{ij} R(i,i+k)\right)_{ij=0}^{\infty}, \tag{2.30}$$

where $R(i, i+k) = (R_{i+s,i+t})_{st=0}^{k}$, and $U : (\ell_r^2)^{k+1} \to \ell_{r(k+1)}^2$ is the unitary transformation defined by (2.22). Moreover, assume that $G(I, S, \dots, S^k)$ is invertible and put

$$G(I,S,\dots,S^k)^{-1} = (\Gamma_{st}^k)_{st=0}^{k}, \tag{2.31}$$

then $R(i, i+k)$ is invertible ($i = 0, 1, \dots$) and

$$\Gamma_{st}^k = (\delta_{ij}\gamma_{st}^i)_{ij=0}^{\infty}, \tag{2.32}$$

where γ_{ts}^i are $r \times r$ matrices defined by

$$R(i,i+k)^{-1} = (\gamma_{st}^i)_{st=0}^{k}. \tag{2.33}$$

PROOF. In M_R we have

$$\langle S^t, S^s\rangle_R = \operatorname{diag}(S^{s*}RS^t) = \operatorname{diag}(R_{i+s,j+t})_{ij=0}^{\infty} = (\delta_{ij}R_{i+s,j+t})_{ij=0}^{\infty}.$$

Hence,

$$G(I,S,\dots,S^k) = \left(\langle S^t, S^s\rangle_R\right)_{st=0}^{k} = \left((\delta_{ij}R_{i+s,j+t})_{ij=0}^{\infty}\right)_{st=0}^{k}.$$

By Lemma 2.3, it follows that

$$\begin{aligned} UG(I,S,\dots,S^k)U^* &= \left((\delta_{ij}R_{i+s,j+t})_{st=0}^{k}\right)_{ij=0}^{\infty} = \left(\delta_{ij}(R_{i+s,j+t})_{st=0}^{k}\right)_{ij=0}^{\infty} \\ &= \left(\delta_{ij}R(i,i+k)\right)_{ij=0}^{\infty}. \end{aligned}$$

Hence, (2.30) holds. Assume now that $G(I, S, \dots, S^k)$ is invertible. Then by the last equality, $R(i, i+k)$ is invertible for $i = 0, 1 \dots$. Define Γ_{st}^k ($s, t = 0, \dots, k$) via (2.31) and put $\Gamma_{st}^k = (g_{ij,st}^k)_{ij=0}^{\infty}$. Then by Lemma 2.3 we have

$$UG(I,S,\dots,S^k)^{-1}U^* = \left((g_{ij,st}^k)_{st=0}^{k}\right)_{ij=0}^{\infty}.$$

On the other hand, it follows from (2.30) that

$$UG(I,S,\dots,S^k)^{-1}U^* = \left(\delta_{ij}R(i,i+k)^{-1}\right)_{ij=0}^{\infty} = \left(\delta_{ij}(\gamma_{st}^i)_{st=0}^{k}\right)_{ij=0}^{\infty},$$

where we also used (2.33). It follows from these equalities and the preceding one that $g_{ij,st}^k = \delta_{ij}\gamma_{st}^i$ ($i, j = 0, 1, \dots;\ s, t = 0, \dots, k$). Since $\Gamma_{st}^k = (g_{ij,st}^k)_{ij=0}^{\infty}$, this proves (2.32). $\square$

3. PRELIMINARIES ON INERTIA THEOREMS FOR UNILATERAL SHIFTS

In this section, we prove certain inertia theorems for two diagonal block operators. In order to apply these results in the next section, it is more convenient to use the following setting. Let S be a unilateral shift of finite multiplicity in a Hilbert space H. We denote the wandering space of S by H_S, thus $H_S = \big(S(H)\big)^{\perp}$, and $H = \bigoplus_{n=0}^{\infty} S^n(H_S)$. An operator T acting in H is called S-diagonal if $T(S^n(H_S)) \subset S^n(H_S)$ $(n = 0, 1, \ldots)$.

THEOREM 3.1. *Set S be a unilateral shift of finite multiplicity on H, and X and D be bounded and S-diagonal operators, with X invertible and self-adjoint. If the following inequalities hold*

$$S^*XS - D^*XD \geq 0 \tag{3.1}$$

and

$$S^{*m}(X - (SD^*)^m X(DS^*)^m)S^m \geq \varepsilon I \tag{3.2}$$

where $\varepsilon > 0$ and m is a positive integer, then the operator $S - D$ is Fredholm and

$$\text{index}\,(S - D) = -\nu_+(X|_{S^n(H_S)})$$

for n sufficiently large. Here, $\nu_+(X|_{S^n(H_S)})$ denotes the number of positive eigenvalues of the restriction of X to the X-invariant finite dimensional space $S^n(H_S)$.

Let us note that the above result holds also in the case when X is not assumed a-priori to be invertible, but we shall not need this here.

PROOF OF THEOREM 3.1. Let us first show that $\nu_+(X|_{S^n(H_S)})$ is constant for n large. Denote the multiplicity of S by r, thus $r = \dim H_S$. We use the decomposition

$$H = \bigoplus_{k=0}^{\infty} S^k(H_S)\,.$$

Choosing suitable orthonormal bases in $S^k(H_S)$, the matrices representing the operators S, X and D relative to this decomposition have the form $S = (\delta_{i,j+1}I_r)_{ij=0}^{\infty}$, $X = (\delta_{ij}X_j)_{ij=0}^{\infty}$ and $D = (\delta_{ij}D_j)_{ij=0}^{\infty}$. Here X_j and D_j are $r \times r$ matrices with X_j selfadjoint $(j = 0, 1, \ldots)$. Thus, (3.1) means that $X_{n+1} \geq D_n^* X_n D_n$ $(n = 0, 1, \ldots)$ and we also have $\nu_+(X_n) = \nu_+(X|_{S^n(H_S)})$. Let N be a positive definite subspace relative to X_n with $\dim N = \nu_+(X_n)$. Put $N' = \{x \in \mathbf{C}^r : D_n x \in N\}$. Then

$\dim N' \geq \dim N = \nu_+(X_n)$ and N' is nonnegative definite relative to X_{n+1}. Thus, $\nu_+(X_{n+1}) + \dim \operatorname{Ker}(X_{n+1}) \geq \dim N' \geq \nu_+(X_n)$. However X is invertible, and therefore, X_{n+1} is also invertible. Hence, $\nu_+(X_{n+1}) \geq \nu_+(X_n)$. Now S is of finite multiplicity and therefore this monotonicity condition shows that $\nu_+(X_n)$ is constant for n sufficiently large. We denote the asymptotic value of $\nu_+(X_n)$ (n-large), by ν_+. Thus, $\nu_+ = \nu_+(X|_{S^n(H_S)})$ for n large.

We now divide the proof into three parts.

Part a) Here we assume $m = 1$ and D is invertible. In this case, using the above block decomposition, the inequality (3.2) means that $X_{n+1} - D_n^* X_n D_n \geq \varepsilon I$. Moreover, since D is invertible, each D_n is invertible and $\sup_n \|D_n^{-1}\| < \infty$. Thus, we obtain

$$D_n^{-1*} X_{n+1} D_n^{-1} - X_n \geq \varepsilon D_n^{-1*} D_n^{-1} \geq \varepsilon' I \quad (n = 0, 1, \ldots)$$

where $\varepsilon' = \varepsilon / \|D\|^2 > 0$.

Now denote $Y_n = -X_n$, $A_n = D_n^{-1}$ $(n = 0, 1, \ldots)$, to obtain

$$Y_n - A_n^* Y_{n+1} A_n \geq \varepsilon' I \quad (n = 0, 1, \ldots) .$$

Put $G = (\delta_{i,j+1} A_j)_{ij=0}^{\infty}$. By Theorem 1.1 of [BG2], these inequalities imply that $I - G$ is Fredholm and index $(I - G) = -\nu_-(Y_n)$ for n sufficiently large. Here $\nu_-(Y_n)$ denotes the number of negative eigenvalues of Y_n. Since $Y_n = -X_n$, $\nu_-(Y_n) = \nu_+(X_n)$. Thus,

$$\text{index } (I - G) = -\nu_+ ,$$

where ν_+ is the asymptotic value of $\nu_+(X_n)$, as above.

Now note that $G = (\delta_{i,j+1} D_j^{-1})_{ij=0}^{\infty} = SD^{-1}$. Thus, $S - D = -(I - G)D$. Since D is invertible, this implies that $S - D$ is Fredholm with

$$\text{index } (S - D) = -\nu_+ .$$

Part b) Here we prove the result assuming only $m = 1$. There exists a sequence $(E_k)_{k=1}^{\infty}$ of S-diagonal operators, such that each E_k is invertible in H and $\lim_{k\to\infty} E_k = D$ in the uniform norm. This follows for example by considering the representation $D = (\delta_{ij} D_j)_{ij=0}^{\infty}$ where each D_j acts in the finite dimensional space $S^j(H_S)$, then decomposing $D_j = U_j P_j$ where U_j is unitary in $S^j(H_S)$ and P_j is nonnegative in $S^j(H_S)$, and finally defining $E_k = \left(\delta_{ij} U_j (P_j + k^{-1} I)\right)_{ij=0}^{\infty}$. Inequality (3.2) with $m = 1$ means that

$$S^* X S - D^* X D \geq \varepsilon I . \tag{3.3}$$

Thus, there exists a nonnegative integer k_0 such that

$$S^*XS - E_k^*XE_k \geq \frac{\varepsilon}{2}I \qquad (k \geq k_0).$$

Since E_k is invertible, we can apply Part a) to this inequality. It follows that $S - E_k$ is Fredholm with index $(S - E_k) = -\nu_+$ when $k \geq k_0$.

We now apply inequality (3.3) to show that $S - D$ is Fredholm. Let $u \in H$ with $\|u\| = 1$. By (3.3) we have

$$\begin{aligned}\varepsilon = \varepsilon\|u\|^2 &\leq (XSu, Su)_H - (XDu, Du)_H = \\ &= (XSu, (S-D)u)_H + ((S-D)u, XDu)_H \leq \\ &\leq \|(S-D)u\|\big(\|XS\| + \|XD\|\big) .\end{aligned}$$

Hence, $\|(S-D)u\| \geq \rho \quad (\|u\| = 1)$ where $\rho = \varepsilon(\|X\| + \|XD\|)^{-1} > 0$. Thus, $S - D$ is bounded below, namely

$$\|(S-D)u\| \geq \rho\|u\| \quad (u \in H) .$$

This implies that Ker $(S - D) = \{0\}$, and Im $(S - D)$ is closed. We now prove that $\dim\big[\text{Im }(S-D)\big]^{\perp} \leq \dim H_S$. In fact, assume the contrary, then there exists a vector $x \neq 0$ orthogonal to Im$(S - D)$ and H_S. Since $x \perp H_S$, we have a representation of the form $x = \sum_{j=h}^{\infty} S^j x_j$ where $h \geq 1$, $x_j \in H_S$ $(j = h, h+1, \cdots)$, and $x_h \neq 0$. In addition, $x \perp \text{Im }(S - D)$ implies $(S - D)^*x = 0$, whence $S^*x = D^*x$. However, on one hand $S^*x = \sum_{j=h-1}^{\infty} S^j x_{j+1} \notin \oplus_{j=h}^{\infty} S^j(H_S)$ because $x_h \neq 0$, and on the other hand, since D^* is S-diagonal and $x \in \oplus_{j=h}^{\infty} S^j(H_S)$ we have $D^*x \in \oplus_{j=h}^{\infty} S^j(H_S)$, yielding a contradiction. Thus, $\dim\big[\text{Im }(S-D)\big]^{\perp} \leq \dim H_S < \infty$ and hence, $S - D$ is Fredholm.

Since $\lim_{k\to\infty} S - E_k = S - D$ and index $(S - E_k) = -\nu_+$ for $k \geq k_0$, it follows that

$$\text{index }(S - D) = -\nu_+ .$$

Part c) We now consider the general case. Let z be an arbitrary complex number with $|z| = 1$. We define two operators in H via

$$\Sigma = S^m \quad \text{and} \quad \Delta_z = z^m(DS^*)^m S^m .$$

Then Σ is a unilateral shift of finite multiplicity in H, and X and Δ_z are Σ-diagonal. Moreover, (3.2) leads to

$$\Sigma^*X\Sigma - \Delta_z^*X\Delta_z \geq \varepsilon I .$$

We apply the previous part b) to this inequality. It follows that $\Sigma - \Delta_z$ is Fredholm with index $(\Sigma - \Delta_z) = -\nu_+(X|_{\Sigma^k(H_\Sigma)})$ for k sufficiently large. Note that $\Sigma = S^m$ leads to

$H_\Sigma = H_S \oplus S(H_S) \oplus \cdots \oplus S^{m-1}(H_S)$, whence $\Sigma^k(H_\Sigma) = S^{mk}(H_S) \oplus \cdots \oplus S^{mk+m-1}(H_S)$. Since $\nu_+(X|_{S^n(H_S)}) = \nu_+$ for n large, it follows that $\nu_+(X|_{\Sigma^k(H_\Sigma)}) = m\nu_+$ for k large, and hence, index $(\Sigma - \Delta_z) = -m\nu_+$. Recalling the definitions of Σ and Δ_z, we have that

$$\Sigma - \Delta_z = S^m - z^m(DS^*)^m S^m = (I - z^m(DS^*)^m)S^m \ .$$

Therefore, the operator

$$(I - z^m(DS^*)^m)S^m S^{*m} = (\Sigma - \Delta_z)S^{*m}$$

is also Fredholm with index given by

$$\text{index } \left[(I - z^m(DS^*)^m)S^m S^{*m}\right] = -m\nu_+ + mr$$

where $r = \dim H_S$ is the multiplicity of S. However, $I - S^m S^{*m}$ is compact, and consequently $I - z^m(DS^*)^m$ is Fredholm with

$$\text{index } \left(I - z^m(DS^*)^m\right) = -m\nu_+ + mr \quad (|z| = 1) \ . \tag{3.4}$$

We now use the following commutative factorization

$$I - z^m(DS^*)^m = \prod_{k=0}^{m-1} \left(I - e^{\frac{2\pi ik}{m}} zDS^*\right) \ . \tag{3.5}$$

First, this leads to

$$\text{Ker}(I - zDS^*) \subset \text{Ker}(I - z^m(DS^*)^m) \ ,$$

and

$$\text{Im } (I - zDS^*) \supset \text{Im } \left(I - z^m(DS^*)^m\right) \ .$$

Since $I - z^m(DS^*)^m$ is Fredholm, it follows that $I - zDS^*$ is also Fredholm for $|z| = 1$. Moreover, the operator $I - zDS^*$ depends continuously on z in the uniform operator topology. Hence, index $(I - zDS^*)$ is independent of z $(|z| = 1)$. Denote

$$\text{index } (I - zDS^*) = \mu \quad (|z| = 1) \ . \tag{3.6}$$

Then we have index $\left(I - e^{\frac{2\pi ik}{m}} DS^*\right) = \mu$ $(k = 0, 1, \ldots, m-1)$. Applying (3.5) with $z = 1$ we obtain

$$\text{index}(I - (DS^*)^m) = m\mu \ .$$

Comparing this with (3.4) with $z = 1$, it follows that $\mu = -\nu_+ + r$. Hence, (3.6) with $z = 1$ leads to

$$\text{index } (I - DS^*) = -\nu_+ + r \ .$$

Finally, since $I - DS^*$ is Fredholm, $S - D = (I - DS^*)S$ is Fredholm too, and

$$\text{index } (S - D) = \text{index } (I - DS^*) + \text{index } S = -\nu_+ + r - r = -\nu_+ \ . \quad \square$$

4. THE MAIN RESULT

The proof of Theorem 1.2 of the introduction appears after the proof of Theorem 4.1 below. Throughout this section, we let $R = (R_{ij})_{ij=0}^{\infty}$ be a bounded self-adjoint operator in ℓ_r^2, where R_{ij} are $r \times r$ complex matrices. We denote by M_R the corresponding Hilbert module over the C^*-algebra B of all block diagonal operators in ℓ_r^2. We also denote by $S = (\delta_{i,j+1} I_r)_{ij=0}^{\infty}$ the block shift in ℓ_r^2, and the Gramian of $I, S, \ldots, S^k$ by

$$G(I, S, \ldots, S^k) = \left(\langle S^j, S^i \rangle_R\right)_{ij=0}^{k} \ . \tag{4.1}$$

THEOREM 4.1. *Assume that for some positive integer* m, $G(I, S, \ldots, S^m)$ *and* $G(I, S, \ldots, S^{m-1})$ *are invertible and denote* $G(I, S, \ldots, S^m)^{-1} = (\Gamma_{ij})_{ij=0}^{m}$. *Put* $W_m = \sum_{i=0}^{m} S^i \Gamma_{i,m}$. *If the following inequality holds*

$$\Gamma_{m,m} \geq 0 \ , \tag{4.2}$$

then $\Gamma_{m,m} \geq \varepsilon I$ *for some positive* ε, W_m *is left invertible and*

$$\operatorname{codim} \operatorname{Im} W_m = \nu_+\left((R_{ij})_{ij=n}^{n+m-1}\right) \tag{4.3}$$

for n *sufficiently large.*

PROOF. We will use the following notation,

$$X = G(I, S, \ldots, S^{m-1}) = \left(\langle S^j, S^i \rangle_R\right)_{ij=0}^{m-1} \ , \tag{4.4}$$

$$Y = \left(\langle S^j, S^i \rangle_R\right)_{ij=1}^{m} \ , \tag{4.5}$$

and

$$Z = \left(\langle S^j, S^i \rangle_R\right)_{i=0,j=1}^{m-1,m} \ . \tag{4.6}$$

Note that

$$\langle S^{j+1}, S^{i+1}\rangle_R = \operatorname{diag}\,(S^{*(i+1)}RS^{j+1}) = S^* \operatorname{diag}\,(S^{*i}RS^j)S = S^*\langle S^j, S^i\rangle_R S\ .$$

Thus, the following equality holds

$$Y = \sigma^* X \sigma\ , \tag{4.7}$$

where

$$\sigma = (\delta_{ij}S)_{ij=0}^{m-1}\ . \tag{4.8}$$

Let us now remark that since $[G(I, S, \dots, S^m)]^{-1} = (\Gamma_{ij})_{ij=0}^{m}$ we have

$$\begin{cases} G(I, S, \dots, S^{m-1}) \begin{pmatrix} \Gamma_{0,m} \\ \vdots \\ \Gamma_{m-1,m} \end{pmatrix} + K\Gamma_{m,m} = 0\ , \\ K^* \begin{pmatrix} \Gamma_{0,m} \\ \vdots \\ \Gamma_{m-1,m} \end{pmatrix} + \langle S^m, S^m\rangle_R \Gamma_{m,m} = I\ , \end{cases} \tag{4.9}$$

where

$$K = \begin{pmatrix} \langle S^m, I\rangle_R \\ \langle S^m, S\rangle_R \\ \vdots \\ \langle S^m, S^{m-1}\rangle_R \end{pmatrix}\ . \tag{4.10}$$

Since $G(I, S, \dots, S^{m-1})$ is invertible, these equalities lead to

$$\left[\langle S^m, S^m\rangle_R - K^*G(I, S, \dots, S^{m-1})^{-1}K\right]\Gamma_{m,m} = I\ .$$

The term in square brackets is self-adjoint and therefore it is invertible with inverse $\Gamma_{m,m}$. Taking into account the notation $X = G(I, S, \dots, S^{m-1})$, we obtain that $\langle S^m, S^m\rangle_R - K^*X^{-1}K$ is invertible with

$$\left[\langle S^m, S^m\rangle_R - K^*X^{-1}K\right]^{-1} = \Gamma_{m,m}\ . \tag{4.11}$$

It also follows that $\Gamma_{m,m}$ is invertible. Since $\Gamma_{m,m} \geq 0$ by (4.2) we have

$$\Gamma_{m,m} \geq \varepsilon I, \quad \Gamma_{m,m}^{-1} \geq \varepsilon I\ , \tag{4.12}$$

for some $\varepsilon > 0$.

We now consider the following companion type matrix in $B_{m\times m}$

$$D = \begin{pmatrix} 0 & \cdots & 0 & -\Gamma_{0,m}\Gamma_{m,m}^{-1} \\ I & \cdots & 0 & -\Gamma_{1,m}\Gamma_{m,m}^{-1} \\ \cdot & \cdots & \cdot & \cdot \\ \cdot & \cdots & \cdot & \cdot \\ 0 & \cdots & I & -\Gamma_{m-1,m}\Gamma_{m,m}^{-1} \end{pmatrix}. \tag{4.13}$$

Note that the upper equality in (4.9), equality (4.10) and definition (4.4) lead to

$$X \begin{pmatrix} -\Gamma_{0,m}\Gamma_{m,m}^{-1} \\ \vdots \\ -\Gamma_{m-1,m}\Gamma_{m,m}^{-1} \end{pmatrix} = K = \begin{pmatrix} \langle S^m, I\rangle_R \\ \vdots \\ \langle S^m, S^{m-1}\rangle_R \end{pmatrix}.$$

On the other hand, it is clear that

$$X \begin{pmatrix} 0 & \cdots & 0 & 0 \\ I & \cdots & 0 & 0 \\ \cdot & \cdots & \cdot & \cdot \\ 0 & \cdots & I & 0 \\ 0 & \cdots & 0 & I \end{pmatrix} = \left(\langle S^j, S^i\rangle_R\right)_{ij=0}^{m-1} \begin{pmatrix} 0 & \cdots & 0 & 0 \\ I & \cdots & 0 & 0 \\ \cdot & \cdots & \cdot & \cdot \\ 0 & \cdots & I & 0 \\ 0 & \cdots & 0 & I \end{pmatrix} = \left(\langle S^j, S^i\rangle_R\right)_{i=0,j=1}^{m-1,m-1}.$$

Combining these two equalities it follows that

$$XD = \left(\langle S^j, S^i\rangle_R\right)_{i=0,j=1}^{m-1,m} = Z\,, \tag{4.14}$$

where we used the notation of (4.6). From the special structure of D given in (4.13), it follows that D^*XD has the following form

$$D^*XD = \begin{pmatrix} \left(\langle S^j, S^i\rangle_R\right)_{i=1,j=1}^{m-1,m} \\ * \;\; * \;\; * \;\; * \end{pmatrix}.$$

This equality and (4.5) show that D^*XD and Y agree on their first $m-1$ rows. In addition, D^*XD and Y are self-adjoint because $\langle S^j, S^i\rangle_R^* = \langle S^i, S^j\rangle_R$. Since D^*XD and Y agree on their first $m-1$ rows, they also agree on their first $m-1$ columns. Consequently, we have

$$Y - D^*XD = \left(\delta_{i,m-1}\delta_{j,m-1}L\right)_{ij=0}^{m-1} \tag{4.15}$$

for some element L in B. To determine L note that (4.14) leads to $D = X^{-1}Z$. Inserting this in (4.15) we obtain

$$Y - Z^*X^{-1}Z = \left(\delta_{i,m-1}\delta_{j,m-1}L\right)_{ij=0}^{m-1}. \tag{4.16}$$

Now comparing the definitions (4.6) and (4.10) it follows that the rightmost column of Z is K. Hence (4.16) and the definition (4.5) of Y imply that

$$\langle S^m, S^m\rangle_R - K^*X^{-1}K = L .$$

By (4.11) this shows that $L = \Gamma_{m,m}^{-1}$, and therefore (4.12) implies that

$$L \geq \varepsilon I . \tag{4.17}$$

Now put

$$E = (\delta_{i,m-1}\delta_{j,m-1}L)_{ij=0}^{m-1} . \tag{4.18}$$

Then equalities (4.7), (4.15) and (4.17) imply

$$\sigma^* X\sigma - D^* XD = E \geq 0 . \tag{4.19}$$

We first use this equality in order to show that

$$\operatorname{Ker}(\sigma - D) = \{0\} . \tag{4.20}$$

In fact, assume that $x = (x_i)_{i=0}^{m-1} \in \operatorname{Ker}(\sigma - D)$ where $x_i \in \ell_r^2$ $(i = 0, \ldots, m-1)$. Then $\sigma x = Dx$, and hence (4.18) and (4.19) imply that

$$\begin{aligned}(Lx_{m-1}, x_{m-1})_{\ell_r^2} &= (Ex, x)_{(\ell_r^2)^m} = ((\sigma^* X\sigma - D^* XD)x, x)_{(\ell_r^2)^m} = \\ &= (X\sigma x, \sigma x)_{(\ell_r^2)^m} - (XDx, Dx)_{(\ell_r^2)^m} = 0 ,\end{aligned}$$

where we denote by $(\cdot,\cdot)_{(\ell_r^2)^m}$ the usual Hilbert inner product in $(\ell_r^2)^m$. Since $L \geq \varepsilon$ by (4.17) these equalities show that $x_{m-1} = 0$. Hence, the definition (4.13) of D leads to

$$Dx = \begin{pmatrix} 0 & \cdots & 0 & 0 \\ I & \cdots & 0 & 0 \\ \cdot & \cdots & \cdot & \cdot \\ 0 & \cdots & I & 0 \end{pmatrix} x .$$

Finally, from this equality, the definition (4.8) of σ and our assumption $(\sigma - D)x = 0$ we obtain

$$\begin{pmatrix} S & 0 & \cdots & 0 & 0 \\ -I & S & \cdots & 0 & 0 \\ \cdot & \cdot & \cdots & \cdot & \cdot \\ 0 & 0 & \cdots & S & 0 \\ 0 & 0 & \cdots & -I & S \end{pmatrix} x = 0 .$$

However, the operator on the left hand side of this equality is injective and hence $x = 0$. This proves (4.20).

We now proceed to obtain strict inequalities by iterating (4.19). Denote

$$\pi_k = \sigma^k \sigma^{*k} (D\sigma^*)^{m-k} \sigma^m \quad (k = 0, \ldots, m) . \tag{4.21}$$

Taking into account the definition (4.8) of σ we have $\sigma^* \sigma = I$ whence

$$\pi_0 = (D\sigma^*)^m \sigma^m, \quad \pi_m = \sigma^m , \tag{4.22}$$

and

$$\sigma^* \pi_{k+1} = \sigma^k \sigma^{*(k+1)} (D\sigma^*)^{m-(k+1)} \sigma^m \quad (k = 0, \ldots, m-1) . \tag{4.23}$$

The last equalities also imply that

$$\sigma \sigma^* \pi_{k+1} = \pi_{k+1} \quad (k = 0, \ldots, m-1) . \tag{4.24}$$

We now multiply (4.19) by $\sigma^* \pi_{k+1}$ on the right and $\pi_{k+1} \sigma$ on the left. Taking into account (4.24) we obtain

$$\pi_{k+1}^* X \pi_{k+1} - \pi_{k+1}^* \sigma D^* X D \sigma^* \pi_{k+1} = E_k \geq 0 \quad (k = 0, \ldots, m-1) , \tag{4.25}$$

where

$$E_k = \pi_{k+1}^* \sigma E \sigma^* \pi_{k+1} \geq 0 \quad (k = 0, \ldots, m-1) . \tag{4.26}$$

Now note that σ given by (4.8) is a unilateral shift, and the entries of D given by (4.13) belong to the algebra B of block diagonal operators in ℓ_r^2. Hence, D is σ-diagonal and D commutes with $\sigma^k \sigma^{*k}$ $(k = 0, 1, \cdots)$. Thus, (4.23) implies that

$$D\sigma^* \pi_{k+1} = \sigma^k \sigma^{*k} D \sigma^* (D\sigma^*)^{m-(k+1)} \sigma^m = \pi_k \quad (k = 0, \ldots, m-1) .$$

Hence, (4.25) yields

$$\pi_{k+1}^* X \pi_{k+1} - \pi_k^* X \pi_k = E_k \geq 0 \quad (k = 0, \ldots, m-1) .$$

Adding for $k = 0, \ldots, m-1$ we obtain

$$\pi_m^* X \pi_m - \pi_0^* X \pi_0 = \Lambda , \tag{4.27}$$

where

$$\Lambda = E_0 + \cdots + E_{m-1} \geq 0 . \tag{4.28}$$

By the previous inequality, Λ is nonnegative. It turns out that Λ is positive definite, namely

$$\Lambda \geq \varepsilon' I , \tag{4.29}$$

for some $\varepsilon' > 0$. We postpone the proof of this inequality to later on, and proceed first to conclude the proof of the theorem based upon (4.29). In view of (4.22), equality (4.27) leads to

$$\sigma^{*m} X \sigma^m - \sigma^{*m} (\sigma D^*)^m X (D\sigma^*)^m \sigma^m = \Lambda .$$

Combining this with (4.29) we obtain the following inequality

$$\sigma^{*m} (X - (\sigma D^*)^m X (D\sigma^*)^m) \sigma^m \geq \varepsilon' I . \tag{4.30}$$

Now recall that σ is a unilateral shift, D is σ-diagonal and X, which is defined by (4.4), is self-adjoint and has entries in B. Thus, X is also σ-diagonal. Furthermore, X is invertible by the assumptions of the theorem and therefore we may apply Theorem 3.1 to the inequalities (4.19) and (4.30). It follows that $\sigma - D$ is a Fredholm operator with

$$\text{index}\,(\sigma - D) = -\nu_+ (X|_{\sigma^n(H_\sigma)}) \tag{4.31}$$

for n sufficiently large.

Recall that $X = (\langle S^j, S^i \rangle_R)_{ij=0}^{m-1}$ by (4.4), and by the definition (4.8) of σ we have $\sigma^n(H_\sigma) = S^n(H_S) \oplus S^n(H_S) \oplus \cdots \oplus S^n(H_S)$, with m summands. Therefore, $X|_{\sigma^n(H_\sigma)}$ is unitarily equivalent with $\left(\langle S^j, S^i \rangle_R|_{S^n(H_S)}\right)_{ij=0}^{m-1}$. Now , $\langle S^j, S^i \rangle_R = \text{diag}\,(S^{*i} R S^j) = (\delta_{ts} R_{t+i,s+j})_{ts=0}^{\infty}$, and therefore, $\langle S^j, S^i \rangle_R|_{S^n(H_S)}$ is unitarily equivalent with $R_{n+i,n+j}$. It follows that $X|_{\sigma^n(H_\sigma)}$ is unitarily equivalent with $(R_{n+i,n+j})_{ij=0}^{m-1} = (R_{ij})_{ij=n}^{n+m-1}$. Thus, (4.31) leads to

$$\text{index}\,(\sigma - D) = -\nu_+ \left((R_{ij})_{ij=n}^{n+m-1}\right) \tag{4.32}$$

for n sufficiently large.

Now let M_1 be the following element of $B_{m\times m}$

$$M_1 = \begin{pmatrix} I & 0 & \cdots & 0 & 0 \\ 0 & I & \cdots & 0 & 0 \\ \cdot & \cdot & \cdots & \cdot & \cdot \\ 0 & 0 & \cdots & I & 0 \\ 0 & 0 & \cdots & 0 & \Gamma_{m,m} \end{pmatrix} .$$

By (4.12), M_1 is invertible and the definitions (4.8) and (4.13) of σ and D leads to

$$(\sigma - D)M_1 = \begin{pmatrix} S & 0 & \cdots & 0 & \Gamma_{0,m} \\ -I & S & \cdots & 0 & \Gamma_{1,m} \\ \cdot & \cdot & \cdots & \cdot & \cdot \\ 0 & 0 & \cdot & S & \Gamma_{m-2,m} \\ 0 & 0 & \cdot & -I & \Gamma_{m-1,m} + S\Gamma_{m,m} \end{pmatrix}$$

Now denote

$$M_2 = \begin{pmatrix} I & S & \cdots & S^{m-1} \\ 0 & I & \cdots & S^{m-2} \\ \cdot & \cdot & \cdots & \cdot \\ 0 & 0 & \cdots & S \\ 0 & 0 & \cdots & I \end{pmatrix}; \quad M_3 = \begin{pmatrix} I & 0 & \cdots & 0 & \sum_{i=0}^{m-1} S^i\Gamma_{i+1,m} \\ 0 & I & \cdots & 0 & \sum_{i=0}^{m-2} S^i\Gamma_{i+2,m} \\ \cdot & \cdot & \cdots & \cdot & \cdot \\ 0 & 0 & \cdots & I & \sum_{i=0}^{1} S^i\Gamma_{i+m-1,m} \\ 0 & 0 & \cdots & 0 & I \end{pmatrix}.$$

Note that

$$-\sum_{i=0}^{m-k} S^i\Gamma_{i+k,m} + S\sum_{i=0}^{m-(k+1)} S^i\Gamma_{i+k+1,m} + \Gamma_{k,m} = 0 \quad (k = 1, \ldots, m-1),$$

and therefore,

$$(\sigma - D)M_1M_3 = \begin{pmatrix} S & \cdots & 0 & W_m \\ -I & \cdots & 0 & 0 \\ \cdot & \cdots & \cdot & \cdot \\ 0 & \cdots & S & 0 \\ 0 & \cdots & -I & 0 \end{pmatrix}.$$

Here $W_m = \sum_{i=0}^{m} S^i\Gamma_{i,m}$ is as in the statement of the theorem. Consequently we obtain

$$M_2(\sigma - D)M_1M_3 = \begin{pmatrix} 0 & \cdots & 0 & W_m \\ -I & \cdots & 0 & 0 \\ \cdot & \cdots & \cdot & \cdot \\ 0 & \cdots & 0 & 0 \\ 0 & \cdots & -I & 0 \end{pmatrix}. \tag{4.33}$$

Now recall that M_1 is invertible by (4.12) and note that also M_2 and M_3 are invertible. Since $\sigma - D$ is Fredholm, this implies that W_m is Fredholm. Moreover, equalities (4.32) and (4.33) lead to

$$\text{index } W_m = -\nu_+\big((R_{ij})_{ij=n}^{n+m-1}\big) \tag{4.34}$$

for n sufficiently large. Furthermore, it follows from (4.20) and (4.33) that Ker $(W_m) = \{0\}$. Since W_m is Fredholm, it follows from this that W_m is left invertible. Moreover,

$\operatorname{Ker}(W_m) = \{0\}$ and (4.34) implies (4.3). Finally, $\Gamma_{m,m} \geq \varepsilon I$ follows from (4.12). This and the above properties of W_m prove Theorem 4.1.

We still have to prove that inequality (4.29) holds for some $\varepsilon' > 0$. Note that Λ is nonnegative by (4.28), and Λ acts in $(\ell_r^2)^m$. Hence, in order to prove (4.29) for some ε' it is enough to show that if $(x_n)_{n=0}^{\infty}$ is a sequence of vectors in $(\ell_r^2)^m$ such that

$$\|x_n\| = 1 \quad (n = 0, 1, \ldots) \tag{4.35}$$

then

$$\limsup_{n\to\infty} (\Lambda x_n, x_n)_{(\ell_r^2)^m} > 0 . \tag{4.36}$$

Here $(\cdot,\cdot)_{(\ell_r^2)^m}$ denotes the usual inner product in $(\ell_r^2)^m$. We now prove this statement. Let $(x_n)_{n=0}^{\infty}$ be a sequence in $(\ell_r^2)^m$ satisfying (4.35). Each $x_n \in (\ell_r^2)^m$ is a vector of the form

$$x_n = \begin{pmatrix} x_{n,0} \\ \vdots \\ x_{n,m-1} \end{pmatrix} \quad (n = 0, 1, \ldots) ,$$

where $x_{n,0}, \ldots, x_{n,m-1}$ belong to ℓ_r^2. It follows from condition (4.35) that there exists an integer $t \in \{0, \ldots, m-1\}$ such that

$$\limsup_{n\to\infty} \|x_{n,t}\| > 0 \tag{4.37}$$

and

$$\lim_{n\to\infty} \|x_{n,t+1}\| = \cdots = \lim_{n\to\infty} \|x_{n,m-1}\| = 0 . \tag{4.38}$$

Here if $t = m-1$ then condition (4.38) is void. Now define a sequence $(y_n)_{n=0}^{\infty}$ in $(\ell_r^2)^m$ via

$$y_n = \begin{pmatrix} x_{n,0} \\ \vdots \\ x_{n,t} \\ 0 \\ \vdots \\ 0 \end{pmatrix} \quad (n = 0, 1, \ldots) . \tag{4.39}$$

Then by (4.38) we have

$$\lim_{n\to\infty} \|x_n - y_n\| = 0 . \tag{4.40}$$

Thus, inequality (4.36) will follow from (4.35), the boundedness of Λ, and the inequality

$$\limsup_{n\to\infty}(\Lambda y_n, y_n)_{(\ell_r^2)^m} > 0 \tag{4.41}$$

which we now prove. To prove (4.41) we compute the inner product $(E_t y_n, y_n)_{(\ell_r^2)^m}$, where $E_t = \pi_{t+1}^* \sigma E \sigma^* \pi_{t+1}$ is given by (4.26), $\sigma^*\pi_{t+1} = \sigma^t \sigma^{*(t+1)}(D\sigma^*)^{m-t-1}\sigma^m$ is given by (4.23), and $E = (\delta_{i,m-1}\delta_{j,m-1}L)_{ij=0}^{m-1}$ is defined by (4.18). Let us first define an operator $P : (\ell_r^2)^m \to \ell_r^2$ via $P(z_1, \ldots, z_m) = z_m$ ($z_i \in \ell_r^2$; $i = 1, \ldots, m$). Namely, P is the projection on the last entry, and is represented by the matrix

$$P = (0 \;\cdots\; 0 \; I)\,. \tag{4.42}$$

The diagonal structure of σ given in (4.8) leads to

$$SP = P\sigma, \quad S^*P = P\sigma^*\,. \tag{4.43}$$

Thus, it follows from (4.23) that

$$P\sigma^*\pi_{t+1} = P\sigma^t\sigma^{*(t+1)}(D\sigma^*)^{m-t-1}\sigma^m = S^tS^{*(t+1)}P(D\sigma^*)^{m-t-1}\sigma^m\,. \tag{4.44}$$

We now proceed to compute $P\sigma^*\pi_{t+1}y_n$. Note that the operator $D\sigma^*$ is given by

$$D\sigma^* = \begin{pmatrix} 0 & \cdots & 0 & -\Gamma_{0,m}\Gamma_{m,m}^{-1}S^* \\ S^* & \cdots & 0 & -\Gamma_{1,m}\Gamma_{m,m}^{-1}S^* \\ \cdot & \cdots & \cdot & \cdot \\ \cdot & \cdots & \cdot & \cdot \\ 0 & \cdots & S^* & -\Gamma_{m-1,m}\Gamma_{m,m}^{-1}S^* \end{pmatrix}.$$

In addition, the matrix representing the operator $P(D\sigma^*)^{m-t-1}$ is given by the last row of $(D\sigma^*)^{m-t-1}$. By the above expression for $D\sigma^*$ we obtain

$$P(D\sigma^*)^{m-t-1} = (\overbrace{0 \cdots 0}^{t} \; S^{*(m-t-1)} \; \overbrace{* \cdots *}^{m-t-1})\,,$$

where $\overbrace{* \cdots *}^{m-t-1}$ denote $m-t-1$ unspecified operators in ℓ_r^2. Since $S^tS^{*(t+1)}S^{*(m-t-1)}S^m = S^t$ we obtain that

$$S^tS^{*(t+1)}P(D\sigma^*)^{m-t-1}\sigma^m = (\overbrace{0 \cdots 0}^{t} \; S^t \; \overbrace{* \cdots *}^{m-t-1})\,.$$

By (4.44) this leads to

$$P\sigma^*\pi_{t+1} = (\overbrace{0 \cdots 0}^{t} \; S^t \; \overbrace{* \cdots *}^{m-t-1})\,.$$

Hence, formula (4.39) of y_n shows that

$$P\sigma^*\pi_{t+1}y_n = S^t x_{n,t} \quad (n = 0, 1, \ldots) . \tag{4.45}$$

Let us now remark that the definitions (4.18) of E and (4.42) of P show that

$$E = P^*LP .$$

Hence, (4.26) leads to $E_t = \pi_{t+1}^*\sigma P^*LP\sigma^*\pi_{t+1}$, and therefore

$$(E_t y_n, y_n)_{(\ell_r^2)^m} = (LP\sigma^*\pi_{t+1}y_n, P\sigma^*\pi_{t+1}y_n)_{\ell_r^2} .$$

By (4.45), this leads to

$$(E_t y_n, y_n)_{(\ell_r^2)^m} = (LS^t x_{n,t}, S^t x_{n,t})_{\ell_r^2} ,$$

and since $L \geq \varepsilon I$ by (4.17) we finally obtain

$$(E_t y_n, y_n)_{(\ell_r^2)^m} \geq \varepsilon \|S^t x_{n,t}\|^2 = \varepsilon \|x_{n,t}\|^2 \quad (n = 0, 1, \ldots) .$$

By (4.37) this inequality implies

$$\limsup_{n\to\infty} (E_t y_n, y_n)_{(\ell_r^2)^m} > 0 .$$

However $\Lambda \geq E_t$ by (4.28) and (4.26), and therefore (4.41) holds. As stated above, (4.41) leads to (4.36) which implies in turn that $\Lambda \geq \varepsilon' I$ for some $\varepsilon' > 0$. □

PROOF OF THEOREM 1.2. Since the system $I, S, \ldots, S^m$ admits an orthogonalization with invertible squares in M_R, it follows from Theorem 2.1 that the Gramians $G(I, S, \ldots, S^k)$ $(k = 0, \ldots, m)$ are invertible. Denote $(\Gamma_{ij}^k)_{ij=0}^k = G(I, S, \ldots, S^k)^{-1}$ $(k = 0, \ldots, m)$. By Corollary 2.2, inequality (1.7) implies $\Gamma_{m,m}^m \geq 0$. In view of these two properties we may apply Theorem 4.1 with $\Gamma_{ij} = \Gamma_{ij}^m$. Define $W_k = \sum_{j=0}^k S^j \Gamma_{jk}^k$ $(k = 0, \ldots, m)$. By Theorem 4.1, W_m is left invertible and (4.3) holds. On the other hand, by Theorem 2.1, $W_0, \ldots, W_m$ is an orthogonalization with invertible squares of $I, S, \ldots, S^m$. Since $V_0, \ldots, V_m$ is also an orthogonalization with invertible squares of $I, S, \ldots, S^m$, it follows again from Theorem 2.1 that exists an invertible element b_m in B such that $V_m = W_m b_m$. Since W_m is left invertible by Theorem 4.1, V_m is left invertible, and (1.8) follows from (4.3). □

Theorem 1.2 admits the following generalization.

THEOREM 4.2. *Let $R = (R_{ij})_{ij=0}^{\infty}$ be a bounded self-adjoint operator in ℓ_r^2 where R_{ij} are complex $r \times r$ matrices, and let m be a positive integer. If all the minors $R(n, n+k) = (R_{ij})_{i=n}^{n+k}$ $(n = 0, 1, \dots;\ k = 0, \dots, m)$ are invertible with*

$$\sup\{\|R(n, n+k)^{-1}\| :\ n = 0, 1, \dots;\ k = 0, \dots, m\} < \infty \ , \tag{4.46}$$

then the Gramians $G(I, S, \dots, S^k)$ $(k = 0, \dots, m)$ are invertible, where $S = (\delta_{i,j+1} I_r)_{ij=0}^{\infty}$ is the block shift, and the system $I, S, \dots, S^m$ admits an orthogonalization with invertible squares. Assume also that $\left[R(n, n+m)^{-1}\right]_{m,m} \geq 0$ $(n = n_0, n_0 + 1, \dots)$ for some integer n_0, where $\left[R(n, n+m)^{-1}\right]_{m,m}$ denotes the $r \times r$ matrix appearing in the right lower corner of $R(n, n+m)$. Then for any orthogonalization $V_0, \dots, V_m$ with invertible squares of $I, S, \dots, S^m$, the operator V_m is Fredholm and

$$\text{index } V_m = -\nu_+\big(R(n, n+m-1)\big) \tag{4.47}$$

for n sufficiently large. Finally, if $n_0 = 0$ then

$$\text{Ker } V_m = \{0\} \ , \tag{4.48}$$

and

$$\langle V_m, V_m \rangle_R \geq \varepsilon I, \quad \Gamma_{m,m} \geq \varepsilon I \tag{4.49}$$

for some positive number ε, where $\Gamma_{m,m}$ is defined via

$$(\Gamma_{ij})_{ij=0}^{m} = G(I, S, \dots, S^m)^{-1} \ . \tag{4.50}$$

PROOF. First remark that the equality (2.30) of Corollary 2.4 shows that the operator $G(I, S, \dots, S^k)$ and the block diagonal operator $\big(\delta_{ij} R(i, i+k)\big)_{ij=0}^{\infty}$ are unitarily equivalent $(k = 0, 1, \dots)$. Hence, if $R(n, n+k)$ is invertible $(n = 0, 1, \dots;$ $k = 0, \dots, m)$ and inequality (4.46) holds, then $\big(\delta_{ij} R(i, i+k)\big)_{ij=0}^{\infty}$ is invertible and therefore, also $G(I, S, \dots, S^k)$ is invertible $(k = 0, \dots, m)$. By Theorem 2.1, this implies that $I, S, \dots, S^m$ admits an orthogonalization with invertible squares. This proves the first part of the theorem.

We denote in the sequel

$$G(I, S, \dots, S^k)^{-1} = (\Gamma_{st}^k)_{st=0}^{k} \quad (k = 0, \dots, m) \ ,$$

and remark that Γ_{ij} as defined in equality (4.50) of the statement is given by

$$\Gamma_{ij} = \Gamma_{ij}^m \quad (i,j = 0,\dots,m) . \tag{4.51}$$

We now proceed with the remaining part of the theorem, and begin by assuming $n_0 = 0$. That is we assume

$$\left[R(n,n+m)^{-1}\right]_{m,m} \geq 0 \quad (n = 0,1,\dots) , \tag{4.52}$$

and let $V_0,\dots,V_m$ be an arbitrary orthogonalization with invertible squares of the system $I, S, \dots, S^m$. Denote

$$R(n,n+m)^{-1} = (\gamma_{st}^n)_{st=0}^m .$$

Then (4.52) means that

$$\gamma_{m,m}^n \geq 0 \quad (n = 0,1,\dots) .$$

However, by equality (2.32) of Corollary 2.4 with $k = s = t = m$ we have $\Gamma_{m,m}^m = (\delta_{ij}\gamma_{m,m}^i)_{ij=0}^\infty$. Therefore, the preceding inequality and (4.51) imply $\Gamma_{m,m} = \Gamma_{m,m}^m \geq 0$. It follows from this inequality and Corollary 2.2 that the inequalities (4.49) hold. Hence, inequality (1.7) holds, and therefore Theorem 1.2 shows that V_m is Fredholm, and (4.47) and (4.48) hold. This proves Theorem 4.2 in the case when $n_0 = 0$.

We now consider the general case where $n_0 \geq 0$. Let us remark that by the first part of the proof, the system $I, S, \dots, S^m$ admits an orthogonalization with invertible squares. Let $V_0,\dots,V_m$ be an arbitrary orthogonalization of $I, S, \dots, S^m$ with invertible squares. We will show that V_m is Fredholm and that (4.47) holds.

Define operators

$$R^{(0)} = S^{*n_0} R S^{n_0} \tag{4.53}$$

and

$$V_i^{(0)} = S^{*n_0} V_i S^{n_0} \quad (i = 0,\dots,m) . \tag{4.54}$$

We now consider the module $M_{R^{(0)}}$, where $R^{(0)}$ is the self-adjoint operator defined by (4.53). Denote the entries of $R^{(0)}$ via

$$R^{(0)} = (R_{ij}^{(0)})_{ij=0}^\infty ,$$

and put

$$R^{(0)}(n,n+k) = (R_{ij}^{(0)})_{ij=n}^{n+k} \quad (n = 0,1,\dots;\ k = 0,\dots,m) .$$

By (4.53) we have $R_{ij}^{(0)} = R_{i+n_0,j+n_0}$. Consequently,

$$R^{(0)}(n,n+k) = R(n+n_0, n+n_0+k) \; . \tag{4.55}$$

Since the matrices $R(n,n+k)$ are invertible and (4.46) holds, it follows that $R^{(0)}(n,n+k)$ is invertible ($n = 0,1,\dots;\ k = 0,\dots,m$) and

$$\sup\{\|R^{(0)}(n,n+k)^{-1}\| :\ n = 0,1,\dots;\ k = 0,\dots,m\} < \infty \; .$$

Moreover, (4.55) and the assumptions of the theorem lead to

$$\left[R^{(0)}(n,n+m)^{-1}\right]_{m,m} \geq 0 \quad (n = 0,1,\dots) \; .$$

Thus, we may apply the theorem with $n_0 = 0$ to the module $M_{R^{(0)}}$. It follows that for any orthogonalization $V_0', \dots, V_m'$ with invertible squares of $I, S, \dots, S^m$ in the module $M_{R^{(0)}}$, the operator V_m' is Fredholm and index $V_m' = -\nu_+\big(R^{(0)}(n,n+m-1)\big)$ for n large. By (4.55) this also leads to index $V_m' = -\nu_+\big(R(n,n+m-1)\big)$ for n large. We will show below that $V_0^{(0)}, \dots, V_m^{(0)}$ as defined in (4.54) is an orthogonalization with invertible squares of $I, S, \dots, S^m$ in the module $M_{R^{(0)}}$. Then, the above statement implies that $V_m^{(0)}$ is Fredholm with

$$\text{index } V_m^{(0)} = -\nu_+\big(R(n,n+m-1)\big) \tag{4.56}$$

for n large. By the definition (4.54) this shows that V_m is Fredholm, and since (4.54) also leads to index $V_m^{(0)} =$ index V_m we obtain (4.47) from equality (4.56).

We now show that $V_0^{(0)}, \dots, V_m^{(0)}$ is an orthogonalization with invertible squares of $I, S, \dots, S^m$ in the module $M_{R^{(0)}}$. Let us first note that since B consists of all block diagonal operators in ℓ_r^2 we have

$$bS^{n_0} = S^{n_0}S^{*n_0}bS^{n_0} \quad (b \in B) \; . \tag{4.57}$$

Now recall that $V_0, \dots, V_m$ is an orthogonalization with invertible squares of $I, S, \dots, S^m$. Therefore, $V_i \in \operatorname{span}_B(V_0, \dots, V_i) = \operatorname{span}_B(I, S, \dots, S^i)$ ($i = 0, \dots, m$), and hence, there are elements $b_{ti} \in B$ ($i = 0, \dots, m;\ t = 0, \dots, i$) such that

$$V_i = \sum_{t=0}^{i} S^t b_{ti} \quad (i = 0, \dots, m) \; .$$

Taking into account (4.57) we have

$$V_i S^{n_0} = \sum_{t=0}^{i} S^t b_{ti} S^{n_0} = \sum_{t=0}^{i} S^{t+n_0} S^{*n_0} b_{ti} S^{n_0} \quad (i = 0, \dots, m) \; . \tag{4.58}$$

Put

$$b_{ti}^{(0)} = S^{*n_o} b_{ti} S^{n_o} \quad (i = 0, \ldots, m;\ t = 0, \ldots, i)\,. \tag{4.59}$$

Then $b_{ti}^{(0)} \in B$ and (4.58) leads to

$$V_i S^{n_o} = \sum_{t=0}^{i} S^{t+n_o} b_{ti}^{(0)} \quad (i = 0, \ldots, m)\,. \tag{4.60}$$

Hence, the definition (4.54) of $V_i^{(0)}$ leads to

$$V_i^{(0)} = \sum_{t=0}^{i} S^t b_{ti}^{(0)}\,. \tag{4.61}$$

This equality and (4.60) also imply that

$$V_i S^{n_o} = S^{n_o} V_i^{(0)} \quad (i = 0, \ldots, m)\,, \tag{4.62}$$

whence

$$S^{*n_o} V_i^* = V_i^{(0)*} S^{*n_o} \quad (i = 0, \ldots, m)\,. \tag{4.63}$$

We now compute inner products in $M_{R^{(0)}}$ and denote such an inner product by $\langle \cdot, \cdot \rangle_{R^{(0)}}$. We also denote by $\langle \cdot, \cdot \rangle_R$ the inner product in M_R. We have by definition

$$\langle V_j^{(0)}, V_i^{(0)} \rangle_{R^{(0)}} = \operatorname{diag}\,(V_i^{(0)*} R^{(0)}\, V_j^{(0)}) = \operatorname{diag}\,(V_i^{(0)*} S^{*n_o} R S^{n_o} V_j^{(0)})\,,$$

where we used (4.53). Taking into account (4.62) and (4.63) it follows that

$$\langle V_j^{(0)}, V_i^{(0)} \rangle_{R^{(0)}} = \operatorname{diag}\,(S^{*n_o} V_i^* R V_j S^{n_o})\,.$$

Since $\operatorname{diag}\,(S^{*n_o} V_i^* R V_j S^{n_o}) = S^{*n_o} \operatorname{diag}\,(V_i^* R V_j) S^{n_o}$, this leads to

$$\langle V_j^{(0)}, V_i^{(0)} \rangle_{R^{(0)}} = S^{*n_o} \langle V_j, V_i \rangle_R S^{n_o} \quad (i, j = 0, \ldots, m)\,. \tag{4.64}$$

We now show that $V_0^{(0)}, \ldots, V_m^{(0)}$ satisfies conditions (1.3) – (1.5) of the introduction in the module $M_{R^{(0)}}$, with $T_i = S^i$ $(i = 0, \ldots, m)$. Since $V_0, \ldots, V_m$ is an orthogonalization with invertible squares of $I, S, \ldots, S^m$ in M_R, then $V_0, \ldots, V_m$ satisfy (1.3) – (1.5) in M_R. Hence,

$$\langle V_j, V_i \rangle_R = 0 \quad (i \neq j) \tag{4.65}$$

$$\text{(4.66)} \qquad \operatorname{span}_B(I, S, \ldots, S^k) = \operatorname{span}_B(V_0, \ldots, V_k) \quad (k = 0, \ldots, m)\ ,$$

and

$$\text{(4.67)} \qquad \langle V_i, V_i \rangle_R \text{ is invertible} \quad (i = 0, \ldots, m)\ .$$

By (4.64) and (4.65) we obtain

$$\text{(4.68)} \qquad \langle V_j^{(0)}, V_i^{(0)} \rangle_{R^{(0)}} = 0 \quad (i \neq j)\ .$$

Moreover, since $\langle V_i, V_i \rangle_R \in B$, $\langle V_i, V_i \rangle_R$ is diagonal. Hence, it follows from equality (4.67) that $S^{*n_0} \langle V_i, V_i \rangle_R S^{n_0}$ is invertible. By (4.64) we obtain

$$\text{(4.69)} \qquad \langle V_i^{(0)}, V_i^{(0)} \rangle_{R^{(0)}} \text{ is invertible} \quad (i = 0, \ldots, m)\ .$$

Moreover, since $b_{\ell i}^{(0)} \in B$ equality (4.61) implies

$$\text{(4.70)} \qquad V_k^{(0)} \in \operatorname{span}_B(I, S, \ldots, S^k) \quad (k = 0, \ldots, m)\ .$$

Finally, by (4.66) there are elements $c_{k\ell} \in B$ $(k = 0, \ldots, m;\ \ell = 0, \ldots, k)$ such that

$$S^k = \sum_{\ell=0}^{k} V_\ell c_{k\ell} \quad (k = 0, \ldots, m)\ .$$

By (4.57) we have $c_{k\ell} S^{n_0} = S^{n_0} S^{*n_0} c_{k\ell} S^{n_0}$. Thus, the preceding equality leads to

$$S^k = S^{*n_0} S^k S^{n_0} = \sum_{\ell=0}^{k} S^{*n_0} V_\ell c_{k\ell} S^{n_0} = \sum_{\ell=0}^{k} S^{*n_0} V_\ell S^{n_0} S^{*n_0} c_{k\ell} S^{n_0}\ .$$

Put $c_{k\ell}^{(0)} = S^{*n_0} c_{k\ell} S^{n_0}$. Taking into account the definition (4.54) of $V_i^{(0)}$ we obtain

$$S^k = \sum_{\ell=0}^{k} V_\ell^{(0)} c_{k\ell}^{(0)} \quad (k = 0, \ldots, m;\ \ell = 0, \ldots, k)\ .$$

Since $c_{k\ell} \in B$, we have $c_{k\ell}^{(0)} \in B$ and therefore,

$$S^k \in \operatorname{span}_B(V_0^{(0)}, \ldots, V_k^{(0)}) \quad (k = 0, \ldots, m)\ .$$

Combining this with (4.70) we obtain

$$\operatorname{span}_B(V_0^{(0)}, \ldots, V_k^{(0)}) = \operatorname{span}_B(I, S, \ldots, S^k) \quad (k = 0, \ldots, m)\ .$$

This equality and (4.68) – (4.69) show that $V_0^{(0)}, \ldots, V_m^{(0)}$ is an orthogonalization with invertible squares of $I, S, \ldots, S^m$ in $M_{R^{(0)}}$. As explained above, this proves Theorem 4.2.

□

5. REFERENCES

[AG] D. Alpay and I. Gohberg, On orthogonal matrix polynomials, Operator Theory: Advances and Applications, Vol. 34, 25-46, (1988), Birkhäuser Verlag.

[BG1] A. Ben-Artzi and I. Gohberg, Extension of a theorem of M.G. Krein on orthogonal polynomials to the nonstationary case, Operator Theory: Advances and Applications, Vol. 34, 65-78, (1988), Birkhäuser Verlag.

[BG2] A. Ben-Artzi and I. Gohberg, Inertia theorems for block weighted shifts and applications, Operator Theory: Advances and Applications, Vol. 56, 120-152, (1992), Birkhäuser Verlag.

[BG3] A. Ben-Artzi and I. Gohberg, On time dependent orthogonal polynomials on the unit circle, to appear.

[I] S. Itoh, Reproducing kernels in modules over C^*-algebras and their applications. Bull. Kyushu Inst. Tech. (Math. Natur. Sci.) No. 37, 1-20, (1990).

[P] W.L. Paschke, Inner product modules over B^*-algebras, Trans. Amer. Math. Soc., Vol. 182, 443-468, (1973).

School of Mathematical Sciences
Raymond and Beverly Sackler Faculty of Exact Sciences
Tel Aviv University
Ramat Aviv 69978, Israel

AMS subject classifications: 33D45, 47A53

Operator Theory:
Advances and Applications, Vol. 73

RELATIONS OF LINKING AND DUALITY BETWEEN SYMMETRIC GAUGE FUNCTIONS

Rajendra Bhatia [1]

Chandler Davis [2]

Dedicated to Moshe Livšic, in appreciation of his inspiring leadership

1. INTRODUCTION

Unitarily invariant norms, and the symmetric gauge functions which represent them, appear increasingly often in recent literature on matrices and other compact Hilbert-space operators. The variety of norms which can be of interest is greater than might have been anticipated. We will recall later in this Section some of the processes by which new u.i. norms or s.g.f. can be concocted, and some of the motivations for taking advantage of these possibilities.

The class of all s.g.f. being a wide and variegated landscape, we feel an invitation to chart it. There are a few relationships between the functions already: inequality and duality are the clearly important ones. The main matter of the present paper is the introduction of two related concepts: 'linked' symmetric gauge functions, and the 'quotient' of two symmetric gauge functions.

These are defined and their theory developed in Sections 2 and 3. As explained in detail there, the linking concept generalizes a type of inequality which is already prominent in the subject, and our Theorem 2.1 can be quoted from the literature. Nevertheless the concept has not hitherto been systematically investigated. The concept of

[1] This author thanks D.A.E. (India) for support.
[2] This author thanks NSERC (Canada) and the Indian Statistical Institute for making possible a visit to Delhi during which most of this work was done.

quotient generalizes duality; it may be regarded as a sort of relative duality, though its properties depart somewhat from the model case. There is a certain amount of interest in a subclass of u.i. norms, the Q-norms [2], [4], [8]. They fit into the development of the quotient operation, as we will explain in Section 4.

Throughout the paper, the norms $|||\ |||$ considered are norms on the complex vector space of complex $n \times n$ matrices, and they are unitarily invariant (u.i.): that is, they satisfy

$$|||UAV||| = |||A|||$$

for any A and any unitary U and V. The normalization used is that $|||P||| = 1$ for any rank-1 orthoprojector. Such norms correspond one-one to symmetric gauge functions f on the set of n-tuples of nonnegative numbers,

$$||\ ||_f \leftrightarrow f,$$

the correspondence being by way of the n-tuple $(s_1(A), \ldots, s_n(A))$ of singular values of A: by definition

$$||A||_f = f(s(A)).$$

It must be kept in mind throughout that if f is a s.g.f. and, μ, ν are n-tuples, then $\mu \prec_w \nu$ implies $f(\mu) \leq f(\nu)$; here $\prec_w$ is the relation of weak majorization. For details see [6], [3], or [1].

Though we treat only the matrix case, so that we can keep n finite, we prefer to leave n arbitrary. Consequently, every f we admit has as arguments n-tuples $\mu = (\mu_1, \ldots, \mu_n)$ for every positive n. Also, every n-tuple can be regarded also as an $(n+1)$-tuple by adjoining a 0. Some of our functions have a cut-off: for some f there may exist m such that even if $n > m$, the value of $f(\mu)$ does not depend on μ_k for $k > m$. Then we say f has 'scope' m. Many of our f, on the other hand, do not have finite scope in this sense.

Our functions being symmetric, we are free to rearrange the components of an argument in the usual (decreasing) order when convenient. (Throughout the paper, decreasing means nonincreasing.) Notation: The set of nonnegative n-tuples is written $\mathbf{R}^n_+$; $\mathbf{R}^n_\downarrow$ is the subset of nonnegative decreasing n-tuples.

We begin, as is customary, with the examples of the bound norm, the c_p norms, and the Ky Fan norms; and we will return to them repeatedly. The bound norm

$$||A|| = \max\{||Ax|| : ||x|| = 1\}$$

is, in the above notation, $\|\ \|_{\max}$. The familiar c_p norm $(1 \leq p \leq \infty)$ corresponds to taking, for f, the function c_p, i.e., the ℓ_p mean; we will abbreviate it simply $\|\ \|_p$. Then

$$\|\ \| = \|\ \|_\infty .$$

The k-th Ky Fan norm, which we write $|||\ |||_k$ $(k = 1, 2, ..., n)$, is the sum of the largest k singular values. The corresponding s.g.f. will be denoted by κ_k. Then

$$|||\ |||_1 = \|\ \|, \quad |||\ |||_n = \|\ \|_1 .$$

Various inequalities hold between these:

$$c_p \geq c_r \quad \text{for} \quad p \leq r,$$
$$\kappa_k \leq \kappa_\ell \quad \text{for} \quad k \leq \ell.$$

The duality relation – which, we have mentioned, we will be generalizing – is as follows [10], [6]. If f is a s.g.f., then a new s.g.f. can be defined by setting (here λ and μ are in $\mathbf{R}^n_+$)

$$h(\mu) = \max\{\Sigma_j \lambda_j \mu_j / f(\lambda)\} .$$

This h is called the dual of f. One verifies that duality is an involutive and order-reversing function. It is familiar that c_p is dual to c_q if p and q are conjugate exponents; especially striking cases are $p = 2$ (c_2 is unique in being self-dual) and $p = 1$ (because c_1 is maximal and c_∞ minimal in the set of all s.g.f.).

The s.g.f. form a convex set. It is also an order-convex set in the sense that if functions satisfy $f \leq g \leq h$ and f and h are s.g.f. then g is a s.g.f.

The big merit of the c_∞ norm is what it ignores: only the maximum is taken into account. But sometimes this simplicity is not appropriate. In some contexts, we would want to consider a matrix with nine large singular values to be "bigger" than one with one large singular value. Then we would prefer a u.i. norm with significant dependence upon the largest nine arguments: simplest would be κ_9; or, if we wanted, we could choose the root mean square of the top nine arguments and get computational simplification. These few remarks already exemplify some of the operations by which, from given s.g.f., new ones may be manufactured. Here is a more systematic list.

(M1) Chopping: given f we construct, for given integer $k \leq 1$, the function $(Tf)(\mu) = f(\mu_1, ..., \mu_k)$ – disregarding any further μ_j. (The Ky Fan norms arise in this way from c_1.)

(M2) Diploidy: given f we construct the function $(Tf)(\mu) = f(\mu_1, \mu_1, \mu_2, \mu_2, \mu_3, ...)$ and then normalize; more generally, polyploidy.

(M3) Given two s.g.f. we form their maximum.

(M4) Given two s.g.f. we form a convex combination of them.

We hope to return in a later paper to a still larger class of methods for generating new s.g.f.

2. LINKED SYMMETRIC GAUGE FUNCTIONS

DEFINITION 2.1. *For symmetric gauge functions f, g, h, we say that f 'links' g and h, or symbolically that $f\mathcal{L}(g,h)$, in case for all decreasing tuples λ, μ we have*

$$f(\lambda\mu) \leq g(\lambda)h(\mu). \tag{2.1}$$

Here the product $\lambda\mu$ is the pointwise product; it is automatically a decreasing sequence.

We must note for future reference these immediate consequences of the definition:

PROPOSITION 2.1. *If $f\mathcal{L}(g,h)$ then*

(i) $f\mathcal{L}(h,g)$;

(ii) $f\mathcal{L}(g,k)$ *for any s.g.f.* $k \geq h$;

(iii) $k\mathcal{L}(g,h)$ *for any s.g.f.* $k \leq f$.

THEOREM 2.1. *The following are equivalent:*

(a) $f\mathcal{L}(g,h)$, *i.e.,* $f(\lambda\mu) \leq g(\lambda)h(\mu)$ *for* $\lambda, \mu \in \mathbf{R}^n_\downarrow$.

(b) $f(\lambda\mu) \leq g(\lambda)h(\mu)$ *for all* $\lambda, \mu \in \mathbf{R}^n_+$.

(c) *For all matrices A and B,* $\|AB\|_f \leq \|A\|_g\|B\|_h$.

(d) *For all matrices A and B,* $\|A \circ B\|_f \leq \|A\|_g\|B\|_h$.

Here $A \circ B$ means the Schur product (elementwise product) of matrices [7].

The theorem is known. Especially important is the equivalence of (a) and (c). Knowing (c) we deduce (b) at once by taking A and B to be commuting normals; and it is trivial that (b) implies (a). The point is the reverse implication from (a) to (c). Let the tuples λ, μ, ν be the decreasing sequences of singular values of A, B, AB respectively. What we need is to infer $\nu \prec_w \lambda\mu$. This is a key fact due to A. Horn ([6], Cor. II.4.1).

As to (d), again it immediately implies (b). The reverse implication is a theorem of K. Okubo [9]. □

Two major inequalities of matrix theory can be stated in terms of the relation of linking.

First example. Recall that, for any u.i. norm $|||\ |||$, we have $|||AB||| \leq ||A||\ |||B|||$ for all A, B. In the terms introduced here, for any s.g.f. f, $f\mathcal{L}(c_\infty, f)$.

Second example. One often uses the inequality involving the c_1 and c_2 norms $||AB||_1 \leq ||A||_2||B||_2$. In the present terminology, $c_1\mathcal{L}(c_2, c_2)$.

Some other, less standard examples were established by Okubo [9]. In the course of the development below we will mention these and other examples.

3. QUOTIENT OF SYMMETRIC GAUGE FUNCTIONS

DEFINITION 3.1. *Let f and g be symmetric gauge functions. By the 'quotient' $f \div g$ we mean the function h given on $\mathbf{R}^n_\downarrow$ by*

$$h(\mu) \;=\; \max\{f(\lambda\mu)/g(\lambda) : \lambda \in \mathbf{R}^n_\downarrow \setminus \{0\}\}. \tag{3.1}$$

PROPOSITION 3.1. *$f \div g$ is defined, and extends to a symmetric gauge function.*

PROOF. A little care is needed in handling the domains of the functions. Remember that all s.g.f. are defined on n-tuples for arbitrary finite n, and some of them are (in the terminology proposed in Section 1) of infinite scope, i.e., they genuinely depend on all the components of their arguments. Is there any loss here in fixing n? The argument μ is to be finitely non-zero in any event, say $\mu_k = 0$ for $k > m$. Such μ can be considered as an element of $\mathbf{R}^n_+$ for any $n \geq m$; we need to make sure that the choice of such n does not affect the value given by (3.1). It does not, for adjoining additional non-zero components of λ beyond the m-th can only increase denominators of ratios under the maximum sign, without changing numerators.

Fix n, then. $h(0) = 0$ is obvious, as is $h(\mu) > 0$ for $\mu \neq 0$. Clearly h is positively homogeneous. Without loss of generality, we confine attention for the moment to elements μ, λ of $\mathbf{R}^n_\downarrow$ having first component 1. This makes the domain over which λ varies compact, so the maximum exists (an explicit upper bound for the values on the right is $f(\mu)$). The normalization of h, $h(1,0,0,...) = 1$, follows easily from normalization of f and g.

The main point is left for last. It must be verified either that the extension of h by symmetry to all of $\mathbf{R}^n$ satisfies $h(\mu+\nu) \leq h(\mu) + h(\nu)$, or equivalently, that on $\mathbf{R}^n_\downarrow$ h satisfies

$$\mu \prec_w \nu \quad \text{implies} \quad h(\mu) \leq h(\nu).$$

For this it will be more than enough to show that for such μ and ν, the quotients in the right-hand side of (3.1) satisfy the desired inequality for each λ separately. Now if all the tuples are in $\mathbf{R}^n_\downarrow$ then $\mu \prec_w \nu$ implies $\lambda\mu \prec_w \lambda\nu$; this is a familiar exercise. Then because f is a s.g.f., we do indeed have

$$f(\lambda\mu) \leq f(\lambda\nu). \qquad \square$$

The relationship between the quotient and linking is fundamental.

PROPOSITION 3.2. *$f\mathcal{L}(g,h)$ if and only if $h \geq f \div g$. In particular, $f\mathcal{L}(g, f \div g)$.*

PROOF. If $f\mathcal{L}(g,h)$, then each quotient under consideration in the right-hand expression in (3.1) is h, hence so is their maximum. Now suppose for the converse that $h(\mu) \geq f \div g(\mu)$ given by (3.1); then perforce $h(\mu) \geq f(\lambda\mu)/g(\lambda)$ for all λ, and that is exactly (2.1). $\square$

PROPOSITION 3.3. *The mapping $f \to f \div g$ is order-preserving. The mapping $g \to f \div g$ is order-reversing.*

This is clear from the definition.

PROPOSITION 3.4. *The mapping $g \to f \div (f \div g))$ is order-preserving and decreasing.*

PROOF. The first statement is immediate from Proposition 3.3. To show that $f \div (f \div g)) \leq g$ we invoke Proposition 3.2 twice: first to affirm that $f\mathcal{L}(g, f \div g)$, then to infer from this that $g \geq f \div (f \div g))$. $\square$

The next result has a similar role in the theory to the second assertion of Proposition 3.2. That showed that, among h with $f\mathcal{L}(g,h)$, the choice $f \div g$ is minimal (remember Proposition 2.1(ii)). Now we can find a way to reduce either function on the right in the linking relation (or both!):

PROPOSITION 3.5. *If $f\mathcal{L}(g,h)$ then $f\mathcal{L}(f \div (f \div g), h)$.*

Note that the conclusion directly strengthens the hypothesis, according to Propositions 2.1(ii), 3.4.

PROOF. Using Proposition 3.2, translate the hypothesis into the statement $h \geq f \div g$ and the conclusion into the statement $f \div h \leq f \div (f \div g)$. The first statement does imply the second, by Proposition 3.3. □

Propositions 3.3 and 3.4 say that we have an instance of what is called a "Galois correspondence" [5]. As in any Galois correspondence, we have the following simplification.

PROPOSITION 3.6. $f \div (f \div (f \div g)) = f \div g$.

PROOF. From $g \geq f \div (f \div g)$ (Proposition 3.4) follows $f \div g \leq f \div (f \div (f \div g))$ by Proposition 3.3. On the other hand, the reverse inequality follows by applying Proposition 3.4 to $f \div g$. □

All of this is true in particular in the special case $f = c_1$. In that case we see that $c_1 \div g$ is exactly von Neumann's dual of g; so think of the general $f \div g$ as a relative dual. Proposition 3.6 tells us that the third relative dual is the same as the first. To judge by the special case, we might venture to ask whether a strengthening simultaneously of Propositions 3.4 and 3.6 might hold: whether always $f \div (f \div g) = g$.

A simple obstruction to this is the scope. If f has scope m (i.e., depends only upon the m largest components of a tuple), then it is clear that $f \div k$ has the same limitation. More interesting departures of the second relative dual of g from g do occur, as witness the following.

Example 3.1. Let $f = c_2$ on $\mathbf{R}^2$, and define g on $\mathbf{R}^2_\downarrow$ by $g(\lambda) = \lambda_1 + \frac{1}{3}\lambda_2$. We compute $f \div g$ for $1 \geq \mu \geq 0$:

$$(f \div g)(1,\mu) = \begin{cases} 1 & (\mu^2 \leq 7/9) \\ \frac{3}{4}(1+\mu^2)^{1/2} & (\mu^2 \geq 7/9). \end{cases}$$

From this we compute $f \div (f \div g)) \neq g$: namely, for $1 \geq \nu \geq 0$,

$$(f \div (f \div g))(1,\nu) = (1 + \frac{7}{9}\nu^2)^{1/2}.$$

The reader is invited to verify in this example that $f \div (f \div g) \leq g$ as Proposition 3.4 insists.

Thus information is lost, in general, in going to the second relative dual, and the equations written do not admit cancellation. Points (iii) and (iv) of the following make this explicit.

PROPOSITION 3.7.

(i) *$f \div f = c_\infty$ for all f.*

(ii) *$f \div c_\infty = f$ for all f.*

(iii) *$c_\infty \div g = c_\infty$ for all g; hence knowledge of f and $f \div g$ can be insufficient to determine g.*

(iv) *Knowledge of g and $f \div g$ can be insufficient to determine f.*

PROOF. The first three assertions are proved from the definition in analogous ways; we illustrate with (iii). Without loss of generality, assume that $\mu_1 = 1$, and maximize only over those λ with $\lambda_1 = 1$. Then the dividend in each quotient on the right in (3.1) is $c_\infty(1,...) = 1$, independent of μ. The maximum will thus be attained when the divisor is minimized; but this occurs for $\lambda = (1,0,0,...)$ and the value is $g(\lambda) = 1$. This proves (iii).

As to (iv) – the third alternative in the following Proposition will provide a glaring example of the phenomenon. □

PROPOSITION 3.8. *$c_r \div c_p = c_q$, where*

(i) *$1/q = 1/r - 1/p$ if $r < p < \infty$;*

(ii) *$q = r$ if $p = \infty$;*

(iii) *$q = \infty$ if $r \geq p$.*

PROOF. This is covered by the cases already proved in the preceding Proposition, except for r and p both finite and unequal; so assume this. We seek, for arbitrary $\mu \in \mathbf{R}^n_+$,

$$(c_r \div c_p)(\mu) = \max\{c_r(\lambda\mu)/c_p(\lambda) : \lambda \in \mathbf{R}^n_+\}.$$

But if we rewrite this in terms of new variables $t = \lambda^r$, $a = \mu^r$, we recognize a familiar problem:

$$(c_r \div c_p)(\mu)^r = \max\left\{\frac{\Sigma_j t_j a_j}{(\Sigma_j t_j^{p/r})^{p/r}} : t \in \mathbf{R}^n_+\right\} \quad (a \in \mathbf{R}^n_+).$$

We define $P = p/r$, and we use the notation s' for the conjugate exponent of any s, $1/s' = 1 - 1/s$. By Hölder's inequality, the maximum is attained, if $P < 1$, for an n-tuple t with only one non-zero component; while if $P > 1$, the maximum value is $(\Sigma_j a_j^{P'})^{1/P'}$. Now $P < 1$ or $p < r$ is case (iii) and we have confirmed the stated conclusion. If $P > 1$ or $p > r$, some arithmetic remains. What we have found (as $(p/r)' = p/(p-r)$) is

$$(c_r \div c_p)(\mu)^r = \left(\Sigma_j((\mu_j)^r)^{p/(p-r)}\right)^{(p-r)/p}$$

This agrees with the announced formula for case (i). □

Let us recapitulate. We have found that sometimes $f \div g$ is the same quotient as could have been obtained with a somewhat different f or a somewhat different g. This raises the question of finding critical quotients in which g is as large as possible without diminishing the quotient, or as small as possible without increasing the quotient. The answer to the latter question flows directly from the above relations.

PROPOSITION 3.9. *Fix f; the following conditions on g are equivalent.*

(a) $g = f \div (f \div g)$.

(b) $g = f \div h$ *for some s.g.f.* h.

(c) *Every s.g.f.* k *with* $f \div k = f \div g$ *satisfies* $k \geq g$.

PROOF. Trivially (a) implies (b); the converse follows from Proposition 3.6.

Assume (b), and $f \div k = f \div g = f \div (f \div h)$; for (c), it is to be proved that $k \geq f \div h$. We get $f \div h = f \div (f \div k) \leq k$ by Proposition 3.6 followed by Proposition 3.4.

Finally, assume (c). Make the particular choice $k = f \div (f \div g)$ (which works, by Proposition 3.6). Then by the hypothesis $k \geq g$; and $k \leq g$ by Proposition 3.4. This establishes (a). □

4. Q-NORMS

The notion of Q-norm arose in 1981. We had noted that some inequalities involving c_p norms hold only for $p \geq 2$, and we wondered if the crucial thing might not be that these are exactly the p for which c_p can be obtained "as a square root" in the sense of the following definition.

DEFINITION. *A symmetric gauge function g gives a 'Q-norm' if and only if there is another s.g.f. f such that for all λ,*

$$g(\lambda) = (f(\lambda^2))^{1/2}.$$

In this case we say g is the Q-function associated with f, and write $g = Qf$.

Thus in particular c_{2q} is Qc_q, and the p with $\| \, \|_p$ a Q-norm are exactly those ≥ 2. The program was to extend properties they enjoy to all Q-norms, and it has met with some success.

We now want to show the relationship between this notion and those we are introducing in the present paper. Bhatia proved [2] that for any s.g.f. f and $\lambda, \mu \in \mathbf{R}^n_+$,

$$|f(\lambda\mu)| \leq (f(\lambda^2)f(\mu^2))^{1/2}. \qquad \text{(C–S)}$$

In the notation used here, this says that $f\mathcal{L}(Qf, Qf)$. The special case $f = c_1$ of this result is the usual Cauchy-Schwarz inequality or $c_1\mathcal{L}(c_2, c_2)$. The result is accordingly called the Cauchy-Schwarz inequality for s.g.f. (We remark in passing that, more generally, the Hölder inequality for s.g.f. [9] is also an assertion about the linking relation.)

PROPOSITION 4.1. *The following relations between symmetric gauge functions f and g are equivalent:*

(a) $g = Qf$, *i.e., g is the Q-function associated with f;*

(b) $g = f \div g$.

So here is a better sense in which to think of Q-norms as square roots!

PROOF. The first half of this is a strengthening of the known inequality (C-S) and will be proved using it. So assume (a), and set $h = f \div (f \div g)$. The first goal now is to show that $h = g$. By Proposition 3.4, $h \leq g$. By (C-S), $f\mathcal{L}(g, g)$, and according to Proposition 3.5 this entails $f\mathcal{L}(h, h)$. Now substituting $\mu = \lambda$ in the Definition 2.1,

$$f(\lambda^2) \leq h(\lambda)^2 \leq g(\lambda)^2 = f(\lambda^2).$$

Hence these inequalities must both be equalities; we have proved $g = h$ as desired. This will be used in proving the conclusion, $g = f \div g$.

First, by Proposition 3.2, $f\mathcal{L}(g, g)$ entails $g \geq f \div g$. What remains is (like Proposition 3.6) general principles of Galois correspondences: from $g \geq f \div g$ follows $f \div g \leq f \div (f \div g)$ by Proposition 3.3, but the last function has just been shown equal to g, so both inequalities between g and $f \div g$ hold. The proof that (a) implies (b) is complete.

The converse deploys the same ideas. Assume (b). By Proposition 3.2, it is the same to say $f\mathcal{L}(g, g)$ or $f(\lambda\mu) \leq g(\lambda)g(\mu)$. Taking $\lambda = \mu$ we get $f(\lambda^2) \leq g(\lambda)^2$, which is to say $Qf \leq g$. Now the reverse inequality between these two is derived by

$$Qf = f \div Qf \geq f \div g = g;$$

the first equality here comes from the first half of the proof, and the inequality relies on Proposition 3.3. We have deduced (a) from (b). □

In the special case $f = c_1$, the Proposition gives the familiar assertion (which we recalled in Section 1) that c_2 is the only self-dual s.g.f. The Proposition also contains the assertion that $c_{2p} = c_p \div c_{2p}$, in accord with Proposition 3.8.

REFERENCES

1. T. Ando, Majorization, doubly stochastic matrices and comparison of eigenvalues, Linear Algebra Appl. 118 (1989), 163-248.

2. R. Bhatia, Some inequalities for norm ideals, Commun. Math. Phys. 111 (1987), 33-39.

3. R. Bhatia, Perturbation Bounds for Matrix Eigenvalues, Pitman Research Notes in Mathematics 162, Longman, Harlow and New York, 1987.

4. R. Bhatia, Perturbation inequalities for the absolute value map in norm ideals of operators, J. Operator Theory 19 (1988), 129-136.

5. G. Birkhoff, Lattice Theory, 2d edn., American Mathematical Society, Providence, 1948.

6. I.C. Gohberg & M.G. Krein, Introduction to the Theory of Linear Nonselfadjoint Operators, American Mathematical Society, Providence, 1969.

7. R.A. Horn, The Hadamard product, Proc. Symp. Appl. Math. 40 (1990), 87-169.

8. R. Horn, R. Mathias, & Y. Nakamura, Inequalities for unitarily invariant norms and bilinear matrix products, Linear and Multilinear Algebra 30 (1991), 303-314.

9. K. Okubo, Hölder-type inequalities for Schur products of matrices, Linear Algebra Appl. 91 (1987), 13-28.

10. J. von Neumann, Some matrix inequalities and metrization of matric space, Izv. Mat. Mekh. Tomsk. Univ. 1 (1937), 286- 300; also in Collected Works, vol. IV, Pergamon Press, 1961, pp. 205-214.

Rajendra Bhatia
Indian Statistical Institute
SJS Sansanwal Marg
New Delhi, 110016 India

Chandler Davis
Department of Mathematics
University of Toronto
Toronto, M5S 1A1 Canada

AMS(MOS) subject classifications: 15A60, 15A45

Operator Theory:
Advances and Applications, Vol. 73

Julia Operators and Coefficient Problems

GENE CHRISTNER, KIN Y. LI, AND JAMES ROVNYAK†

To M. S. Livšic, with best wishes on his retirement.

The form of a Julia operator for a given Kreĭn space operator is determined when an invariant subspace and certain factorizations are known. The result is related to contractive triangular operator matrices and the Schur algorithm. In the context of one-step extension problems for power series, the method yields closed-form expressions for the centers and radii of disks which characterize interpolation. The main example is a coefficient interpolation problem for normalized Riemann mappings $B(z)$ of the unit disk into itself. If $B(z)^\nu$ is expanded in powers $z^\nu, z^{\nu+1}, \ldots$ for any real ν, then the n-th coefficient belongs to a closed disk whose center and radius depend on ν and lower order coefficients.

1. Introduction

Julia operators arise in Kreĭn space operator theory in extension problems for contraction operators. Let $T \in \mathbf{B}(\mathcal{H}, \mathcal{K})$, where $\mathcal{H}$ and $\mathcal{K}$ are Kreĭn spaces. By a **Julia operator** for T we mean a unitary operator U of the form

$$U = \begin{pmatrix} T & D \\ \tilde{D}^* & L \end{pmatrix} \in \mathbf{B}(\mathcal{H} \oplus \mathcal{D}, \mathcal{K} \oplus \tilde{\mathcal{D}}),$$

where $\mathcal{D}$ and $\tilde{\mathcal{D}}$ are Kreĭn spaces, $D \in \mathbf{B}(\mathcal{D}, \mathcal{K})$ and $\tilde{D} \in \mathbf{B}(\tilde{\mathcal{D}}, \mathcal{H})$ have zero kernels, and $L \in \mathbf{B}(\mathcal{D}, \tilde{\mathcal{D}})$. A Julia operator always exists [**3**]. In important cases such as T a contraction, Julia operators are essentially unique [**18, 19**]. Other sources for Julia operators and unitary colligations include [**15, 16, 17**].

The study of Julia operators can be motivated in a number of ways. They arise in the theory of canonical models and characteristic operator functions, and a prototype example [**20**] of a defect space $\mathcal{D}$ in a Julia operator can be taken to be a Hilbert space $\mathfrak{H}(B)$ with reproducing kernel of the form $[1 - B(z)\bar{B}(w)]/(1 - z\bar{w})$ for some function $B(z)$ which is analytic and bounded by one in the unit disk [**14**].

† Research supported by the National Science Foundation.

In this view, it is natural to ask if the multiplication theorem for characteristic operator functions of Livšic and Potapov **[31]** has an analog for Julia operators? We can repose the question to ask, what is the influence of an invariant subspace on the structure of a Julia operator? The answer lies in a construction of Julia operators for triangular operator matrices which have a suitable structure. A special case is the product construction of unitary colligations **[15, 25, 6]**. Corollaries characterize contractive triangular operator matrices. Related results appear in **[1, 2]**, where Hilbert spaces are assumed. In examples, various analogs of the Hilbert spaces $\mathfrak{H}(B)$ appear.

Two concrete examples are considered. For the purpose of illustration, we first take up a one-step interpolation problem which is related to the Schur algorithm. Here we are concerned with contractive multiplication operators

$$T : f(z) \to B(z)f(z),$$

where $B(z) = B_0 + B_1 z + B_2 z^2 + \cdots$ is a formal power series. The operators act on Hilbert spaces $\mathcal{C}_\sigma(z)$ whose elements are (equivalence classes of) power series

$$f(z) = \sum_{n=0}^{\infty} a_n z^n$$

with

$$\|f(z)\|^2 = \sum_{n=0}^{r} \sigma_n |a_n|^2,$$

where $\sigma = \{\sigma_0, \ldots, \sigma_r\}$ are positive weights. The data of the problem consists of the coefficients $B_0, \ldots, B_r$. It is required to determine all numbers B_{r+1} for which the operator remains contractive relative to a one-dimensional extension of the Hilbert space. Well-known methods exist to solve such problems. Monographic accounts which emphasize operator methods are given in **[5, 24, 21]**. An approach through reproducing kernel spaces may be found in **[4]**. The work of I. Schur is reprinted in English translations in a volume dedicated to Schur's ideas and their impact in operator theory and signal processing **[26]**. We use complementation properties of contractively contained Hilbert spaces to derive closed-form expressions for the center and radius of the disk of values B_{r+1} for which interpolation holds. These results appear in §3.

The main example is similar but involves transformations of the form

$$T : f(z) \to f(B(z))$$

acting on Kreĭn spaces $\mathfrak{G}_\sigma^\nu$ whose elements are (equivalence classes of) generalized power series

$$f(z) = \sum_{n=1}^{\infty} a_n z^{\nu+n}.$$

Here $B(z) = B_1 z + B_2 z^2 + \dots$ is a formal power series with constant term zero and coefficient of z positive. The space $\mathfrak{G}_\sigma^\nu$ depends on a real parameter ν and positive numbers $\sigma = \{\sigma_1, \dots, \sigma_r\}$. For any element $f(z) = \sum_{n=1}^{\infty} a_n z^{\nu+n}$ of the space,

$$\langle f(z), f(z) \rangle_{\mathfrak{G}_\sigma^\nu} = \sum_{n=1}^{r} (\nu + n)\sigma_n |a_n|^2 .$$

Such spaces $\mathfrak{G}_\sigma^\nu$ are called Grunsky spaces. Algebraic properties of generalized power series relating to the operation of substitution are discussed in an appendix.

The extension problem for Grunsky spaces is treated in §4. Given $B_1, \dots, B_r$ such that $T : f(z) \to f(B(z))$ is a contraction on some space $\mathfrak{G}_\sigma^\nu$, it is required to characterize the coefficients B_{r+1} for which T remains contractive when the space is extended by one dimension. The set of all such numbers is again a closed disk, and we derive formulas for the radius and center. The calculation involves scalar quantities that play a role analogous to the classical Schur parameters. Now, however, the quantities are functions of a real parameter ν. A Kreĭn space generalization of complementation theory is used in the development. This is due to de Branges [**11**], and an account by different methods may be found in [**20**]. In principle, our results give an iterative method to determine numerically coefficients B_{r+1} with the property that the contractive substitution condition holds for the $(r+1)^{\mathrm{st}}$ stage if it holds for the rth stage.

Examples arise in geometric function theory. Let $B(z)$ be a normalized Riemann mapping of the unit disk into itelf (this means that $B(z)$ is analytic and univalent on the unit disk, bounded by one, vanishes at the origin and has positive derivative at the origin). Then substitution by $B(z)$ is contractive in $\mathfrak{G}_\sigma^\nu$ for every real ν and unit weights $\sigma = \{1, \dots, 1\}$ of arbitrary length r. By a theorem of de Branges [**13**], this property characterizes the class of normalized Riemann mappings of the unit disk into itelf. The theorem assumes that the infinite sequence of coefficients of $B(z)$ is given. It is an open problem to characterize initial segments $B_1, \dots, B_r$ of the coefficients of such mappings. In particular, it is unknown if the contractive substitution property is characteristic of such segments [**30**]. We do not solve this problem†, but our results yield an estimate for the coefficients in the expansion

$$B(z)^\nu = \sum_{n=0}^{\infty} P_n(\nu) z^{\nu+n}.$$

† *Added in proof:* The interpolation procedure described in this paper has been programmed, and a numerical example has been found which shows that the property does not characterize segments of length $r = 3$. Details will appear elsewhere.

If $\nu \geq -1$ and $n \geq 2$, then

$$\left|\frac{P_n(\nu)}{\nu}\right| \leq \sqrt{\frac{1-B'(0)^{2\nu}}{\nu}}\ \sqrt{\frac{1-B'(0)^{2\nu+2n}}{\nu+n}}\ .$$

The approach to coefficient problems for Riemann mappings via contractive substitution transformations in Grunsky spaces is due to de Branges [**9, 10, 12**] and played a role in the proof of the Bieberbach conjecture [**8**]. See [**29, 32, 34**] for related results and extensions.

2. Julia operators for triangular matrices

All underlying spaces in this section are assumed to be Kreĭn spaces [**7**]. The inner product of vectors f and g in a Kreĭn space $\mathcal{H}$ is written $\langle f,g\rangle_{\mathcal{H}}$. Let $\mathbf{B}(\mathcal{H})$ and $\mathbf{B}(\mathcal{H},\mathcal{K})$ be the continuous operators on $\mathcal{H}$ to itself and on $\mathcal{H}$ to $\mathcal{K}$. The **adjoint** of an operator $A \in \mathbf{B}(\mathcal{H},\mathcal{K})$ is the unique operator $A^* \in \mathbf{B}(\mathcal{K},\mathcal{H})$ such that $\langle Af,g\rangle_{\mathcal{K}} = \langle f,A^*g\rangle_{\mathcal{H}}$ for all f in $\mathcal{H}$ and g in $\mathcal{K}$. An operator $A \in \mathbf{B}(\mathcal{H},\mathcal{K})$ is a **contraction** if

$$\langle Af,Af\rangle_{\mathcal{K}} \leq \langle f,f\rangle_{\mathcal{H}}, \qquad f \in \mathcal{H}.$$

If both A and A^* are contractions, A is said to be a **bicontraction**. If $\mathcal{H}$ and $\mathcal{K}$ have equal finite negative indices, every contraction $A \in \mathbf{B}(\mathcal{H},\mathcal{K})$ is a bicontraction.

A **defect operator** for a given $T \in \mathbf{B}(\mathcal{H},\mathcal{K})$ is any operator $\tilde{D} \in \mathbf{B}(\tilde{\mathcal{D}},\mathcal{H})$, where $\tilde{\mathcal{D}}$ is a Kreĭn space, such that $\tilde{D}$ has zero kernel and $1 - T^*T = \tilde{D}\tilde{D}^*$. We call $\tilde{\mathcal{D}}$ a **defect space** for T. In this case,

$$V = \begin{pmatrix} T \\ \tilde{D}^* \end{pmatrix} \in \mathbf{B}(\mathcal{H},\mathcal{K}\oplus\tilde{\mathcal{D}})$$

is an isometric extension of T. By a **Julia operator** for T we mean a unitary operator U having the form

$$U = \begin{pmatrix} T & D \\ \tilde{D}^* & L \end{pmatrix} \in \mathbf{B}(\mathcal{H}\oplus\mathcal{D},\mathcal{K}\oplus\tilde{\mathcal{D}}),$$

where $\mathcal{D}$ and $\tilde{\mathcal{D}}$ are Kreĭn spaces, $D \in \mathbf{B}(\mathcal{D},\mathcal{K})$ and $\tilde{D} \in \mathbf{B}(\tilde{\mathcal{D}},\mathcal{H})$ have zero kernels, and $L \in \mathbf{B}(\mathcal{D},\tilde{\mathcal{D}})$. In this situation, $\tilde{D}$ is a defect operator for T, and D is a defect operator for T^*. The six relations

$$\begin{aligned} T^*T + \tilde{D}\tilde{D}^* &= 1, & TT^* + DD^* &= 1, \\ T^*D + \tilde{D}L &= 0, & \tilde{D}^*T^* + LD^* &= 0, \\ D^*D + L^*L &= 1, & \tilde{D}^*\tilde{D} + LL^* &= 1, \end{aligned}$$

express the fact that U is unitary. For details and basic properties, see [**19**].

The construction of Julia operators requires choices of defect spaces $\tilde{\mathcal{D}}$ and $\mathcal{D}$ for a given operator $T \in \mathbf{B}(\mathcal{H}, \mathcal{K})$ and its adjoint. When the kernel of T is a regular subspace of $\mathcal{H}$, natural choices can be made using the theory of complementation in Kreĭn spaces [**11, 20**].

If $\mathcal{P}$ and $\mathcal{H}$ are Kreĭn spaces with $\mathcal{P}$ a vector subspace of $\mathcal{H}$, we say that $\mathcal{P}$ is **contained continuously** in $\mathcal{H}$ if the inclusion mapping is continuous. The inclusion of $\mathcal{P}$ in $\mathcal{H}$ is called **contractive** if the inclusion mapping is both continuous and contractive. If $\mathcal{P}$ is contained continuously in $\mathcal{H}$ and A is the inclusion mapping, then $P = AA^*$ is a selfadjoint operator on $\mathcal{H}$ whose action coincides with the adjoint of the inclusion of $\mathcal{P}$ in $\mathcal{H}$. Conversely, given any selfadjoint operator $P \in \mathbf{B}(\mathcal{H})$, there is a Kreĭn space $\mathcal{P}$ which is contained continuously in $\mathcal{H}$ such that the adjoint of the inclusion coincides with P. Such a space can be constructed from any factorization $P = EE^*$, where $E \in \mathbf{B}(\mathcal{E}, \mathcal{H})$ for some Kreĭn space $\mathcal{E}$ and $\ker E = \{0\}$. A space $\mathcal{P}$ with the required properties is obtained as the range of E in the inner product which makes E an isomorphism from $\mathcal{E}$ onto $\mathcal{P}$. Under suitable conditions [**20**], the space $\mathcal{P}$ is uniquely determined by P. These conditions are always met in our applications. For example, uniqueness is automatic if $\mathcal{H}$ is finite dimensional.

These notions can be used to construct a Julia operator for any operator $T \in \mathbf{B}(\mathcal{H}, \mathcal{K})$ such that $\mathfrak{N} = \ker T$ is a regular subspace of $\mathcal{H}$. Let $\mathfrak{M}(T)$ be the range of T in the inner product which makes T a partial isometry from $\mathcal{H}$ onto $\mathfrak{M}(T)$ with initial space $\mathcal{H} \ominus \mathfrak{N}$. Then $\mathfrak{M}(T)$ is a Kreĭn space which is contained continuously in $\mathcal{K}$ such that the adjoint of the inclusion coincides with TT^*. Let $\mathfrak{H}(T)$ be a Kreĭn space which is contained continuously in $\mathcal{K}$ such that the adjoint of the inclusion coincides with $1 - TT^*$.

We construct a Kreĭn space $\mathfrak{L}(T)$ which is contained continuously in $\mathcal{H}$ such that the adjoint of the inclusion coincides with $1-T^*T$. Start with the intersection $\mathfrak{R}(T)$ of $\mathfrak{M}(T)$ and $\mathfrak{H}(T)$ viewed as a Kreĭn space in the inner product

$$\langle f, g \rangle_{\mathfrak{R}(T)} = \langle f, g \rangle_{\mathfrak{H}(T)} + \langle f, g \rangle_{\mathfrak{M}(T)}$$

for any elements f, g of the space [**20, Th. 3**]. Define

$$\mathfrak{L}(T) = \mathfrak{N} \oplus \mathfrak{L}_0(T),$$

where $\mathfrak{L}_0(T)$ is the Kreĭn space of elements f of $\mathcal{H} \ominus \mathfrak{N}$ such that Tf is in $\mathfrak{R}(T)$ in the inner product which makes T an isomorphism from $\mathfrak{L}_0(T)$ onto $\mathfrak{R}(T)$. One sees easily that $\mathfrak{L}(T)$ is contained continuously in $\mathcal{H}$ and that the adjoint of the inclusion coincides with $1 - T^*T$. In fact, for any elements $f = u + f_1$ of $\mathcal{H}$ ($u \in \mathfrak{N}$

and $f_1 \in \mathcal{H} \ominus \mathfrak{N}$) and $g = v + g_1$ of $\mathfrak{L}(T)$ ($v \in \mathfrak{N}$ and $g_1 \in \mathfrak{L}_0(T)$),

$$\begin{aligned}\langle g, (1 - T^*T)f\rangle_{\mathfrak{L}(T)} &= \langle v, u\rangle_{\mathcal{H}} + \langle Tg_1, T(1 - T^*T)f_1\rangle_{\mathfrak{R}(T)} \\ &= \langle v, u\rangle_{\mathcal{H}} + \langle Tg_1, (1 - TT^*)Tf_1\rangle_{\mathfrak{H}(T)} \\ &\qquad + \langle Tg_1, T(1 - T^*T)f_1\rangle_{\mathfrak{M}(T)} \\ &= \langle v, u\rangle_{\mathcal{H}} + \langle Tg_1, Tf_1\rangle_{\mathcal{K}} + \langle g_1, (1 - T^*T)f_1\rangle_{\mathcal{H}} \\ &= \langle g, f\rangle_{\mathcal{H}},\end{aligned}$$

proving that the adjoint of the inclusion of $\mathfrak{L}(T)$ in $\mathcal{H}$ coincides with $1 - T^*T$.

With $\mathfrak{H}(T)$ and $\mathfrak{L}(T)$ so defined, a Julia operator for T is given by

$$\begin{pmatrix} T & D \\ \tilde{D}^* & L \end{pmatrix} \in \mathbf{B}(\mathcal{H} \oplus \mathfrak{H}(T), \mathcal{K} \oplus \mathfrak{L}(T)),$$

where D, $\tilde{D}$ are inclusion mappings and $L \in \mathbf{B}(\mathfrak{H}(T), \mathfrak{L}(T))$ is an operator such that $L^*f = -Tf$ for every f in $\mathfrak{L}(T)$. This assertion is [**20, Th. 13**] when $\ker T = \{0\}$, and the proof is essentially the same in the general case. An important special case is when T and T^* are contractions. Then $\mathfrak{H}(T)$ and $\mathfrak{L}(T)$ are Hilbert spaces. In this case, an element f of $\mathcal{K}$ belongs to $\mathfrak{H}(T)$ if and only if [**11, 19**]

$$\sup_{g \in \mathcal{H}} \left[\langle f + Tg, f + Tg\rangle_{\mathcal{K}} - \langle g, g\rangle_{\mathcal{H}}\right] < \infty,$$

in which case the value of the supremum is $\|f\|^2_{\mathfrak{H}(T)}$.

We are concerned with operators $T \in \mathbf{B}(\mathcal{H}, \mathcal{K})$ which have triangular form

$$T = \begin{pmatrix} T_1 & 0 \\ Q & T_2 \end{pmatrix} \tag{2-1}$$

relative to some orthogonal decompositions $\mathcal{H} = \mathcal{H}_1 \oplus \mathcal{H}_2$ and $\mathcal{K} = \mathcal{K}_1 \oplus \mathcal{K}_2$. We consider the problem to describe a Julia operator for T when Julia operators

$$\begin{pmatrix} T_j & D_j \\ \tilde{D}_j^* & L_j \end{pmatrix} \in \mathbf{B}(\mathcal{H}_j \oplus \mathcal{D}_j, \mathcal{K}_j \oplus \tilde{\mathcal{D}}_j), \qquad j = 1, 2, \tag{2-2}$$

for the diagonal entries are given. The solution has a simple form when the off-diagonal entry Q in (2-1) has a factorization.

Theorem 2.1. *Let $T \in \mathbf{B}(\mathcal{H},\mathcal{K})$ have the form (2-1) for the decompositions $\mathcal{H} = \mathcal{H}_1 \oplus \mathcal{H}_2$ and $\mathcal{K} = \mathcal{K}_1 \oplus \mathcal{K}_2$, and let (2-2) be Julia operators for the diagonal entries T_1 and T_2. Assume that Q has a factorization $Q = D_2 Y \tilde{D}_1^*$ for some operator $Y \in \mathbf{B}(\tilde{\mathcal{D}}_1, \mathcal{D}_2)$, and let*

$$(2-3) \qquad \begin{pmatrix} Y & D_Y \\ \tilde{D}_Y^* & L_Y \end{pmatrix} \in \mathbf{B}(\tilde{\mathcal{D}}_1 \oplus \mathcal{D}_Y, \mathcal{D}_2 \oplus \tilde{\mathcal{D}}_Y)$$

be a Julia operator for Y. Then a Julia operator for T is given by

$$(2-4) \qquad \begin{pmatrix} T & D_T \\ \tilde{D}_T^* & L_T \end{pmatrix} \in \mathbf{B}(\mathcal{H} \oplus \mathcal{D}_T, \mathcal{K} \oplus \tilde{\mathcal{D}}_T),$$

where

$$D_T = \begin{pmatrix} D_1 & 0 \\ 0 & D_2 \end{pmatrix} \begin{pmatrix} 1 & 0 \\ Y L_1 & D_Y \end{pmatrix} \in \mathbf{B}(\mathcal{D}_1 \oplus \mathcal{D}_Y, \mathcal{K}_1 \oplus \mathcal{K}_2),$$

$$\tilde{D}_T = \begin{pmatrix} \tilde{D}_1 & 0 \\ 0 & \tilde{D}_2 \end{pmatrix} \begin{pmatrix} Y^* L_2^* & \tilde{D}_Y \\ 1 & 0 \end{pmatrix} \in \mathbf{B}(\tilde{\mathcal{D}}_2 \oplus \tilde{\mathcal{D}}_Y, \mathcal{H}_1 \oplus \mathcal{H}_2),$$

$$L_T = \begin{pmatrix} L_2 & 0 \\ 0 & 1 \end{pmatrix} \begin{pmatrix} Y & D_Y \\ \tilde{D}_Y^* & L_Y \end{pmatrix} \begin{pmatrix} L_1 & 0 \\ 0 & 1 \end{pmatrix} \in \mathbf{B}(\mathcal{D}_1 \oplus \mathcal{D}_Y, \tilde{\mathcal{D}}_2 \oplus \tilde{\mathcal{D}}_Y).$$

Such factorizations have a long history which goes back to the multiplication theorem for characteristic operator functions of Livšic and Potapov [**31**] and counterparts in the relationship between invariant subspaces and factorization in canonical models. In colligation form [**15, 25, 6**], this case arises when $\tilde{\mathcal{D}}_1 = \mathcal{D}_2$ and $Y = 1$. Other cases appear in [**1, 2**].

The hypotheses of Theorem 2.1 are met in our applications. It is an interesting problem to calculate Julia operators of triangular matrices that do not meet these conditions.

Michael Dritschel has pointed out that the Julia operator (2-4) has the factored form

$$\begin{pmatrix} T & D_T \\ \tilde{D}_T^* & L_T \end{pmatrix} = \begin{pmatrix} 1 & 0 & 0 & 0 \\ 0 & T_2 & D_2 & 0 \\ 0 & \tilde{D}_2^* & L_2 & 0 \\ 0 & 0 & 0 & 1 \end{pmatrix} \begin{pmatrix} 1 & 0 & 0 & 0 \\ 0 & 1 & 0 & 0 \\ 0 & 0 & Y & D_Y \\ 0 & 0 & \tilde{D}_Y^* & L_Y \end{pmatrix} \begin{pmatrix} T_1 & 0 & D_1 & 0 \\ 0 & 1 & 0 & 0 \\ \tilde{D}_1^* & 0 & L_1 & 0 \\ 0 & 0 & 0 & 1 \end{pmatrix}$$

as an operator on the Kreĭn space

$$\mathcal{H} \oplus \mathcal{D}_T = (\mathcal{H}_1 \oplus \mathcal{H}_2) \oplus (\mathcal{D}_1 \oplus \mathcal{D}_Y)$$

to the Kreĭn space

$$\mathcal{K} \oplus \tilde{\mathcal{D}}_T = (\mathcal{K}_1 \oplus \mathcal{K}_2) \oplus (\tilde{\mathcal{D}}_2 \oplus \tilde{\mathcal{D}}_Y).$$

Corollaries 2.2 and 2.3 below are similar to Theorems 4.8 and 4.9 of [**18**].

Proof. Clearly D_T and $\tilde{D}_T$ have zero kernels. To prove the theorem we must verify the identities

$$\begin{aligned} T^*T + \tilde{D}_T\tilde{D}_T^* &= 1, & TT^* + D_TD_T^* &= 1, \\ T^*D_T + \tilde{D}_TL_T &= 0, & \tilde{D}_T^*T^* + L_TD_T^* &= 0, \\ D_T^*D_T + L_T^*L_T &= 1, & \tilde{D}_T^*\tilde{D}_T + L_TL_T^* &= 1. \end{aligned}$$

We use the corresponding identities for the Julia operators for T_1, T_2, and Y. Thus

$$\begin{aligned} 1 - T^*T &= \begin{pmatrix} 1 - T_1^*T_1 - Q^*Q & -Q^*T_2 \\ -T_2^*Q & 1 - T_2^*T_2 \end{pmatrix} \\ &= \begin{pmatrix} \tilde{D}_1\tilde{D}_1^* - \tilde{D}_1Y^*D_2^*D_2Y\tilde{D}_1^* & -\tilde{D}_1Y^*D_2^*T_2 \\ -T_2^*D_2Y\tilde{D}_1^* & \tilde{D}_2\tilde{D}_2^* \end{pmatrix} \\ &= \begin{pmatrix} \tilde{D}_1(1 - Y^*Y)\tilde{D}_1^* + \tilde{D}_1Y^*L_2^*L_2Y\tilde{D}_1^* & \tilde{D}_1Y^*L_2^*\tilde{D}_2^* \\ \tilde{D}_2L_2Y\tilde{D}_1^* & \tilde{D}_2\tilde{D}_2^* \end{pmatrix} \\ &= \tilde{D}_T\tilde{D}_T^* \end{aligned}$$

and

$$\begin{aligned} 1 - TT^* &= \begin{pmatrix} 1 - T_1T_1^* & -T_1Q^* \\ -QT_1^* & 1 - T_2T_2^* - QQ^* \end{pmatrix} \\ &= \begin{pmatrix} D_1D_1^* & -T_1\tilde{D}_1Y^*D_2^* \\ -D_2Y\tilde{D}_1^*T_1^* & D_2D_2^* - D_2Y\tilde{D}_1^*\tilde{D}_1Y^*D_2^* \end{pmatrix} \\ &= \begin{pmatrix} D_1D_1^* & D_1L_1^*Y^*D_2^* \\ D_2YL_1D_1^* & D_2(1 - YY^*)D_2^* + D_2YL_1L_1^*Y^*D_2^* \end{pmatrix} \\ &= D_TD_T^*. \end{aligned}$$

The identity $\tilde{D}_T^*T^* + L_TD_T^* = 0$ asserts that

$$\begin{pmatrix} L_2Y\tilde{D}_1^* & \tilde{D}_2^* \\ \tilde{D}_Y^*\tilde{D}_1^* & 0 \end{pmatrix} \begin{pmatrix} T_1^* & Q^* \\ 0 & T_2^* \end{pmatrix} = - \begin{pmatrix} L_2YL_1 & L_2D_Y \\ \tilde{D}_Y^*L_1 & L_Y \end{pmatrix} \begin{pmatrix} D_1^* & L_1^*Y^*D_2^* \\ 0 & D_Y^*D_2^* \end{pmatrix}.$$

The four equalities are readily checked using the six relations for the entries of the Julia operators for T_1, T_2, and Y. For the 11- and 21-entries, this is just one step. Equality of the 22-entries asserts that

$$\tilde{D}_Y^*\tilde{D}_1^*Q^* = -\tilde{D}_Y^*L_1L_1^*Y^*D_2^* - L_YD_Y^*D_2^*.$$

This follows from the identities $L_Y D_Y^* = -\tilde{D}_Y^* Y^*$ and $\tilde{D}_1^* \tilde{D}_1 = -L_1 L_1^* + 1$. For equality of the 12-entries we must show that

$$L_2 Y \tilde{D}_1^* Q^* + \tilde{D}_2^* T_2^* = -L_2 Y L_1 L_1^* Y^* D_2^* - L_2 D_Y D_Y^* D_2^*.$$

For this use in addition $D_Y D_Y^* = 1 - YY^*$ and $\tilde{D}_2^* T_2^* = -L_2 D_2^*$.

We have verified three of the six identities. The remaining three can be verified in a similar manner. Alternatively, they can be deduced from the first three and the fact that $D_T, \tilde{D}_T$ have zero kernels as in **[19, p. 297]**. ■

Corollary 2.2. *A Julia operator for* $\begin{pmatrix} X \\ Y\tilde{D}_X^* \end{pmatrix} \in \mathbf{B}(\mathcal{H}, \mathcal{K} \oplus \mathcal{L})$ *is given by*

$$\begin{pmatrix} \begin{pmatrix} X \\ Y\tilde{D}_X^* \end{pmatrix} & \begin{pmatrix} D_X & 0 \\ YL_X & D_Y \end{pmatrix} \\ (\tilde{D}_Y^* \tilde{D}_X^*) & (\tilde{D}_Y^* L_X \quad L_Y) \end{pmatrix} = \begin{pmatrix} 1 & 0 & 0 \\ 0 & Y & D_Y \\ 0 & \tilde{D}_Y^* & L_Y \end{pmatrix} \begin{pmatrix} X & D_X & 0 \\ \tilde{D}_X^* & L_X & 0 \\ 0 & 0 & 1 \end{pmatrix}$$

as an operator on $\mathcal{H} \oplus (\mathcal{D}_X \oplus \mathcal{D}_Y)$ *to* $(\mathcal{K} \oplus \mathcal{L}) \oplus \tilde{\mathcal{D}}_Y$*, assuming that*

$$\begin{pmatrix} X & D_X \\ \tilde{D}_X^* & L_X \end{pmatrix} \in \mathbf{B}(\mathcal{H} \oplus \mathcal{D}_X, \mathcal{K} \oplus \tilde{\mathcal{D}}_X),$$

$$\begin{pmatrix} Y & D_Y \\ \tilde{D}_Y^* & L_Y \end{pmatrix} \in \mathbf{B}(\tilde{\mathcal{D}}_X \oplus \mathcal{D}_Y, \mathcal{L} \oplus \tilde{\mathcal{D}}_Y)$$

are Julia operators for $X \in \mathbf{B}(\mathcal{H}, \mathcal{K})$ *and* $Y \in \mathbf{B}(\tilde{\mathcal{D}}_X, \mathcal{L})$.

Proof. This follows from Theorem 2.1 on choosing $\mathcal{H}_2 = \{0\}$ and changing notation. ■

Corollary 2.3. *A Julia operator for* $(X \quad D_X Y) \in \mathbf{B}(\mathcal{H} \oplus \mathcal{K}, \mathcal{L})$ *is given by*

$$\begin{pmatrix} (X \quad D_X Y) & (D_X D_Y) \\ \begin{pmatrix} \tilde{D}_X^* & L_X Y \\ 0 & \tilde{D}_Y^* \end{pmatrix} & \begin{pmatrix} L_X D_Y \\ L_Y \end{pmatrix} \end{pmatrix} = \begin{pmatrix} X & D_X & 0 \\ \tilde{D}_X^* & L_X & 0 \\ 0 & 0 & 1 \end{pmatrix} \begin{pmatrix} 1 & 0 & 0 \\ 0 & Y & D_Y \\ 0 & \tilde{D}_Y^* & L_Y \end{pmatrix}$$

as an operator on $(\mathcal{H} \oplus \mathcal{K}) \oplus \mathcal{D}_Y$ *to* $\mathcal{L} \oplus (\tilde{\mathcal{D}}_X \oplus \tilde{\mathcal{D}}_Y)$*, assuming that*

$$\begin{pmatrix} X & D_X \\ \tilde{D}_X^* & L_X \end{pmatrix} \in \mathbf{B}(\mathcal{H} \oplus \mathcal{D}_X, \mathcal{L} \oplus \tilde{\mathcal{D}}_X),$$

$$\begin{pmatrix} Y & D_Y \\ \tilde{D}_Y^* & L_Y \end{pmatrix} \in \mathbf{B}(\mathcal{K} \oplus \mathcal{D}_Y, \mathcal{D}_X \oplus \tilde{\mathcal{D}}_Y)$$

are Julia operators for $X \in \mathbf{B}(\mathcal{H}, \mathcal{L})$ *and* $Y \in \mathbf{B}(\mathcal{K}, \mathcal{D}_X)$.

Proof. First apply Corollary 2.2 to the adjoint of $(X \quad D_X Y)$. Then change notation and take adjoints to get the result. ■

Theorem 2.4. *Let $T \in \mathbf{B}(\mathcal{H},\mathcal{K})$ have the form (2-1) for the decompositions $\mathcal{H} = \mathcal{H}_1 \oplus \mathcal{H}_2$ and $\mathcal{K} = \mathcal{K}_1 \oplus \mathcal{K}_2$. Let $\tilde{D}_1 \in \mathbf{B}(\tilde{\mathcal{D}}_1, \mathcal{H}_1)$ and $D_2 \in \mathbf{B}(\mathcal{D}_2, \mathcal{K}_2)$ be defect operators for T_1 and T_2^* respectively, and let $Q = D_2 Y \tilde{D}_1^*$ for some operator $Y \in \mathbf{B}(\tilde{\mathcal{D}}_1, \mathcal{D}_2)$. Then*

(i) T is a contraction if and only if T_2 and Y are contractions,

(ii) T^* is a contraction if and only if T_1^* and Y^* are contractions, and

(iii) T is a bicontraction if and only if T_1^* and T_2 are contractions and Y is a bicontraction.

Proof. Choose Julia operators (2-2) for T_1 and T_2 which contain the given defect operators [**19**] and a Julia operator (2-3) for Y. Compute a Julia operator (2-4) for T as in Theorem 2.1. The adjoints of (2-3) and (2-4) are Julia operators for Y^* and T^*.

By [**19, Th. 1.4.1**], T is a contraction if and only if $\tilde{\mathcal{D}}_T = \tilde{\mathcal{D}}_2 \oplus \tilde{\mathcal{D}}_Y$ is a Hilbert space. This holds if and only if T_2 and Y are contractions. This proves part (i). In a similar way, T^* is a contraction if and only if $\mathcal{D}_T = \mathcal{D}_1 \oplus \mathcal{D}_Y$ is a Hilbert space, or equivalently, T_1^* and Y^* are contractions. Hence (ii) follows. We obtain (iii) by combining the first two parts. ■

A similar result holds for 3×3 matrices.

Theorem 2.5. *Let $T \in \mathbf{B}(\mathcal{H},\mathcal{K})$ have the form*

$$(2-5) \qquad T = \begin{pmatrix} T_0 & 0 & 0 \\ P & T_1 & 0 \\ R & Q & T_2 \end{pmatrix}$$

relative to decompositions $\mathcal{H} = \mathcal{H}_0 \oplus \mathcal{H}_1 \oplus \mathcal{H}_2$ and $\mathcal{K} = \mathcal{K}_0 \oplus \mathcal{K}_1 \oplus \mathcal{K}_2$. Let

$$(2-6) \qquad \begin{pmatrix} T_j & D_j \\ \tilde{D}_j^* & L_j \end{pmatrix} \in \mathbf{B}(\mathcal{H}_j \oplus \mathcal{D}_j, \mathcal{K}_j \oplus \tilde{\mathcal{D}}_j), \qquad j = 0,1,2,$$

be Julia operators for the diagonal entries in (2-5). Assume that

$$P = D_1 X \tilde{D}_0^* \in \mathbf{B}(\mathcal{H}_0, \mathcal{K}_1), \qquad X \in \mathbf{B}(\tilde{\mathcal{D}}_0, \mathcal{D}_1),$$
$$Q = D_2 Y \tilde{D}_1^* \in \mathbf{B}(\mathcal{H}_1, \mathcal{K}_2), \qquad Y \in \mathbf{B}(\tilde{\mathcal{D}}_1, \mathcal{D}_2),$$
$$R = D_2 Z \tilde{D}_0^* \in \mathbf{B}(\mathcal{H}_0, \mathcal{K}_2), \qquad Z \in \mathbf{B}(\tilde{\mathcal{D}}_0, \mathcal{D}_2),$$

and

$$Z = Y L_1 X + D_Y U \tilde{D}_X^*, \qquad U \in \mathbf{B}(\tilde{\mathcal{D}}_X, \mathcal{D}_Y),$$

for some defect operators $\tilde{D}_X \in \mathbf{B}(\tilde{\mathcal{D}}_X, \tilde{\mathcal{D}}_0)$ and $D_Y \in \mathbf{B}(\mathcal{D}_Y, \mathcal{D}_2)$ for X and Y^*. The following assertions are equivalent:

(i) T is a contraction;

(ii) T_2, Y, and U are contractions;

(iii) T_2 and $\begin{pmatrix} X & 0 \\ Z - YL_1X & Y \end{pmatrix}$ are contractions.

Proof. Write T in block form,

$$T = \begin{pmatrix} \begin{pmatrix} T_0 & 0 \\ P & T_1 \end{pmatrix} & (0) \\ (R \quad Q) & (T_2) \end{pmatrix}.$$

Choose Julia operators

$$\begin{pmatrix} X & D_X \\ \tilde{D}_X^* & L_X \end{pmatrix} \in \mathbf{B}(\tilde{\mathcal{D}}_0 \oplus \mathcal{D}_X, \mathcal{D}_1 \oplus \tilde{\mathcal{D}}_X),$$
$$\begin{pmatrix} Y & D_Y \\ \tilde{D}_Y^* & L_Y \end{pmatrix} \in \mathbf{B}(\tilde{\mathcal{D}}_1 \oplus \mathcal{D}_Y, \mathcal{D}_2 \oplus \tilde{\mathcal{D}}_Y),$$

and construct a Julia operator for $A = \begin{pmatrix} T_0 & 0 \\ P & T_1 \end{pmatrix}$ as in Theorem 2.1. Then

$$(2-7) \qquad (Y \quad D_YU) \in \mathbf{B}(\tilde{\mathcal{D}}_A, \mathcal{D}_2)$$

and

$$\begin{aligned} D_2(Y \quad D_YU)\tilde{D}_A^* &= D_2(Y \quad D_YU)\begin{pmatrix} L_1X\tilde{D}_0^* & \tilde{D}_1^* \\ \tilde{D}_X^*\tilde{D}_0^* & 0 \end{pmatrix} \\ &= (D_2YL_1X\tilde{D}_0^* + D_2D_YU\tilde{D}_X^*\tilde{D}_0^* \quad D_2Y\tilde{D}_1^*) \\ &= (R \quad Q). \end{aligned}$$

By Theorem 2.4, T is a contraction if and only if T_2 and $(Y \quad D_YU)$ are contractions. By Corollary 2.3, a defect space for $(Y \quad D_YU)$ is $\tilde{\mathcal{D}}_Y \oplus \tilde{\mathcal{D}}_U$ for any defect space $\tilde{\mathcal{D}}_U$ for U. By [**19, Th. 1.4.1**], $(Y \quad D_YU)$ is a contraction if and only if $\tilde{\mathcal{D}}_Y \oplus \tilde{\mathcal{D}}_U$ is a Hilbert space, that is, Y and U are contractions. This proves the equivalence of (i) and (ii). Since

$$\begin{pmatrix} X & 0 \\ Z - YL_1X & Y \end{pmatrix} = \begin{pmatrix} X & 0 \\ D_YU\tilde{D}_X^* & Y \end{pmatrix},$$

the equivalence of (ii) and (iii) follows by applying Theorem 2.1 again. ■

The proof of Theorem 2.5 can be carried out using a different block form for T:

$$T = \begin{pmatrix} (T_0) & (0) \\ \begin{pmatrix} P \\ R \end{pmatrix} & \begin{pmatrix} T_1 & 0 \\ Q & T_2 \end{pmatrix} \end{pmatrix}.$$

Construct a Julia operator for $B = \begin{pmatrix} T_1 & 0 \\ Q & T_2 \end{pmatrix}$ as in Theorem 2.1. We then have

$$D_B \begin{pmatrix} X \\ U\tilde{D}_X^* \end{pmatrix} \tilde{D}_0^* = \begin{pmatrix} D_1 & 0 \\ D_2 Y L_1 & D_2 D_Y \end{pmatrix} \begin{pmatrix} X\tilde{D}_0^* \\ U\tilde{D}_X^* \tilde{D}_0^* \end{pmatrix} = \begin{pmatrix} P \\ R \end{pmatrix},$$

where

$$(2-8) \qquad \begin{pmatrix} X \\ U\tilde{D}_X^* \end{pmatrix} \in \mathbf{B}(\tilde{\mathcal{D}}_0, \mathcal{D}_B).$$

The same result is obtained. We remark that a Julia operator for T can be constructed by repeated use of Theorem 2.1 with either of the block forms of T. In the construction, we use Corollaries 2.2 and 2.3 to find Julia operators for (2-7) and (2-8).

Related results hold for positive two-by-two matrices. Briefly, in place of an operator of the form $1 - T^*T$, we can consider any selfadjoint operator

$$(2-9) \qquad R = \begin{pmatrix} A & B \\ B^* & D \end{pmatrix} \in \mathbf{B}(\mathcal{K}_1 \oplus \mathcal{K}_2),$$

and ask for conditions for positivity in the partial ordering of selfadjoint operators on a Kreĭn space. The answer depends on the existence of certain factorizations, of a type which have been used, for example, in **[23, 27]**. One result will indicate the possibilities.

Theorem 2.6. *Let (2-9) be a selfadjoint operator on a direct sum of Kreĭn spaces $\mathcal{K}_1$ and $\mathcal{K}_2$.*

(i) *If $B = AX$ for some operator $X \in \mathbf{B}(\mathcal{K}_2, \mathcal{K}_1)$, then $R \geq 0$ if and only if $A \geq 0$ and $D - X^*AX \geq 0$.*

(ii) *If $B = YD$ for some operator $Y \in \mathbf{B}(\mathcal{K}_2, \mathcal{K}_1)$, then $R \geq 0$ if and only if $A - YDY^* \geq 0$ and $D \geq 0$.*

Proof. The assertions follow from the factorizations

$$R = \begin{pmatrix} 1 & 0 \\ X^* & 1 \end{pmatrix} \begin{pmatrix} A & 0 \\ 0 & D - X^*AX \end{pmatrix} \begin{pmatrix} 1 & X \\ 0 & 1 \end{pmatrix},$$

$$R = \begin{pmatrix} 1 & Y \\ 0 & 1 \end{pmatrix} \begin{pmatrix} A - YDY^* & 0 \\ 0 & D \end{pmatrix} \begin{pmatrix} 1 & 0 \\ Y^* & 1 \end{pmatrix}$$

by elementary arguments. ■

The analysis of positive three-by-three matrices is more complicated, but the approach is a viable alternative to the one that we adopted here using Julia operators.

3. Multiplication transformations on power series

Our main example in §4 treats substitution transformations on Kreĭn spaces of generalized power series. It may be helpful first to see what the results look like in a more familiar setting. Accordingly, here we discuss the simpler situation of multiplication transformations on Hilbert spaces of formal power series.

If $\sigma = \{\sigma_0, \ldots, \sigma_r\}$ are positive numbers, let $\mathcal{C}_\sigma(z)$ be the Hilbert space of all formal power series

$$(3-1) \qquad f(z) = \sum_{n=0}^{\infty} a_n z^n$$

with complex coefficients in the norm

$$\|f(z)\|^2_{\mathcal{C}_\sigma(z)} = \sum_{n=0}^{r} \sigma_n |a_n|^2.$$

Two series of the form (3-1) are identified in the Hilbert space if their difference has zero norm. If $B(z) = B_0 + B_1 z + B_2 z^2 + \cdots$ is a formal power series, define a linear transformation of $\mathcal{C}_\sigma(z)$ into itself by

$$(3-2) \qquad T : f(z) \to B(z) f(z)$$

for any element $f(z)$ of the space. This transformation depends only on the coefficients $B_0, \ldots, B_r$ of $B(z)$.

A formal symmetry connects multiplication by $B(z)$ on $\mathcal{C}_\sigma(z)$ and multiplication by $B^*(z) = \bar{B}_0 + \bar{B}_1 z + \bar{B}_2 z^2 + \cdots$ on a related space.

Theorem 3.1. *Let $B(z) = B_0 + B_1 z + B_2 z^2 + \cdots$ be a formal power series, and define transformations $T : f(z) \to B(z)f(z)$ on $\mathcal{C}_\sigma(z)$ and $T' : f(z) \to B^*(z)f(z)$ on $\mathcal{C}_\tau(z)$, where $\tau = \{\sigma_r^{-1}, \ldots, \sigma_0^{-1}\}$. An isometry W of $\mathcal{C}_\sigma(z)$ onto $\mathcal{C}_\tau(z)$ is defined by $W : f(z) \to \tilde{f}(z)$, where if*

$$f(z) = \sum_{n=0}^{r} a_n z^n, \qquad \text{then} \qquad \tilde{f}(z) = \sum_{n=0}^{r} \sigma_{r-n} a_{r-n} z^n .$$

We have $T^* = W^{-1}T'W$. In particular, T is a contraction on $\mathcal{C}_\sigma(z)$ if and only if T' is a contraction on $\mathcal{C}_\tau(z)$.

Proof. The mapping W is an isometry by a straightforward calculation of norms. If $f(z) = \sum_{n=0}^{r} a_n z^n$ and $g(z) = \sum_{n=0}^{r} b_n z^n$, then

$$\begin{aligned}
\langle B(z)f(z), g(z)\rangle_{\mathcal{C}_\sigma(z)} &= \sigma_0 B_0 a_0 \bar{b}_0 + \sigma_1 (B_0 a_1 + B_1 a_0)\bar{b}_1 \\
&\qquad + \cdots + \sigma_r (B_0 a_r + \cdots + B_r a_0)\bar{b}_r \\
&= a_0 \left(\sigma_0 B_0 \bar{b}_0 + \sigma_1 B_1 \bar{b}_1 + \cdots + \sigma_r B_r \bar{b}_r\right) \\
&\quad + a_1 \left(\sigma_1 B_0 \bar{b}_1 + \sigma_2 B_1 \bar{b}_2 + \cdots + \sigma_r B_{r-1} \bar{b}_r\right) \\
&\quad + \cdots + a_{r-1}\left(\sigma_{r-1} B_0 \bar{b}_{r-1} + \sigma_r B_1 \bar{b}_r\right) + a_r \left(\sigma_r B_0 \bar{b}_r\right) \\
&= \left\langle f(z), W^{-1}T'Wg(z)\right\rangle_{\mathcal{C}_\sigma(z)}.
\end{aligned}$$

Hence $T^* = W^{-1}T'W$. ■

Definition 3.2. *Let $B(z)$ be a formal power series, and let T be the transformation (3-2) on a space $\mathcal{C}_\sigma(z)$.*

(i) By $\mathfrak{M}_\sigma(B)$ we mean the unique Hilbert space which is contained in $\mathcal{C}_\sigma(z)$ such that the adjoint of the inclusion coincides with TT^*.

(ii) By $\mathfrak{H}_\sigma(B)$ we mean the unique Kreĭn space which is contained in $\mathcal{C}_\sigma(z)$ such that the adjoint of the inclusion coincides with $1 - TT^*$.

(iii) By $\mathfrak{L}_\sigma(B)$ we mean the unique Kreĭn space which is contained in $\mathcal{C}_\sigma(z)$ such that the adjoint of the inclusion coincides with $1 - T^*T$.

In the notation described in §2 for general operators,

$$\mathfrak{M}_\sigma(B) = \mathfrak{M}(T), \qquad \mathfrak{H}_\sigma(B) = \mathfrak{H}(T),$$

and

$$\mathfrak{L}_\sigma(B) = \mathfrak{L}(T).$$

Uniqueness of these spaces is automatic because $\mathcal{C}_\sigma(z)$ is finite dimensional.

If T is contractive, so is T^* by the finite-dimensionality of $\mathcal{C}_\sigma(z)$. In this case $\mathfrak{H}_\sigma(B)$ and $\mathfrak{L}_\sigma(B)$ are Hilbert spaces. The space $\mathfrak{H}_\sigma(B)$ is the set of all $f(z)$ of $\mathcal{C}_\sigma(z)$ such that (see §2)

$$(3-3) \qquad \|f(z)\|^2_{\mathfrak{H}_\sigma(B)} = \sup \left[\|f(z) + B(z)g(z)\|^2_{\mathcal{C}_\sigma(z)} - \|g(z)\|^2_{\mathcal{C}_\sigma(z)} \right] < \infty,$$

where the supremum is over all $g(z)$ in $\mathcal{C}_\sigma(z)$.

Theorem 3.3. *Let $B(z)$ be a formal power series, and let T be the transformation (3-2) on a space $\mathcal{C}_\sigma(z)$. If $\sigma = \{\sigma_0, \ldots, \sigma_r\}$, let $\tau = \{\sigma_r^{-1}, \ldots, \sigma_0^{-1}\}$. Then a Julia operator for T is given by*

$$\begin{pmatrix} T & D \\ \tilde{D}^* & L \end{pmatrix} \in \mathbf{B}(\mathcal{C}_\sigma(z) \oplus \mathcal{D}, \mathcal{C}_\sigma(z) \oplus \tilde{\mathcal{D}}),$$

where

(i) $\mathcal{D} = \mathfrak{H}_\sigma(B)$ and D is inclusion in $\mathcal{C}_\sigma(z)$,
(ii) $\tilde{\mathcal{D}} = \mathfrak{H}_\tau(B^*)$ and $\tilde{D} : f(z) \to W^{-1}f(z)$ for every $f(z)$ in $\tilde{\mathcal{D}}$, and
(iii) $L : f(z) \to -B^*(z)Wf(z)$ for every $f(z)$ in $\mathcal{D}$.

Proof. By the discussion in §2, a Julia operator for T is given by

$$\begin{pmatrix} T & D_0 \\ \tilde{D}_0^* & L_0 \end{pmatrix} \in \mathbf{B}(\mathcal{C}_\sigma(z) \oplus \mathcal{D}_0, \mathcal{C}_\sigma(z) \oplus \tilde{\mathcal{D}}_0),$$

where

$$\mathcal{D}_0 = \mathfrak{H}_\sigma(B), \qquad \tilde{\mathcal{D}}_0 = \mathfrak{L}_\sigma(B),$$

$D_0, \tilde{D}_0$ are inclusion mappings, and $L_0^* : f(z) \to -B(z)f(z)$ for every $f(z)$ in $\tilde{\mathcal{D}}_0$. Write $T^* = W^{-1}T'W$ as in Theorem 3.1. Then

$$W(1 - T^*T) = W - W(W^{-1}T'W)(W^{-1}T'^*W) = (1 - T'T'^*)W.$$

Since $\mathfrak{L}_\sigma(B)$ coincides with the range of $1 - T^*T$ and $\mathfrak{H}_\tau(B^*)$ coincides with the range of $1 - T'T'^*$, W maps $\mathfrak{L}_\sigma(B)$ onto $\mathfrak{H}_\tau(B^*)$. The mapping is an isometry, since for any $f(z)$ in $\mathcal{C}_\sigma(z)$,

$$\begin{aligned}
&\langle W(1 - T^*T)f(z), W(1 - T^*T)f(z)\rangle_{\mathfrak{H}_\tau(B^*)} \\
&\quad = \langle (1 - T'T'^*)Wf(z), (1 - T'T'^*)Wf(z)\rangle_{\mathfrak{H}_\tau(B^*)} \\
&\quad = \langle (1 - T'T'^*)Wf(z), Wf(z)\rangle_{\mathcal{C}_\tau(z)} \\
&\quad = \langle W(1 - T^*T)f(z), Wf(z)\rangle_{\mathcal{C}_\tau(z)} \\
&\quad = \langle (1 - T^*T)f(z), (1 - T^*T)f(z)\rangle_{\mathfrak{L}_\sigma(B)}.
\end{aligned}$$

Therefore a second Julia operator for T is given by

$$\begin{pmatrix} T & D \\ \tilde{D}^* & L \end{pmatrix} = \begin{pmatrix} 1 & 0 \\ 0 & W \end{pmatrix} \begin{pmatrix} T & D_0 \\ \tilde{D}_0^* & L_0 \end{pmatrix} = \begin{pmatrix} T & D_0 \\ (\tilde{D}_0 W^{-1})^* & WL_0 \end{pmatrix}.$$

By construction, (i) holds. We obtain (ii) from the fact that W acts as a Kreĭn space isomorphism of $\mathfrak{L}_\sigma(B)$ onto $\mathfrak{H}_\tau(B^*)$. Since the identity $T^*D_0 + \tilde{D}_0 L_0 = 0$ holds by the unitarity of a Julia operator and D_0 and $\tilde{D}_0$ are inclusion mappings, $L_0 f = -T^* f$ for every f in $\mathcal{D}_0 = \mathcal{D}$. Hence

$$Lf = WL_0 f = -WT^* f = -T'Wf$$

for every f in $\mathcal{D}$, and (iii) follows. ■

Theorem 3.4. *Let $T : f(z) \to B(z)f(z)$ on $\mathcal{C}_\sigma(z)$, where $B(z)$ is a formal power series and $\sigma = \{\sigma_0, \dots, \sigma_r\}$ are positive numbers. If T is contractive, then T remains contractive if σ_0 and σ_r are replaced by any positive numbers $\sigma_0' \geq \sigma_0$ and $\sigma_r' \leq \sigma_r$.*

Proof. The hypotheses imply $|B_0| \leq 1$. Write $\sigma_0' = \sigma_0 + \delta$, $\delta \geq 0$. Since T is a contraction, if $f(z) = \sum_{n=0}^\infty a_n z^n$ and $B(z)f(z) = \sum_{n=0}^\infty b_n z^n$, then

$$\sum_{n=0}^{r} \sigma_n |b_n|^2 \leq \sum_{n=0}^{r} \sigma_n |a_n|^2.$$

Since $b_0 = B_0 a_0$, adding $\delta|b_0|^2$ to the left side and $\delta|a_0|^2$ to the right side does not change the inequality, and so T remains contractive when σ_0 is replaced by σ_0'. The second part is reduced to the first using Theorem 3.1. ■

Let $\mathcal{H} = \mathcal{C}_{\{\sigma_0,\dots,\sigma_{r+1}\}}$, where the weights $\sigma_0, \dots, \sigma_{r+1}$ are given positive numbers. A natural decomposition

$$(3-4) \qquad \mathcal{H} = \mathcal{H}_0 \oplus \mathcal{H}_1 \oplus \mathcal{H}_2$$

is obtained with

$$(3-5) \qquad \mathcal{H}_0 = \{a : a \text{ complex}\},$$

$$(3-6) \qquad \mathcal{H}_1 = \left\{ \sum_{n=1}^{r} a_n z^n : a_1, \dots, a_r \text{ complex} \right\},$$

$$(3-7) \qquad \mathcal{H}_2 = \left\{ a z^{r+1} : a \text{ complex} \right\}.$$

Write $\{\tau_0, \dots, \tau_{r+1}\} = \{\sigma_{r+1}^{-1}, \dots, \sigma_0^{-1}\}$ and

$$(3-8) \qquad \sigma = \{\sigma_1, \dots, \sigma_r\}, \qquad \tau = \{\tau_1, \dots, \tau_r\}.$$

Let W be the isometry from $\mathcal{C}_\sigma(z)$ onto $\mathcal{C}_\tau(z)$ defined as in Theorem 3.1. We examine what it means for multiplication by

$$(3-9) \qquad B(z) = B_0 + B_1 z + B_2 z^2 + \cdots$$

to be contractive, first in $\mathcal{H}_0 \oplus \mathcal{H}_1$ and $\mathcal{H}_1 \oplus \mathcal{H}_2$, and later in $\mathcal{H}$.

Theorem 3.5. *Assume the setting described in (3-4) – (3-9).*

(i) Multiplication by $B(z)$ is contractive in $\mathcal{H}_0 \oplus \mathcal{H}_1$ if and only if

(a) multiplication by $B(z)$ is contractive in $\mathcal{C}_\sigma(z)$, and

(b) $[B(z) - B(0)]/z$ belongs to $\mathfrak{H}_\sigma(B)$ and

$$\left\| \frac{B(z) - B(0)}{z} \right\|^2_{\mathfrak{H}_\sigma(B)} \le \sigma_0 \left(1 - |B_0|^2\right). \tag{3-10}$$

(ii) Multiplication by $B(z)$ is contractive in $\mathcal{H}_1 \oplus \mathcal{H}_2$ if and only if

(a) multiplication by $B^*(z)$ is contractive in $\mathcal{C}_\tau(z)$, and

(b) $[B^*(z) - B^*(0)]/z$ belongs to $\mathfrak{H}_\tau(B^*)$ and

$$\left\| \frac{B^*(z) - B^*(0)}{z} \right\|^2_{\mathfrak{H}_\tau(B^*)} \le \sigma_{r+1}^{-1} \left(1 - |B_0|^2\right). \tag{3-11}$$

Proof. If multiplication by $B(z)$ is contractive in $\mathcal{H}_0 \oplus \mathcal{H}_1$, its restriction to $\mathcal{H}_1$ is contractive. Hence multiplication by $B(z)$ is contractive in $\mathcal{C}_\sigma(z)$. For any

$$g(z) = \sum_{n=0}^{\infty} b_n z^n$$

in $\mathcal{C}_\sigma(z)$,

$$f(z) = 1 + zg(z)$$

defines an element of $\mathcal{H}_0 \oplus \mathcal{H}_1 = \mathcal{C}_{\{\sigma_0,\ldots,\sigma_r\}}(z)$. By hypothesis,

$$\|B(z)f(z)\|^2_{\mathcal{C}_{\{\sigma_0,\ldots,\sigma_r\}}(z)} \le \|f(z)\|^2_{\mathcal{C}_{\{\sigma_0,\ldots,\sigma_r\}}(z)},$$

and hence

$$\left\| \frac{B(z) - B(0)}{z} + B(z)g(z) \right\|^2_{\mathcal{C}_\sigma(z)} - \|g(z)\|^2_{\mathcal{C}_\sigma(z)} \le \left(1 - |B_0|^2\right) \sigma_0.$$

The necessity part of (i) follows from the formula (3-3) for the norm in $\mathfrak{H}_\sigma(B)$. Sufficiency is essentially proved by reversing these steps. Part (ii) follows from (i) and Theorem 3.1. ■

Our main result of this section exhibits the center and radius of the disk which characterizes when interpolation is possible.

Theorem 3.6. *Assume the setting described in (3-4) – (3-8), but in place of (3-9) let*

$$B(z) = B_0 + B_1 z + \cdots + B_r z^r$$

be a polynomial with $|B_0| < 1$. *Assume that multiplication by* $B(z)$ *is contractive in each of the spaces* $\mathcal{H}_0 \oplus \mathcal{H}_1$ *and* $\mathcal{H}_1 \oplus \mathcal{H}_2$. *If* B_{r+1} *is a complex number, then multiplication by*

$$C(z) = B(z) + B_{r+1} z^{r+1}$$

is contractive in $\mathcal{H} = \mathcal{H}_0 \oplus \mathcal{H}_1 \oplus \mathcal{H}_2$ *if and only if*

$$\left| B_{r+1} + \left\langle B^*(z) W \frac{B(z) - B(0)}{z}, \frac{B^*(z) - B^*(0)}{z} \right\rangle_{\mathfrak{H}_r(B^*)} \right|^2$$
$$\leq \left(\sigma_0 (1 - |B_0|^2) - \left\| \frac{B(z) - B(0)}{z} \right\|^2_{\mathfrak{H}_\sigma(B)} \right) \times$$
$$\times \left(\sigma_{r+1}^{-1} (1 - |B_0|^2) - \left\| \frac{B^*(z) - B^*(0)}{z} \right\|^2_{\mathfrak{H}_r(B^*)} \right).$$

Proof. Write the operator $T : f(z) \to C(z) f(z)$ on $\mathcal{H}$ in triangular form

$$(3-12) \qquad T = \begin{pmatrix} T_0 & 0 & 0 \\ P & T_1 & 0 \\ R & Q & T_2 \end{pmatrix}$$

relative to the decomposition $\mathcal{H} = \mathcal{H}_0 \oplus \mathcal{H}_1 \oplus \mathcal{H}_2$. The blocks

$$\begin{pmatrix} T_0 & 0 \\ P & T_1 \end{pmatrix} \quad \text{and} \quad \begin{pmatrix} T_1 & 0 \\ Q & T_2 \end{pmatrix}$$

represent the contractive transformations of multiplication by $B(z)$ in $\mathcal{H}_0 \oplus \mathcal{H}_1$ and $\mathcal{H}_1 \oplus \mathcal{H}_2$, respectively. The entries T_0, T_1, T_2, P, Q are compressions of multiplication by $B(z)$, and R has the action

$$Ra = B_{r+1} a z^{r+1}$$

for any number a.

Assume first that T is a contraction. By Theorem 3.4, the hypotheses of the theorem remain satisfied and T is still a contraction if the original weights are replaced by $\{\sigma_0', \sigma_1, \ldots, \sigma_r, \sigma_{r+1}'\}$, where $\sigma_0' > \sigma_0$ and $0 < \sigma_{r+1}' < \sigma_{r+1}$. We proceed with the modified weights but otherwise keep the same notation.

We choose Julia operators

$$\begin{pmatrix} T_j & D_j \\ \tilde{D}_j^* & L_j \end{pmatrix} \in \mathbf{B}(\mathcal{H}_j \oplus \mathcal{D}_j, \mathcal{H}_j \oplus \tilde{\mathcal{D}}_j), \qquad j = 0, 1, 2,$$

for the diagonal entries in (3-12). Factor $T_1 = S T_\sigma S^{-1}$, where T_σ is multiplication by $B(z)$ on $\mathcal{C}_\sigma(z)$ and S is multiplication by z on $\mathcal{C}_\sigma(z)$ to $\mathcal{H}_1$. Choose a Julia operator for T_σ using Theorem 3.3, and then modify it to obtain a Julia operator for T_1 such that

$$\mathcal{D}_1 = \mathfrak{H}_\sigma(B) \qquad \text{and} \qquad \tilde{\mathcal{D}}_1 = \mathfrak{H}_\tau(B^*),$$

$$D_1 : f(z) \to S f(z),$$

$$\tilde{D}_1 : g(z) \to S W^{-1} g(z),$$

$$L_1 : f(z) \to -B^*(z) W f(z),$$

for any $f(z)$ in $\mathfrak{H}_\sigma(B)$ and $g(z)$ in $\mathfrak{H}_\tau(B^*)$. For $j = 0, 2$, choose the defect spaces for T_0 and T_2 to be the complex numbers as vector spaces, with

$$\langle a, b\rangle_{\mathcal{D}_0} = \sigma_0' \frac{a\bar{b}}{1 - |B_0|^2} \qquad \text{and} \qquad \langle a, b\rangle_{\tilde{\mathcal{D}}_0} = \sigma_0'^{-1} \frac{a\bar{b}}{1 - |B_0|^2},$$

$$D_0\, a = a \qquad \text{and} \qquad D_0^*\, a = \left(1 - |B_0|^2\right) a,$$

$$\tilde{D}_0\, a = \sigma_0'^{-1} a \qquad \text{and} \qquad \tilde{D}_0^*\, a = \sigma_0' \left(1 - |B_0|^2\right) a,$$

$$L_0\, a = -\bar{B}_0 \sigma_0' a \qquad \text{and} \qquad \tilde{L}_0^*\, a = -B_0 \sigma_0'^{-1} a,$$

and

$$\langle a, b\rangle_{\mathcal{D}_2} = \sigma_{r+1}' \frac{a\bar{b}}{1 - |B_0|^2} \qquad \text{and} \qquad \langle a, b\rangle_{\tilde{\mathcal{D}}_2} = \sigma_{r+1}'^{-1} \frac{a\bar{b}}{1 - |B_0|^2},$$

$$D_2\, a = a z^{r+1} \qquad \text{and} \qquad D_2^*\, a z^{r+1} = \left(1 - |B_0|^2\right) a,$$

$$\tilde{D}_2\, a = \sigma_{r+1}'^{-1} a z^{r+1} \qquad \text{and} \qquad \tilde{D}_2^*\, a z^{r+1} = \sigma_{r+1}' \left(1 - |B_0|^2\right) a,$$

$$L_2\, a = -\bar{B}_0 \sigma_{r+1}' a \qquad \text{and} \qquad L_2^*\, a = -B_0 \sigma_{r+1}'^{-1} a$$

for any complex numbers a and b. Notice that the inequalities (3-10) and (3-11) in Theorem 3.5 are strict for the modified weights,

$$\left\| \frac{B(z) - B(0)}{z} \right\|_{\mathcal{D}_1}^2 < \sigma_0' \left(1 - |B_0|^2\right), \tag{3-13}$$

$$\left\| \frac{B^*(z) - B^*(0)}{z} \right\|_{\tilde{\mathcal{D}}_1}^2 < \sigma_{r+1}'^{-1} \left(1 - |B_0|^2\right), \tag{3-14}$$

because $\sigma_0' > \sigma_0$, $\sigma_{r+1}' < \sigma_{r+1}$, and $|B_0| < 1$.

The operator $X \in \mathbf{B}(\tilde{\mathcal{D}}_0, \mathcal{D}_1)$ defined by

$$X = \langle \cdot, 1 \rangle_{\tilde{\mathcal{D}}_0} \frac{B(z) - B(0)}{z}$$

satisfies $P = D_1 X \tilde{D}_0^*$, since for any number a,

$$\begin{aligned} D_1 X \tilde{D}_0^* a &= D_1 X \{\sigma_0' (1 - |B_0|^2) a\} = D_1 \langle \sigma_0' (1 - |B_0|^2) a, 1 \rangle_{\tilde{\mathcal{D}}_0} \frac{B(z) - B(0)}{z} \\ &= [B(z) - B(0)]\, a = P a . \end{aligned}$$

The operator $Y \in \mathbf{B}(\tilde{\mathcal{D}}_1, \mathcal{D}_2)$ defined by

$$Y = \left\langle \cdot, \frac{B^*(z) - B^*(0)}{z} \right\rangle_{\tilde{\mathcal{D}}_1}$$

satisfies $Q = D_2 Y \tilde{D}_1^*$. For if $a_1, \dots, a_r$ are any numbers,

$$\begin{aligned} &D_2 Y \tilde{D}_1^* (a_1 z + \cdots + a_r z^r) \\ &\quad = D_2 \left\langle \tilde{D}_1^* (a_1 z + \cdots + a_r z^r), \frac{B^*(z) - B^*(0)}{z} \right\rangle_{\tilde{\mathcal{D}}_1} \\ &\quad = D_2 \left\langle a_1 z + \cdots + a_r z^r, S W^{-1} (\bar{B}_1 + \bar{B}_2 z + \cdots + \bar{B}_r z^{r-1}) \right\rangle_{\mathcal{H}_1} \\ &\quad = D_2 \left\langle a_1 z + \cdots + a_r z^r, \sigma_1^{-1} \bar{B}_r z + \sigma_2^{-1} \bar{B}_{r-1} z^2 + \cdots + \sigma_r^{-1} \bar{B}_1 z^r \right\rangle_{\mathcal{H}_1} \\ &\quad = D_2 \, (B_r a_1 + B_{r-1} a_2 + \cdots + B_1 a_r) \\ &\quad = (B_r a_1 + B_{r-1} a_2 + \cdots + B_1 a_r)\, z^{r+1} \\ &\quad = Q \, (a_1 z + \cdots + a_r z^r) . \end{aligned}$$

Define $Z \in \mathbf{B}(\tilde{\mathcal{D}}_0, \mathcal{D}_2)$ by

$$Z = \langle \cdot, 1 \rangle_{\tilde{\mathcal{D}}_0} B_{r+1}.$$

Then Z satisfies $R = D_2 Z \tilde{D}_0^*$, since if a is any number,

$$D_2 Z \tilde{D}_0^* a = D_2 \langle \sigma_0' (1 - |B_0|^2) a, 1 \rangle_{\tilde{\mathcal{D}}_0} B_{r+1} = B_{r+1} a z^{r+1} = R a.$$

Straightforward calculations yield

$$\begin{aligned} (3-15) \qquad (1 - X^* X) a &= a - X^* \left[\langle a, 1 \rangle_{\tilde{\mathcal{D}}_0} \frac{B(z) - B(0)}{z} \right] \\ &= a - X^* \left[\sigma_0'^{-1} \frac{a}{1 - |B_0|^2} \frac{B(z) - B(0)}{z} \right] \\ &= \left(1 - \frac{1}{\sigma_0' (1 - |B_0|^2)} \left\| \frac{B(z) - B(0)}{z} \right\|_{\mathcal{D}_1}^2 \right) a , \end{aligned}$$

$$(3-16)\quad (1-YY^*)a = a - Y\left[\langle a,1\rangle_{\mathcal{D}_2}\frac{B^*(z)-B^*(0)}{z}\right]$$

$$= a - Y\left[\sigma'_{r+1}\frac{a}{1-|B_0|^2}\frac{B^*(z)-B^*(0)}{z}\right]$$

$$= \left(1 - \frac{1}{\sigma'^{-1}_{r+1}(1-|B_0|^2)}\left\|\frac{B^*(z)-B^*(0)}{z}\right\|^2_{\tilde{\mathcal{D}}_1}\right)a,$$

$$(3-17)\quad (Z-YL_1X)a = Za - YL_1\left[\langle a,1\rangle_{\tilde{\mathcal{D}}_0}\frac{B(z)-B(0)}{z}\right]$$

$$= Za - YL_1\left[\sigma'^{-1}_0\frac{a}{1-|B_0|^2}\frac{B(z)-B(0)}{z}\right]$$

$$= Za - \left\langle L_1\frac{B(z)-B(0)}{z}, \frac{B^*(z)-B^*(0)}{z}\right\rangle_{\tilde{\mathcal{D}}_1}\frac{a}{\sigma'_0(1-|B_0|^2)}$$

$$= Za + \left\langle B^*(z)W\frac{B(z)-B(0)}{z}, \frac{B^*(z)-B^*(0)}{z}\right\rangle_{\tilde{\mathcal{D}}_1}\frac{a}{\sigma'_0(1-|B_0|^2)}$$

$$= \left(\frac{B_{r+1}}{\sigma'_0(1-|B_0|^2)} + \frac{1}{\sigma'_0(1-|B_0|^2)}\left\langle B^*(z)W\frac{B(z)-B(0)}{z}, \frac{B^*(z)-B^*(0)}{z}\right\rangle_{\tilde{\mathcal{D}}_1}\right)a,$$

for any number a.

By (3-13) and (3-14), $1-X^*X$ and $1-YY^*$ are invertible. For any defect operators $\tilde{D}_X \in \mathbf{B}(\tilde{\mathcal{D}}_X, \tilde{\mathcal{D}}_0)$ for X and $D_Y \in \mathbf{B}(\mathcal{D}_Y, \mathcal{D}_2)$ for Y^*, $\tilde{\mathcal{D}}_X$ and $\mathcal{D}_Y$ have dimension one and we may define $U \in \mathbf{B}(\tilde{\mathcal{D}}_X, \mathcal{D}_Y)$ by

$$U = D_Y^{-1}(Z-YL_1X)\tilde{D}_X^{*-1}.$$

Then $Z = YL_1X + D_YU\tilde{D}_X^*$. By Theorem 2.5, since T is a contraction, so is U. Hence for any number a,

$$\left\langle U\tilde{D}_X^*a, U\tilde{D}_X^*a\right\rangle_{\mathcal{D}_Y} \le \langle D_X^*a, D_X^*a\rangle_{\tilde{\mathcal{D}}_X}.$$

This says

$$\langle D_Y^{-1}(Z-YL_1X)a, D_Y^{-1}(Z-YL_1X)a\rangle_{\mathcal{D}_Y} \le \left\langle \tilde{D}_X\tilde{D}_X^*a, a\right\rangle_{\tilde{\mathcal{D}}_0}.$$

Since $D_YD_Y^* = 1-YY^*$ and $\tilde{D}_X\tilde{D}_X^* = 1-X^*X$, we get

$$\langle (1-YY^*)^{-1}(Z-YL_1X)a, (Z-YL_1X)a\rangle_{\mathcal{D}_2} \le \langle (1-X^*X)a, a\rangle_{\tilde{\mathcal{D}}_0}.$$

Using (3-15), (3-16), and (3-17), we transform this inequality to the form

$$\left| B_{r+1} + \left\langle B^*(z) W \frac{B(z) - B(0)}{z}, \frac{B^*(z) - B^*(0)}{z} \right\rangle_{\tilde{\mathcal{D}}_1} \right|^2$$

$$\leq \left(\sigma_0'(1 - |B_0|^2) - \left\| \frac{B(z) - B(0)}{z} \right\|^2_{\mathcal{D}_1} \right) \times$$

$$\times \left(\sigma_{r+1}'^{-1}(1 - |B_0|^2) - \left\| \frac{B^*(z) - B^*(0)}{z} \right\|^2_{\tilde{\mathcal{D}}_1} \right).$$

The inequality in the theorem is obtained in the limit as $\sigma_0' \downarrow \sigma_0$ and $\sigma_{r+1}' \uparrow \sigma_{r+1}$.

Conversely, if B_{r+1} satisfies the inequality in the theorem, it does so as well for any weights $\{\sigma_0', \sigma_1, \dots, \sigma_r, \sigma_{r+1}'\}$, where $\sigma_0' > \sigma_0$ and $0 < \sigma_{r+1}' < \sigma_{r+1}$. The argument can be reversed to show that U is a contraction, and Theorem 2.4 implies that T_2 and Y are contractions. Hence by Theorem 2.5, T is contractive relative to the weights $\{\sigma_0', \sigma_1, \dots, \sigma_r, \sigma_{r+1}'\}$, and T remains contractive in the limit as $\sigma_0' \downarrow \sigma_0$ and $\sigma_{r+1}' \uparrow \sigma_{r+1}$. ∎

4. Extension problem for substitution transformations

If ν is a real number and $\sigma = \{\sigma_1, \dots, \sigma_r\}$ are positive numbers, define the **Grunsky space** $\mathfrak{G}_\sigma^\nu = \mathfrak{G}_{\{\sigma_1,\dots,\sigma_r\}}^\nu$ to be the finite-dimensional Kreĭn space of generalized power series

$$f(z) = \sum_{n=1}^{\infty} a_n z^{\nu+n}$$

in the inner product such that

$$\langle f(z), f(z) \rangle_{\mathfrak{G}_\sigma^\nu} = \sum_{n=1}^{r} (\nu + n) \sigma_n |a_n|^2.$$

Two elements of the space,

$$f(z) = \sum_{n=1}^{\infty} a_n z^{\nu+n} \qquad \text{and} \qquad g(z) = \sum_{n=1}^{\infty} b_n z^{\nu+n},$$

are identified if $a_n = b_n$ whenever $1 \leq n \leq r$ and $\nu + n \neq 0$. Thus constants, if present, are identified to zero. Given a formal power series $B(z) = B_1 z + B_2 z^2 + \cdots$ with $B_1 > 0$, define a linear transformation of $\mathfrak{G}_\sigma^\nu$ into itself by

$$T : f(z) \to f(B(z))$$

whenever $f(z)$ is in the space. It is easy to see that T is well defined ($f(z)$ equivalent to zero implies $f(B(z))$ equivalent to zero), one-to-one, and depends just on the coefficients $B_1, \ldots, B_r$ of $B(z)$. Set $B^*(z) = \bar{B}_1 z + \bar{B}_2 z^2 + \cdots$.

Theorem 4.1. *Let $\mathfrak{G}_\sigma^\nu = \mathfrak{G}_{\{\sigma_1,\ldots,\sigma_r\}}^\nu$ and $\mathfrak{G}_\tau^\mu = \mathfrak{G}_{\{\tau_1,\ldots,\tau_r\}}^\mu$ be Grunsky spaces connected by*

$$\mu = -\nu - r - 1 \quad \text{and} \quad \{\tau_1, \cdots, \tau_r\} = \{1/\sigma_r, \cdots, 1/\sigma_1\}.$$

A Kreĭn space isomorphism from $\mathfrak{G}_\sigma^\nu$ onto the anti-space of $\mathfrak{G}_\tau^\mu$ is defined by $W : f(z) \to \tilde{f}(z)$, where if

$$f(z) = \sum_{n=1}^{r} a_n z^{\nu+n}, \quad \text{then} \quad \tilde{f}(z) = \sum_{n=1}^{r} \sigma_n a_n z^{-\nu-n}.$$

The transformations

$$\begin{aligned} T &: f(z) \to f(B(z)) \quad \text{on} \quad \mathfrak{G}_\sigma^\nu, \\ S &: f(z) \to f(B^*(z)) \quad \text{on} \quad \mathfrak{G}_\tau^\mu, \end{aligned}$$

are connected by $T^* = W^{-1} S^{-1} W$. Thus S is a contraction if and only if T is a contraction.

We call W the **tilde transformation** from $\mathfrak{G}_\sigma^\nu$ to $\mathfrak{G}_\tau^\mu$. Its inverse is the tilde transformation from $\mathfrak{G}_\tau^\mu$ to $\mathfrak{G}_\sigma^\nu$. Since W is an isomorphism from $\mathfrak{G}_\sigma^\nu$ onto the anti-space of $\mathfrak{G}_\tau^\mu$, $W^* = -W^{-1}$ and so

$$\left\langle \tilde{f}(z), \tilde{f}(z) \right\rangle_{\mathfrak{G}_\tau^\mu} = -\langle f(z), f(z) \rangle_{\mathfrak{G}_\sigma^\nu}$$

for every $f(z)$ in $\mathfrak{G}_\sigma^\nu$. Theorem 4.1 is proved in [**30, Th. 2.5**]. It generalizes a result in [**29**] and is related to the Lagrange-Bürmann identity (Theorem B in the Appendix).

Definition 4.2. *Let $T : f(z) \to f(B(z))$ on $\mathfrak{G}_\sigma^\nu$ for some formal power series $B(z) = B_1 z + B_2 z^2 + \cdots$ $(B_1 > 0)$.*

(i) By $\mathfrak{M}_\sigma^\nu(B)$ we mean the unique Kreĭn space which is contained in $\mathfrak{G}_\sigma^\nu$ such that the adjoint of the inclusion coincides with TT^*.

(ii) By $\mathfrak{G}_\sigma^\nu(B)$ we mean the unique Kreĭn space which is contained in $\mathfrak{G}_\sigma^\nu$ such that the adjoint of the inclusion coincides with $1 - TT^*$.

(iii) Define $\mathfrak{L}_\sigma^\nu(B)$ as the Kreĭn space of elements $f(z)$ of $\mathfrak{G}_\sigma^\nu$ such that $f(B(z))$ belongs to $\mathfrak{G}_\sigma^\nu(B)$ in the inner product such that

$$\langle f(z), g(z)\rangle_{\mathfrak{L}_\sigma^\nu(B)} = \langle f(z), g(z)\rangle_{\mathfrak{G}_\sigma^\nu} + \langle f(B(z)), g(B(z))\rangle_{\mathfrak{G}_\sigma^\nu(B)}$$

for all $f(z)$ and $g(z)$ in $\mathfrak{L}_\sigma^\nu(B)$.

These spaces coincide with those of §2: $\mathfrak{M}_\sigma^\nu(B) = \mathfrak{M}(T)$, $\mathfrak{G}_\sigma^\nu(B) = \mathfrak{H}(T)$, and $\mathfrak{L}_\sigma^\nu(B) = \mathfrak{L}(T)$. When T is contractive, T^* is also contractive and $\mathfrak{G}_\sigma^\nu(B)$ and $\mathfrak{L}_\sigma^\nu(B)$ are Hilbert spaces. Then $\mathfrak{G}_\sigma^\nu(B)$ is the set of elements $f(z)$ of $\mathfrak{G}_\sigma^\nu$ such that

$$\|f(z)\|^2_{\mathfrak{G}_\sigma^\nu(B)} = \sup\left[\langle f(z) + g(B(z)), f(z) + g(B(z))\rangle_{\mathfrak{G}_\sigma^\nu} - \langle g(z), g(z)\rangle_{\mathfrak{G}_\sigma^\nu}\right] < \infty,$$

where the supremum is over all $g(z)$ in $\mathfrak{G}_\sigma^\nu$.

Theorem 4.3. *Let $T : f(z) \to f(B(z))$ on $\mathfrak{G}_\sigma^\nu = \mathfrak{G}_{\{\sigma_1,\ldots,\sigma_r\}}^\nu$ for some formal power series $B(z) = B_1 z + B_2 z^2 + \cdots$ $(B_1 > 0)$. Let $\mathfrak{G}_\sigma^\nu$ and $\mathfrak{G}_\tau^\mu$ be related as in Theorem 4.1, and let W be the tilde transformation from $\mathfrak{G}_\sigma^\nu$ to $\mathfrak{G}_\tau^\mu$. Then a Julia operator for T is given by*

$$\begin{pmatrix} T & D \\ \tilde{D}^* & L \end{pmatrix} \in \mathbf{B}(\mathfrak{G}_\sigma^\nu \oplus \mathcal{D}, \mathfrak{G}_\sigma^\nu \oplus \tilde{\mathcal{D}}),$$

where

(i) $\mathcal{D}$ is $\mathfrak{G}_\sigma^\nu(B)$ and D is inclusion in $\mathfrak{G}_\sigma^\nu$,

(ii) $\tilde{\mathcal{D}}$ is $\mathfrak{G}_\tau^\mu(B^*)$ and $\tilde{D} = T^* W^{-1}|\tilde{\mathcal{D}}$, and

(iii) $L : f(z) \to -\tilde{f}(z)$ for every $f(z)$ in $\mathcal{D}$.

Proof. Start with the Julia operator for T constructed in §2 such that

$$\begin{pmatrix} T & D \\ \tilde{D}_1^* & L_1 \end{pmatrix} \in \mathbf{B}(\mathfrak{G}_\sigma^\nu \oplus \mathcal{D}, \mathfrak{G}_\sigma^\nu \oplus \tilde{\mathcal{D}}_1),$$

where $\mathcal{D} = \mathfrak{G}_\sigma^\nu(B)$ and D is inclusion in $\mathfrak{G}_\sigma^\nu$, $\tilde{\mathcal{D}}_1 = \mathfrak{L}_\sigma^\nu(B)$ and $\tilde{D}_1$ is inclusion in $\mathfrak{G}_\sigma^\nu$, and $L_1^* : f(z) \to -f(B(z))$ for every $f(z)$ in $\tilde{\mathcal{D}}_1$. In particular, $1 - T^*T = \tilde{D}_1\tilde{D}_1^*$.

Define S on $\mathfrak{G}^{\mu}_{\tau}$ as in Theorem 4.1, and let E be the inclusion mapping from $\mathfrak{G}^{\mu}_{\tau}(B^*)$ into $\mathfrak{G}^{\mu}_{\tau}$. By the definition of $\mathfrak{G}^{\mu}_{\tau}(B^*)$, the adjoint of E coincides with $1 - SS^*$, and hence $1 - SS^* = EE^*$. Since

$$T^* = W^{-1}S^{-1}W = -W^*S^{-1}W$$

by Theorem 4.1,

$$\begin{aligned} \tilde{D}_1\tilde{D}_1^* &= 1 - T^*T = 1 - W^*S^{-1}WW^*S^{*-1}W \\ &= 1 + W^*S^{-1}S^{*-1}W = W^*S^{-1}(-SS^* + 1)S^{*-1}W \\ &= T^*W^{-1}(1 - SS^*)W^{*-1}T = T^*W^{-1}EE^*W^{*-1}T. \end{aligned}$$

By [**20, Th. 2**], the operator

$$\tilde{V} = T^*W^{-1}E$$

acts as an isometry from $\tilde{\mathcal{D}} = \mathfrak{G}^{\mu}_{\tau}(B^*)$ onto $\tilde{\mathcal{D}}_1 = \mathfrak{L}^{\nu}_{\sigma}(B)$. It follows that a second Julia operator for T is given by

$$\begin{pmatrix} T & D \\ \tilde{D}^* & L \end{pmatrix} = \begin{pmatrix} 1 & 0 \\ 0 & \tilde{V}^* \end{pmatrix} \begin{pmatrix} T & D \\ \tilde{D}_1^* & L_1 \end{pmatrix} \in \mathbf{B}(\mathfrak{G}^{\nu}_{\sigma} \oplus \mathcal{D}, \mathfrak{G}^{\nu}_{\sigma} \oplus \tilde{\mathcal{D}}).$$

Clearly (i) is satisfied. Since $\tilde{D}_1$ and E are inclusion mappings,

$$\tilde{D}f = \tilde{D}_1\tilde{V}f = \tilde{D}_1T^*W^{-1}Ef = T^*W^{-1}f$$

for every $f(z)$ in $\tilde{\mathcal{D}}$, and so (ii) holds.

To check (iii), first notice that since $\tilde{V}$ is a Kreĭn space isomorphism, $\tilde{V}^* = \tilde{V}^{-1}$. The identity $T^*D + \tilde{D}_1L_1 = 0$ follows from the definition of a Julia operator. Since $\tilde{D}$ and $\tilde{D}_1$ are inclusion mappings, $L_1f = -T^*f$ for every $f(z)$ in $\mathcal{D}$. Therefore for any $f(z)$ in $\mathcal{D}$,

$$Lf = \tilde{V}^*L_1f = -\tilde{V}^{-1}T^*f = -Wf,$$

and this verifies (iii). ■

Theorem 4.4. *Let $B(z)$ be a formal power series with constant term zero and coefficient of z positive such that $T : f(z) \to f(B(z))$ is contractive on $\mathfrak{G}^{\nu}_{\sigma} = \mathfrak{G}^{\nu}_{\{\sigma_1,\ldots,\sigma_r\}}$. Then T remains contractive if σ_1 and σ_r are replaced by any positive numbers $\sigma_1' \geq \sigma_1$ and $\sigma_r' \leq \sigma_r$.*

Proof. In the degenerate case $r = 1$ and $\nu = -1$, $\mathfrak{G}^{\nu}_{\sigma}$ contains no nonzero element, and there is nothing to prove. We exclude this case in what follows.

The hypotheses imply that $B_1 \leq 1$. To see this, let $z^{\nu+n}$ be the highest order monomial in $\mathfrak{G}_\sigma^\nu$ which is not equivalent to zero (this is $z^{\nu+r}$ if $\nu+r \neq 0$ and $z^{\nu+r-1}$ otherwise). Since T is a contraction,

$$\left\langle B(z)^{\nu+n}, B(z)^{\nu+n}\right\rangle_{\mathfrak{G}_\sigma^\nu} \leq \left\langle z^{\nu+n}, z^{\nu+n}\right\rangle_{\mathfrak{G}_\sigma^\nu},$$

and hence $(\nu+n)B_1^{2\nu+2n} \leq \nu+n$. Since $\nu+n \neq 0$, $B_1 \leq 1$.

Write $\sigma_1' = \sigma_1 + \delta$, where $\delta \geq 0$. Consider an element $f(z) = \sum_{n=1}^\infty a_n z^{\nu+n}$ in $\mathfrak{G}_\sigma^\nu$, and set $f(B(z)) = \sum_{n=1}^\infty b_n z^{\nu+n}$. Since T is a contraction,

$$\sum_{n=1}^{r} (\nu+n)\sigma_n |b_n|^2 \leq \sum_{n=1}^{r} (\nu+n)\sigma_n |a_n|^2.$$

We have also $b_1 = B_1^{\nu+1} a_1$, and therefore

$$(\nu+1)\delta|b_1|^2 \leq (\nu+1)\delta|a_1|^2,$$

because $B_1 \leq 1$ and $\delta \geq 0$. Adding the two inequalities, we get an inequality which asserts that T remains contractive if σ_1 is replaced by σ_1'.

To prove the second part, consider the space $\mathfrak{G}_\tau^\mu$, where $\mu = -\nu - r - 1$ and $\tau = \{1/\sigma_r, \ldots, 1/\sigma_1\}$. Since $T : f(z) \to f(B(z))$ is contractive on $\mathfrak{G}_\sigma^\nu$, $S : f(z) \to f(B^*(z))$ is contractive on $\mathfrak{G}_\tau^\mu$ by Theorem 4.1. Applying what was proved above to S, we see that S remains contractive if σ_r is replaced by any positive number $\sigma_r' \leq \sigma_r$. Then another application of Theorem 4.1 shows that T remains contractive if σ_r is replaced by σ_r'. ■

The setting for the extension problem is a Grunsky space $\mathcal{H}$ of generalized power series

$$f(z) = \sum_{n=0}^{r+1} a_n z^{\nu+n}$$

in an inner product such that

$$\langle f(z), f(z)\rangle_{\mathcal{H}} = \sum_{n=0}^{r+1} (\nu+n)\sigma_n |a_n|^2$$

for some real number ν, positive integer r, and positive numbers $\sigma_0, \ldots, \sigma_{r+1}$. Write

$$\mathcal{H} = \mathcal{H}_0 \oplus \mathcal{H}_1 \oplus \mathcal{H}_2, \tag{4-1}$$

where

$$(4-2)\qquad \begin{cases} \mathcal{H}_0 = \{az^\nu : a \text{ complex}\}\,, \\ \mathcal{H}_1 = \left\{\sum_{n=1}^{r} a_n z^{\nu+n} : a_1,\dots,a_r \text{ complex}\right\}, \\ \mathcal{H}_2 = \{az^{\nu+r+1} : a \text{ complex}\}\,. \end{cases}$$

Set $\mu = -\nu - r - 1$ and $\{\tau_0, \tau_1, \dots, \tau_r, \tau_{r+1}\} = \{1/\sigma_{r+1}, 1/\sigma_r, \dots, 1/\sigma_1, 1/\sigma_0\}$. Write $\sigma = \{\sigma_1, \dots, \sigma_r\}$ and $\tau = \{\tau_1, \dots, \tau_r\}$, so that

$$(4-3)\qquad \mathfrak{G}^\nu_\sigma = \mathfrak{G}^\nu_{\{\sigma_1,\dots,\sigma_r\}} \quad \text{and} \quad \mathfrak{G}^\mu_\tau = \mathfrak{G}^\mu_{\{\tau_1,\dots,\tau_r\}}\,.$$

All of the spaces in (4-1) are Grunsky spaces:

$$(4-4)\qquad \begin{cases} \mathcal{H} \;= \mathfrak{G}^{\nu-1}_{\{\sigma_0,\sigma_1,\dots,\sigma_r,\sigma_{r+1}\}}\,, \\ \mathcal{H}_0 = \mathfrak{G}^{\nu-1}_{\{\sigma_0\}}\,, \\ \mathcal{H}_1 = \mathfrak{G}^{\nu}_{\sigma}\,, \\ \mathcal{H}_2 = \mathfrak{G}^{\nu+r}_{\{\sigma_{r+1}\}}\,. \end{cases}$$

So are

$$(4-5)\qquad \begin{cases} \mathcal{K} \;= \mathfrak{G}^{\mu-1}_{\{\tau_0,\tau_1,\dots,\tau_r,\tau_{r+1}\}}\,, \\ \mathcal{K}_0 = \mathfrak{G}^{\mu+r}_{\{\tau_{r+1}\}}\,, \\ \mathcal{K}_1 = \mathfrak{G}^{\mu}_{\tau}\,, \\ \mathcal{K}_2 = \mathfrak{G}^{\mu-1}_{\{\tau_0\}}\,, \end{cases}$$

and

$$(4-6)\qquad \mathcal{K} = \mathcal{K}_2 \oplus \mathcal{K}_1 \oplus \mathcal{K}_0\,.$$

The tilde transformation from $\mathcal{H}$ onto $\mathcal{K}$ is diagonal with respect to the decompositions (4-1) and (4-6), and the diagonal entries are the tilde transformations for the summands taken pairwise.

We determine the conditions for a substitution transformation to be contractive in $\mathcal{H}_0 \oplus \mathcal{H}_1$ and $\mathcal{H}_1 \oplus \mathcal{H}_2$. A similar result appears in [**30, Th. 2.3**], but we need a more precise statement and include the proof.

Theorem 4.5. *Let $B(z)$ be a formal power series with constant term zero and coefficient of z positive. Assume that Grunsky spaces are given as above.*

(i) If $\nu \neq 0$, substitution by $B(z)$ is contractive in $\mathcal{H}_0 \oplus \mathcal{H}_1$ if and only if

(a) it is contractive in $\mathfrak{G}_\sigma^\nu$, and

(b) the element $[B(z)^\nu - B'(0)^\nu z^\nu]/\nu$ of $\mathfrak{G}_\sigma^\nu$ belongs to $\mathfrak{G}_\sigma^\nu(B)$ and

$$\left\| \frac{B(z)^\nu - B'(0)^\nu z^\nu}{\nu} \right\|^2_{\mathfrak{G}_\sigma^\nu(B)} \leq \frac{1 - B'(0)^{2\nu}}{\nu} \sigma_0. \tag{4-7}$$

(ii) If $\nu \neq -r-1$, substitution by $B(z)$ is contractive in $\mathcal{H}_1 \oplus \mathcal{H}_2$ if and only if

(a) substitution by $B^*(z)$ is contractive in $\mathfrak{G}_\tau^\mu$, and

(b) the element $[B^*(z)^\mu - B'(0)^\mu z^\mu]/\mu$ of $\mathfrak{G}_\tau^\mu$ belongs to $\mathfrak{G}_\tau^\mu(B^*)$ and

$$\left\| \frac{B^*(z)^\mu - B'(0)^\mu z^\mu}{\mu} \right\|^2_{\mathfrak{G}_\tau^\mu(B^*)} \leq \frac{1 - B'(0)^{2\mu}}{\mu} \frac{1}{\sigma_{r+1}}. \tag{4-8}$$

Proof. If substitution by $B(z)$ is contractive on $\mathcal{H}_0 \oplus \mathcal{H}_1$, its restriction to the invariant subspace $\mathcal{H}_1 = \mathfrak{G}_\sigma^\nu$ is also contractive. In this case, $\mathfrak{G}_\sigma^\nu(B)$ is a Hilbert space. For any element $g(z) = \sum_{n=1}^\infty b_n z^{\nu+n}$ of $\mathfrak{G}_\sigma^\nu$,

$$f(z) = \frac{1}{\nu} z^\nu + g(z)$$

defines an element of $\mathfrak{G}^{\nu-1}_{\{\sigma_0,\sigma_1,\ldots,\sigma_r\}}$, and

$$f(B(z)) = \frac{1}{\nu} B'(0)^\nu z^\nu + \frac{B(z)^\nu - B'(0)^\nu z^\nu}{\nu} + g(B(z)).$$

By hypothesis,

$$\langle f(B(z)), f(B(z)) \rangle_{\mathfrak{G}^{\nu-1}_{\{\sigma_0,\sigma_1,\ldots,\sigma_r\}}} \leq \langle f(z), f(z) \rangle_{\mathfrak{G}^{\nu-1}_{\{\sigma_0,\sigma_1,\ldots,\sigma_r\}}}.$$

An equivalent inequality is

$$\left\langle \frac{B(z)^\nu - B'(0)^\nu z^\nu}{\nu} + g(B(z)), \frac{B(z)^\nu - B'(0)^\nu z^\nu}{\nu} + g(B(z)) \right\rangle_{\mathfrak{G}_\sigma^\nu} - \langle g(z), g(z) \rangle_{\mathfrak{G}_\sigma^\nu}$$
$$\leq \frac{1 - B'(0)^{2\nu}}{\nu} \sigma_0.$$

The necessity part of (i) follows on taking the supremum over all $g(z)$ in $\mathfrak{G}_\sigma^\nu$ and using the supremum formula for the norm in $\mathfrak{G}_\sigma^\nu(B)$. Sufficiency is proved by reversing these steps.

Substitution by $B(z)$ is contractive on $\mathcal{H}_1 \oplus \mathcal{H}_2$ if and only if substitution by $B^*(z)$ is contractive on $\mathcal{K}_0 \oplus \mathcal{K}_1$, by Theorem 4.1. Therefore part (ii) of the theorem follows from part (i). ■

The next result is the main extension theorem.

Theorem 4.6. *Assume the situation described in (4-1) – (4-6), and let*

$$B(z) = B_1 z + \cdots + B_{r+1} z^{r+1}$$

be a polynomial such that $0 < B'(0) < 1$. *Assume that substitution by* $B(z)$ *is contractive in each of the spaces* $\mathcal{H}_0 \oplus \mathcal{H}_1$ *and* $\mathcal{H}_1 \oplus \mathcal{H}_2$. *Let* B_{r+2} *be a complex number, and set*

$$C(z) = B(z) + B_{r+2} z^{r+2}.$$

(i) If $\nu = 0$ or $\nu = -r-1$, *substitution by* $C(z)$ *is contractive in* $\mathcal{H} = \mathcal{H}_0 \oplus \mathcal{H}_1 \oplus \mathcal{H}_2$ *for any choice of complex number* B_{r+2}.
(ii) If $\nu \neq 0$ *and* $\nu \neq -r-1$, *substitution by* $C(z)$ *is contractive in* $\mathcal{H} = \mathcal{H}_0 \oplus \mathcal{H}_1 \oplus \mathcal{H}_2$ *if and only if the coefficient* $Q_{r+1}(\nu)$ *in the expansion*

$$C(z)^\nu = \sum_{n=0}^{\infty} Q_n(\nu)\, z^{\nu+n} \tag{4-9}$$

satisfies

$$\left| \frac{Q_{r+1}(\nu)}{\nu B'(0)^{\nu+r+1}} - \left\langle L_1 \frac{B(z)^\nu - B'(0)^\nu z^\nu}{\nu}, \frac{B^*(z)^\mu - B'(0)^\mu z^\mu}{\mu} \right\rangle_{\mathfrak{G}_\tau^\mu(B^*)} \right|^2 \tag{4-10}$$
$$\leq \left(\frac{1 - B'(0)^{2\nu}}{\nu} \sigma_0 - \left\| \frac{B(z)^\nu - B'(0)^\nu z^\nu}{\nu} \right\|^2_{\mathfrak{G}_\sigma^\nu(B)} \right) \times$$
$$\times \left(\frac{1 - B'(0)^{2\mu}}{\mu} \frac{1}{\sigma_{r+1}} - \left\| \frac{B^*(z)^\mu - B'(0)^\mu z^\mu}{\mu} \right\|^2_{\mathfrak{G}_\tau^\mu(B^*)} \right),$$

where L_1 *is the operator on* $\mathfrak{G}_\sigma^\nu(B)$ *to* $\mathfrak{G}_\tau^\mu(B^*)$ *such that* $L_1 : f(z) \to -\tilde{f}(z)$ *relative to the tilde transformation from* $\mathfrak{G}_\sigma^\nu$ *to* $\mathfrak{G}_\tau^\mu$.

All quantities in (4-10) depend only on ν and $B_1, \ldots, B_{r+1}$, except $Q_{r+1}(\nu)$. This has the form

$$Q_{r+1}(\nu) = \nu B_1^{\nu-1} B_{r+2} + \varphi(\nu, B_1, \ldots, B_{r+1}). \tag{4-11}$$

The term $\varphi(\nu, B_1, \ldots, B_{r+1})$ can be computed recursively using Theorem C of the Appendix. Combining (4-10) and (4-11), we are able to determine the center and radius of the closed disk to which B_{r+2} must belong for a contractive extension to exist. For example, with $r = 1$ and $\sigma_0 = \sigma_1 = \sigma_2 = 1$, the inequality (4-10) can be expanded and brought to the equivalent form

$$(4-12)\quad \left| B_3 + \left(\frac{\nu - 1}{2} B_1^{-1} + B_1^{2\nu+1} \frac{\nu+1}{1 - B_1^{2\nu+2}} \right) B_2^2 \right|^2$$

$$\le B_1^{2\nu} \left(\frac{\nu+1}{1 - B_1^{2\nu+2}} \right)^2 \left(B_1^{-2\nu+2} \frac{1 - B_1^{2\nu}}{\nu} \frac{1 - B_1^{2\nu+2}}{\nu+1} - |B_2|^2 \right) \times$$

$$\times \left(B_1^{-2\nu} \frac{1 - B_1^{2\nu+2}}{\nu+1} \frac{1 - B_1^{2\nu+4}}{\nu+2} - |B_2|^2 \right).$$

Proof of Theorem 4.6. Let us dispose of the degenerate case (i). If $\nu = 0$, then $\mathcal{H}_0 = \{0\}$ and substitution by $C(z)$ in $\mathcal{H}$ reduces to substitution by $B(z)$ in $\mathcal{H}_1 \oplus \mathcal{H}_2$, which is independent of B_{r+2} and is contractive by hypothesis. If $\nu = -r - 1$, then $\mathcal{H}_2 = \{0\}$ and a similar situation occurs. In the rest of the proof we assume that we are in case (ii), that is, $\nu \neq 0$ and $\nu \neq -r - 1$. This insures that $\mathcal{H}_0$ and $\mathcal{H}_2$ are one-dimensional spaces.

Let $T : f(z) \to f(C(z))$ on $\mathcal{H}$, and let

$$(4-13)\qquad T = \begin{pmatrix} T_0 & 0 & 0 \\ P & T_1 & 0 \\ R & Q & T_2 \end{pmatrix}$$

be its matrix relative to the decomposition $\mathcal{H} = \mathcal{H}_0 \oplus \mathcal{H}_1 \oplus \mathcal{H}_2$. The upper left and lower right 2×2 blocks in (4-13) represent substitution by $B(z)$ in $\mathcal{H}_0 \oplus \mathcal{H}_1$ and $\mathcal{H}_1 \oplus \mathcal{H}_2$, respectively, and these transformations are contractive by hypothesis. Except for R, the entries of (4-13) are compressions of substitution by $B(z)$, and

$$R : az^\nu \to aQ_{r+1}(\nu) z^{\nu+r+1}.$$

We are to show that T is a contraction if and only if (4-10) holds.

First assume that T is a contraction. To begin, replace σ_0 and σ_{r+1} by positive numbers $\sigma_0' > \sigma_0$ and $\sigma_{r+1}' < \sigma_{r+1}$. By Theorem 4.4, the hypotheses of the theorem are satisfied and T is still a contraction with these substitutions.

We choose Julia operators

$$\begin{pmatrix} T_j & D_j \\ \tilde{D}_j^* & L_j \end{pmatrix} \in \mathbf{B}(\mathcal{H}_j \oplus \mathcal{D}_j, \mathcal{H}_j \oplus \tilde{\mathcal{D}}_j), \qquad j = 0, 1, 2,$$

for the diagonal entries in (4-13). For T_1, use Theorem 4.3 to do this so that

$$\mathcal{D}_1 = \mathfrak{G}_\sigma^\nu(B) \qquad \text{and} \qquad \tilde{\mathcal{D}}_1 = \mathfrak{G}_\tau^\mu(B^*),$$

$$D_1 : f(z) \to f(z),$$

$$\tilde{D}_1 : g(z) \to T_1^* W_1^{-1} g(z),$$

$$L_1 : f(z) \to -W_1 f(z),$$

for any $f(z)$ in $\mathfrak{G}_\sigma^\nu(B)$ and $g(z)$ in $\mathfrak{G}_\tau^\mu(B^*)$. Here $W_1 : f(z) \to \tilde{f}(z)$ is the tilde transformation from $\mathfrak{G}_\sigma^\nu$ to $\mathfrak{G}_\tau^\mu$. Let the defect spaces for T_0 be $\mathcal{D}_0 = \{az^\nu : a \text{ complex}\}$ and $\tilde{\mathcal{D}}_0 = \{az^{-\nu} : a \text{ complex}\}$, with

$$\begin{aligned}
\langle az^\nu, bz^\nu \rangle_{\mathcal{D}_0} &= \frac{\nu}{1 - B_1^{2\nu}} \sigma_0' a\bar{b}, \\
\langle az^{-\nu}, bz^{-\nu} \rangle_{\tilde{\mathcal{D}}_0} &= \frac{-\nu}{1 - B_1^{-2\nu}} \sigma_0'^{-1} a\bar{b}, \\
D_0 az^\nu &= az^\nu, \\
D_0^* az^\nu &= (1 - B_1^{2\nu}) az^\nu, \\
\tilde{D}_0 az^{-\nu} &= B_1^\nu \sigma_0'^{-1} az^\nu, \\
\tilde{D}_0^* az^\nu &= B_1^{-\nu} \sigma_0' (1 - B_1^{2\nu}) az^{-\nu}.
\end{aligned}$$

Let the defect spaces for T_2 be $\mathcal{D}_2 = \{az^{\nu+r+1} : a \text{ complex}\}$ and $\tilde{\mathcal{D}}_2 = \{az^{-\nu-r-1} : a \text{ complex}\}$, with

$$\begin{aligned}
\langle az^{\nu+r+1}, bz^{\nu+r+1} \rangle_{\mathcal{D}_2} &= \frac{\nu + r + 1}{1 - B_1^{2\nu+2r+2}} \sigma_{r+1}' a\bar{b}, \\
\langle az^{-\nu-r-1}, bz^{-\nu-r-1} \rangle_{\tilde{\mathcal{D}}_2} &= \frac{-\nu - r - 1}{1 - B_1^{-2\nu-2r-2}} \sigma_{r+1}'^{-1} a\bar{b}, \\
D_2\, az^{\nu+r+1} &= az^{\nu+r+1}, \\
D_2^*\, az^{\nu+r+1} &= (1 - B_1^{2\nu+2r+2})\, az^{\nu+r+1}, \\
\tilde{D}_2\, az^{-\nu-r-1} &= B_1^{\nu+r+1} \sigma_{r+1}'^{-1} az^{\nu+r+1}, \\
\tilde{D}_2^*\, az^{\nu+r+1} &= B_1^{-\nu-r-1} \sigma_{r+1}' (1 - B_1^{2\nu+2r+2})\, az^{-\nu-r-1}.
\end{aligned}$$

By Theorem 4.5, we have the strict inequalities

$$(4-14) \qquad \left\| \frac{B(z)^\nu - B'(0)^\nu z^\nu}{\nu} \right\|_{\mathcal{D}_1}^2 < \frac{1 - B'(0)^{2\nu}}{\nu} \sigma_0',$$

and

$$(4-15) \qquad \left\| \frac{B^*(z)^\mu - B'(0)^\mu z^\mu}{\mu} \right\|_{\tilde{\mathcal{D}}_1}^2 < \frac{1 - B'(0)^{2\mu}}{\mu} \frac{1}{\sigma_{r+1}'}.$$

Define an operator $X \in \mathbf{B}(\tilde{\mathcal{D}}_0, \mathcal{D}_1)$ by

$$X = \langle \cdot, B'(0)^{-\nu} z^{-\nu} \rangle_{\tilde{\mathcal{D}}_0} \frac{B(z)^\nu - B'(0)^\nu z^\nu}{\nu}.$$

Since P is a compression of substitution by $B(z)$,

$$P\,az^\nu = \mathrm{Pr}_{\mathcal{H}_1}\, aB(z)^\nu = a\,[B(z)^\nu - B'(0)^\nu z^\nu]\,.$$

It follows that $P = D_1 X \tilde{D}_0^*$.

Define $Y \in \mathbf{B}(\tilde{\mathcal{D}}_1, \mathcal{D}_2)$ by

$$Y = \left\langle \cdot\,, \frac{B^*(z)^\mu - B'(0)^\mu z^\mu}{\mu} \right\rangle_{\tilde{\mathcal{D}}_1} B'(0)^{\nu+r+1} z^{\nu+r+1}\,.$$

We show that $Q = D_2 Y \tilde{D}_1^*$. Write

$$T^* = W^{-1} S^{-1} W \qquad \text{and} \qquad T_1^* = W_1^{-1} S_1^{-1} W_1\,,$$

where W is the tilde transformation on $\mathcal{H}$ to $\mathcal{K}$ and W_1 is the tilde transformation on $\mathcal{H}_1 = \mathfrak{G}_\sigma^\nu$ to $\mathcal{K}_1 = \mathfrak{G}_\tau^\mu$, S is substitution by $C^*(z)$ on $\mathcal{K}$, and S_1 is substitution by $B^*(z)$ on $\mathcal{K}_1$. An easy calculation shows that

$$W_1 \mathrm{Pr}_{\mathcal{H}_1} = \mathrm{Pr}_{\mathcal{K}_1} W.$$

The identity $Q = D_2 Y \tilde{D}_1^*$ will be proved by showing that

$$S_1 W_1 Q^* az^{\nu+r+1} = S_1 W_1 \tilde{D}_1 Y^* D_2^* az^{\nu+r+1} \tag{4-16}$$

for any element $az^{\nu+r+1}$ of $\mathcal{H}_2$. Let $A(z)$ be the compositional inverse of $C(z)$. By Theorem B in the Appendix, $A(z)$ coincides with the compositional inverse of $B(z)$ up to terms of order $r+1$. Therefore for the left side of (4-16), we obtain

$$\begin{aligned} S_1 W_1 Q^* az^{\nu+r+1} &= S_1 W_1 \mathrm{Pr}_{\mathcal{H}_1} T^* az^{\nu+r+1} \\ &= S_1 W_1 \mathrm{Pr}_{\mathcal{H}_1} W^{-1} S^{-1} \sigma'_{r+1} az^{-\nu-r-1} \\ &= S_1 \mathrm{Pr}_{\mathcal{K}_1} \sigma'_{r+1} aA^*(z)^\mu \\ &= S_1 \sigma'_{r+1} a\,[A^*(z)^\mu - A'(0)^\mu z^\mu] \\ &= -B'(0)^{-\mu}\,[B^*(z)^\mu - B'(0)^\mu z^\mu]\,\sigma'_{r+1} a\,, \end{aligned}$$

where equality is in the sense of elements of $\mathcal{K}_1 = \mathfrak{G}_\tau^\mu$. The last step is justified by Theorem A in the Appendix. For the right side of (4-16), we have

$$\begin{aligned} S_1 W_1 \tilde{D}_1 Y^* D_2^* az^{\nu+r+1} &= S_1 W_1 T_1^* W_1^{-1} Y^* \left(1 - B_1^{2\nu+2r+2}\right) az^{\nu+r+1} \\ &= \left(1 - B_1^{2\nu+2r+2}\right) Y^* az^{\nu+r+1} \\ &= \left(1 - B_1^{2\nu+2r+2}\right) \left\langle az^{\nu+r+1}, B'(0)^{\nu+r+1} z^{\nu+r+1} \right\rangle_{\mathcal{D}_2} \times \\ &\qquad\qquad \times \frac{B^*(z)^\mu - B'(0)^\mu z^\mu}{\mu} \\ &= -B'(0)^{-\mu}\,[B^*(z)^\mu - B'(0)^\mu z^\mu]\,\sigma'_{r+1} a\,, \end{aligned}$$

where again equality is in the sense of elements of $\mathcal{K}_1 = \mathfrak{G}^\mu_\tau$. We have verified (4-16), and this completes the proof of the identity $Q = D_2 Y \tilde{D}_1^*$. (A different proof of the identity $Q = D_2 Y \tilde{D}_1^*$ can be given parallel to the argument for the same formula in Theorem 3.6. This form of the argument also uses Theorems A and B of the Appendix.)

Define $Z \in \mathbf{B}(\tilde{\mathcal{D}}_0, \mathcal{D}_2)$ by

$$Z = \left\langle \cdot, B'(0)^{-\nu} z^{-\nu} \right\rangle_{\tilde{\mathcal{D}}_0} \frac{Q_{\tau+1}(\nu)}{\nu} z^{\nu+\tau+1}.$$

The identity $R = D_2 Z \tilde{D}_0^*$ is verified by checking the action of each side on any element az^ν of $\mathcal{H}_0$ using (4-9) and the definition of R as a compression of substitution by $C(z)$.

A short calculation gives

$$Z - YL_1X = \left(\frac{Q_{\tau+1}(\nu)}{\nu B'(0)^{\nu+\tau+1}} - \left\langle L_1 \frac{B(z)^\nu - B'(0)^\nu z^\nu}{\nu}, \frac{B^*(z)^\mu - B'(0)^\mu z^\mu}{\mu} \right\rangle_{\tilde{\mathcal{D}}_1} \right) \times$$
$$\times \left\langle \cdot, B'(0)^{-\nu} z^{-\nu} \right\rangle_{\tilde{\mathcal{D}}_0} B'(0)^{\nu+\tau+1} z^{\nu+\tau+1}.$$

In a similar way,

$$1 - X^*X = \left(\frac{1 - B'(0)^{2\nu}}{\nu} \sigma_0' - \left\| \frac{B(z)^\nu - B'(0)^\nu z^\nu}{\nu} \right\|^2_{\mathcal{D}_1} \right) \times$$
$$\times \left\langle \cdot, B'(0)^{-\nu} z^{-\nu} \right\rangle_{\tilde{\mathcal{D}}_0} B'(0)^{-\nu} z^{-\nu}$$

and

$$1 - YY^* = \left(\frac{1 - B'(0)^{2\mu}}{\mu} \frac{1}{\sigma_{\tau+1}'} - \left\| \frac{B^*(z)^\mu - B'(0)^\mu z^\mu}{\mu} \right\|^2_{\tilde{\mathcal{D}}_1} \right) \times$$
$$\times \left\langle \cdot, B'(0)^{\nu+\tau+1} z^{\nu+\tau+1} \right\rangle_{\mathcal{D}_2} B'(0)^{\nu+\tau+1} z^{\nu+\tau+1}.$$

These operators have rank one by (4-14) – (4-15). Hence

$$\dim \tilde{\mathcal{D}}_X = \dim \mathcal{D}_Y = 1$$

for any defect operator $\tilde{D}_X \in \mathbf{B}(\tilde{\mathcal{D}}_X, \tilde{\mathcal{D}}_0)$ for X and any defect operator $D_Y \in \mathbf{B}(\mathcal{D}_Y, \mathcal{D}_2)$ for Y^*. This allows us to define $U \in \mathbf{B}(\tilde{\mathcal{D}}_X, \mathcal{D}_Y)$ by

$$U = D_Y^{-1}(Z - YL_1X)\tilde{D}_X^{*-1}.$$

The identity $Z = YL_1X + D_YU\tilde{D}_X^*$ holds by construction. The hypotheses of Theorem 2.5 are thus satisfied for the operator (4-13).

By Theorem 2.5, $U^*U \le 1$. Hence for any element $az^{-\nu}$ in $\tilde{\mathcal{D}}_0$,

$$\left\langle U\tilde{D}_X^* az^{-\nu}, U\tilde{D}_X^* az^{-\nu}\right\rangle_{\mathcal{D}_Y} \le \left\langle \tilde{D}_X^* az^{-\nu}, \tilde{D}_X^* az^{-\nu}\right\rangle_{\tilde{\mathcal{D}}_X}$$

and

$$\langle D_Y^{-1}(Z - YL_1X)az^{-\nu}, D_Y^{-1}(Z - YL_1X)az^{-\nu}\rangle_{\mathcal{D}_Y} \le \left\langle \tilde{D}_X^* az^{-\nu}, \tilde{D}_X^* az^{-\nu}\right\rangle_{\tilde{\mathcal{D}}_X}.$$

Since $D_YD_Y^* = 1 - YY^*$ and $\tilde{D}_X\tilde{D}_X^* = 1 - X^*X$,

$$\begin{aligned}\langle (1 - YY^*)^{-1}(Z - YL_1X)az^{-\nu}, (Z - YL_1X)az^{-\nu}\rangle_{\mathcal{D}_2}& \\ \le \langle (1 - X^*X)az^{-\nu}, az^{-\nu}\rangle_{\tilde{\mathcal{D}}_0}&.\end{aligned}$$

By our previous calculations, this inequality becomes

$$\tag{4-17} \begin{aligned}&\left| \frac{Q_{r+1}(\nu)}{\nu B'(0)^{\nu+r+1}} - \left\langle L_1 \frac{B(z)^\nu - B'(0)^\nu z^\nu}{\nu}, \frac{B^*(z)^\mu - B'(0)^\mu z^\mu}{\mu}\right\rangle_{\tilde{\mathcal{D}}_1}\right|^2 \\ &\le \left(\frac{1 - B'(0)^{2\nu}}{\nu}\sigma_0' - \left\| \frac{B(z)^\nu - B'(0)^\nu z^\nu}{\nu}\right\|_{\mathcal{D}_1}^2\right) \times \\ &\times \left(\frac{1 - B'(0)^{2\mu}}{\mu}\frac{1}{\sigma_{r+1}'} - \left\| \frac{B^*(z)^\mu - B'(0)^\mu z^\mu}{\mu}\right\|_{\tilde{\mathcal{D}}_1}^2\right).\end{aligned}$$

Letting $\sigma_0' \downarrow \sigma_0$ and $\sigma_{r+1}' \uparrow \sigma_{r+1}$ in (4-17), we obtain (4-10).

Conversely, (4-10) implies (4-17). The steps in the argument reverse to show that U is a contraction. The operators Y and T_2 are contractions by Theorem 2.4. By Theorem 2.5, T is contractive relative to the weights $\sigma_0', \sigma_1, \dots, \sigma_r, \sigma_{r+1}'$, and this remains true in the limit for the original weights $\sigma_0, \sigma_1, \dots, \sigma_r, \sigma_{r+1}$. ■

Remark. The operators U which appear in the proof of Theorem 4.6 play a role analogous to the Schur parameters [**5, 24, 21**]. As in the classical case, they are scalar quantities because the underlying spaces are one-dimensional. However, whereas the Schur parameters are constants bounded by 1, their counterparts in Theorem 4.6 are functions of ν which are bounded by 1.

Unit weights are of interest in geometric function theory. For background, see §1. Write $\mathfrak{D}_r^\nu$ for $\mathfrak{G}_\sigma^\nu$ with

$$\sigma = \{\overbrace{1, \ldots, 1}^{r}\}$$

We say that given numbers $B_1, \ldots, B_r$ $(B_1 > 0)$ have the **contractive substitution property** if substitution by $B_1 z + \cdots + B_r z^r$ is contractive in $\mathfrak{D}_r^\nu$ for all real ν. It is sufficient to check the property for $\nu > -(r+1)/2$ [**30**]. If $B_1, \ldots, B_r$ have the contractive substitution property, so do $B_1, \ldots, B_s$ whenever $1 \leq s \leq r$. The contractive substitution property holds if $B_1, \ldots, B_r$ are an initial segment of coefficients of a normalized Riemann mapping of the unit disk into itself.

Open Problem. *If $B_1, \ldots, B_r$ $(B_1 > 0)$ satisfy the contractive substitution property, is there always a number B_{r+1} such that $B_1, \ldots, B_r, B_{r+1}$ satisfy the contractive substitution property?*

If the answer is affirmative, it follows from de Branges' theorem (see §1) that the contractive substitution property characterizes initial segments of coefficients of a normalized Riemann mapping of the unit disk into itself.

The answer is affirmative for $r = 1, 2$. For $r = 1$, the contractive substitution property simply says that $B_1 \leq 1$. For such B_1, the contractive substitution property holds for B_1, B_2 if and only if

$$|B_2|^2 \leq B_1^{-2\nu} \frac{1 - B_1^{2\nu+2}}{\nu+1} \frac{1 - B_1^{2\nu+4}}{\nu+2}$$

for every real number ν. The right side is a minimum when $\nu = -3/2$ [**29**], and taking this value we reduce the condition to the inequality

$$(4-18) \qquad |B_2| \leq 2B_1(1 - B_1).$$

Equality holds if and only if there is a Koebe mapping whose first two coefficients coincide with B_1, B_2 (cf. Pick's Theorem [**33, 22, p. 74**]). By a **Koebe mapping** we mean a normalized Riemann mapping of the unit disk into itself of the form

$$C = f_b^{-1} \circ f_a,$$

where for positive t, $f_t(z) = tz/(1 - wz)^2$ for some fixed w, $|w| = 1$, and $0 < a \leq b < \infty$. If $c = a/b$, then

$$(4-19) \qquad C(z) = cz + 2c(1-c)wz^2 + c(1-c)(3-5c)w^2 z^3 + \cdots.$$

The case $r = 2$ can be handled by showing that if B_1, B_2 $(B_1 > 0)$ satisfy $B_1 \leq 1$ and $|B_2| \leq 2B_1(1 - B_1)$, there exist Koebe mappings $C_1(z)$ and $C_2(z)$ such that B_1, B_2 coincide with the first two coefficients of $C_1(C_2(z))$ (cf. [**29**]). If B_3 is taken to be the third coefficient of $C_1(C_2(z))$, then B_1, B_2, B_3 satisfy the contractive substitution property.

We show another way to do the case $r = 2$, using Theorem 4.6. Suppose that B_1, B_2 $(B_1 > 0)$ satisfy the contractive substitution property and we wish to find B_3 such that B_1, B_2, B_3 satisfy the property. Theorem 4.6 asserts that such a number is in the intersection of the closed disks (4-12). The centers of the disks are candidates for points in the intersection. Calculations suggest that the value $\nu = -1/2$ is special. Solving

$$B_3 + \left(\frac{\nu - 1}{2} B_1^{-1} + B_1^{2\nu+1} \frac{\nu + 1}{1 - B_1^{2\nu+2}} \right) B_2^2 = 0$$

with $\nu = -1/2$, we get

$$B_3 = \frac{3 - 5B_1}{4B_1(1 - B_1)} B_2^2. \tag{4-20}$$

When $B_1 = 1$, in accordance with (4-18) we interpret the formula as $B_3 = 0$.

Theorem 4.7. *If B_1, B_2 $(B_1 > 0)$ satisfy the contractive substitution property and B_3 is defined by (4-20), then B_1, B_2, B_3 satisfy the property.*

Proof. If $B_1 = 1$, then $B_2 = B_3 = 0$ and the conclusion is trivial. Assume $B_1 < 1$. By Theorem 4.6, it is enough to show that (4-12) holds for all real ν. Let

$$C(z) = C_1 z + C_2 z^2 + C_3 z^3 + \cdots$$

be the Koebe mapping (4-19) with $c = B_1$ and $w = 1$. By (4-12) applied to C(z),

$$\begin{aligned} &\left| C_3 + \left(\frac{\nu - 1}{2} C_1^{-1} + C_1^{2\nu+1} \frac{\nu + 1}{1 - C_1^{2\nu+2}} \right) C_2^2 \right|^2 \\ &\leq C_1^{2\nu} \left(\frac{\nu + 1}{1 - C_1^{2\nu+2}} \right)^2 \left(C_1^{-2\nu+2} \frac{1 - C_1^{2\nu}}{\nu} \frac{1 - C_1^{2\nu+2}}{\nu + 1} - |C_2|^2 \right) \times \\ &\times \left(C_1^{-2\nu} \frac{1 - C_1^{2\nu+2}}{\nu + 1} \frac{1 - C_1^{2\nu+4}}{\nu + 2} - |C_2|^2 \right). \end{aligned} \tag{4-21}$$

We are to show that

$$\begin{aligned} &\left| B_3 + \left(\frac{\nu - 1}{2} B_1^{-1} + B_1^{2\nu+1} \frac{\nu + 1}{1 - B_1^{2\nu+2}} \right) B_2^2 \right|^2 \\ &\leq B_1^{2\nu} \left(\frac{\nu + 1}{1 - B_1^{2\nu+2}} \right)^2 \left(x(\nu - 1) - |B_2|^2 \right) \left(x(\nu) - |B_2|^2 \right), \end{aligned}$$

where

$$x(\nu) = B_1^{-2\nu} \frac{1 - B_1^{2\nu+2}}{\nu + 1} \frac{1 - B_1^{2\nu+4}}{\nu + 2}.$$

By (4-20), this is trivial if $B_2 = 0$. If $B_2 \neq 0$, the inequality is the same as

$$\left| \frac{3 - 5B_1}{4B_1(1 - B_1)} + \left(\frac{\nu - 1}{2} B_1^{-1} + B_1^{2\nu+1} \frac{\nu + 1}{1 - B_1^{2\nu+2}} \right) \right|^2 \leq B_1^{2\nu} \left(\frac{\nu + 1}{1 - B_1^{2\nu+2}} \right)^2 \left(\frac{x(\nu - 1)}{|B_2|^2} - 1 \right) \left(\frac{x(\nu)}{|B_2|^2} - 1 \right).$$

In view of (4-18), it is enough to prove

$$\left| \frac{3 - 5B_1}{4B_1(1 - B_1)} + \left(\frac{\nu - 1}{2} B_1^{-1} + B_1^{2\nu+1} \frac{\nu + 1}{1 - B_1^{2\nu+2}} \right) \right|^2 \leq B_1^{2\nu} \left(\frac{\nu + 1}{1 - B_1^{2\nu+2}} \right)^2 \left(\frac{x(\nu - 1)}{4B_1^2(1 - B_1)^2} - 1 \right) \left(\frac{x(\nu)}{4B_1^2(1 - B_1)^2} - 1 \right).$$

This holds by (4-21) because $B_1 = C_1$ and

$$C_3 = \frac{3 - 5C_1}{4C_1(1 - C_1)} C_2^2$$

by (4-19). ■

Theorem 4.6 implies an inequality for the coefficients of powers of Riemann mappings.

Theorem 4.8. *Let $B(z)$ be a normalized Riemann mapping of the unit disk into itself, and let*

$$B(z)^\nu = \sum_{n=0}^{\infty} P_n(\nu) z^{\nu+n}.$$

If $\nu \geq -1$ and $n \geq 2$, then

$$\left| \frac{P_n(\nu)}{\nu} \right| \leq \sqrt{\frac{1 - B'(0)^{2\nu}}{\nu}} \sqrt{\frac{1 - B'(0)^{2\nu+2n}}{\nu + n}}.$$

Proof. The conclusion is trivial if $B_1 = 1$. Assume $B_1 < 1$, and write $n = r + 1$, $r \geq 1$. We apply Theorem 4.6 with unit weights $\sigma_n \equiv 1$. Set

$$a = \sqrt{\frac{1 - B_1^{2\nu}}{\nu}}, \qquad b = \sqrt{\frac{1 - B_1^{2\mu}}{\mu}}$$

and

$$u(z) = \frac{B(z)^\nu - B'(0)^\nu z^\nu}{\nu},$$
$$v(z) = \frac{B^*(z)^\mu - B'(0)^\mu z^\mu}{\mu},$$

where $\mu = -\nu - r - 1$. Then by (4-10),

$$\left| \frac{P_n(\nu)}{\nu B_1^{\nu+n}} - \langle L_1 u, v\rangle_{\tilde{\mathcal{D}}_1} \right|^2 \le \left(a^2 - \|u\|^2_{\mathcal{D}_1}\right)\left(b^2 - \|v\|^2_{\tilde{\mathcal{D}}_1}\right),$$

where L_1 acts from $\mathcal{D}_1 = \mathfrak{G}^\nu_\sigma(B)$ to $\tilde{\mathcal{D}}_1 = \mathfrak{G}^\mu_\tau(B^*)$ as described in Theorem 4.6. Hence

$$\left| \frac{P_n(\nu)}{\nu B_1^{\nu+n}} \right| \le \|L_1 u\|_{\tilde{\mathcal{D}}_1} \|v\|_{\tilde{\mathcal{D}}_1} + \sqrt{a^2 - \|u\|^2_{\mathcal{D}_1}}\,\sqrt{b^2 - \|v\|^2_{\tilde{\mathcal{D}}_1}}\,.$$

By Theorem 4.3,

$$L_1^* L_1 = 1 - D_1^* D_1,$$

where D_1 is the inclusion of $\mathcal{D}_1$ in $\mathfrak{G}^\nu_\sigma$. Since we assume that $\nu \ge -1$, $\mathfrak{G}^\nu_\sigma$ is a Hilbert space. Therefore

$$\begin{aligned} \|L_1 u\|^2_{\tilde{\mathcal{D}}_1} &= \langle L_1^* L_1 u, u\rangle_{\mathcal{D}_1} \\ &= \langle (1 - D_1^* D_1) u, u\rangle_{\mathcal{D}_1} \\ &= \|u\|^2_{\mathcal{D}_1} - \|u\|^2_{\mathfrak{G}^\nu_\sigma} \\ &\le \|u\|^2_{\mathcal{D}_1}, \end{aligned}$$

and so

$$\left| \frac{P_n(\nu)}{\nu B_1^{\nu+n}} \right| \le \|u\|_{\mathcal{D}_1} \|v\|_{\tilde{\mathcal{D}}_1} + \sqrt{a^2 - \|u\|^2_{\mathcal{D}_1}}\,\sqrt{b^2 - \|v\|^2_{\tilde{\mathcal{D}}_1}}\,.$$

The maximum of

$$f(x,y) = xy + \sqrt{a^2 - x^2}\,\sqrt{b^2 - y^2}$$

on the rectangle $0 \le x \le a$, $0 \le y \le b$ occurs on the diagonal $y = bx/a$, $0 \le x \le a$. For all points on the diagonal,

$$f(x,y) = ab.$$

We obtain

$$\left| \frac{P_n(\nu)}{\nu B_1^{\nu+n}} \right| \le ab = \sqrt{\frac{1 - B_1^{2\nu}}{\nu}}\,\sqrt{\frac{1 - B_1^{-2\nu-2r-2}}{-\nu - r - 1}},$$

which is equivalent to the stated result. ■

Appendix. Formal algebra

We view generalized power series $f(z) = \sum_{n=1}^{\infty} a_n z^{\nu+n}$ as formal objects depending on a real parameter ν. Elementary properties can be derived by writing $f(z) = z^\nu f_0(z)$ where $f_0(z)$ is a formal power series and using properties of formal power series. An infinite series $\sum_{n=1}^{\infty} f_n(z)$ of generalized power series has **formal sum** $f(z)$ if the partial sums converge formally to $f(z)$. Formal convergence of a sequence of generalized power series means convergence of coefficients.

Exponentiation and composition require care. Given a formal power series $B(z) = B_1 z + B_2 z^2 + \cdots$ $(B_1 > 0)$, define

$$\begin{aligned} B(z)^\nu &= B_1^\nu z^\nu \left[1 + (B_1^{-1} B_2 z + B_1^{-1} B_3 z^2 + \cdots)\right]^\nu \\ &= B_1^\nu z^\nu \sum_{n=0}^{\infty} \binom{\nu}{n} (B_1^{-1} B_2 z + B_1^{-1} B_3 z^2 + \cdots)^n . \end{aligned}$$

The sum has the form

$$B(z)^\nu = \sum_{n=0}^{\infty} P_n(\nu)\, z^{\nu+n},$$

where for each n, $P_n(\nu)/B_1^\nu$ is a polynomial of degree n in ν depending on the coefficients $B_1, \dots, B_{n+1}$. Special cases are $B(z)^0 = 1$ and $B(z)^1 = B(z)$. For all real μ and ν,

$$\begin{aligned} B(z)^\mu B(z)^\nu &= B_1^{\mu+\nu} z^{\mu+\nu} \sum_{m=0}^{\infty} \binom{\mu}{m} \left[\frac{B(z) - B_1 z}{B_1 z}\right]^m \sum_{n=0}^{\infty} \binom{\nu}{n} \left[\frac{B(z) - B_1 z}{B_1 z}\right]^n \\ &= B_1^{\mu+\nu} z^{\mu+\nu} \sum_{p=0}^{\infty} \sum_{\substack{m+n=p \\ m,n\geq 0}} \binom{\mu}{m}\binom{\nu}{n} \left[\frac{B(z) - B_1 z}{B_1 z}\right]^p \\ &= B_1^{\mu+\nu} z^{\mu+\nu} \sum_{p=0}^{\infty} \binom{\mu+\nu}{p} \left[\frac{B(z) - B_1 z}{B_1 z}\right]^p \\ &= B(z)^{\mu+\nu}, \end{aligned}$$

by properties of formal power series [28, Ch. 1]. If $f(z) = \sum_{n=1}^{\infty} a_n z^{\nu+n}$, define $f(B(z)) = \sum_{n=1}^{\infty} a_n B(z)^{\nu+n}$. Substitution is well behaved with respect to addition, multiplication, and formal differentiation: if $f(z) = \sum_{n=1}^{\infty} a_n z^{\nu+n}$, then $f'(z) = \sum_{n=1}^{\infty} (\nu+n)\, a_n z^{\nu+n-1}$. For example, if $g(z) = f(B(z))$, then $g'(z) = f'(B(z))B'(z)$. Observe that z^{-1} can appear in $f'(z)$ only with coefficient zero.

It is important to know that the expression $A(B(z))^\nu$ is unambiguous for any $A(z)$ and $B(z)$.

Theorem A. *Let $A(z) = A_1 z + A_2 z^2 + \cdots$ and $B(z) = B_1 z + B_2 z^2 + \cdots$ be formal power series such that $A_1 > 0$ and $B_1 > 0$. If $C(z) = A(B(z))$ and $D(z) = A(z)^\nu$, then $C(z)^\nu = D(B(z))$.*

Proof. The result holds for $\nu + 1$ if it holds for ν. Since the theorem is clearly true for $\nu = 1$, it is true for every positive integer ν. By the definition of composition, $D(B(z))$ and $C(z)^\nu$ have the form

$$D(B(z)) = (A_1 B_1)^{\nu-1} \left[\varphi_0(\nu) z^\nu + \varphi_1(\nu) z^{\nu+1} + \cdots\right],$$
$$C(z)^\nu = (A_1 B_1)^{\nu-1} \left[\psi_0(\nu) z^\nu + \psi_1(\nu) z^{\nu+1} + \cdots\right],$$

where $\varphi_0, \varphi_1, \ldots$ and $\psi_0, \psi_1, \ldots$ are polynomials in ν. Since the result holds for positive integer values of ν, $\varphi_n(\nu)$ and $\psi_n(\nu)$ are equal identically in ν for all n. Therefore the generalized power series coincide. ∎

The **compositional inverse** of $B(z) = B_1 z + B_2 z^2 + \cdots$ $(B_1 > 0)$ is the unique series $A(z) = A_1 z + A_2 z^2 + \cdots$ $(A_1 > 0)$ such that $A(B(z)) = B(A(z)) = z$ (see [**28, Ch. 1**]). We give a formal proof of the Lagrange-Bürmann identity for generalized power series [**34**].

Theorem B. *Let $B(z) = B_1 z + B_2 z^2 + \cdots$ be a formal power series such that $B_1 > 0$, and let $A(z)$ be its compositional inverse. If*

$$A(z)^\nu = \sum_{n=0}^{\infty} Q_n(\nu)\, z^{\nu+n} \qquad \text{and} \qquad B(z)^\nu = \sum_{n=0}^{\infty} P_n(\nu)\, z^{\nu+n},$$

then for all nonnegative integers n,

$$Q_n(\nu) = \frac{\nu}{\nu+n} P_n(-\nu - n).$$

Proof. By Theorem A,

$$z^\nu = B(A(z))^\nu = \sum_{n=0}^{\infty} P_n(\nu) A(z)^{\nu+n}$$
$$= \sum_{n=0}^{\infty} P_n(\nu) \sum_{j=0}^{\infty} Q_j(\nu+n) z^{\nu+n+j} = \sum_{p=0}^{\infty} \left\{ \sum_{n=0}^{p} P_n(\nu) Q_{p-n}(\nu+n) \right\} z^{\nu+p},$$

and so

$$(A-1) \qquad \sum_{n=0}^{p} P_n(\nu) Q_{p-n}(\nu+n) = \delta_{p0}, \qquad p = 0, 1, 2, \ldots .$$

For any $p = 0, 1, 2, \dots$, the coefficient of z^{-1} in the expression

$$B(z)^{-\nu-p}\frac{d}{dz}B(z)^{\nu} = \sum_{j=0}^{\infty} P_j(-\nu-p)z^{-\nu-p+j}\sum_{k=0}^{\infty}(\nu+k)P_k(\nu)z^{\nu+k-1}$$

is equal to $\sum_{n=0}^{p}(\nu+n)P_{p-n}(-\nu-p)P_n(\nu)$. When $p = 0$, the coefficient of z^{-1} is ν because

$$B(z)^{-\nu}\frac{d}{dz}B(z)^{\nu} = \nu\, B(z)^{-1}B'(z).$$

When $p > 0$, the coefficient of z^{-1} is 0, because in this case,

$$B(z)^{-\nu-p}\frac{d}{dz}B(z)^{\nu} = \nu\, B(z)^{-p-1}B'(z) = -\frac{\nu}{p}\frac{d}{dz}B(z)^{-p}$$

is the derivative of a generalized power series. It follows that

$$(A-2) \qquad \sum_{n=0}^{p}(\nu+n)P_{p-n}(-\nu-p)P_n(\nu) = (\nu+p)\delta_{p0}, \qquad p = 0, 1, 2, \dots .$$

Comparison of (A-1) and (A-2) shows that $Q_n(\nu)$ and $P_n(-\nu-n)\,\nu/(\nu+n)$ satisfy a system of linear equations which has a unique solution, yielding the result. ■

The coefficients in the expansion $B(z)^{\nu} = \sum_{n=0}^{\infty} P_n(\nu)\, z^{\nu+n}$ are

$$P_0(\nu) = B_1^{\nu}$$

$$P_1(\nu) = B_1^{\nu}\left[B_1^{-1}B_2\,\frac{\nu}{1}\right]$$

$$P_2(\nu) = B_1^{\nu}\left[B_1^{-1}B_3\,\frac{\nu}{1} + B_1^{-2}B_2^2\,\frac{\nu}{1}\frac{\nu-1}{2}\right]$$

$$P_3(\nu) = B_1^{\nu}\left[B_1^{-1}B_4\,\frac{\nu}{1} + 2B_1^{-2}B_2B_3\,\frac{\nu}{1}\frac{\nu-1}{2} + B_1^{-3}B_2^3\,\frac{\nu}{1}\frac{\nu-1}{2}\frac{\nu-2}{3}\right]$$

$$\dots$$

Other explicit and recursive formulas for the coefficients may be given. Henrici [28, Ch. 1] calls one of these identities the Euler-J. C. P. Miller formula.

Theorem C. *Let $B(z) = B_1 z + B_2 z^2 + \cdots$ be a formal power series such that $B_1 > 0$. If $B(z)^{\nu} = \sum_{n=0}^{\infty} P_n(\nu)\, z^{\nu+n}$, then $P_0(\nu) = B_1^{\nu}$ and for all $n \geq 1$,*

$$P_n(\nu) = \frac{1}{nB_1}\sum_{j=0}^{n-1}[(n-j)\nu - j]B_{n-j+1}P_j(\nu).$$

Proof. The relation $P_0(\nu) = B_1^\nu$ has already been noted as a consequence of the definition of substitution. In the identity

$$B(z)\, z\frac{d}{dz}B(z)^\nu = \nu\, B(z)^\nu\, z\frac{d}{dz}B(z),$$

compare coefficients of $z^{\nu+n+1}$ to get

$$\sum_{j=0}^{n}(\nu+j)B_{n-j+1}P_j(\nu) = \sum_{j=0}^{n}\nu\,(n-j+1)B_{n-j+1}P_j(\nu).$$

This relation is equivalent to the stated result. ■

References

[1] Gr. Arsene, Z. Ceauşescu, and C. Foiaş, "On intertwining dilations. VIII," J. Operator Theory 4 (1980), 55–91. MR 82d:47013.

[2] Gr. Arsene, Z. Ceauşescu, and T. Constantinescu, "Multiplicative properties in the structure of contractions," Oper. Theory: Adv. Appl., Vol. 19, pp. 9–22, Birkhäuser, Basel-Boston, 1986. MR 89a:47025.

[3] Gr. Arsene, T. Constantinescu, and A. Gheondea, "Lifting of operators and prescribed numbers of negative squares," Michigan J. Math. 34 (1987), 201–216. MR 88j:47008.

[4] D. Alpay and H. Dym, "On applications of reproducing kernel spaces to the Schur algorithm and rational J unitary factorization," Oper. Theory: Adv. Appl., Vol. 18, pp. 89–159, Birkhäuser, Basel-Boston, 1986. MR 89g:46051.

[5] M. Bakonyi and T. Constantinescu, *Schur's Algorithm and Several Applications,* Monografii Matematice, Universitatea din Timişoara, Timişoara, 1989.

[6] J. A. Ball and N. Cohen, "de Branges-Rovnyak operator models and systems theory: a survey," Oper. Theory: Adv. Appl., Vol. 50, pp. 93-136, Birkhäuser, Basel-Boston, 1991.

[7] J. Bognár, *Indefinite Inner Product Spaces,* Springer, Berlin-New York, 1974. MR 57#7125.

[8] L. de Branges, " A proof of the Bieberbach conjecture," Acta Math. 154 (1985), 137–152. MR 86h:30026.

[9] L. de Branges, "Unitary linear systems whose transfer functions are Riemann mapping functions," Integral Equations and Operator Theory 19 (1986), 105–124. MR 88k:47011.

[10] L. de Branges, "Powers of Riemann mapping functions," The Bieberbach Conjecture (West Lafayette, Ind., 1985), pp. 51–67, Math. Surveys Monographs 21, Amer. Math. Soc., Providence, 1986. MR 88j:30034.

[11] L. de Branges, "Complementation in Krein spaces," Trans. Amer. Math. Soc. 305 (1988), 277–291. MR 89c:46034.

[12] L. de Branges, "Underlying concepts in the proof of the Bieberbach conjecture," Proceedings of the International Congress of Mathematicians 1986, pp.25–42, Berkeley, California, 1986. MR 89f:30029.

[13] L. de Branges, *Square Summable Power Series,* in preparation.

[14] L. de Branges and J. Rovnyak, *Square Summable Power Series*, Holt, Rinehart and Winston, New York, 1966. MR 35 # 5909.

[15] M. S. Brodskiĭ, "Unitary operator colligations and their characteristic functions," Uspehi Mat. Nauk 33, no. 4 (202) (1978), 141–168, 256; English transl. Russian Math. Surveys 33, no. 4 (1978), 159–191. MR 80e:47010.

[16] B. Ćurgus, A. Dijksma, H. Langer, and H. S. V. de Snoo, "Characteristic functions of unitary colligations and of bounded operators in Krein spaces," Oper. Theory: Adv. Appl., Vol. 41, pp. 125–152, Birkhäuser, Basel-Boston, 1989. MR 91c:47020.

[17] A. Dijksma, H. Langer, and H. S. V. de Snoo, "Unitary colligations in Krein spaces and their role in the extension theory of isometries and symmetric linear relations in Hilbert spaces," Functional Analysis, II (Dubrovnic 1985), pp. 1–42, Lecture Notes in Math., Vol. 1242, Springer, Berlin-New York, 1987; MR 89a:47055.

[18] M. A. Dritschel, "The essential uniqueness property for linear operators in Kreĭn spaces," preprint, 1990.

[19] M. A. Dritschel and J. Rovnyak, "Extension theorems for contraction operators on Kreĭn spaces," Oper. Theory: Adv. Appl., Vol. 47, pp. 221–305, Birkhäuser, Basel-Boston, 1990; MR 92m:47068.

[20] M. A. Dritschel and J. Rovnyak, "Julia operators and complementation in Kreĭn spaces," Indiana U. Math. J. 40 (1991), 885–901.

[21] V. K. Dubovoj, B. Fritsche, and B. Kirstein, *Matricial Version of the Classical Schur Problem,* Teubner-Texte zur Mathematik, Bd. 129, B. G. Teubner Verlagsgesellschaft, Leipzig, 1992.

[22] P. Duren, *Univalent Functions,* Springer-Verlag, New York-Berlin, 1983. MR 85j:30034.

[23] H. Dym, *J Contractive Matrix Functions, Reproducing Kernel Hilbert Spaces, and Interpolation,* CBMS Regional Conference Series in Mathematics, Vol. 71, Amer. Math. Soc., Providence, 1989. MR 90g:47003.

[24] C. Foiaş and A. Frazho, *The Commutant Lifting Approach to Interpolation Problems,* Oper. Theory: Adv. Appl., Vol. 44, Birkhäuser, Basel-Boston, 1990.

[25] Y. Genin, P. Van Dooren, T. Kailath, J.-M. Delosme, and M. Morf, "On Σ-lossless transfer functions and related questions," Linear Algebra Appl. 50 (1983), 251–275. MR 85g:93024.

[26] I. Gohberg et al., *I. Schur methods in operator theory and signal processing,* Oper. Theory: Adv. Appl., Vol. 18, Birkhäuser, Basel-Boston, 1986. MR 88d:00006.

[27] L. A. Harris, "Linear fractional transformations of circular domains in operator spaces," Indiana U. Math. J. 41 (1992), 125-147.

[28] P. Henrici, *Applied and Computational Complex Analysis,* Vol. 1, John Wiley & Sons, New York, 1974. MR 51 # 8378.

[29] Kin Y. Li, "de Branges' conjecture on bounded Riemann mapping coefficients," J. Analyse Math. 54 (1990), 289–296. MR 91a:30013.

[30] Kin Y. Li and James Rovnyak, "On the coefficients of Riemann mappings of the unit disk into itself," Oper. Theory: Adv. Appl., to appear.
[31] M. S. Livšic and V. P. Potapov, "A theorem on the multiplication of characteristic matrix functions," Dokl. Akad. Nauk SSSR (N.S.) 72 (1950), 625–628. MR 11, 669.
[32] N. K. Nikol'skiĭ and V. I. Vasyunin, "Quasiorthogonal decompositions in complementary metrics and estimates of univalent functions," Algebra and Analysis 2 (4) (1990), 1–81; Engl. transl., Leningrad Math. J. 2 (1991), 691–764, and "Operator-valued measures and coefficients of univalent functions", Algebra and Analysis 3 (6) (1991), 1-75; Engl. transl., St. Petersburg Math. J. 3 (6) (1992).
[33] G. Pick, "Über die konforme Abbildung eines Kreises auf ein schlichtes und zugleich beschränktes Gebiet," S.-B. Kaiserl. Akad. Wiss. Wien 126 (1917), 247–263.
[34] J. Rovnyak, "Coefficient estimates for Riemann mapping functions," J. Analyse Math. 52 (1989), 53–93. MR 90f:30030.

Gene Christner and James Rovnyak
Department of Mathematics
Mathematics-Astronomy Building
University of Virginia
Charlottesville, Virginia 22903-3199
U. S. A.

Kin Y. Li
Department of Mathematics
Hong Kong University of Science and Technology
Clear Water Bay
Kowloon
Hong Kong

Operator Theory:
Advances and Applications, Vol. 73

SHIFTS, REALIZATIONS AND INTERPOLATION, REDUX

Harry Dym*

Dedicated to M.S. Livšic with respect, admiration and affection.

Finite dimensional spaces of vector valued meromorphic functions which are invariant with respect to the generalized backward shift operator play a central, though oft unrecognized, role in many problems of analysis. In this paper we shall review a number of applications of such spaces: to realization theory, to finite dimensional reproducing kernel Hilbert (and Krein) spaces, to interpolation theory, and to the Livšic theory of characteristic functions. A list of basic formulas in which the shift intervenes is also provided.

TABLE OF CONTENTS

*The author wishes to thank Renee and Jay Weiss for endowing the chair which supported this research.

1. INTRODUCTION

The objective of this paper is to summarize in one convenient place a number of the many useful properties and applications of the generalized backward shift operator R_α, which is defined on matrix valued functions $f(\lambda)$ of the complex variable λ by the rule

$$(R_\alpha f)(\lambda) = \begin{cases} \dfrac{f(\lambda) - f(\alpha)}{\lambda - \alpha} & \text{if } \lambda \neq \alpha , \\ f'(\alpha) & \text{if } \lambda = \alpha . \end{cases}$$

Here it is assumed that f is analytic in some open nonempty subset of $\mathbb{C}$, which includes the point α. The operator R_α is a generalization of the familiar backwards shift operator

$$(R_0 f)(\lambda) = \frac{f(\lambda) - f(0)}{\lambda} \qquad (\lambda \neq 0) ,$$

which appears extensively in the theory of the Hardy space $H_2(\mathbb{D})$ over the open unit disc $\mathbb{D}$, but is of little use for spaces of functions which are not analytic at zero, such as the Hardy spaces $H_2(\mathbb{C}_+)$ and $H_2(\Pi_+)$, over the open upper halfplane $\mathbb{C}_+$ and the open right halfplane Π_+, respectively. The main theme is the central role played by finite dimensional R_α invariant subspaces in the theory of realization and interpolation. The scope is pretty clear from the table of contents.

By its very nature this paper is largely expository. Most of the material under discussion is known in one form or another; the innovation if any, is in the viewpoint, although in a number of spots new proofs had to be invented to make things fit together. Particular effort has been invested to clarify the connections between the author's own approach to interpolation (via structured reproducing kernel Hilbert spaces of the de Branges type) as developed in [D2] and [D3] with methods based on realization theory, as in the monograph [BGR] of Ball, Gohberg and Rodman. Enroute the equivalence between the de Branges structural identity, specialized to finite dimensional R_α invariant reproducing kernel Hilbert spaces and the Liapunov/Stein equation for the Grammian of the space, which was established in [D3], is given a simpler proof. A number of other calculations from that paper are also simplified.

Finally, in the last section, we touch upon the theory of characteristic functions which was invented by M.S. Livšic almost fifty years ago. Given the nature of this volume, this seems a particularly fitting place to pause. There is much more to be said, but this lies in the future: [D8], if the forces of evil can be overcome.

Throughout this paper we shall work with a flexible notation which allows us to work with functions defined on each of the regions $\mathbb{D}$, $\mathbb{C}_+$, and Π_+, more or less simultaneously. The rules are spelled out in the following table:

Ω_+	$\mathbb{D}$	$\mathbb{C}_+$	Π_+
$\rho_\omega(\lambda)$	$1-\lambda\omega^*$	$-2\pi i(\lambda-\omega^*)$	$2\pi(\lambda+\omega^*)$
$\langle f,g\rangle$	$\frac{1}{2\pi}\int_0^{2\pi} g(e^{i\theta})^* f(e^{i\theta})d\theta$	$\int_{-\infty}^{\infty} g(x)^* f(x)dx$	$\int_{-\infty}^{\infty} g(iy)^* f(iy)dy$
$b_\omega(\lambda)$	$(\lambda-\omega)/(1-\lambda\omega^*)$	$(\lambda-\omega)/(\lambda-\omega^*)$	$(\lambda-\omega)/(\lambda+\omega^*)$
Ω_0	$\mathbb{T}$	$\mathbb{R}$	$i\mathbb{R}$
λ°	$1/\lambda^*$	λ^*	$-\lambda^*$
$f^\#(\lambda)$	$f(\lambda^\circ)^*$	$f(\lambda^\circ)^*$	$f(\lambda^\circ)^*$
$a(\lambda)$	1	$\sqrt{\pi}(1-i\lambda)$	$\sqrt{\pi}(1+\lambda)$
$b(\lambda)$	λ	$\sqrt{\pi}(1+i\lambda)$	$\sqrt{\pi}(1-\lambda)$
$ab'-ba'$	1	$2\pi i$	-2π
$\varphi_{j,\omega}(\lambda)$	$\lambda^j/(1-\lambda\omega^*)^{j+1}$	$-1/2\pi i(\lambda-\omega^*)^{j+1}$	$(-1)^j/2\pi(\lambda+\omega^*)^{j+1}$
$(R_\alpha\rho_\omega^{-1})(\lambda)$	$\omega^*/\rho_\omega(\alpha)\rho_\omega(\lambda)$	$2\pi i/\rho_\omega(\alpha)\rho_\omega(\lambda)$	$-2\pi/\rho_\omega(\alpha)\rho_\omega(\lambda)$

TABLE 1.1

For each of the three listed choices of the kernel function $\rho_\omega(\lambda)$,

$$\Omega_+ = \{\omega \in \mathbb{C} : \rho_\omega(\omega) > 0\}$$

and

$$\Omega_0 = \{\omega \in \mathbb{C} : \rho_\omega(\omega) = 0\} .$$

We shall take

$$\Omega_- = \{\omega \in \mathbb{C} : \rho_\omega(\omega) < 0\} .$$

The use of such a flexible notation to cover problems in both $\mathbb{D}$ and $\mathbb{C}_+$ more or less simultaneously was promoted in [AD1] and [D2]. The observation that the kernels $\rho_\omega(\lambda)$ which intervene in these problems can be expressed in terms of a pair of polynomials $a(\lambda)$ and $b(\lambda)$ as

$$\rho_\omega(\lambda) = a(\lambda)a(\omega)^* - b(\lambda)b(\omega)^* \tag{1.1}$$

is due Lev-Ari and Kailath [LAK]. (A.A. Nudelman reported on some related use of the form (1.1) in the former Soviet Union [Nu3], but I do not have exact references.) They noticed that certain fast algorithms in which the term $\rho_\omega(\lambda)$ intervenes will work if and only if $\rho_\omega(\lambda)$ can be expressed in the form (1.1). A general theory of reproducing kernels with denominators of this form and their applications was developed in [AD4], [AD5] and [AD6].

The notation is fairly standard: The symbol A^* denotes the adjoint of an operator A on a Hilbert space, with respect to the inner product of the space. If A is a finite matrix, then the adjoint will always be computed with respect to the standard inner product so that, in this case, A^* will be the Hermitian transpose, or just the complex conjugate if A is a number. We shall let $H_2^{m\times n}(\Omega_+)$ denote the set of $m \times n$ matrix valued functions with entries which belong to the Hardy space $H_2(\Omega_+)$, and shall abbreviate $H_2^{m\times 1}$ by H_2^m. The symbol $\underline{\underline{p}}$ [resp. $\underline{\underline{q}}$] will designate the orthogonal projection of $L_2^k(\Omega_0)$ onto $H_2^k(\Omega_+)$ [resp. $H_2^k(\Omega_-)$] without regard to the value of k; we shall also set

$$\underline{\underline{q}}' = I - \underline{\underline{p}} ,$$

the orthogonal projection onto the orthogonal complement $\{H_2^k(\Omega_+)\}^\perp$ of $H_2^k(\Omega_+)$ in $L_2^k(\Omega_0)$. Clearly $q' = \underline{\underline{q}}$ if $\Omega_+ = \mathbb{C}_+$ or $\Omega_+ = \Pi_+$, but not if $\Omega_+ = \mathbb{D}$.

The symbol $\mathcal{S}^{p\times q}(\Omega_+)$ designates the Schur class of $p \times q$ matrix valued functions which are both analytic and contractive in Ω_+.

2. FORMULAS AND FACTS

In this section we present a number of formulas and facts involving R_α.

2.1. Formulas

Let f and g be matrix valued functions which are analytic in some open nonempty subset Δ of $\mathbb{C}$, and let α, β be points in Δ. Then:

The j'th iterate, $j = 1, 2, \ldots$,

$$(R_\alpha^j f)(\lambda) = \frac{f(\lambda) - f(\alpha) - \cdots - \frac{f^{(j-1)}(\alpha)}{(j-1)!}(\lambda - \alpha)^{j-1}}{(\lambda - \alpha)^j} \quad \text{if } \lambda \neq \alpha , \tag{2.1}$$

and

$$(R_\alpha^j f)(\alpha) = \frac{f^{(j)}(\alpha)}{j!} . \tag{2.2}$$

If the multiplication fg is meaningful, then

$$\{R_\alpha(fg)\}(\lambda) = f(\lambda)(R_\alpha g)(\lambda) + (R_\alpha f)(\lambda)g(\alpha) \tag{2.3}$$

and

$$\{R_\alpha(fg)\}(\lambda) = (R_\alpha f)(\lambda)g(\lambda) + f(\alpha)(R_\alpha g)(\lambda) . \tag{2.4}$$

If f is invertible at α, then

$$(R_\alpha f^{-1})(\lambda) = -f^{-1}(\lambda)(R_\alpha f)(\lambda)f^{-1}(\alpha) = -f^{-1}(\alpha)(R_\alpha f)(\lambda)f^{-1}(\lambda) \,. \tag{2.5}$$

The resolvent identity

$$R_\alpha f - R_\beta f = (\alpha - \beta)R_\alpha R_\beta f \tag{2.6}$$

holds, and may be used to verify that

$$R_\alpha R_\beta f = R_\beta R_\alpha f \,. \tag{2.7}$$

2.2. Evaluations in $H_2^m(\Omega_+)$

LEMMA 2.1. *The Hardy space $H_2^m(\Omega_+)$ is R_α invariant for every point $\alpha \in \Omega_+$, and*

$$\|R_\alpha f\| \leq \{d(\alpha, \Omega_0)\}^{-1}\|f\| \,, \tag{2.8}$$

wherein $d(\alpha, \Omega_0)$ designates the distance from the point $\alpha \in \Omega_+$ to Ω_0, i.e., $1 - |\alpha|$, $(\alpha - \alpha^*)/2i$, or $(\alpha + \alpha^*)/2$, according as Ω_+ is $\mathbb{D}$, $\mathbb{C}_+$ or Π_+.

The next set of evaluations depend upon the fact that

$$k_\omega(\lambda) = I_m/\rho_\omega(\lambda) \tag{2.9}$$

is a reproducing kernel for the Hardy space $H_2^m(\Omega_+)$. This means that

$$(1) \quad k_\omega x \in H_2^m(\Omega_+) \,, \qquad \text{and} \qquad (2) \quad \langle f, k_\omega x\rangle = x^* f(\omega) \,, \tag{2.10}$$

for every choice of $\omega \in \Omega_+$, $x \in \mathbb{C}^m$ and $f \in H_2^m(\Omega_+)$. The verification of (1) is easy, whereas (2) is equivalent to Cauchy's formula for these spaces. Similarly, the kernels based on the functions

$$\varphi_{j,\omega}(\lambda) = \left\{\frac{1}{j!}\left(\frac{\partial}{\partial\omega^*}\right)^j \frac{1}{\rho_\omega}\right\}(\lambda) \tag{2.11}$$

serve to evaluate derivatives:

$$\langle f, \varphi_{j,\omega} x\rangle = \frac{x^* f^{(j)}(\omega)}{j!} \,, \tag{2.12}$$

for every choice of $\omega \in \Omega_+$, $x \in \mathbb{C}^m$ and $f \in H_2^m(\Omega_+)$. Explicit formulas for $\varphi_{j,\omega}(\lambda)$ for each of the three cases of interest are given in Table 1.1.

LEMMA 2.2. *If* $f \in H_\infty^m(\Omega_+)$, *then* $(R_\alpha f) \in H_2^m(\Omega_+)$ *for every choice of* $\alpha \in \Omega_+$.

PROOF. The proof is straightforward, and is left to the reader. ∎

LEMMA 2.3. *If* $F \in H_\infty^{p\times q}(\Omega_+)$, *then*

$$\{\underline{\underline{p}}F^*\varphi_{k,\omega}\xi\}(\lambda) = \sum_{j=0}^{k}(R_\omega^{k-j}F)(\omega)^*\varphi_{j,\omega}(\lambda)\xi \tag{2.13}$$

for $k = 0, 1, \ldots$, *and*

$$\{\underline{\underline{p}}F\frac{\eta}{(\lambda-\omega)^k}\}(\lambda) = (R_\omega^k F)(\lambda)\eta \tag{2.14}$$

for $k = 1, 2, \ldots$, *for every choice of* $\omega \in \Omega_+$, $\xi \in \mathbb{C}^p$ *and* $\eta \in \mathbb{C}^q$. *In (2.13) we adopt the convention that*

$$(R_\omega^{k-j}F)(\omega) = \frac{F^{(k-j)}(\omega)}{(k-j)!} \tag{2.15}$$

for $j = k$ *too, which is consistent with formula (2.2).*

PROOF. To obtain (2.13), let $\alpha \in \Omega_+$ and observe that

$$\begin{aligned}\eta^*(\underline{\underline{p}}F^*\varphi_{k,\omega}\xi)(\alpha) &= \langle F^*\varphi_{k,\omega}\xi, \frac{\eta}{\rho_\alpha}\rangle \\ &= \langle \frac{F\eta}{\rho_\alpha}, \varphi_{k,\omega}\xi\rangle^* \\ &= \left\{\xi^*\frac{1}{k!}\left(\frac{F}{\rho_\alpha}\right)^{(k)}(\omega)\eta\right\}^*\end{aligned}$$

for every choice of $\alpha \in \Omega_+$ and $\eta \in \mathbb{C}^q$. By Leibnitz's formula

$$\left(\frac{F}{\rho_\alpha}\right)^{(k)}(\omega) = \sum_{j=0}^{k}\binom{k}{j}F^{(k-j)}(\omega)\left(\frac{1}{\rho_\alpha}\right)^{(j)}(\omega)\,.$$

Therefore, since

$$\left\{\frac{1}{j!}\left(\frac{1}{\rho_\alpha}\right)^{(j)}(\omega)\right\}^* = \varphi_{j,\omega}(\alpha)\,, \tag{2.16}$$

the rest of the proof of (2.13) reduces to a routine calculation.

Formula (2.14) is clear from the decomposition

$$\frac{F(\lambda)\eta}{(\lambda-\omega)^k} = (R_\omega^k F)(\lambda)\eta + \frac{1}{(\lambda-\omega)^k}\sum_{j=0}^{k-1}(\lambda-\omega)^j\frac{F^{(j)}(\omega)}{j!}\eta\,, \tag{2.17}$$

since the first term on the right belongs to $H_2^m(\Omega_+)$, thanks to Lemma 2.1, and the second belongs to $(H_2^\perp)^m(\Omega_+)$. ∎

COROLLARY. *If* $F \in H_\infty^{p\times q}(\Omega_+)$, *then*

$$\underline{\underline{p}}F^*\frac{\xi}{\rho_\omega} = F(\omega)^*\frac{\xi}{\rho_\omega}\ , \tag{2.18}$$

and

$$\left\{\underline{\underline{q}}'F\frac{\eta}{(\lambda-\omega)^k}\right\}(\lambda) = \frac{1}{(\lambda-\omega)^k}\sum_{j=0}^{k-1}(\lambda-\omega)^j(R_\omega^jF)(\omega)\eta \tag{2.19}$$

for every choice of $\xi \in \mathbb{C}^p$, $\eta \in \mathbb{C}^q$, $k = 1, 2, \ldots$, *and* $\omega \in \Omega_+$.

PROOF. The first assertion is just (2.13) with $k = 0$ (which we have isolated because of its frequent use); the second is immediate from (2.17). ∎

2.3. Operators which commute with R_α

THEOREM 2.1. *Let* $F \in H_\infty^{p\times q}(\Omega_+)$. *Then the operator* R_α *commutes with the operator* $\underline{\underline{p}}F^*|_{H_2^p(\Omega_+)}$ *for every choice of* $\alpha \in \Omega_+$.

PROOF. It is readily seen via (2.18) and the bottom row of Table 1.1 that

$$\underline{\underline{p}}F^*R_\alpha f = R_\alpha \underline{\underline{p}}F^*f$$

for $\alpha \in \Omega_+$ and finite sums $f = \Sigma\xi_j/\rho_{\omega_j}$ with $\xi_j \in \mathbb{C}^p$ and $\omega_j \in \Omega_+$. The same holds for general $f \in H_2^p(\Omega_+)$ since such sums are dense in $H_2^p(\Omega_+)$. Lemma 2.1 serves to justify the requisite estimates. ∎

THEOREM 2.2. *Let* $F \in H_\infty^{p\times q}(\Omega_+)$, *and let* $\mathcal{M}$ *be any subspace of* $H_2^p(\Omega_+)$ *which is* R_α *invariant for at least one (and hence every) point* $\alpha \in \Omega_+$. *Then the space*

$$\{\underline{\underline{p}}F^*f:\ f \in \mathcal{M}\}$$

is R_α *invariant for every* $\alpha \in \Omega_+$. *In particular*

$$\ker\ \underline{\underline{p}}F^*|_{H_2^p(\Omega_+)} = \{f \in H_2^p(\Omega_+) : \underline{\underline{p}}F^*f = 0\}$$

is R_α *invariant for every* $\alpha \in \Omega_+$.

Theorem 2.2 is an immediate corollary of Theorem 2.1. The identification of the indicated kernel as an R_α invariant subspace is extremely useful in the study of assorted root distribution problems; see e.g., [D1], [D4] and [D7] ([D6] supplements the first two). The next result is a converse to Theorem 2.1.

THEOREM 2.3. *Let T be a bounded linear operator from $H_2^p(\Omega_+)$ into $H_2^q(\Omega_+)$ such that*

$$R_\alpha T = TR_\alpha \tag{2.20}$$

for some point $\alpha \in \Omega_+$. Then there exists a matrix valued function $S \in H_\infty^{p\times q}(\Omega_+)$ such that

$$Tf = \underline{\underline{p}}S^* f \tag{2.21}$$

for every $f \in H_2^p(\Omega_+)$, and

$$\|T\| = \sup_{\omega\in\Omega_+} |S(\omega)| \ , \tag{2.22}$$

where $|S(\omega)|$ stands for the maximum singular value of the matrix S at the point ω.

PROOF. By formula (2.4),

$$\begin{aligned} 0 &= \{R_\alpha(\rho_\omega \cdot \rho_\omega^{-1})\}(\lambda) \\ &= (R_\alpha\rho_\omega)(\lambda)\rho_\omega^{-1}(\lambda) + \rho_\omega(\alpha)(R_\alpha\rho_\omega^{-1})(\lambda) \end{aligned}$$

for every point α at which $\rho_\omega(\alpha) \neq 0$. This exhibits ρ_ω^{-1} as an eigenfunction of R_α for each such α with eigenvalue

$$\mu = -\frac{(R_\alpha\rho_\omega)(\lambda)}{\rho_\omega(\alpha)} \ ,$$

which is independent of λ since $\rho_\omega(\lambda)$ is linear in λ:

$$(R_\alpha\rho_\omega^{-1})(\lambda) = \mu\rho_\omega^{-1}(\lambda) \ .$$

Now let

$$F_\omega(\lambda)\xi = \left\{T\left(\frac{\xi}{\rho_\omega}\right)\right\}(\lambda)$$

for every choice of λ and ω in Ω_+ and $\xi \in \mathbb{C}^p$. Then, since R_α commutes with T, we also have

$$(R_\alpha F_\omega)(\lambda) = \mu F_\omega(\lambda) \ .$$

Therefore, by another application of (2.4),

$$\begin{aligned} \{R_\alpha(\rho_\omega F_\omega)\}(\lambda) &= (R_\alpha\rho_\omega)(\lambda)F_\omega(\lambda) + \rho_\omega(\alpha)(R_\alpha F_\omega)(\lambda) \\ &= \{(R_\alpha\rho_\omega) + \mu\rho_\omega(\alpha)\}F_\omega(\lambda) \\ &= 0 \ . \end{aligned}$$

This proves that $\rho_\omega(\lambda)F_\omega(\lambda)$ is independent of λ. Let

$$S(\omega)^* = \rho_\omega(\lambda)F_\omega(\lambda) ,$$

which is the same as to say that,

$$\left\{T\left(\frac{I_p}{\rho_\omega}\right)\right\}(\lambda) = \frac{S(\omega)^*}{\rho_\omega(\lambda)} .$$

Thus, for any choice of $f \in H_2^q(\Omega_+)$, $\xi \in \mathbb{C}^p$ and $\omega \in \Omega_+$,

$$\begin{aligned}\xi^*(T^*f)(\omega) &= \langle f, T\frac{\xi}{\rho_\omega}\rangle \\ &= \langle S(\omega)f, \frac{\xi}{\rho_\omega}\rangle \\ &= \xi^*(Sf)(\omega) .\end{aligned}$$

The choice $f = \eta/\rho_\beta$ with $\eta \in \mathbb{C}^q$ and $\beta \in \Omega_+$ yields the identity

$$\frac{\xi^* S(\omega)\eta}{\rho_\beta(\omega)} = \xi^*(T^*f)(\omega) ,$$

which serves to exhibit S as an analytic matrix valued function in Ω_+. Moreover, by Schwarz's inequality

$$\begin{aligned}\left|\frac{\xi^* S(\omega)\eta}{\rho_\beta(\omega)}\right| &\le \left\|\frac{\eta}{\rho_\beta}\right\| \; \|T\| \; \left\|\frac{\xi}{\rho_\omega}\right\| \\ &= \frac{|\eta| \; \|T\| \; |\xi|}{\{\rho_\beta(\beta)\rho_\omega(\omega)\}^{\frac{1}{2}}} .\end{aligned}$$

Thus upon setting $\beta = \omega$ and $\xi = S(\omega)\eta$, we see that

$$|S(\omega)\eta|^2 \le |\eta| \; \|T\| \; |S(\omega)\eta| ,$$

and hence that

$$|S(\omega)| \le \|T\|$$

for every point $\omega \in \Omega_+$.

On the other hand, the fact that for $f \in H_2^p(\Omega_+)$

$$\|f\| = \lim_{\varepsilon \downarrow 0} \|f_\varepsilon\| ,$$

where

$$f_\varepsilon(\lambda) = \begin{cases} f((1-\varepsilon)\lambda) & \text{if } \Omega_+ = \mathbb{D} \\ f(\lambda + i\varepsilon) & \text{if } \Omega_+ = \mathbb{C}_+ \\ f(\lambda+\varepsilon)) & \text{if } \Omega_+ = \Pi_+ , \end{cases}$$

leads to the supplementary bound

$$\begin{aligned} \|T^* f\| &= \|Sf\| \\ &= \lim_{\varepsilon \downarrow 0} \|S_\varepsilon f_\varepsilon\| \\ &\leq \sup_{\omega \in \Omega_+} |S(\omega)| \, \|f\| . \end{aligned}$$

This serves to complete the proof. ∎

The results of this section are more or less well known. Usually they are presented in adjoint form for $\Omega_+ = \mathbb{D}$, i.e., in terms of the shift and the operator M_F of multiplication by F, alias the adjoints of R_0 and $\underline{\underline{p}}F^*|_{H_2^p(\mathbb{D})}$, respectively. The present formulation has the advantage of uniformity of method, and relatively painless proofs which are easily adapted to the more general setting of [AD5]; see Theorem 3.3 therein.

3. R_α INVARIANCE

In this section we shall study finite dimensional spaces of vector valued functions which are invariant under the action of R_α for at least one appropriately chosen point $\alpha \in \mathbb{C}$.

3.1. The main observation

THEOREM 3.1. *Let $\mathcal{M}$ be an n dimensional vector space of $m \times 1$ vector valued functions which are meromorphic in some open nonempty set $\Delta \subset \mathbb{C}$ and suppose further that $\mathcal{M}$ is R_α invariant for some point $\alpha \in \Delta$ in the domain of analyticity of $\mathcal{M}$. Then $\mathcal{M}$ is spanned by the columns of a rational $m \times n$ matrix valued function of the form*

$$F(\lambda) = V\{M - \lambda N\}^{-1} , \tag{3.1}$$

where $V \in \mathbb{C}^{m\times n}$, $M, N \in \mathbb{C}^{n \times n}$,

$$MN = NM \quad \textit{and} \quad M - \alpha N = I_n . \tag{3.2}$$

Moreover, $\lambda \in \Delta$ is a point of analyticity of F if and only if the $n \times n$ matrix $M - \lambda N$ is invertible.

PROOF. Let $f_1, \ldots, f_n$ be a basis for $\mathcal{M}$ and let

$$F(\lambda) = [f_1(\lambda) \cdots f_n(\lambda)]$$

be the $m \times n$ matrix valued function with columns $f_1(\lambda), \ldots, f_n(\lambda)$. Then, because of the presumed R_α invariance of the columns of F,

$$R_\alpha F(\lambda) = \frac{F(\lambda) - F(\alpha)}{\lambda - \alpha} = F(\lambda)E_\alpha$$

for some $n \times n$ matrix E_α which is independent of λ. Thus

$$F(\lambda)\{I_n - (\lambda - \alpha)E_\alpha\} = F(\alpha) ,$$

and hence, since $\det\{I_n - (\lambda - \alpha)E_\alpha\} \not\equiv 0$,

$$F(\lambda) = F(\alpha)\{I_n + \alpha E_\alpha - \lambda E_\alpha\}^{-1}$$

which is of the form (3.1) with $V = F(\alpha)$, $M = I_n + \alpha E_\alpha$ and $N = E_\alpha$.

Suppose next that F is analytic at a point $\omega \in \Delta$ and that $u \in \ker(M - \omega N)$. Then

$$F(\lambda)(M - \lambda N)u = Vu = 0 ,$$

first for $\lambda = \omega$, and then for every $\lambda \in \Delta$ in the domain of analyticity of F. Thus, for all such λ,

$$(\omega - \lambda)F(\lambda)Nu = F(\lambda)\{M - \lambda N - (M - \omega N)\}u = 0 .$$

Therefore, since the columns of $F(\lambda)$ are linearly independent functions of λ, $Nu = 0$. But this in conjunction with the prevailing assumption $(M - \omega N)u = 0$ implies that

$$u \in \ker M \cap \ker N \Longrightarrow u = 0 \Longrightarrow M - \omega N \text{ is invertible.}$$

Thus we have shown that if F is analytic at ω, then $M - \omega N$ is invertible. Since the opposite implication is easy, this serves to complete the proof. ∎

COROLLARY 1. *If* $\det(M - \lambda N) \not\equiv 0$ *and* $F(\lambda) = V(M - \lambda N)^{-1}$ *is a rational $m \times n$ matrix valued function with n linearly independent columns, then:*

(1) M is invertible if and only if F is analytic at zero.

(2) N is invertible if and only if F is analytic at infinity and $F(\infty) = 0$.

Moreover, in case (1) F can be expressed in the form

$$F(\lambda) = C(I_n - \lambda A)^{-1} , \tag{3.3}$$

whereas in case (2) F can be expressed in the form

$$F(\lambda) = C(A - \lambda I_n)^{-1} . \tag{3.4}$$

PROOF. The first assertion is contained in the theorem, the second is obtained in much the same way. More precisely, if $\lim_{\lambda\to\infty} F(\lambda) = 0$ and $u \in \ker N$, then

$$F(\lambda)Mu = F(\lambda)(M - \lambda N)u = Vu .$$

But now, upon letting $\lambda \to \infty$ it follows that

$$Vu = 0 \Longrightarrow F(\lambda)Mu = 0 \Longrightarrow Mu = 0 \Longrightarrow u \in \ker M \cap \ker N \Longrightarrow u = 0 .$$

Thus N is invertible. The other direction is easy, as are formulas (3.3) and (3.4). Just take $C = VM^{-1}$ and $A = NM^{-1}$ in the first case, and $C = VN^{-1}$ and $A = MN^{-1}$ in the second. ∎

COROLLARY 2. *Let f be an* $m \times 1$ *vector valued function which is meromorphic in some open nonempty set* $\Delta \subset \mathbb{C}$ *and let* $\alpha \in \Delta$ *be a point of analyticity of f. Then f is an eigenfunction of* R_α *if and only if it can be expressed in the form*

$$f(\lambda) = \frac{v}{\rho_\omega(\lambda)}$$

for one or more choices of $\rho_\omega(\lambda)$ *in Table 1.1 with* $\rho_\omega(\alpha) \neq 0$ *and some nonzero constant vector* $v \in \mathbb{C}^m$.

3.2. Linear independence

It seems well to emphasize that herein the n columns of an $m \times n$ matrix valued function $F(\lambda)$ are said to be linearly independent if they are linearly independent in the vector space of continuous $m \times 1$ vector valued functions on the domain of analyticity of F. If

$$F(\lambda) = C(I_n - \lambda A)^{-1} \quad \text{or} \quad F(\lambda) = C(A - \lambda I_n)^{-1} ,$$

this is easily seen to be equivalent to the statement that

$$\bigcap_{j=0}^{n-1} \ker CA^{j-1} = 0 ,$$

i.e., that the pair (C, A) is observable. Such a realization for F is minimal in the sense of Kalman because (in the usual terminology, see e.g., Kailath [K]) the pair (A, B) is automatically controllable:

$$\bigcap_{j=0}^{n-1} \ker B^* A^{*j} = \{0\} \quad \text{(equivalently, } \operatorname{rank}[B\ AB \cdots A^{n-1}B] = n) ,$$

since $B = I_n$.

3.3. R_α invariant subspaces of $H_2^m(\Omega_+)$

Let $\mathcal{M}$ be an n dimensional R_α invariant subspace of $H_2^m(\Omega_+)$. Then, by Theorem 2.1, $\mathcal{M}$ is spanned by the columns of a rational $m \times n$ matrix valued function $F(\lambda)$ of the form (3.1). But more is true, because if $\Omega_+ = \mathbb{D}$, then $F(\lambda)$ is analytic at zero and hence (3.3) prevails, whereas if $\Omega_+ = \mathbb{C}_+$ or Π_+, then $\lim_{\lambda\to\infty} F(\lambda) = 0$ and hence (3.4) prevails.

In this subsection we shall obtain more concrete forms of F by reexpressing the matrix A in terms of its Jordan form.

THEOREM 3.2. *Let $\mathcal{M}$ be an n dimensional R_α invariant subspace of $H_2^m(\Omega_+)$. Then $\mathcal{M}$ is spanned by the columns of an $m \times n$ matrix valued function of the form*

$$F(\lambda) = V \begin{bmatrix} \Phi_{n_1,\omega_1} & & \\ & \ddots & \\ & & \Phi_{n_k,\omega_k} \end{bmatrix} , \tag{3.5}$$

where the matrix displayed in (3.5) is block diagonal, each block

$$\Phi_{j,\alpha} = \begin{bmatrix} \varphi_{0,\alpha} & \cdots & & \varphi_{j-1,\alpha} \\ 0 & \varphi_{0,\alpha} & \cdots & \varphi_{j-2,\alpha} \\ & & \ddots & \\ 0 & & & \varphi_{0,\alpha} \end{bmatrix} \tag{3.6}$$

is a $j \times j$ upper triangular Toeplitz matrix with entries

$$\varphi_{j,\alpha}(\lambda) = \left\{ \frac{1}{j!} \left(\frac{\partial}{\partial\alpha^*} \right)^j \frac{1}{\rho_\alpha} \right\} (\lambda) ,$$

which are specified explicitly in Table 1.1,

$$n_1 + \cdots + n_k = n \ ,$$

and

$$\omega_1, \ldots, \omega_k \in \Omega_+ \ .$$

PROOF. Suppose first that $\Omega_+ = \mathbb{D}$. Then by Theorem 3.1, $\mathcal{M}$ is spanned by the columns of an $m \times n$ matrix valued function of the form $C(I_n - \lambda A)^{-1}$. Therefore, upon writing A in the Jordan form

$$A = Q\mathcal{J}Q^{-1} \ ,$$

we see that $\mathcal{M}$ is also spanned by the columns of

$$F(\lambda) = CQ(I_n - \lambda\mathcal{J})^{-1} \ .$$

Formulas (3.5) and (3.6) are plain from the sequence of identifications

$$\begin{aligned}
\mathcal{J} &= \mathcal{J}_1 \oplus \cdots \oplus \mathcal{J}_k \ , \\
\mathcal{J}_j &= \begin{bmatrix} \omega_j^* & 1 & & \\ & & \ddots & \\ & & \ddots & 1 \\ & & & \omega_j^* \end{bmatrix} \\
&= \omega_j^* I_{n_j} + Z_{n_j}
\end{aligned}$$

for $j = 1, \ldots, k$, and

$$\begin{aligned}
(I_{n_j} - \lambda\mathcal{J}_j)^{-1} &= \{\rho_{\omega_j}(\lambda)I_{n_j} - \lambda Z_{n_j}\}^{-1} \\
&= \frac{1}{\rho_{\omega_j}(\lambda)}\left\{I_{n_j} - \frac{\lambda}{\rho_{\omega_j}(\lambda)}Z_{n_j}\right\}^{-1} \\
&= \Phi_{n_j,\omega_j}(\lambda) \ ,
\end{aligned}$$

where Z_k denotes the $k \times k$ matrix with ones on the first super diagonal (i.e., $(Z_k)_{i,i+1} = 1$ for $i = 1, \ldots, k-1$) and zeros elsewhere.

Moreover, upon writing $F = [f_1 \cdots f_n]$ and $CQ = V = [v_1 \cdots v_n]$, for the moment, it is readily seen that

$$f_1 = \frac{v_1}{\rho_{\omega_1}} \ , \quad \text{and} \quad f_{n_j+1} = \frac{v_{n_j+1}}{\rho_{\omega_{j+1}}} \ , \quad j = 1, \ldots, k-1 \ .$$

Because of the presumed linear independence of the columns of F, none of these functions can be zero. Therefore, since they are also presumed to belong to $H_2^m(\mathbb{D})$, the points $\omega_1, \ldots, \omega_k$ must all belong to $\mathbb{D}$.

Suppose next that $\Omega_+ = \mathbb{C}_+$. Then, since f is analytic in $\mathbb{C}_+$ and $f(\lambda) \to 0$ as $\lambda \to \infty$ for every $f \in \mathcal{M}$, it follows from the preceding corollary that $\mathcal{M}$ is spanned by the columns of an $m \times n$ matrix valued function of the form $C(A - \lambda I)^{-1}$. Upon writing $A = Q\mathcal{J}Q^{-1}$ as before, this implies that $\mathcal{M}$ is spanned by the columns of the $m \times n$ matrix valued function

$$F(\lambda) = \frac{1}{2\pi i} V(\mathcal{J} - \lambda I_n)^{-1}$$

with $V = 2\pi i\, CQ$. The final formula drops out upon writing

$$\begin{aligned}(\mathcal{J}_j - \lambda I_{n_j}) &= \{(\omega_j^* - \lambda)I_{n_j} + Z_{n_j}\}^{-1} \\ &= 2\pi i \Phi_{n_j,\omega_j}(\lambda) \ ,\end{aligned}$$

with $\omega_j \in \mathbb{C}_+$. It remains only to check that the points $\omega_1, \ldots, \omega_k \in \Omega_+$, but this goes through just as before.

The verification of $\Omega_+ = \mathbb{\Pi}_+$ is much the same as for $\Omega_+ = \mathbb{C}_+$. The details are left to the reader. ▮

We remark that $\mathcal{M}$ may also be identified in terms of a Blaschke-Potapov product B of n elementary inner factors (see e.g., (1.28) of [D2] for one of the standard forms) as

$$\mathcal{M} = H_2^m(\Omega_+) \ominus BH_2^m(\Omega_+) \ . \tag{3.7}$$

In fact, if P designates the $n \times n$ matrix with ij entry

$$p_{ij} = \langle f_j, f_i \rangle \ , \qquad i,j = 1, \ldots, n \ ,$$

and μ is any point in Ω_0, then B is uniquely specified by the formula

$$B(\lambda) = I_m - \rho_\mu(\lambda)F(\lambda)P^{-1}F(\mu)^* \ , \tag{3.8}$$

up to a unitary constant factor on the right, and

$$F(\lambda)P^{-1}F(\omega)^* = \frac{I_m - B(\lambda)B(\omega)^*}{\rho_\omega(\lambda)} \tag{3.9}$$

for every choice of λ and ω in Ω_+.

The identification (3.7) is a finite dimensional version of the Beurling-Lax theorem.

4. REALIZATIONS

In this section we shall use R_α invariance to extract a few sample realization formulas for matrix valued rational functions.

THEOREM 4.1. *Let $P(\lambda)$ be an $m \times n$ matrix polynomial of degree k. Then P can be expressed in the form*

$$P(\lambda) = D + \lambda C(I - \lambda A)^{-1}B \tag{4.1}$$

for suitably sized matrices A, B, C and D.

PROOF. By assumption

$$P(\lambda) = \sum_{j=0}^{k} P_j \lambda^j .$$

Therefore,

$$(R_\alpha P)(\lambda) = \sum_{j=1}^{k} P_j \frac{\lambda^j - \alpha^j}{\lambda - \alpha}$$

$$= \sum_{j=1}^{k} P_j \sum_{i=0}^{j-1} \lambda^i \alpha^{j-1-i}$$

is clearly a polynomial of degree at most $k-1$. Thus

$$\mathcal{M} = \text{span}\{(R_\alpha^j P)(\lambda)u : u \in \mathbb{C}^n \ , \quad j = 1, \dots, k\}$$

is a vector space of $m \times 1$ vector valued polynomials of dimension $\nu \le kn$. Moreover, by the resolvent identity (2.6), $\mathcal{M}$ is R_β invariant for every $\beta \in \mathbb{C}$. Let

$$G(\lambda) = [g_1(\lambda) \cdots g_\nu(\lambda)]$$

be a $m \times \nu$ matrix valued function whose columns form a basis for $\mathcal{M}$. Then, since the columns of $(R_0 P)(\lambda)$ belong to $\mathcal{M}$, there exists a $\nu \times n$ constant matrix B such that

$$\frac{P(\lambda) - P(0)}{\lambda} = G(\lambda)B .$$

But this does the trick since $G(\lambda)$ can be expressed in the form

$$G(\lambda) = C(I_\nu - \lambda A)^{-1}$$

for a suitable choice of constant matrices C and A, thanks to the first corollary to Theorem 3.1. ∎

THEOREM 4.2. *Let $F(\lambda)$ be a strictly proper, rational, $m\times n$ matrix valued function. Then F can be expressed in the form*

$$F(\lambda) = C(A-\lambda I)^{-1}B \tag{4.2}$$

for suitable chosen constant matrices A, B and C.

PROOF. By assumption

$$F(\lambda) = P(\lambda)Q(\lambda)^{-1} ,$$

where P and Q are matrix polynomials of sizes $m\times n$ and $n\times n$, respectively, such that $\det Q(\lambda) \not\equiv 0$ and $F(\lambda)\to 0$ as $\lambda\to\infty$. Let $\alpha\in\mathbb{C}$ be such that $Q(\alpha)$ is invertible. Then by (2.3) and (2.5),

$$(R_\alpha F)(\lambda) = -F(\lambda)(R_\alpha Q)(\lambda)Q(\alpha)^{-1} + (R_\alpha P)(\lambda)Q(\alpha)^{-1} . \tag{4.3}$$

Let

$$\mathcal{M}_Q = \operatorname{span}\left\{R_\alpha^k Qu :\ k=1,2,\dots\ ,\ u\in\mathbb{C}^n\right\}$$

and

$$\mathcal{M}_P = \operatorname{span}\left\{R_\alpha^k Pu :\ k=1,2,\dots\ ,\ u\in\mathbb{C}^n\right\} .$$

By the resolvent identity (2.6), both of these spaces are R_β invariant for every choice of $\beta\in\mathbb{C}$ and hence are independent of the initial choice of α. We now proceed in steps.

STEP 1. *The linear space*

$$\mathcal{M} = \{-Fg+h :\ g\in\mathcal{M}_Q\ ,\ \ h\in\mathcal{M}_P\}$$

is R_α invariant.

PROOF OF STEP 1. By (2.3) and (4.3),

$$\begin{aligned} R_\alpha\{-Fg+h\} &= -F(\lambda)(R_\alpha g)(\lambda) - (R_\alpha F)(\lambda)g(\alpha) + (R_\alpha h)(\lambda) \\ &= -F(\lambda)(R_\alpha g)(\lambda) + \{F(\lambda)(R_\alpha Q)(\lambda)Q(\alpha)^{-1} - (R_\alpha P)(\lambda)Q(\alpha)^{-1}\}g(\alpha) \\ &\quad + (R_\alpha h)(\lambda) \\ &= F(\lambda)\{-(R_\alpha g)(\lambda) + (R_\alpha Q)(\lambda)Q(\alpha)^{-1}g(\alpha)\} \\ &\quad - (R_\alpha P)(\lambda)Q(\alpha)^{-1}g(\alpha) + (R_\alpha h)(\lambda) , \end{aligned}$$

which belongs to the indicated set.

STEP 2. *F admits a representation of the form (4.2).*

PROOF OF STEP 2. Let

$$\mathcal{M}_F = \operatorname{span}\{R_\alpha^j F u : j = 1, 2, \dots , u \in \mathbb{C}^n\} .$$

Then, since $\mathcal{M}_F \subset \mathcal{M}$ and $\mathcal{M}$ is finite dimensional, $\mathcal{M}_F$ is clearly a finite dimensional R_α invariant subspace of $m \times 1$ vector valued functions which are analytic in all of $\mathbb{C}$ except at the zeros of Q. Let $G(\lambda) = [g_1(\lambda) \cdots g_\nu(\lambda)]$ be an $m \times \nu$ matrix valued function whose columns form a basis for $\mathcal{M}_F$. Since $G(\lambda) \to 0$ as $\lambda \to \infty$, G admits a representation of the form

$$G(\lambda) = C(A - \lambda I_\nu)^{-1} ,$$

by the corollary to Theorem 3.1. Thus

$$F(\lambda) = F(\alpha) + (\lambda - \alpha)C(A - \lambda I_\nu)^{-1} E_\alpha$$

for some $\nu \times n$ matrix E_α which is independent of λ. But now, upon letting $\lambda \to \infty$, it follows that

$$0 = F(\alpha) - CE_\alpha ,$$

and hence that

$$\begin{aligned} F(\lambda) &= C(A - \lambda I_\nu)^{-1}\{(\lambda - \alpha)I_\nu + A - \lambda I_\nu\}E_\alpha \\ &= C(A - \lambda I_\nu)^{-1}(A - \alpha I_\nu)E_\alpha , \end{aligned}$$

which is clearly of the form (4.2). ∎

COROLLARY. *Let $F(\lambda)$ be a proper, rational, $m \times n$ matrix valued function. Then F can be expressed in the form*

$$F(\lambda) = D + C(A - \lambda I)^{-1} B$$

for suitably chosen constant matrices A, B, C, and $D = F(\infty)$.

5. REPRODUCING KERNEL SPACES

Recall that a Hilbert space of $m \times 1$ vector valued functions which are defined in a subset Δ of $\mathbb{C}$, is said to be a RKHS (reproducing kernel Hilbert space) if there exists an $m \times m$ matrix valued function $K_\omega(\lambda)$ such that

(1) $K_\omega u \in \mathcal{H}$, and

(2) $\langle f, K_\omega u\rangle_{\mathcal{H}} = u^* f(\omega)$, (5.1)

for every choice of $\omega \in \Delta$, $f \in \mathcal{H}$ and $u \in \mathbb{C}^m$. It is both well known and readily checked that:

(1) The reproducing kernel is unique.

(2) $K_\alpha(\beta) = K_\beta(\alpha)^*$, (5.2)

for every choice of α, β in Δ.

(3) $\sum_{i,j=1}^n u_i^* K_{\alpha_j}(\alpha_i) u_j \geq 0$, (5.3)

for every choice of $\alpha_1, \ldots, \alpha_n$ in Δ and $u_1, \ldots, u_n$ in $\mathbb{C}^m$.

Present interest focuses on spaces of analytic functions for which the reproducing kernel can be expressed in a special form which will be described below. The main theorem is an elaboration of a fundamental result which is due to de Branges [dB]. It is formulated in terms of the pair of polynomials $a(\lambda)$ and $b(\lambda)$ which are given in Table 1.1, and symmetric sets Δ. A set Δ is said to be symmetric with respect to Ω_0 (or $\rho_\omega(\lambda)$) if for every $\lambda \in \Delta$ (except 0 for $\Omega_0 = \mathbb{T}$) the point $\lambda^o \in \Delta$; note that $\rho_\omega(\omega^o) = 0$. Recall that $\rho_\omega(\lambda) = a(\lambda)a(\omega)^* - b(\lambda)b(\omega)^*$.

THEOREM 5.1. *Let $\mathcal{H}$ be a RKHS of $m \times 1$ vector valued functions which are analytic in an open nonempty subset Δ of $\mathbb{C}$ which is symmetric with respect to Ω_0. Then the reproducing kernel $K_\omega(\lambda)$ can be expressed in the form*

$$K_\omega(\lambda) = \frac{J - \Theta(\lambda)J\Theta(\omega)^*}{\rho_\omega(\lambda)} , \tag{5.4}$$

for some choice of $m \times m$ matrix valued function $\Theta(\lambda)$ which is analytic in Δ and some signature matrix J, if and only if the following two conditions hold:

(1) $\mathcal{H}$ is R_α invariant for every $\alpha \in \Delta$.

(2) The structural identity

$$\langle R_\alpha(bf), R_\beta(bg)\rangle_{\mathcal{H}} - \langle R_\alpha(af), R_\beta(ag)\rangle_{\mathcal{H}} = |ab' - ba'|^2 g(\beta)^* J f(\alpha) \tag{5.5}$$

holds for every choice of α, β in Δ and f, g in $\mathcal{H}$.

Moreover, in this case, the function Θ which appears in (5.4) is unique up to a J unitary constant factor on the right. If there exists a point $\mu \in \Delta \cap \Omega_0$, it can be taken equal to

$$\Theta(\lambda) = I_m - \rho_\mu(\lambda)K_\mu(\lambda)J . \tag{5.6}$$

This formulation is adapted from [AD6]; see especially Theorems 4.1, 4.3 and 4.4. The restriction to the three choices of $a(\lambda)$ and $b(\lambda)$ specified earlier, permits some simplification in the presentation, because the terms $r(a,b;\alpha)f$ and $r(b,a;\alpha)f$ which intervene there are constant multiples of $R_\alpha(af)$ and $R_\alpha(bf)$, respectively.

For the three cases of interest, the structural identity (5.5) can be reexpressed as:

$$\langle (I+\alpha R_\alpha)f,(I+\beta R_\beta)g\rangle_{\mathcal{H}} - \langle R_\alpha f, R_\beta g\rangle_{\mathcal{H}} = g(\beta)^* Jf(\alpha) \tag{5.7}$$

if $\Omega_+ = \mathbb{D}$,

$$\langle R_\alpha f, g\rangle_{\mathcal{H}} - \langle f, R_\beta g\rangle_{\mathcal{H}} - (\alpha-\beta^*)\langle R_\alpha f, R_\beta g\rangle_{\mathcal{H}} = 2\pi i\ g(\beta)^* Jf(\alpha) \tag{5.8}$$

if $\Omega_+ = \mathbb{C}_+$, and

$$\langle R_\alpha f, g\rangle_{\mathcal{H}} + \langle f, R_\beta g\rangle_{\mathcal{H}} + (\alpha+\beta^*)\langle R_\alpha f, R_\beta g\rangle_{\mathcal{H}} = -2\pi\ g(\beta)^* Jf(\alpha) \tag{5.9}$$

if $\Omega_+ = \Pi_+$.

Formula (5.8) appears in de Branges [dB]; formula (5.7) is equivalent to a formula which appears in Ball [Ba], who adapted de Branges' work to the disc, including a technical improvement due to Rovnyak [Rov].

The role of the two conditions in Theorem 5.1 becomes particularly transparent when $\mathcal{H}$ is finite dimensional. Indeed we have already observed in Theorem 3.1 that if $\mathcal{M}$ is an n dimensional vector space of meromorphic functions which is R_α invariant for some point α in the domain of analyticity of any basis, then $\mathcal{M}$ is spanned by the columns of an $m \times n$ matrix valued function $F(\lambda)$ which can be expressed in the form

$$F(\lambda) = V(M-\lambda N)^{-1} \tag{5.10}$$

with M and N satisfying (3.2). Thus R_α invariance forces the elements of $\mathcal{M}$ to be rational of the indicated form.

Suppose further that $\mathcal{M}$ is endowed with an indefinite inner product based on an arbitrary $n \times n$ invertible Hermitian matrix P according to the rule

$$\langle Fu, Fv\rangle_{\mathcal{M}} = v^* Pu \tag{5.11}$$

for every choice of u and $v \in \mathbb{C}^n$. Then it is readily checked that

$$K_\omega(\lambda) = F(\lambda)P^{-1}F(\omega)^* \tag{5.12}$$

is a reproducing kernel for $\mathcal{M}$:

(1) $K_\omega v \in \mathcal{M}$, and

(2) $\langle Fu, K_\omega v\rangle_{\mathcal{M}} = v^*F(\omega)u$

for every choice of ω in $\mathcal{A}_F$, the domain of analyticity of F, and $u, v \in \mathbb{C}^n$. If $P > 0$, then $\mathcal{M}$ is a RKHS; if P is just Hermitian and invertible, then $\mathcal{M}$ is a finite dimensional reproducing kernel Krein space. The analysis in [AD2] and [AD6] covers this indefinite case also and implies in particular that for F of the form (5.10), the reproducing kernel in (5.12) can be expressed in the form (5.4) if and only if the structural identity (5.5) is met. Since

$$(R_\beta F)(\lambda) = F(\lambda)N(M - \beta N)^{-1}$$

for every point β at which the matrix $M - \beta N$ is invertible, i.e., for every $\beta \in \mathcal{A}_F$, it is readily checked that

$$\begin{aligned}\langle R_\alpha Fu, Fv\rangle_{\mathcal{M}} &= \langle FN(M - \alpha N)^{-1}u, Fv\rangle_{\mathcal{M}} \\ &= v^*PN(M - \alpha N)^{-1}u\ ,\end{aligned} \tag{5.13}$$

and similarly that

$$\langle Fu, R_\beta Fv\rangle_{\mathcal{M}} = v^*(M^* - \beta^*N^*)^{-1}N^*Pu\ , \tag{5.14}$$

and

$$\langle R_\alpha Fu, R_\beta v\rangle_{\mathcal{M}} = v^*(M^* - \beta^*N^*)^{-1}N^*PN(M - \alpha N)^{-1}u \tag{5.15}$$

for every choice of α, β in $\mathcal{A}_F$ and u, v in $\mathbb{C}^n$. For each of the three special choices of Ω_+ under consideration, it is now readily checked that the structural identity (5.5) reduces to a matrix equation for P by working out (5.7)-(5.9) with the aid of (5.13)-(5.15). Or, in other words, in a finite dimensional R_α invariant space $\mathcal{M}$ with Gram matrix P, the de Branges structural identity is equivalent to a Liapunov/Stein equation for P. This was first established explicitly in [D3] by a considerably lengthier calculation. If F is analytic at zero, then we may presume that $M = I_n$ in (5.10) and take $\alpha = \beta = 0$ in the structural identity (5.5). The fact that this reduced structural identity is equivalent to a Liapunov/Stein equation for the Gram matrix P appears in [AGo], in a somewhat different language.

THEOREM 5.2. *Let* $F(\lambda) = V(M - \lambda N)^{-1}$ *be an* $m \times n$ *matrix valued function with* $\det(M - \lambda N) \not\equiv 0$ *and linearly independent columns, and let the vector space*

$$\mathcal{M} = \{F(\lambda)u :\ u \in \mathbb{C}^n\}$$

be endowed with the indefinite inner product

$$\langle Fu, Fv\rangle_{\mathcal{M}} = v^*Pu \ ,$$

which is based on an $n\times n$ invertible Hermitian matrix P. Then $\mathcal{M}$ is a finite dimensional reproducing kernel Krein space with reproducing kernel $K_\omega(\lambda)$ given by (5.12).

The reproducing kernel can be expressed in the form

$$K_\omega(\lambda) = \frac{J - \Theta(\lambda)J\Theta(\omega)^*}{\rho_\omega(\lambda)}$$

with $\rho_\omega(\lambda)$ as in Table 1.1 if and only if P is a solution of the equation

$$M^*PM - N^*PN = V^*JV \quad for \ \ \Omega_+ = \mathbb{D} \ , \tag{5.16}$$

$$M^*PN - N^*PM = 2\pi i V^*JV \quad for \ \ \Omega_+ = \mathbb{C}_+ \ , \tag{5.17}$$

$$M^*PN + N^*PM = -2\pi V^*JV \quad for \ \ \Omega_+ = \Pi_+ \ . \tag{5.18}$$

Moreover, in each of these cases Θ is uniquely specified up to a J unitary constant multiplier on the right by the formula

$$\Theta(\lambda) = I_m - \rho_\mu(\lambda)F(\lambda)P^{-1}F(\mu)^*J \tag{5.19}$$

for any choice of the point $\mu \in \Omega_0 \cap \mathcal{A}_F$.

$\mathcal{M}$ is a RKHS if P is positive definite.

PROOF. This is an easy consequence of Theorem 5.1 and the discussion preceding the statement of this theorem. The basic point is that because of the special form of F, (5.7) [resp. (5.8), (5.9)] holds if and only if P is a solution of (5.16) [resp. (5.17), (5.18)]. ∎

It is well to note that formula (5.19) is a realization formula for $\Theta(\lambda)$, and that in the usual notation of (3.3) and (3.4) it depends only upon A, C and P. It can be reexpressed in one of the standard A, B, C, D forms by elementary manipulations.

Formulas (5.6) and (5.19) for $\Theta(\lambda)$ are obtained by matching the right hand sides of (5.4) and (5.12). This leads to the formula

$$\Theta(\lambda)J\Theta(\omega)^* = J - \rho_\omega(\lambda)F(\lambda)P^{-1}F(\omega)^* \ ,$$

which is clearly a necessary constraint on $\Theta(\lambda)$ since $\mathcal{M}$ has only one reproducing kernel, and hence any two recipes for it must agree. The final formula emerges upon setting

$\omega = \mu \in \Omega_0 \cap \mathcal{A}_F$ and then discarding J unitary constant factors on the right such as $\Theta(\mu)^{-1}$ and J. Thus the general theory of "structured" reproducing kernel spaces as formulated in Theorem 5.1 yields formula (5.19). However, once the formula is available, it can be used to check that

$$F(\lambda)P^{-1}F(\omega)^* = \frac{J - \Theta(\lambda)J\Theta(\omega)^*}{\rho_\omega(\lambda)}$$

for every pair of points λ, ω in $\mathcal{A}_F$ by straightforward calculation, using only the fact that P is a solution of one of the equations (5.16)-(5.18), according to the choice of Ω_+.

Formulas (5.16)-(5.18) can be expressed in a more unified way by introducing the notation used in [AD4]-[AD6]: Upon writing

$$M - \lambda N = a(\lambda)A - b(\lambda)B \tag{5.20}$$

in terms of the polynomials $a(\lambda)$ and $b(\lambda)$ which determine Ω_+, it is readily checked that A and B are given by the formulas in Table 5.1, and hence that (5.16)-(5.18) hold if and only if

$$A^*PA - B^*PB = V^*JV\ . \tag{5.21}$$

Ω_+	$\mathbb{D}$	$\mathbb{C}_+$	Π_+
A	M	$(M - iN)/2\sqrt{\pi}$	$(M - N)/2\sqrt{\pi}$
B	N	$-(M + iN)/2\sqrt{\pi}$	$-(M + N)/2\sqrt{\pi}$

TABLE 5.1

Finally we remark that connections between rational matrix functions which are J unitary on Ω_0 and Liapunov/Stein equations may be found in many places with assorted degrees of generality; see e.g., [AGo], [BR], [G–M], and [G]. The point here is to exhibit their equivalence to the structural identity (5.5) in the finite dimensional setting of Theorem 5.2. For infinite dimensional applications of Theorem 5.1 see e.g., [D7], and for another point of view [S].

6. $\mathcal{H}(S)$ SPACES

For each $S \in \mathcal{S}^{p\times q}(\Omega_+)$, the kernel

$$\Lambda_\omega(\lambda) = \frac{I_p - S(\lambda)S(\omega)^*}{\rho_\omega(\lambda)} \tag{6.1}$$

is positive in the sense exhibited in inequality (5.3). Perhaps the easiest way to see this is to observe that

$$\sum_{i,j=1}^{n} \xi_i^* \Lambda_{\alpha_j}(\alpha_i)\xi_j = \langle g, g\rangle - \langle \underline{\underline{p}}S^*g, \underline{\underline{p}}S^*g\rangle$$

with $g = \sum_{j=1}^n \xi_j/\rho_{\alpha_j}$; see (2.18) for help with the evaluation, if need be.

Because of this positivity, it follows on general grounds (see e.g., [Arn]) that $\Lambda_\omega(\lambda)$ is the reproducing kernel of exactly one RKHS which we shall designate by $\mathcal{H}(S)$. The following beautiful characterization of $\mathcal{H}(S)$ is due to de Branges and Rovnyak [dBR].

THEOREM 6.1. *Let $S \in \mathcal{S}^{p\times q}(\Omega_+)$, and for $f \in H_2^p(\Omega_+)$ let*

$$\kappa(f) = \sup\{\|f + \dot{S}g\|^2 - \|g\|^2 : g \in H_2^q(\Omega_+)\} . \tag{6.2}$$

Then

$$\mathcal{H}(S) = \{f \in H_2^p(\Omega_+) : \kappa(f) < \infty\} ,$$

and

$$\|f\|_{\mathcal{H}(S)}^2 = \kappa(f) .$$

PROOF. A proof for $\Omega_+ = \mathbb{D}$ can be found in [dBR], but it goes through for the other two cases in just the same way. ∎

We list a number of useful implications.

COROLLARY 1. *If $S \in \mathcal{S}^{p\times q}(\Omega_+)$ and $f \in H_2^p(\Omega_+)$, then $(I - S\underline{\underline{p}}S^*)f \in \mathcal{H}(S)$, and*

$$\|(I - S\underline{\underline{p}}S^*)f\|_{\mathcal{H}(S)}^2 = \langle (I - S\underline{\underline{p}}S^*)f, f\rangle . \tag{6.3}$$

COROLLARY 2. *If $S \in \mathcal{S}^{p\times q}(\Omega_+)$ is isometric a.e. on Ω_0, then $p \geq q$,*

$$\mathcal{H}(S) = H_2^p(\Omega_+) \ominus SH_2^q(\Omega_+) ,$$

and

$$\|f\|_{\mathcal{H}(S)} = \|f\| .$$

LEMMA 6.1. *If* $S \in \mathcal{S}^{p\times q}(\Omega_+)$, $u \in L_2^q(\Omega_0) \ominus H_2^q(\Omega_+)$, *and*

$$h = [I_p \quad -S]\underline{\underline{q}}' \begin{bmatrix} Su \\ u \end{bmatrix} = -\underline{\underline{p}}Su ,$$

then

$$\|h + Sg\|^2 - \|g\|^2 \leq \|u\|^2 - \|\underline{\underline{q}}'Su\|^2 . \tag{6.4}$$

The proofs of the last three statements can be found e.g., on p.27 of [D2]; the Supplementary Notes to that section discusses additional references. We supplement these facts with the following identity, which serves to simplify a number of the calculations which appear in [D2] and [D3].

LEMMA 6.2. *If* $f \in H_2^p(\Omega_+)$ *and* $g \in \mathcal{H}(S)$, *then*

$$\langle g, (I - S\underline{\underline{p}}S^*)f\rangle_{\mathcal{H}(S)} = \langle g, f\rangle . \tag{6.5}$$

PROOF. Let $f_1, f_2, \ldots$, be a sequence of finite sums of the form ξ_j/ρ_{ω_j} with $\xi_j \in \mathbb{C}^p$ and $\omega_j \in \Omega_+$ such that

$$\|f_n - f\| \to 0$$

as $n \uparrow \infty$. Then

$$\begin{aligned}\|(I - S\underline{\underline{p}}S^*)(f_n - f)\|^2_{\mathcal{H}(S)} &= \langle (I - S\underline{\underline{p}}S^*)(f_n - f), (f_n - f)\rangle \\ &= \|f_n - f\|^2 - \|\underline{\underline{p}}S^*(f_n - f)\|^2 \\ &\leq \|f_n - f\|^2 .\end{aligned}$$

Thus

$$\begin{aligned}\langle g, (I - S\underline{\underline{p}}S^*)f\rangle_{\mathcal{H}(S)} &= \lim_{n\uparrow\infty} \langle g, (I - S\underline{\underline{p}}S^*)f_n\rangle_{\mathcal{H}(S)} \\ &= \lim_{n\uparrow\infty} \langle g, f_n\rangle \\ &= \langle g, f\rangle ,\end{aligned}$$

because of the special form of the f_n. To amplify: since f_n is a finite sum of the form $f_n = \Sigma\xi_j/\rho_{\omega_j}$ with $\omega_j \in \Omega_+$,

$$(I - S\underline{\underline{p}}S^*)f_n = \Sigma\Lambda_{\omega_j}\xi_j ,$$

by (2.18). Thus

$$\begin{aligned}\langle g,(I - S\underline{p}S^*)f_n\rangle_{\mathcal{H}(S)} &= \Sigma\langle g, \Lambda_{\omega_j}\xi_j\rangle_{\mathcal{H}(S)} \\ &= \Sigma\xi_j^* g(\omega_j) \\ &= \langle g, f_n\rangle ,\end{aligned}$$

as claimed. ∎

A number of other useful properties of $\mathcal{H}(S)$ are summarized in Theorem 2.3 of [D2] and Theorem 9.1 of [D3]. In particular, it is shown there that $\mathcal{H}(S)$ is R_α invariant, that $R_\alpha S\eta \in \mathcal{H}(S)$, and that

$$\left\|\sum_{j=1}^{n} R_{\alpha_j} S\eta_j\right\|_{\mathcal{H}(S)}^2 \leq k \sum_{i,j=1}^{n} \eta_i^* \left\{\frac{I_q - S(\alpha_i)^* S(\alpha_j)}{\rho_{\alpha_i}(\alpha_j)}\right\} \eta_j \tag{6.6}$$

with a constant

$$k = |a(\lambda)b'(\lambda) - b(\lambda)a'(\lambda)|^2 ,$$

for every choice of $\alpha, \alpha_1, \ldots, \alpha_n$ in Ω_+ and $\eta, \eta_1, \ldots, \eta_n$ in $\mathbb{C}^q$. This, in conjuction with the selfevident inequality

$$\left\|\sum_{j=1}^{\mu} \Lambda_{\omega_j}\xi_j + \sum_{t=1}^{\nu} R_{\alpha_t} S\eta_t\right\|_{\mathcal{H}(S)}^2 \geq 0 , \tag{6.7}$$

leads readily to the conclusion that the block matrix

$$\left[\begin{array}{c|c} \dfrac{I_p - S(\omega_i)S(\omega_j)^*}{\rho_{\omega_j}(\omega_i)} & \dfrac{S(\omega_i) - S(\alpha_t)}{\omega_i - \alpha_t} \\ \hline \dfrac{S(\omega_j)^* - S(\alpha_r)^*}{\omega_j^* - \omega_r^*} & k\dfrac{I_q - S(\alpha_r)^* S(\alpha_t)}{\rho_{\alpha_r}(\alpha_t)} \end{array}\right] \geq 0 , \tag{6.8}$$

wherein $i, j = 1, \ldots, \mu$; $r, t = 1, \ldots, \nu$ and all the indicated points $\omega_1, \ldots, \omega_\mu$ and $\alpha_1, \ldots, \alpha_\nu$ belong to Ω_+. The inequality (6.8) (and variations thereof) is usually referred to as the Potapov-Ginzburg inequality.

If $\omega_i = \alpha_t$ in one of the corner blocks of the block matrix in (6.8), then we set

$$\frac{S(\omega_i) - S(\alpha_t)}{\omega_i - \alpha_t} = S'(\omega_i) .$$

Terms involving higher order derivatives intervene if the second sum on the left hand side of (6.7) includes powers of R_α.

7. A BASIC INTERPOLATION PROBLEM

In this section we first formulate a basic interpolation problem and establish necessary conditions for its solution. We then show that if the space $\mathcal{M}$ generated by the data of the problem is R_α invariant, then these conditions are sufficient. The verification rests first on the observation that the R_α invariance forces the elements of $\mathcal{M}$ to be spanned by the columns of a function F of the form (3.1), and then on the observation that for a space based on such an F, the structural identity (5.5) reduces to a Liapunov/Stein equation which must be met by the Gram matrix P of the space; see Theorem 5.2. This is basically the strategy which was followed in [D3], though the present analysis is a shade more general and less computationally heavy. A quick introduction to a simpler case is sketched in [D5].

7.1. The BIP and necessary conditions for its solution

We shall say that $S \in \mathcal{S}^{p\times q}(\Omega_+)$ is a solution of the BIP (basic interpolation problem) based on a given set of $m \times 1$ vector valued functions

$$f_j = \begin{bmatrix} g_j \\ h_j \end{bmatrix} , \qquad j = 1, \ldots, n ,$$

with components

$$g_j \in H_2^p \text{ and } h_j \in H_2^q , \text{ for } j = 1, \ldots, \mu ,$$

and

$$g_j \in (H_2^p)^\perp \text{ and } h_j \in (H_2^q)^\perp , \text{ for } j = \mu+1, \ldots, n ,$$

if

$$\underline{\underline{p}} S^* g_j = h_j , \text{ for } j = 1, \ldots, \mu , \tag{7.1}$$

and

$$g_j = \underline{\underline{q}}' S h_j , \text{ for } j = \mu+1, \ldots, n . \tag{7.2}$$

We shall assume throughout that $g_1, \ldots, g_\mu$ are linearly independent in H_2^p, and that $h_{\mu+1}, \ldots, h_n$ are linearly independent in $(H_2^q)^\perp$. The assumption that $f_j \in H_2^m$ for $j = 1, \ldots, \mu$ and $f_j \in (H_2^m)^\perp$ for $j = \mu+1, \ldots, n$ can be relaxed. The fundamental requirement is that $(S^* g_j - h_j) \in (H_2^q)^\perp$ for $j = 1, \ldots, \mu$ and $(Sh_j - g_j) \in H_2^p$ for

$j = \mu + 1, \ldots, n$; see Theorem 2.7 of [D1]. The present formulation is adopted because it simplifies the exposition and still covers a number of problems of interest. The extreme cases $\mu = 0$ and $\mu = n$ are permitted, with the obvious conventions.

The terminology BIP stems from the fact that the bitangential interpolation problems of Nevanlinna-Pick, Carathéodory-Fejér, mixtures of the two as well as the more general interpolation problem of Nudelman emerge upon specializing the choice of the f_j. Thus for example, if

$$\begin{bmatrix} g_1 \cdots g_\mu \\ h_1 \cdots h_\mu \end{bmatrix} = \begin{bmatrix} \xi_1 \cdots \xi_\mu \\ \eta_1 \cdots \eta_\mu \end{bmatrix} \Phi_{\mu,\alpha}$$

with $\Phi_{\mu,\alpha}$ as in (3.6), and

$$\begin{bmatrix} g_{\mu+1} \cdots g_n \\ h_{\mu+1} \cdots h_n \end{bmatrix} = \begin{bmatrix} \xi_{\mu+1} \cdots \xi_n \\ \eta_{\mu+1} \cdots \eta_n \end{bmatrix} \{(\lambda - \beta)I_\nu - Z_\nu\}^{-1}$$

with Z_ν as in the last part of Section 3 and α, β in Ω_+, then a straightforward calculation based on (2.13), (2.15) and (2.19) shows that S is a solution of the BIP problem if and only if

$$\eta_j = \sum_{t=0}^{j-1} \frac{S^{(t)}(\alpha)^*}{t!} \xi_{j-t} \quad \text{for} \quad j = 1, \ldots, \mu \ ,$$

and

$$\xi_{\mu+j} = \sum_{t=0}^{j-1} \frac{S^{(t)}(\beta)}{t!} \eta_{\mu+j-t} \quad \text{for} \quad j = 1, \ldots, \nu \ .$$

Numerous other examples in this formalism may be found in [D2] and [D3].

Throughout this section we shall fix the signature matrix

$$J = \begin{bmatrix} I_p & 0 \\ 0 & -I_q \end{bmatrix} , \qquad p + q = m \ .$$

This is the form which is particularly suited to problems in the Schur class $\mathcal{S}^{p \times q}(\Omega_+)$.

THEOREM 7.1. *If S is a solution of the BIP, then the $n \times n$ Hermitian matrix P with ij entry*

$$p_{ij} = \begin{cases} \langle J f_j, f_i \rangle & \textit{for} \quad i,j = 1, \ldots, \mu \ , \\ -\langle S h_j, g_i \rangle \ , & \textit{for} \quad i = 1, \ldots, \mu; \ \ j = \mu + 1, \ldots, n \ , \\ -\langle J f_j, f_i \rangle \ , & \textit{for} \quad i,j = \mu + 1, \ldots, n \ , \end{cases} \tag{7.3}$$

is positive semidefinite: $P \geq 0$.

PROOF. Let $X = [I_p \quad -S]$ and let

$$q_{ij} = \langle Xf_j, Xf_i \rangle_{\mathcal{H}(S)} \cdot$$

Then, since S is a solution of the BIP,

$$Xf_j = \begin{cases} (I_p - S\underline{\underline{p}}S^*)g_j\,, & \text{for } j = 1, \ldots, \mu\,, \\ -\underline{\underline{p}}Sh_j\,, & \text{for } j = \mu + 1, \ldots, n\,. \end{cases}$$

Therefore, for $i, j = 1, \ldots, \mu$,

$$\begin{aligned} q_{ij} &= \langle (I - S\underline{\underline{p}}S^*)g_j, (I - S\underline{\underline{p}}S^*)g_i \rangle_{\mathcal{H}(S)} \\ &= \langle (I - S\underline{\underline{p}}S^*)g_j, g_i \rangle \\ &= \langle g_j, g_i \rangle - \langle h_j, h_i \rangle \\ &= p_{ij}\,, \end{aligned}$$

by (6.3), whereas, for $i = 1, \ldots, \mu$ and $j = \mu + 1, \ldots, n$,

$$\begin{aligned} q_{ij} &= -\langle \underline{\underline{p}}Sh_j, (I - S\underline{\underline{p}}S^*)g_i \rangle_{\mathcal{H}(S)} \\ &= -\langle Sh_j, g_i \rangle \\ &= p_{ij}\,, \end{aligned}$$

by (6.5).

Thus upon expressing

$$P = \begin{bmatrix} P_{11} & P_{12} \\ P_{21} & P_{22} \end{bmatrix} \quad \text{and} \quad Q = \begin{bmatrix} Q_{11} & Q_{12} \\ Q_{21} & Q_{22} \end{bmatrix} \tag{7.4}$$

in block form with 11 blocks of size $\mu \times \mu$, and 22 blocks of size $\nu \times \nu$, we see that

$$0 \leq \begin{bmatrix} Q_{11} & Q_{12} \\ Q_{21} & Q_{22} \end{bmatrix} = \begin{bmatrix} P_{11} & P_{12} \\ P_{21} & Q_{22} \end{bmatrix} .$$

The desired conclusion is now immediate from the fact that $Q_{22} \leq P_{22}$, which follows easily from (6.4). ∎

7.2. Reformulation of necessary conditions under R_α invariance

To proceed further, it is convenient to set

$$F_{11} = [g_1 \cdots g_\mu] , \qquad F_{12} = [g_{\mu+1} \cdots g_n] ,$$

$$F_{21} = [h_1 \cdots h_\mu] , \qquad F_{22} = [h_{\mu+1} \cdots h_n] ,$$

$$F_1 = \begin{bmatrix} F_{11} \\ F_{21} \end{bmatrix} = [f_1 \cdots f_\mu] , \qquad F_2 = \begin{bmatrix} F_{12} \\ F_{22} \end{bmatrix} = [f_{\mu+1} \cdots f_n] ,$$

and

$$F = [F_1 \quad F_2] .$$

Then the entries in the block decomposition (7.4) of P are given by the formulas

$$v^* P_{11} u = \langle JF_1 u, F_1 v \rangle , \tag{7.5a}$$

$$v^* P_{12} x = -\langle SF_{22} x, F_{11} v \rangle , \tag{7.5b}$$

$$y^* P_{22} x = -\langle JF_2 x, F_2 y \rangle , \tag{7.5c}$$

for every choice of $u, v \in \mathbb{C}^\mu$ and $x, y \in \mathbb{C}^\nu$, where $\nu = n - \mu$.

The next step is to show that if the span of the columns of F_1 [resp. F_2] form an R_α invariant subspace of H_2^m [resp. $(H_2^m)^\perp$] for some point $\alpha \in \Omega_+$ [resp. $\alpha \in \Omega_-$], then the condition $P \geq 0$ is also sufficient for the existence of a solution to the BIP. The proof is developed in a sequence of lemmas. The first clarifies the form of $\mathcal{M}$.

LEMMA 7.1. *If the span of the columns of F_1 [resp. F_2] is a μ [resp. ν] dimensional R_α invariant subspace of H_2^m [resp. $(H_2^m)^\perp$] for some point $\alpha \in \Omega_+$ [resp. $\alpha \in \Omega_-$], then*

$$F_1(\lambda) = \begin{cases} C_1(I_\mu - \lambda A_1)^{-1} \ \ \textit{with} \ \ \sigma(A_1) \subset \Omega_+ \ \textit{if} \ \Omega_+ = \mathbb{D} \\ C_1(A_1 - \lambda I_\mu)^{-1} \ \ \textit{with} \ \ \sigma(A_1) \subset \Omega_- \ \textit{if} \ \ \Omega_+ = \mathbb{C}_+ \ \textit{or} \ \ \Pi_+ , \end{cases} \tag{7.6}$$

and

$$F_2(\lambda) = C_2(A_2 - \lambda I_\nu)^{-1} \ \ \textit{with} \ \ \sigma(A_2) \subset \Omega_+ . \tag{7.7}$$

Moreover, the block diagonal components in the block decomposition (7.4) of the matrix P whose entries are specified by (7.3) are given by the following formulas:

Ω_+	P_{11}	P_{22}
$\mathbb{D}$	$\sum_{j=0}^{\infty}(A_1^*)^j C_1^* J C_1 A_1^j$	$-\sum_{j=0}^{\infty}(A_2^*)^j C_2^* J C_2 A_2^j$
$\mathbb{C}_+$	$2\pi \int_0^{\infty} e^{iA_1^* t} C_1^* J C_1 e^{-iA_1 t} dt$	$-2\pi \int_0^{\infty} e^{-iA_2^* t} C_2^* J C_2 e^{iA_2 t} dt$
Π_+	$2\pi \int_0^{\infty} e^{A_1^* t} C_1^* J C_1 e^{A_1 t} dt$	$-2\pi \int_0^{\infty} e^{-A_2^* t} C_2^* J C_2 e^{-A_2 t} dt$

TABLE 7.1

PROOF. The indicated forms for F_1 and F_2 are immediate from Theorem 2.1. The evaluations of the block diagonal entries of P is divided into three cases, according to the choice of Ω.

CASE 1. Evaluation for $\Omega_+ = \mathbb{D}$.

By (7.5a) and (7.6),

$$v^* P_{11} u = \langle JC_1(I_\mu - \lambda A_1)^{-1} u,\ C_1(I_\mu - \lambda A_1)^{-1} v \rangle$$

for every choice of u and v in $\mathbb{C}^\mu$. But since $\sigma(A_1) \subset \mathbb{D}$, this inner product is easily evaluated either by expressing it as a contour integral over the unit circle, or by writing $(I_\mu - \lambda A_1)^{-1} = \sum_{j=0}^{\infty}(\lambda A_1)^j$ and invoking the Plancherel theorem for Fourier series.

The evaluation of P_{22} goes through in much the same way via (7.5c) and (7.7), since $\sigma(A_2) \subset \mathbb{D}$.

CASE 2. Evaluation for $\Omega_+ = \mathbb{C}_+$.

By (7.5a) and (7.6),

$$v^* P_{11} u = \langle JC_1(A_1 - \lambda I_\mu)^{-1} u,\ C_1(A_1 - \lambda I_\mu)^{-1} v \rangle$$

for every choice of u and v in $\mathbb{C}^\mu$. But now as $\sigma(A_1) \subset \mathbb{C}_-$,

$$(A_1 - \lambda I_\mu)^{-1} u = i \int_0^{\infty} e^{i\lambda t} e^{-iA_1 t} u\, dt ,$$

and the desired formula for P_{11} now drops out by the Plancherel formula for Fourier integrals.

The evaluation of P_{22} is similar except that now since $\sigma(A_2) \subset \mathbb{C}_+$, we use the representation

$$(A_2 - \lambda I_\nu)^{-1}x = -i \int_{-\infty}^{0} e^{i\lambda t} e^{-iA_2 t} x dt$$

for $x \in \mathbb{C}^\nu$ to evaluate (7.5c) via the Plancherel formula.

CASE 3. Evaluation for $\Omega_+ = \Pi_+$.

By (7.5a) and (7.6),

$$\begin{aligned} v^* P_{11} u &= \int_{-\infty}^{\infty} v^*(A_1^* + ibI_\mu)^{-1} C_1^* J C_1 (A_1 - ibI_\mu)^{-1} u db \\ &= \int_{-\infty}^{\infty} v^* \left\{(iA_1)^* - bI_\mu\right\}^{-1} C_1^* J C_1 \left\{(iA_1) - bI_\mu\right\}^{-1} u db , \end{aligned}$$

which can now be evaluated as in Case 2, since $\sigma(iA_1) \subset \mathbb{C}_-$. The evaluation of P_{22} is similar. ∎

We remark that if $F = [F_1 \quad F_2]$ and P are specified as in the lemma and if $P \geq 0$, then the columns $f_1, \ldots, f_n$ are linearly independent if and only if $g_1, \ldots, g_\mu$ are linearly independent in H_2^p and $h_{\mu+1}, \ldots, h_n$ are linearly independent in $(H_2^q)^\perp$. The proof in the nontrivial direction rests on formulas (7.5a) and (7.5c).

LEMMA 7.2. *If F_1 and F_2 are given by (7.6) and (7.7) respectively, then the block diagonal components of P with respect to the block decomposition (7.4) which are specified in Table 7.1 may be characterized as the unique solutions of the equations given in Table 7.2.*

Ω_+	P_{11}	P_{22}
$\mathbb{D}$	$X - A_1^* X A_1 = C_1^* J C_1$	$A_2^* Y A_2 - Y = C_2^* J C_2$
$\mathbb{C}_+$	$A_1^* X - X A_1 = 2\pi i C_1^* J C_1$	$A_2^* Y - Y A_2 = 2\pi i C_2^* J C_2$
Π_+	$A_1^* X + X A_1 = -2\pi C_1^* J C_1$	$A_2^* Y + Y A_2 = -2\pi C_2^* J C_2$

TABLE 7.2

PROOF. It is readily checked that the matrices P_{ii} are solutions of the indicated equations. The uniqueness is easy for $\Omega_+ = \mathbb{D}$; for the other two cases it is a consequence of the well known principle that an equation of the form $BX - XA = C$ is uniquely solvable if and only if $\sigma(B) \cap \sigma(A)$ is empty; see e.g., Rosenblum [Ros] for this and much more. ∎

The block component P_{12} in (7.3) is only partially specified by the data unless $\sigma(A_1^*) \cap \sigma(A_2)$ is empty. Specific calculations which exhibit the nonuniqueness may be found in [D2] and [D3]. [BGR] utilized this freedom to impose extra conditions in the associated interpolation problem. It is also exhibited in the nonuniqueness of the solutions to the equations which appear in the next lemma.

The finer subdivision of C_1 and C_2 into

$$C_1 = \begin{bmatrix} C_{11} \\ C_{21} \end{bmatrix} , \quad \text{and} \quad C_2 = \begin{bmatrix} C_{12} \\ C_{22} \end{bmatrix} ,$$

where the top entry is a matrix with columns of height p and the bottom entry is a matrix with columns of height q, will prove convenient.

LEMMA 7.3. *If F_1 and F_2 are specified by (7.6) and (7.7), respectively, then the block component P_{12} in the block decomposition (7.4) of the matrix P whose entries are specified by (7.3) is a solution of the equation*

$$P_{12}A_2 - A_1^*P_{12} = C_1^*JC_2 , \quad \text{if } \Omega_+ = \mathbb{D} , \tag{7.8}$$

$$A_1^*P_{12} - P_{12}A_2 = 2\pi i C_1^*JC_2 , \quad \text{if } \Omega_+ = \mathbb{C}_+ , \tag{7.9}$$

$$P_{12}A_2 + A_1^*P_{12} = -2\pi C_1^*JC_2 , \quad \text{if } \Omega_+ = \Pi_+ . \tag{7.10}$$

PROOF. The proof is broken into three steps according to the choice of Ω_+.

STEP 1. *P_{12} is a solution of (7.8) if $\Omega_+ = \mathbb{D}$.*

PROOF OF STEP 1. By (7.5b), (7.6) and (7.7),

$$\begin{aligned} v^*(P_{12}A_2 - A_1^*P_{12})x &= -\langle SF_{22}A_2x, F_{11}v\rangle + \langle SF_{22}x, F_{11}A_1v\rangle \\ &= -\langle SF_{22}(A_2 - \lambda I_\nu)x, F_{11}v\rangle \\ &\qquad - \langle \lambda SF_{22}x, F_{11}v\rangle + \langle \lambda SF_{22}x, F_{11}\lambda A_1 v\rangle \\ &= -\langle SC_{22}x, F_{11}v\rangle - \langle SF_{22}x, \lambda^{-1}C_{11}v\rangle \\ &= -\langle C_{22}x, \underline{\underline{p}}S^*F_{11}v\rangle - \langle \underline{\underline{q}}'SF_{22}x, \lambda^{-1}C_{11}v\rangle \\ &= -\langle C_{22}x, F_{21}v\rangle - \langle F_{12}x, \lambda^{-1}C_{11}v\rangle , \end{aligned}$$

since S is a solution of the BIP. The proof is now easily completed by evaluating the last term on the right as

$$-v^*C_{21}^*C_{22}x + v^*C_{11}^*C_{12}x = v^*C_1^*JC_2x \ .$$

STEP 2. *P_{12} is a solution of (7.9) if $\Omega_+ = \mathbb{C}_+$.*

PROOF OF STEP 2. By (7.5b), (7.6) and (7.7),

$$\begin{aligned}
v^*(P_{12}A_2 - A_1^*P_{12})x &= -\langle SF_{22}A_2x, F_{11}v\rangle + \langle SF_{22}x, F_{11}A_1v\rangle \\
&= \lim_{\varepsilon\downarrow 0} \frac{2\pi}{\varepsilon}\left\{-\langle\frac{SF_{22}A_2x}{\rho_{i/\varepsilon}}, F_{11}v\rangle + \langle SF_{22}x, \frac{F_{11}A_1v}{(\rho_{i/\varepsilon})^*}\rangle\right\} ,
\end{aligned}$$

where the terms $\rho_{i/\varepsilon}$ and $(\rho_{i/\varepsilon})^*$ are inserted in order to be able to reexpress the last line as

$$\begin{aligned}
&\lim_{\varepsilon\downarrow 0} \frac{2\pi}{\varepsilon}\left\{-\langle\frac{SF_{22}(A_2 - \lambda I_\nu)x}{\rho_{i/\varepsilon}}, F_{11}v\rangle - \langle SF_{22}x, \frac{F_{11}(A_1 - \lambda I_\mu)v}{\rho_{-i/\varepsilon}}\rangle\right\} \\
&= \lim_{\varepsilon\downarrow 0} \frac{2\pi}{\varepsilon}\left\{-\langle\frac{SC_{22}x}{\rho_{i/\varepsilon}}, F_{11}v\rangle - \langle SF_{22}x, \frac{C_{11}v}{\rho_{-i/\varepsilon}}\rangle\right\} \\
&= \lim_{\varepsilon\downarrow 0} \frac{2\pi}{\varepsilon}\left\{-\langle\frac{C_{22}x}{\rho_{i/\varepsilon}}, \underline{\underline{p}}S^*F_{11}v\rangle - \langle \underline{\underline{q}}'SF_{22}x, \frac{C_{11}v}{\rho_{-i/\varepsilon}}\rangle\right\} \\
&= \lim_{\varepsilon\downarrow 0} \frac{2\pi}{\varepsilon}\left\{-\langle\frac{C_{22}x}{\rho_{i/\varepsilon}}, F_{21}v\rangle - \langle F_{12}x, \frac{C_{11}v}{\rho_{-i/\varepsilon}}\rangle\right\} .
\end{aligned}$$

But now as $1/\rho_\omega$ [resp. $-1/\rho_\omega$] is a reproducing kernel for $H_2(\mathbb{C}_+)$ if $\omega \in \mathbb{C}_+$ [resp. $\{H_2(\mathbb{C}_+)\}^\perp$ if $\omega \in \mathbb{C}_-$], the inner products in the last term can be evaluated to obtain

$$\lim_{\varepsilon\downarrow 0} \frac{2\pi}{\varepsilon}\{-v^*F_{21}(i/\varepsilon)^*C_{22}x + v^*C_{11}^*F_{12}(-i/\varepsilon)x\} ,$$

which is readily seen to reduce to

$$\begin{aligned}
\lim_{\varepsilon\downarrow 0} v^*\left\{-(A_1^* + \frac{i}{\varepsilon}I_\mu)^{-1}C_{21}^*C_{22} + C_{11}^*C_{12}(A_2 + \frac{i}{\varepsilon}I_\nu)^{-1}\right\}x& \\
= -2\pi i v^*C_1^*JC_2x ,&
\end{aligned}$$

as needed.

STEP 3. *P_{12} is a solution of (7.10) if $\Omega_+ = \Pi_+$.*

PROOF OF STEP 3. By (7.5b), (7.6) and (7.7),

$$v^*\{P_{12}A_2 + A_1^*P_{12}\}x = -\langle SF_{22}A_2x, F_{11}v\rangle - \langle SF_{22}x, F_{11}A_1v\rangle$$

$$+ \lim_{\varepsilon\downarrow 0} \frac{2\pi}{\varepsilon}\left\{-\langle\frac{SF_{22}A_2x}{\rho_{1/\varepsilon}}, F_{11}v\rangle + \langle SF_{22}x, \frac{F_{11}A_1v}{\rho_{-1/\varepsilon}}\rangle\right\},$$

$$= \lim_{\varepsilon\downarrow 0} \frac{2\pi}{\varepsilon}\left\{-\langle\frac{SF_{22}(A_2-\lambda I_\nu)x}{\rho_{1/\varepsilon}}, F_{11}v\rangle + \langle SF_{22}x, \frac{F_{11}(A_1-\lambda I_\mu)v}{\rho_{-1/\varepsilon}}\rangle\right\},$$

since

$$\langle\frac{\lambda f}{\rho_{1/\varepsilon}}, g\rangle = \langle f, \frac{\lambda g}{\rho_{-1/\varepsilon}}\rangle$$

for every choice of f and g in $H_2^m(\Pi_+)$. The rest goes through much as before:

$$\lim_{\varepsilon\downarrow 0} \frac{2\pi}{\varepsilon}\left\{-\langle\frac{SC_{22}x}{\rho_{1/\varepsilon}}, F_{11}v\rangle + \langle SF_{22}x, \frac{C_{11}v}{\rho_{-1/\varepsilon}}\rangle\right\}$$

$$= \lim_{\varepsilon\downarrow 0} \frac{2\pi}{\varepsilon}\left\{-\langle\frac{C_{22}x}{\rho_{1/\varepsilon}}, \underline{\underline{p}}S^*F_{11}v\rangle + \langle \underline{\underline{q}}'SF_{22}x, \frac{C_{11}v}{\rho_{-1/\varepsilon}}\rangle\right\},$$

$$= \lim_{\varepsilon\downarrow 0} \frac{2\pi}{\varepsilon}\left\{-\langle\frac{C_{22}x}{\rho_{1/\varepsilon}}, F_{21}v\rangle + \langle F_{12}x, \frac{C_{11}v}{\rho_{-1/\varepsilon}}\rangle\right\},$$

$$= \lim_{\varepsilon\downarrow 0} \frac{2\pi}{\varepsilon}\{-v^*F_{21}(1/\varepsilon)^*C_{22}x - v^*C_{11}^*F_{12}(-1/\varepsilon)x\}$$

since $1/\rho_\omega$ [resp. $-1/\rho_\omega$] is a reproducing kernel for $H_2(\Pi_+)$ [resp. Π_-] when $\omega \in \Pi_+$ [resp. Π_-]. But this is easily seen to be equal to

$$-2\pi v^*C_1^*JC_2x\ ,$$

which yields the desired result. This completes the proof of both Step 3 and the lemma. ∎

Equations of the form exhibited in Table 7.2 for P_{11} appear in Nudelman [Nu1], [Nu2], who formulated a general one-sided interpolation problem (i.e., $\mu = n$ in (7.1) and (7.2)) in terms of contour integrals. This approach was adapted and extended to an extensive collection of two-sided problems by Ball, Gohberg and Rodman [BGR]. The supplementary equations for P_{22} in Table 7.2 and P_{12} in Lemma 7.3 are deduced there too, but by different methods and from a different point of view. The present formulation of the BIP in terms of (7.1) and (7.2) (which can of course also be reexpressed in terms of contour integrals) is a natural outgrowth of the analysis in [D2] and [D3]; see especially Lemma 9.1 of the latter. It is more convenient for working with reproducing kernel Hilbert spaces which are simply related to the Hardy spaces $H_2^m(\Omega_+)$.

7.3. Necessary is sufficient under R_α invariance

To this point we have shown that if S is a solution of the BIP based on the $m \times n$ matrix valued function $F(\lambda) = [F_1(\lambda) \quad F_2(\lambda)]$ with linearly independent columns and block entries $F_1(\lambda)$ and $F_2(\lambda)$ specified by (7.6) and (7.7), respectively, then the $n \times n$ matrix P which is defined by (7.3) is a positive definite solution of one of the equations (5.16) - (5.18), according to the choice of Ω_+, with

$$M = \begin{bmatrix} I_\mu & 0 \\ 0 & A_2 \end{bmatrix} \quad \text{and} \quad N = \begin{bmatrix} A_1 & 0 \\ 0 & I_\nu \end{bmatrix} \quad \text{for } \Omega_+ = \mathbb{D} ,$$

$$M = \begin{bmatrix} A_1 & 0 \\ 0 & A_2 \end{bmatrix} \quad \text{and} \quad N = I_n \quad \text{for } \Omega_+ = \mathbb{C}_+ \quad \text{and} \quad \Omega_+ = \Pi_+ ,$$

and

$$V = [C_1 \quad C_2] .$$

It turns out that for F of this form (which is dictated by the presumed R_α invariance), this condition is also sufficient.

THEOREM 7.2. *If the system of equations specified in Table 7.2 and Lemma 7.3 admits a positive semidefinite solution*

$$P = \begin{bmatrix} P_{11} & P_{12} \\ P_{21} & P_{22} \end{bmatrix} \geq 0 ,$$

then the BIP based on the corresponding $F(\lambda) = [F_1(\lambda) \quad F_2(\lambda)]$ with entries $F_1(\lambda)$ and $F_2(\lambda)$ specified by (7.6) and (7.7), respectively, is solvable.

If $P > 0$, then there exists a rational $m \times m$ matrix valued (J inner) function $\Theta(\lambda)$ such that

$$F(\lambda)P^{-1}F(\omega)^* = \frac{J - \Theta(\lambda)J\Theta(\omega)^*}{\rho_\omega(\lambda)} \tag{7.11}$$

for every pair of points λ, ω in $\mathcal{A}_F$. This function Θ is uniquely specified up to a J unitary constant factor on the right by the formula

$$\Theta(\lambda) = I_m - \rho_\mu(\lambda)F(\lambda)P^{-1}F(\mu)^*J \tag{7.12}$$

where μ is any point in Ω_0. (Two different choices of $\mu \in \Omega_0$ lead to two choices of Θ which agree up to a J unitary constant factor on the right.) Moreover, the set of all solutions to the BIP is equal to

$$\{T_\Theta[S_o]: \; S_o \in \mathcal{S}^{p\times q}(\Omega_+)\}$$

in which

$$T_\Theta[S_o] = (\Theta_{11}S_o + \Theta_{12})(\Theta_{21}S_o + \Theta_{22})^{-1} \tag{7.13}$$

is the linear fractional transformation based on the entries in the block decomposition

$$\Theta = \begin{bmatrix} \Theta_{11} & \Theta_{12} \\ \Theta_{21} & \Theta_{22} \end{bmatrix}$$

which is conformal with J.

Moreover, $S = T_\Theta[S_o]$ is strictly contractive if and only if S_o is.

PROOF. If $P > 0$, then the space

$$\mathcal{M} = \{Fu: \; u \in \mathbb{C}^n\}$$

endowed with the inner product

$$\langle Fu, \; Fv\rangle_{\mathcal{M}} = v^*Pu$$

is a RKHS of $m \times 1$ vector valued functions which are analytic in $\mathcal{A}_F$, the domain of analyticity of F, with reproducing kernel

$$K_\omega(\lambda) = F(\lambda)P^{-1}F(\omega)^* \; .$$

By the chosen form of F, $\mathcal{M}$ is R_α invariant for every $\alpha \in \mathcal{A}_F$ and hence for every $\alpha \in \mathbb{C}$ except for at most a finite number of points. Moreover, by assumption, the structural identity (5.5), as expressed in (5.16) - (5.18), is met; see the discussion preceding the statement of this theorem. Therefore, by Theorem 5.2, there exists an $m \times m$ matrix valued J inner function $\Theta(\lambda)$ such that (7.11) holds. The explicit formula (7.12) emerges upon setting $\omega = \mu \in \Omega_0$ in (7.11) and solving for $\Theta(\lambda)$. Factors of J and $\Theta(\mu)^{\pm 1}$ can be multiplied on the right at will, since they are J unitary.

The rest of the proof can be completed much as in the proofs of Theorems 9.3 and 9.4 of [D3]. (In making comparisons, it is well to note that Theorem 9.2 is equivalent to the present Lemma 7.3. Apparent differences are due only to the fact that the block decompositions are finer in the former and the normalizations are a little different; see Theorems 5.1 and 5.2 and formulas (5.12) - (5.14) in [D3]. Full details will be furnished in [D8]. ∎

It is perhaps worth reiterating that once formula (7.12) for $\Theta(\lambda)$ is known, it is possible to verify (7.11) by direct calculation (with the help of (5.16) - (5.18)) without explicit reference to the theory of reproducing kernel spaces.

7.4. Residues

For ease of comparison with other approaches (notably [Nu1], [Nu2], and [BGR]), we remark that if $\mathcal{M}$ is R_α invariant and if F and C are expressed in four block form as in the proof of Lemma 7.3, then S is a solution of the BIP if and only if

$$\sum_{\Omega_+} \operatorname{res}\{(\lambda I_\mu - \delta_1 A_1^*)^{-1} C_{11}^* S\} = C_{21}^* , \tag{7.13}$$

where $\delta_1 = -1$ if $\Omega_+ = \mathbb{\Pi}_+$ and $\delta_1 = 1$ otherwise, and

$$\sum_{\Omega_+} \operatorname{res}\{S C_{22}(\lambda I_\nu - A_2)^{-1}\} = C_{12} . \tag{7.14}$$

Moreover, if S is a solution of the BIP, then the $\mu \times \nu$ off diagonal block in (7.3) is given by the formula

$$P_{12} = \delta_0 \sum_{\Omega_+} \operatorname{res}\{(\lambda I_\mu - \delta_1 A_1^*)^{-1} C_{11}^* S C_{22}(\lambda I_\nu - A_2)^{-1}\} , \tag{7.15}$$

where

$$(\delta_0, \delta_1) = (1,1),\ (-2\pi i, 1)\ \text{ or }\ (-2\pi, -1) ,$$

according as $\Omega_+ = \mathbb{D}$, $\mathbb{C}$ or $\mathbb{\Pi}_+$, respectively. In all three formulas the residues are summed over the poles in Ω_+ only.

8. FACTORIZATION AND RECURSIVE METHODS

Our present objective is to clarify the fundamental role of R_α invariance in factorization and recursive methods. Throughout the section we shall let

$$F(\lambda) = [f_1(\lambda) \cdots f_n(\lambda)] = C(M - \lambda N)^{-1}$$

be an $m \times n$ matrix valued function with linearly independent columns, where

$$M = \begin{bmatrix} I_\mu & 0 \\ 0 & A_2 \end{bmatrix}, \quad N = \begin{bmatrix} A_1 & 0 \\ 0 & I_\nu \end{bmatrix}, \quad \text{and } \sigma(A_i) \subset \mathbb{D}, \quad \text{if } \Omega_+ = \mathbb{D};$$

$$M = \begin{bmatrix} A_1 & 0 \\ 0 & A_2 \end{bmatrix}, \; N = I_n, \; \sigma(A_1) \subset \Omega_-, \text{ and } \sigma(A_2) \subset \Omega_+, \text{ if } \Omega_+ = \mathbb{C}_+ \text{ or } \Pi_+ .$$

It is well to recall (from Section 3) that the exhibited form of F is equivalent to the assumption that

$$\mathcal{M}_1 = \text{span}\{f_1, \ldots, f_\mu\} \text{ is an } R_\alpha \text{ invariant subspace of } H_2^m(\Omega_+)$$

for some point $\alpha \in \Omega_+$, and

$$\mathcal{M}_2 = \text{span}\{f_{\mu+1}, \ldots, f_n\} \text{ is an } R_\alpha \text{ invariant subspace of } \{H_2^m)(\Omega_+)\}^\perp$$

for some point $\alpha \in \Omega_-$. In particular, $\mathcal{M}$ is R_α invariant for every point α in

$$\mathcal{A}_F = \{\lambda \in \mathbb{C} : \det(M - \lambda N) \neq 0\}, \tag{8.1}$$

the domain of analyticity of F.

LEMMA 8.1. *Let*

$$\mathcal{M}_x = \text{span}\{f_1, \ldots, f_k\}$$

and let

$$M = \begin{bmatrix} M_{xx} & M_{xy} \\ M_{yx} & M_{yy} \end{bmatrix} \quad \text{and} \quad N = \begin{bmatrix} N_{xx} & N_{xy} \\ N_{yx} & N_{yy} \end{bmatrix} \tag{8.2}$$

be block decompositions of M and N with M_{xx} and N_{xx} of size $k \times k$. Then $R_\alpha \mathcal{M}_x \subset \mathcal{M}_x$ for some point $\alpha \in \mathcal{A}_F$ if and only if $M_{yx} = N_{yx} = 0$.

PROOF. Suppose first that $R_\alpha \mathcal{M}_x \subset \mathcal{M}_x$ for some point $\alpha \in \mathcal{A}_F$. Then first k columns of $R_\alpha F$ must belong to the span of the first k columns of F:

$$(R_\alpha F)(\lambda) \begin{bmatrix} I_k \\ 0 \end{bmatrix} = F(\lambda) \begin{bmatrix} E \\ 0 \end{bmatrix}$$

for some $k \times k$ matrix $E = E_\alpha$ which is independent of λ. Thus

$$F(\lambda) N (M - \alpha N)^{-1} \begin{bmatrix} I_k \\ 0 \end{bmatrix} = F(\lambda) \begin{bmatrix} E \\ 0 \end{bmatrix},$$

which implies in turn that

$$N\begin{bmatrix} I_k \\ 0 \end{bmatrix} = (M - \alpha N)\begin{bmatrix} E \\ 0 \end{bmatrix},$$

since the columns of F are linearly independent and $MN = NM$. But this is equivalent to the pair of equations

$$N_{xx}(I_k + \alpha E) = M_{xx}E \tag{8.3}$$

and

$$N_{yx}(I_k + \alpha E) = M_{yx}E\,. \tag{8.4}$$

Now if $\Omega_+ = \mathbb{C}_+$ or Π_+, then $N_{xx} = I_k$ and $N_{yx} = 0$. Thus (8.3) and (8.4) reduce to the pair of equations

$$(M_{xx} - \alpha I_k)E = I_k$$

and

$$M_{yx}E = 0\,,$$

which clearly imply that E is invertible, and hence that $M_{yx} = 0$.

Next, if $\Omega_+ = \mathbb{D}$ and $k \leq \mu$, then $M_{xx} = I_k$ and $M_{yx} = 0$. Thus equations (8.3) and (8.4) reduce to the pair of equations

$$N_{xx}(I_k + \alpha E) = E$$

and

$$N_{yx}(I_k + \alpha E) = 0\,.$$

But these imply that $I_k + \alpha E$ is invertible because $(I_k + \alpha E)\eta = 0 \Longrightarrow E\eta = 0 \Longrightarrow \eta = 0$ for every $\eta \in \mathbb{C}^k$. Therefore $N_{yx} = 0$, as needed.

Finally, (in this direction) it remains to consider the case $\Omega_+ = \mathbb{D}$ with $j = k - \mu > 0$. Then, since

$$F = [F_1 \quad F_2]\,,$$

$$F_1(\lambda) = C_1(I_\mu - \lambda A_1)^{-1}\,, \qquad F_2(\lambda) = C_2(A_2 - \lambda I_\nu)^{-1}\,,$$

and

$$(R_\alpha f)(\lambda) = \left[F_1(\lambda)A_1(I_\mu - \alpha A_1)^{-1} \quad F_2(\lambda)(A_2 - \alpha I_\nu)^{-1}\right]\,,$$

the presumed R_α invariance of $\mathcal{M}_x$ amounts to assuming the existence of a $j \times j$ matrix $E = E_\alpha$ which is independent of λ such that

$$F_2(\lambda)(A_2 - \alpha I_\nu)^{-1}\begin{bmatrix} I_j \\ 0 \end{bmatrix} = F_2(\lambda)\begin{bmatrix} E \\ 0 \end{bmatrix}.$$

But upon writing A_2 in the block form

$$A_2 = \begin{bmatrix} B_{11} & B_{12} \\ B_{21} & B_{22} \end{bmatrix}$$

with B_{11} of size $j \times j$, this is readily seen to imply that

$$(B_{11} - \alpha I_j)E = I_j$$

and

$$B_{21}E = 0 \ .$$

This clearly implies that E is invertible and hence that $B_{21} = 0$. It thus follows that if $\mathcal{M}_x$ is R_α invariant, then both M and N have the requisite upper block triangular form.

Since the converse is simple, the proof may be deemed complete. ∎

8.1. The main factorization facts

The general theory of factorization of J contractive matrix valued functions (rational or not) is due to Potapov [P]. He worked with the functions directly and not with the associated reproducing kernel spaces, as in the present analysis. It should perhaps also be emphasized that in the next theorem $\Theta(\lambda)$ will not be J contractive unless P is positive definite.

THEOREM 8.1. *Let $F = [F_1 \quad F_2] = [f_1 \cdots f_n]$ be an $m \times n$ matrix valued function of the form (7.6)-(7.7) with linearly independent columns, and let P be an $n \times n$ invertible Hermitian matrix solution of one of the equations (5.16)-(5.18), according to the choice of Ω_+. Suppose further that*

$$\mathcal{M}_x = \operatorname{span}\{f_1, \ldots, f_k\} \tag{8.5}$$

is R_α invariant for some point $\alpha \in \mathcal{A}_F$, and that the $k \times k$ matrix P_{xx} which sits in the upper left hand corner of the block decomposition

$$P = \begin{bmatrix} P_{xx} & P_{xy} \\ P_{yx} & P_{yy} \end{bmatrix} \tag{8.6}$$

is invertible. Then for any point $\mu \in \Omega_0$,

$$\Theta(\lambda) = I_m - \rho_\mu(\lambda)F(\lambda)P^{-1}F(\mu)^*J \tag{8.7}$$

admits a factorization of the form

$$\Theta(\lambda) = \Theta_x(\lambda)\Theta_y(\lambda) \ , \tag{8.8}$$

where, in terms of the block decomposition (8.2),

$$\Theta_x(\lambda) = I_m - \rho_\mu(\lambda)F_x(\lambda)(P_{xx})^{-1}F_x(\mu)^*J \ , \tag{8.9}$$

$$\Theta_y(\lambda) = I_m - \rho_\mu(\lambda)W(M_{yy} - \lambda N_{yy})^{-1}Q^{-1}(M_{yy}^* - \mu^* N_{yy}^*)^{-1}W^*J \ , \tag{8.10}$$

$$F_x = [f_1, \dots, f_k] \ , \qquad F_y = [f_{k+1} \cdots f_n] \ ,$$

$$\Pi = \begin{bmatrix} -P_{xx}^{-1}P_{xy} \\ I_{n-k} \end{bmatrix} \ ,$$

and

$$W = F(\mu)\Pi(M_{yy} - \mu N_{yy}) \ . \tag{8.11}$$

Moreover:

$$M_{yy}^* Q M_{yy} - N_{yy}^* Q N_{yy} = W^* J W \quad if \quad \Omega_+ = \mathbb{D} \ , \tag{8.12}$$

$$M_{yy}^* Q N_{yy} - N_{yy}^* Q M_{yy} = 2\pi i W^* J W \quad if \quad \Omega_+ = \mathbb{C}_+ \ , \tag{8.13}$$

and

$$M_{yy}^* Q N_{yy} + N_{yy}^* Q M_{yy} = -2\pi W^* J W \quad if \quad \Omega_+ = \Pi_+ \ . \tag{8.14}$$

PROOF. Because of the presumed R_α invariance, Lemma 8.1 guarantees that both M and N are upper block triangular with respect to the indicated decompositions: $M_{yx} = N_{yx} = 0$. Therefore, everything can be read off from the more general factorization results which are presented in Theorems 4.2 and 4.3 of [AD5]; see formula (5.20) and Table 5.1 below for help in transcribing the notation. Nevertheless, in order to clarify the fundamental role of R_α invariance, we shall sketch the underlying strategy.

To begin with, it is readily checked that

$$\mathcal{M}_x^\perp = \{F\Pi u : \ u \in \mathbb{C}^{n-k}\}$$

is equal to the orthogonal complement of $\mathcal{M}_x$ in $\mathcal{M}$ with respect to the inner product induced by P, and that

$$\langle F\Pi u, F\Pi v\rangle_{\mathcal{M}} = v^*\Pi^* P \Pi u = v^* Q u \ , \tag{8.15}$$

for every choice of u and v in $\mathbb{C}^{n-k}$. In general $\mathcal{M}_x^\perp$ is not R_α invariant. However, the space

$$\widehat{\mathcal{M}} = \{\widehat{F}u : \ u \in \mathbb{C}^{n-k}\}$$

with

$$\widehat{F}(\lambda) = \Theta_x(\lambda)^{-1} F(\lambda)\Pi$$

is R_α invariant, thanks to (4.23) of [AD5] which supplies the important identity

$$\widehat{F}(\lambda) = W(M_{yy} - \lambda N_{yy})^{-1} . \tag{8.16}$$

Moreover, since Q is a solution of one of the equations (8.12)-(8.14), it follows that $\widehat{\mathcal{M}}$ endowed with the indefinite inner product

$$\langle \widehat{F}u, \widehat{F}v \rangle_{\widehat{\mathcal{M}}} = v^* Q u$$

is a finite dimensional reproducing kernel Krein space with reproducing kernel of the form

$$K_\omega^y(\lambda) = \frac{J - \Theta_y(\lambda) J \Theta_y(\omega)^*}{\rho_\omega(\lambda)} .$$

The factorization (8.8) emerges on theoretical grounds upon checking that both

$$\frac{J - \Theta(\lambda) J \Theta(\omega)^*}{\rho_\omega(\lambda)}$$

and

$$\frac{J - \Theta_x(\lambda) J \Theta_x(\omega)^*}{\rho_\omega(\lambda)} + \Theta_x(\lambda) K_\omega^y(\lambda) \Theta_x(\omega)^*$$

$$= \frac{J - \Theta_x(\lambda)\Theta_y(\lambda) J \Theta_y(\omega)^* \Theta_x(\omega)^*}{\rho_\omega(\lambda)}$$

are reproducing kernels for $\mathcal{M}$, or upon just multiplying out. ∎

In the general setting which is described in the first three and a half lines of Theorem 8.1, there is no guarantee that there exists a positive integer $k < n$ such that both the space $\mathcal{M}_x$ in (8.5) is R_α invariant and the block P_{xx} in the decomposition (8.6) is invertible. Three illustrative examples which serve to illustrate the difficulties are furnished on pages 152-154 of [AD2] (they are adapted from [A]); for related results, see also Theorems 5.5, 8.2 and 8.3 of [AD2], 2.6 of [AGo], 4.2 of [D2] and 3.3 of [AD3].

THEOREM 8.2. *Let $F = [F_1 \quad F_2] = [f_1 \cdots f_n]$ be an $m \times n$ matrix valued function of the form (7.6)-(7.7) with linearly independent columns and let P be an $n \times n$ invertible Hermitian matrix solution of one of the equations (5.16)-(5.18), according to the choice of Ω_+. Then the associated function $\Theta(\lambda)$ which is defined by (8.7) can be factored into a product*

$$\Theta(\lambda) = \Theta_1(\lambda) \cdots \Theta_n(\lambda) \tag{8.17}$$

of n elementary Blaschke-Potapov factors if and only if

(1) M and N are both upper triangular, and

(2) P admits a factorization of the form $P = U^*DU$, where U is upper triangular and D is diagonal.

If P is positive-definite, then (2) is automatically met.

PROOF. Let $P_{[k]}$ denote the upper left hand $k \times k$ corner of P, alias P_{xx} in the block decomposition (8.6), let $q_{11} = P_{[1]}$, and let $q_{j+1,j+1}$ denote the top left entry in the Schur complement Q_{jj} of q_{jj} for $j = 1, \ldots, n-1$. Then it is both well known and readily checked that

$$q_{11} \cdots q_{kk} = \det P_{[k]} \ ,$$

and hence that the q_{jj} are all nonzero if and only if $P_{[k]}$ is invertible for every choice of k, $k = 1, \ldots, n$. But this in turn is equivalent to (2). It is therefore also clear that (2) holds if P is positive definite.

The rest follows easily from Lemma 8.1 and from Theorem 8.1, which is then applicable for every integer k, $1 < k \leq n$. ∎

8.2. Factors and recursions

Formulas for constructing the factors in (8.17) can now be read off from Theorem 8.1. In particular, upon writing

$$P = [p_{ij}] \ , \quad M = [m_{ij}] \quad \text{and} \quad N = [n_{ij}]$$

for $i, j = 1, \ldots, n$, and

$$C = [c_1 \cdots c_n] \quad \text{with} \quad c_i \in \mathbb{C}^m \ ,$$

it follows from (8.9) with $k = 1$ that

$$\Theta_1(\lambda) = I_m - \rho_\mu(\lambda) c_1 (m_{11} - \lambda n_{11})^{-1} p_{11}^{-1} (m_{11}^* - \mu^* n_{11}^*)^{-1} c_1^* J \ . \tag{8.18}$$

The next factor $\Theta_2(\lambda)$ is computed in just the same way from the first column of the $m \times (n-1)$ matrix valued function $\widehat{F}(\lambda)$ which is defined in (8.16) and the top left entry of the Schur complement Q of p_{11}. In fact $\Theta_2(\lambda)$ may be obtained from (8.10) with $n = 2$. Since M and N are upper triangular, the first column of $\widehat{F}(\lambda)$ is readily seen to be equal to

$$\{-p_{11}^{-1} p_{12} f_1(\mu) + f_2(\mu)\}(m_{22} - \mu n_{22})(m_{22} - \lambda n_{22})^{-1} \ .$$

This exhibits the important fact that, in spite of the apparent dependence of W in (8.16) on all of $F(\mu)$, $\Theta_2(\lambda)$ depends only upon the first two columns of C, M and N, as it should.

If $f_1, \ldots, f_n$ all belong to $H_2^m(\Omega_+)$, $P > 0$ and M and N are upper triangular so that Theorem 8.2 is in force, then the basic recursive algorithm can be summarized as follows: Choose elementary J inner factors of the form (8.18), or better yet (8.19) with $U_1 = I_m$, such that

$$\underline{\underline{p}}\{\Theta_1 \cdots \Theta_j\}^* J f_j = 0$$

for $j = 1, \ldots, n$. This is done above for $j = 1$. Then at the $(k+1)$'st step, it is readily seen that

$$\begin{aligned}\underline{\underline{p}}\{\Theta_1 \cdots \Theta_{k+1}\}^* J f_j &= \underline{\underline{p}}\{\Theta_{j+1} \cdots \Theta_{k+1}\}^* \underline{\underline{p}}\{\Theta_1 \cdots \Theta_j\}^* f_j \\ &= 0\end{aligned}$$

for $j = 1, \ldots, k$, since the $\Theta_j \in H_\infty^{m \times m}(\Omega_+)$, and hence that

$$\begin{aligned}&\underline{\underline{p}}\{\Theta_1 \cdots \Theta_{k+1}\}^* J f_{k+1} \\ &= \underline{\underline{p}}\Theta_{k+1}^* \underline{\underline{p}}\{\Theta_1 \cdots \Theta_k\}^* J\{f_{k+1} - [f_1 \cdots f_k] P_{xx}^{-1} P_{xy}\} \\ &= \underline{\underline{p}}\Theta_{k+1}^* J\{\Theta_1 \cdots \Theta_k\}^{-1}\{f_{k+1} - [f_1 \cdots f_k] P_{xx}^{-1} P_{xy}\} ,\end{aligned}$$

where the decomposition of P is as in Theorem 8.2 but with $n = k+1$. Then by (8.16),

$$\begin{aligned}\{\Theta_1 \cdots \Theta_k\}^{-1}\{f_{k+1} - [f_1 \cdots f_k] P_{xx}^{-1} P_{xy}\} &= w_{k+1}(m_{k+1,k+1} - \lambda n_{k+1,k+1})^{-1} \\ &= \gamma w_{k+1} \rho_{\omega_{k+1}}(\lambda)^{-1} ,\end{aligned}$$

where $w_{k+1} \in \mathbb{C}^m$; see the more detailed calculation just below Table 8.1. Thus Θ_{k+1} can be chosen via formula (8.10), or what is more convenient, by the formula

$$\Theta_{k+1}(\lambda) = I_m + \{b_{\omega_{k+1}}(\lambda) - 1\} w_{k+1}(w_{k+1}^* J w_{k+1})^{-1} w_{k+1}^* J .$$

The preceding analysis gives a theoretical justification for the recursive construction of a rational J inner function $\Theta(\lambda)$ in terms of its elementary factors. For additional discussion and applications of this circle of ideas see: Section 5 of [AD2], the proof of Theorem 4.2, the corollaries on pages 56-57 and the remarks on recursive methods on pages 65 and 74 of [D2]. Other recursive approaches to interpolation are presented in Kimura [Ki], Limebeer and Anderson [LiA], and Sayed and Kailath [SK].

8.3. Alternative forms for elementary factors

Finally, to complete this section we reexpress $\Theta_1(\lambda)$ in more transparent form. There are basically two cases, according as the first column $f_1 \in H_2^m(\Omega_+)$ or not. Because of the assumptions on the spectrum of A_1 and A_2 it is readily seen that m_{11} and n_{11} can be expressed in terms of a point $\omega_1 \in \Omega_+$ as in the following table:

Ω_+	m_{11} if $f_1 \in H_2^m$	n_{11} if $f_1 \in H_2^m$	m_{11} if $f_1 \in (H_2^m)^\perp$	n_{11} if $f_1 \in (H_2^m)^\perp$
$\mathbb{D}$	1	ω_1^*	ω_1	1
$\mathbb{C}_+$	ω_1^*	1	ω_1	1
Π_+	$-\omega_1^*$	1	ω_1	1

TABLE 8.1

Now if $f_1 \in H_2^m(\Omega_+)$, it is readily checked that

$$f_1(\lambda) = c_1(m_{11} - \lambda n_{11})^{-1} = \gamma c_1 \rho_{\omega_1}(\lambda)^{-1} ,$$

and

$$p_{11} = \langle J f_1, f_1 \rangle = |\gamma|^2 \frac{c_1^* J c_1}{\rho_{\omega_1}(\omega_1)} ,$$

where

$$\gamma = a(\lambda)b'(\lambda) - b(\lambda)a'(\lambda)$$

is a constant which depends on the choice of Ω_+; its values are given in Table 1.1. Thus formula (8.18) can be reexpressed as

$$\Theta_1(\lambda) = I_m - \frac{\rho_\mu(\lambda)\rho_{\omega_1}(\omega_1)}{\rho_{\omega_1}(\lambda)\rho_\mu(\omega_1)} c_1(c_1^* J c_1)^{-1} c_1^* J .$$

Therefore, since

$$\frac{\rho_\mu(\lambda)\rho_{\omega_1}(\omega_1)}{\rho_{\omega_1}(\lambda)\rho_\mu(\omega_1)} = 1 - b_{\omega_1}(\lambda) b_{\omega_1}(\mu)^*$$

for all three of the specified choices of $\rho_\omega(\lambda)$ and $b_\omega(\lambda)$ in Table 1.1 (and more),

$$\Theta_1(\lambda) = [I_m + \{b_{\omega_1}(\lambda) - 1\}c_1(c_1^* J c_1)^{-1} c_1^* J]U_1 \ , \tag{8.19}$$

where

$$U_1 = I_m + \{b_{\omega_1}(\mu)^* - 1\}c_1^*(c_1^* J c_1)^{-1} c_1^* J$$

is a J unitary constant factor, since $|b_{\omega_1}(\mu)| = 1$. The key to the last calculation is the fact that $c_1(c_1^* J c_1)^{-1} c_1^* J$ is a projection; Theorem 1.3 of [D2] may also be helpful.

Next, if $f_1 \in \{H_2^m(\Omega_+)\}^\perp$, then

$$f_1(\lambda) = c_1(m_{11} - \lambda n_{11})^{-1} = c_1(\omega_1 - \lambda)^{-1}$$

with $\omega_1 \in \Omega_+$, and

$$p_{11} = -\langle J f_1, f_1 \rangle = -|\gamma|^2 \frac{c_1^* J c_1}{\rho_{\omega_1}(\omega_1)} \ .$$

Thus formula (8.18) can be rexpressed as

$$\Theta_1(\lambda) = I_m + \frac{\rho_\mu(\lambda)\rho_{\omega_1}(\omega_1)}{(\omega_1 - \lambda)(\omega_1^* - \mu^*)} |\gamma|^{-2} c_1(c_1^* J c_1)^{-1} c_1^* J \ .$$

Therefore, since

$$\frac{\rho_\mu(\lambda)\rho_{\omega_1}(\omega_1)}{(\omega_1 - \lambda)(\omega_1^* - \mu^*)} |\gamma|^{-2} = \frac{b_{\omega_1}(\mu)}{b_{\omega_1}(\lambda)} - 1$$

for all three choices of $\rho_\omega(\lambda)$ and $b_\omega(\lambda)$ which are exhibited in Table 1.1,

$$\Theta_1(\lambda) = \left[I_m + \left\{ \frac{1}{b_{\omega_1}(\lambda)} - 1 \right\} c_1(c_1^* J c_1)^{-1} c_1^* J \right] U_1 \ , \tag{8.20}$$

where

$$U_1 = I_m + \{b_{\omega_1}(\mu) - 1\} c_1(c_1^* J c_1)^{-1} c_1^* J$$

is a J unitary constant.

9. CHARACTERISTIC FUNCTIONS

The theory of characteristic functions was introduced by M. S. Livšic in a sequence of remarkable papers [Li1]-[Li4] to study linear operators T on a complex separable Hilbert space $\mathcal{H}$ which are close to selfadjoint in the sense that the range of

$$\Im T = \frac{T - T^*}{2i}$$

is finite dimensional, and linear partially isometric operators T which are close to unitary in the sense that the ranges of $I - TT^*$ and $I - T^*T$ are finite dimensional, of the same dimension.

In this section we shall derive the Livšic characteristic function from the theory of reproducing kernel spaces which was sketched in Section 5. We shall assume throughout that T is bounded, and shall unify the discussion of the two cases referred to above, and more, by working with the polynomials $a(\lambda)$ and $b(\lambda)$ which are specified in Table 1.1. We shall do this in Subsections 9.2 and 9.4, and shall, in a brief interlude, also consider briefly more general choices of $a(\lambda)$ and $b(\lambda)$ in Subsection 9.3. We begin, however, in Subsection 9.1 with some elementary preliminary calculations.

9.1. Preliminary calculations

Let X be a bounded selfadjoint operator on $\mathcal{H}$ with finite dimensional range $\mathcal{G} = X\mathcal{H}$, let $h_1, \ldots, h_m$ be a basis for $\mathcal{G}$, and let H be the $m \times m$ matrix with ij component

$$H_{ij} = \langle h_j, h_i \rangle_{\mathcal{H}} , \qquad i,j = 1, \ldots, m .$$

LEMMA 9.1. *Let Q be the $m \times m$ matrix which is defined by the rule*

$$Xh_t = \sum_{s=1}^{m} h_s Q_{st} , \qquad t = 1, \ldots, m . \tag{9.1}$$

Then HQ is an invertible Hermitian matrix.

PROOF. Clearly

$$\langle Xh_j, h_i \rangle_{\mathcal{H}} = \langle \sum_{s=1}^{m} h_s Q_{sj}, \ h_i \rangle_{\mathcal{H}} = (HQ)_{ij} ,$$

whereas

$$\begin{aligned} \langle h_j, Xh_i \rangle_{\mathcal{H}} &= \langle h_j, \sum_{s=1}^{m} h_s Q_{si} \rangle_{\mathcal{H}} \\ &= \sum_{s=1}^{m} \{H_{js} Q_{si}\}^* \\ &= \{(HQ)_{ji}\}^* . \end{aligned}$$

Therefore, since X is selfadjoint, this serves to prove that $HQ = (HQ)^*$.

Next, let $c \in \mathbb{C}^m$ be a vector with components $c_1, \ldots, c_m$ such that $Qc = 0$. Then

$$X \sum_{t=1}^{m} c_t h_t = \sum_{s,t=1}^{m} h_s Q_{st} c_t = 0 .$$

But this means that

$$\sum_{t=1}^{m} c_t h_t \in (\text{ran } X^*) \cap (\ker X) = \{0\} ,$$

and hence, since the vectors $h_1, \ldots, h_m$ are linearly independent, that $c_1, \ldots, c_m = 0$. Therefore Q is invertible. This completes the proof, since H is invertible. ∎

It is both well known and easily checked that the orthogonal projection $\Pi_{\mathcal{G}}$ of $\mathcal{H}$ onto $\mathcal{G}$ is given by the formula

$$\Pi_{\mathcal{G}} f = \sum_{s,t=1}^{m} h_s (H^{-1})_{st} \langle f, h_t \rangle_{\mathcal{H}} .$$

Therefore, by (9.1),

$$\begin{aligned}
Xh &= \sum_{s,t=1}^{m} h_s (H^{-1})_{st} \langle h, X h_t \rangle_{\mathcal{H}} \\
&= \sum_{s,t=1}^{m} h_s (H^{-1})_{st} \langle h, \sum_{j=1}^{m} h_j Q_{jt} \rangle_{\mathcal{H}} \\
&= \sum_{s,t,j=1}^{m} h_s (H^{-1})_{st} \{Q_{jt}\}^* \langle h, h_j \rangle_{\mathcal{H}} \\
&= \sum_{s,t,j=1}^{m} h_s \{Q_{jt} (H^{-1})_{ts}\}^* \langle h, h_j \rangle_{\mathcal{H}} \\
&= \sum_{s,j=1}^{m} h_s \{(QH^{-1})_{js}\}^* \langle h, h_j \rangle_{\mathcal{H}} \\
&= \sum_{s,j=1}^{m} h_s (QH^{-1})_{sj} \langle h, h_j \rangle_{\mathcal{H}}
\end{aligned}$$

The passage to the last line rests on the identity $QH^{-1} = H^{-1}(HQ)H^{-1}$ and the fact that both H and HQ are Hermitian.

It is convenient to write

$$QH^{-1} = \gamma W J W^* ,$$

where γ is an arbitrary positive constant which may be adjusted at will and J is an $m \times m$

signature matrix ($J = J^*$ and $JJ^* = I_m$). Then, the last formula can be reexpressed as

$$\begin{aligned} Xh &= \gamma \sum_{s,t=1}^{m} \sum_{\mu,\nu=1}^{m} h_s W_{s\mu} J_{\mu\nu} (W^*)_{\nu t} \langle h, h_t \rangle_{\mathcal{H}} \\ &= \gamma \sum_{\mu,\nu=1}^{m} g_\mu J_{\mu\nu} \langle h, g_\nu \rangle_{\mathcal{H}} , \end{aligned}$$

where

$$g_\mu = \sum_{s=1}^{m} h_s W_{s\mu} , \qquad \mu = 1, \ldots, m .$$

Since W is an invertible matrix, $g_1, \ldots, g_m$ is again a basis for $\mathcal{G}$. We have thus proved the following:

THEOREM 9.1. *For each positive constant γ there exits a basis $g_1, \ldots, g_m$ of $\mathcal{G}$ and an $m \times m$ signature matrix J such that*

$$Xh = \gamma \sum_{s,t=1}^{m} g_s J_{st} \langle h, g_t \rangle_{\mathcal{H}} . \tag{9.2}$$

From now on we shall work exclusively with such a fixed basis, and shall let G denote the $m \times m$ Gram matrix with ij entry

$$G_{ij} = \langle g_j, g_i \rangle_{\mathcal{H}} , \qquad i, j = 1, \ldots, m .$$

Also, we shall let Γ denote the linear operator from $\mathbb{C}^m$ onto $\mathcal{G}$ which is defined in term of its action on the standard basis $e_1, \ldots, e_m$ of $\mathbb{C}^m$ by the rule

$$\Gamma e_j = g_j , \qquad j = 1, \ldots, m . \tag{9.3}$$

A number of elementary consequences of this definition are summarized in the next theorem.

THEOREM 9.2. *The following formulas hold:*

(1) $$\Gamma^* \Gamma = G. \tag{9.4}$$

(2) $$\Gamma^* h = 0 \text{ for } h \in \mathcal{G}^\perp = \ker X. \tag{9.5}$$

(3) $$\Gamma^* g_t = \textstyle\sum_{s=1}^{n} e_s G_{st}. \tag{9.6}$$

(4) $$\Gamma G^{-1} \Gamma^* = \Pi_{\mathcal{G}}. \tag{9.7}$$

(5) $$X = \gamma \Gamma J \Gamma^* . \tag{9.8}$$

PROOF. Formulas (9.4) and (9.5) are immediate from the definition of Γ; (9.6) is just the orthogonal expansion

$$\Gamma^* g_t = \sum_{s=1}^{m} e_s \langle \Gamma^* g_t, e_s \rangle_{\mathbb{C}^m}$$

written out. In view of (9.5) it suffices to show that

$$\langle \Gamma G^{-1} \Gamma^* g_t, e_s \rangle_{\mathcal{H}} = G_{st} , \qquad s,t = 1,\dots,m ,$$

in order to obtain (9.7). But that is an easy consequence of (9.6). Finally, by (9.2) and (9.3),

$$Xh = \gamma \sum_{s,t=1}^{m} \Gamma e_s J_{st} \langle \Gamma^* h, e_t \rangle_{\mathbb{C}^m}$$

$$= \gamma \sum_{s,t=1}^{m} \Gamma e_s J_{st} e_t^* \Gamma^* h ,$$

which is the same as (9.8). ▮

9.2. Three basic cases

We now let

$$X = a(T)a(T)^* - b(T)b(T)^* \tag{9.9}$$

and

$$\rho_T(\lambda) = a(\lambda)a(T)^* - b(\lambda)b(T)^* . \tag{9.10}$$

Then X is a bounded linear selfadjoint operator on $\mathcal{H}$ and, for each of the three choices of $a(\lambda)$ and $b(\lambda)$ which are listed in Table 1.1, it is readily checked that X, $\rho_T(\lambda)$ and $(R_\alpha \rho_T^{-1})(\lambda)$ are given by the entries in Table 9.1.

Ω_+	$\mathbb{D}$	$\mathbb{C}_+$	Π_+
X	$I - TT^*$	$-2\pi i(T - T^*)$	$2\pi(T + T^*)$
$\rho_T(\lambda)$	$I - \lambda T^*$	$2\pi i(T^* - \lambda I)$	$2\pi(T^* + \lambda I)$
$(R_\alpha \rho_T^{-1})(\lambda)$	$\rho_T(\lambda)^{-1} T^*(I - \alpha T^*)^{-1}$	$\rho_T(\lambda)^{-1}(T^* - \alpha I)^{-1}$	$-\rho_T(\lambda)^{-1}(T^* + \alpha I)^{-1}$

TABLE 9.1

Let $\mathcal{G} = X\mathcal{H}$ denote the range of X, and assume that it is m dimensional with basis $g_1, \ldots, g_m$ just as in the last subsection, and carry over the symbols Γ, J, γ, with $\gamma = 1$ so that

$$X = \Gamma J \Gamma^* . \tag{9.11}$$

Assume further that T is simple:

$$cls\{T^n \mathcal{G} : n = 0, 1, \ldots\} = \mathcal{H} . \tag{9.12}$$

Then the space

$$\mathcal{M} = \{\Gamma^* \rho_T(\lambda)^{-1} h : h \in \mathcal{H}\} \tag{9.13}$$

of $m \times 1$ vector valued functions of λ has a natural inner product which is given by the rule

$$\langle \Gamma^* \rho_T(\lambda)^{-1} u, \Gamma^* \rho_T(\lambda)^{-1} v \rangle_{\mathcal{M}} = \langle u, v \rangle_{\mathcal{H}} , \tag{9.14}$$

for every choice of u and v in $\mathcal{H}$. Assumption (9.12) guarantees that it is well defined.

LEMMA 9.2. *If* $\Gamma^* \rho_T(\lambda)^{-1} u = 0$ *for every point* λ *at which the indicated inverse exists, then* $u = 0$.

PROOF. Under the given assumption, it is readily checked that

$$\Gamma^* T^{*n} = 0 \qquad \text{for} \quad n = 0, 1, \ldots ,$$

and hence that

$$\langle u, T^n g \rangle_{\mathcal{H}} = 0$$

for $n = 0, 1, \ldots,$ and every choice of $g \in \mathcal{G}$. Therefore, by (9.12), $u = 0$. ∎

Roughly speaking, Lemma 9.2 serves to prove that the "columns" of

$$F(\lambda) := \Gamma^* \rho_T(\lambda)^{-1} \tag{9.15}$$

are "linearly independent" when (9.12) is in force.

LEMMA 9.3. *$\mathcal{M}$ is a reproducing kernel Hilbert space of $m \times 1$ vector valued functions which are analytic in*

$$\mathcal{D}_{\rho_T} = \{\lambda \in \mathbb{C} : \rho_T(\lambda) \ \ \textit{has a bounded inverse}\} . \tag{9.16}$$

The reproducing kernel

$$K_\omega(\lambda) = F(\lambda)F(\omega)^* = \Gamma^* \rho_T(\lambda)^{-1} \rho_T(\omega)^{-*} \Gamma \tag{9.17}$$

for every choice of λ *and* ω *in* $\mathcal{D}_{\rho_T}$.

PROOF. $\mathcal{M}$ is clearly a complex vector space with a well-defined inner product, thanks to Lemma 9.2. Moreover, if $\{Fu_n\}_{n=1}^{\infty}$ is a Cauchy sequence in $\mathcal{M}$, then by (9.14), $\{u_n\}_{n=1}^{\infty}$ is a Cauchy sequence in $\mathcal{H}$. Thus there exists an element u in $\mathcal{H}$ such that $u_n \to u$ (strongly in $\mathcal{H}$) as $n \to \infty$. Therefore $\|Fu_n - Fu\|_{\mathcal{M}} \to 0$ as $n \to \infty$, and $\mathcal{M}$ is seen to be a Hilbert space.

Next, since $K_\omega\xi \in \mathcal{M}$ for every choice of $\omega \in \mathcal{D}_{\rho_T}$ and $\xi \in \mathbb{C}^m$, it remains only to check that

$$\langle Fu, K_\omega\xi\rangle_{\mathcal{M}} = \xi^* F(\omega)u$$

for every $u \in \mathcal{H}$. But that is a straightforward calculation:

$$\begin{aligned}\langle Fu, FF(\omega)^*\xi\rangle_{\mathcal{M}} &= \langle u, F(\omega)^*\xi\rangle_{\mathcal{H}}\\ &= \langle \rho_T(\omega)^{-1}u, \Gamma\xi\rangle_{\mathcal{H}}\\ &= \xi^*\Gamma^*\rho_T(\omega)^{-1}u\ ,\end{aligned}$$

as needed. The proof is complete. ∎

LEMMA 9.4. *If* $\Omega_+ = \mathbb{D}$ *and* T *is a (nonzero) partial isometry, then*

$$\mathbb{C}\backslash\mathbb{T} \subset \mathcal{D}_{\rho_T}\ .$$

If $\Omega_+ = \mathbb{C}_+$ *or* $\Omega_+ = \Pi_+$ *and* T *is bounded, then*

$$\{\lambda \in \mathbb{C} : |\lambda| > \|T\|\} \subset \mathcal{D}_{\rho_T}\ .$$

PROOF. The first assertion is immediate from the well-known bound

$$\|I - \lambda T^*\| = \|I - \lambda^* T\| \geq |(1 - |\lambda|)|$$

for isometric operators on a Hilbert space; see e.g., Akhiezer and Glazman [AG]. The second is even easier, and is left to the reader. ∎

THEOREM 9.3. *Let* Δ *be an open nonempty subset of* $\mathcal{D}_{\rho_T}$ *which is symmetric with respect to* Ω_0 *(such a* Δ *exists by Lemma 9.4). Then there exists an* $m \times m$ *matrix valued function* $\Theta(\lambda)$ *which is analytic in* Δ *such that*

$$F(\lambda)F(\omega)^* = \frac{J - \Theta(\lambda)J\Theta(\omega)^*}{\rho_\omega(\lambda)} \tag{9.18}$$

for every pair of points λ, ω in Δ. Here J is the $m \times m$ signature matrix which appears in (9.11).

The function Θ which appears in (9.18) is unique up to a multiplicative J unitary constant factor on the right. If there exists a point $\mu \in \Delta \cap \Omega_0$, it can be taken equal to

$$\Theta(\lambda) = I_m - \rho_\mu(\lambda)F(\lambda)F(\mu)^* J \ . \tag{9.19}$$

PROOF. From the calculations furnished in Table 9.1 it is readily seen that $\mathcal{M}$ is R_α invariant for every point $\alpha \in \Delta$. Therefore, in view of Theorem 5.1, it remains only to verify that the structural identity (5.5) is met. To this end we let

$$f(\lambda) = F(\lambda)u \quad \text{and} \quad g(\lambda) = F(\lambda)v \ ,$$

for some choice of u and v in $\mathcal{H}$, and proceed in steps, according to the choice of Ω_+.

STEP 1. *The structural identity (5.5) is met if $\Omega_+ = \mathbb{D}$.*

PROOF OF STEP 1. With the help of the entries in Table 9.1, it is readily checked that

$$(I + \alpha R_\alpha)f = F(I - \alpha T^*)^{-1}u$$

and

$$R_\alpha f = FT^*(I - \alpha T^*)^{-1}u \ .$$

Thus the left hand side of (5.7) is equal to

$$\begin{aligned} &\langle (I + \alpha R_\alpha)f, (I + \beta R_\beta)g\rangle_{\mathcal{M}} - \langle R_\alpha f, R_\beta g\rangle_{\mathcal{M}} \\ &= \langle (I - TT^*)(I - \alpha T^*)^{-1}u, (I - \beta T^*)^{-1}v\rangle_{\mathcal{H}} \\ &= \langle \Gamma J \Gamma^*(I - \alpha T^*)^{-1}u, (I - \beta T^*)^{-1}v\rangle_{\mathcal{H}} \\ &= g(\beta)^* J f(\alpha) \ , \end{aligned}$$

as claimed.

STEP 2. *The structural identity (5.5) is met if $\Omega_+ = \mathbb{C}_+$.*

PROOF OF STEP 2. By Table 9.1, it is readily checked that

$$R_\alpha f = F(T^* - \alpha I)^{-1}u \ ,$$

and hence that the left hand side of (5.8) is equal to

$$\begin{aligned}&\langle R_\alpha f, g\rangle_{\mathcal{M}} - \langle f, R_\beta g\rangle_{\mathcal{M}} - (\alpha-\beta^*)\langle R_\alpha f, R_\beta g\rangle_{\mathcal{M}}\\ &= \langle (T^*-\alpha I)^{-1}u, v\rangle_{\mathcal{H}} - \langle u, (T^*-\beta I)^{-1}v\rangle_{\mathcal{H}}\\ &\qquad - (\alpha-\beta)^*\langle (T^*-\alpha I)^{-1}u, (T^*-\beta I)^{-1}v\rangle_{\mathcal{H}}\\ &= \langle\{(T^*-\alpha I)^{-1} - (T-\beta^* I)^{-1} - (\alpha-\beta^*)(T-\beta^* I)^{-1}(T^*-\alpha I)^{-1}\}u, v\rangle_{\mathcal{H}}\\ &= \langle (T-\beta^* I)^{-1}(T-T^*)(T^*-\alpha I)^{-1}u, v\rangle_{\mathcal{H}}\\ &= 4\pi^2\langle (T-T^*)\rho_T(\alpha)^{-1}u, \rho_T(\beta)^{-1}v\rangle_{\mathcal{H}}\,.\end{aligned}$$

Therefore, since

$$T - T^* = \frac{X}{-2\pi i} = \frac{\Gamma^* J\Gamma}{-2\pi i}\,,$$

this last expression is readily seen to reduce to the right hand side of (5.8), as claimed.

STEP 3. *The structural identity (5.5) is satisfied if* $\Omega_+ = \Pi_+$.

PROOF OF STEP 3. For this choice of ρ, it follows from Table 9.1 that

$$R_\alpha f = -F(T^* + \alpha I)^{-1}u$$

and hence that the left hand side of (5.9) is equal to

$$\begin{aligned}&\langle R_\alpha f, g\rangle_{\mathcal{M}} + \langle f, R_\beta g\rangle_{\mathcal{M}} + (\alpha+\beta^*)\langle R_\alpha f, R_\beta g\rangle_{\mathcal{M}}\\ &= -\langle (T^*+\alpha I)^{-1}u, v\rangle_{\mathcal{H}} - \langle u, (T^*+\beta I)^{-1}v\rangle_{\mathcal{H}}\\ &\qquad + (\alpha+\beta^*)\langle (T^*+\alpha I)^{-1}u, (T^*+\beta I)^{-1}v\rangle_{\mathcal{H}}\\ &= -\langle (T+\beta^* I)^{-1}(T^*+T)(T^*+\alpha I)^{-1}u, v\rangle_{\mathcal{H}}\\ &= -4\pi^2\langle (T^*+T)\rho_T(\alpha)^{-1}u, \rho_T(\beta)^{-1}v\rangle_{\mathcal{H}}\,.\end{aligned}$$

But this is readily seen to reduce to the right hand side of (5.9), since

$$T + T^* = \frac{X}{2\pi} = \frac{\Gamma^* J\Gamma}{2\pi}\,.$$

This completes the proof of Step 3 and the theorem. ∎

9.3. A unified approach

The analysis of the preceding subsection can be both unified and extended by taking advantage of the general formulation presented in [AD6]. In particular, it is readily checked that

$$\begin{aligned}\{r(a,b;\alpha)F\}(\lambda) &:= \frac{a(\lambda)F(\lambda)-a(\alpha)F(\alpha)}{a(\alpha)b(\lambda)-b(\alpha)a(\lambda)}\\ &= F(\lambda)b(T)^*\rho_T(\alpha)^{-1}\end{aligned}$$

and

$$\begin{aligned}\{r(b,a;\alpha)F\}(\lambda) &:= \frac{b(\lambda)F(\lambda)-b(\alpha)F(\alpha)}{b(\alpha)a(\lambda)-a(\alpha)b(\lambda)}\\ &= -F(\lambda)a(T)^*\rho_T(\alpha)^{-1}\ ,\end{aligned}$$

for every point $\alpha \in \Omega$ at which $|a(\alpha)|+|b(\alpha)|>0$ (i.e., at which $\rho_\alpha(\lambda)\not\equiv 0$). Therefore,

$$\begin{aligned}&\langle r(b,a;\alpha)Fu, r(b,a;\beta)Fv\rangle_{\mathcal{M}} - \langle r(a,b;\alpha)Fu, r(a,b;\alpha)Fv\rangle_{\mathcal{M}}\\ &= \langle a(T)^*\rho_T(\alpha)^{-1}u, a(T)^*\rho_T(\beta)^{-1}v\rangle_{\mathcal{H}}\langle b(T)^*\rho_T(\alpha)^{-1}u, b(T)^*\rho_T(\beta)^{-1}v\rangle_{\mathcal{H}}\\ &= \langle X\rho_T(\alpha)^{-1}u, \rho_T(\beta)^{-1}v\rangle_{\mathcal{H}}\\ &= \langle \Gamma J\Gamma^*\rho_T(\alpha)^{-1}u, \rho_T(\beta)^{-1}v\rangle_{\mathcal{H}}\\ &= \langle JF(\alpha)u, F(\beta)v\rangle_{\mathbb{C}^m}\ .\end{aligned}$$

This is the structural identity in the general setting of [AD6]. The existence of, and a formula for, $\Theta(\lambda)$ is established in Section 4 of that reference.

9.4. Livšic's formulas

In this subsection we shall obtain formulas for the Livšic characteristic function for two classes of bounded linear operators T acting on a Hilbert space $\mathcal{H}$ from (9.19): close to selfadjoint, and (partial) isometries which are close to unitary.

CASE 1. $\dim\{\text{range}\ \Im T\} = m < \infty$.

DISCUSSION. Set $\Omega_+ = \mathbb{C}_+$ and let $\mu \to \infty$ through real values in the formula

$$\Theta_\mu(\lambda) = I_m + \frac{i}{2\pi}(\lambda-\mu)\Gamma^*(\lambda I - T^*)^{-1}(\mu I - T)^{-1}\Gamma J\ ,$$

which is just (9.19) written out for the case at hand. The limit

$$\Theta_\infty(\lambda) = I_m + \frac{i}{2\pi}\Gamma^*(T^* - \lambda I)^{-1}\Gamma J \tag{9.20}$$

agrees with formula (3) of [Li5], up to normalization.

Livšic himself seems to have obtained formula (9.20) by a mystical combination of insight and magic. Once this formula is available, however, it is readily checked by direct calculation with the help of (9.11), that

$$\frac{J - \Theta_\infty(\lambda)J\Theta_\infty(\omega)^*}{\rho_\omega(\lambda)} = F(\lambda)F(\omega)^* , \tag{9.21}$$

for every choice of λ and ω in $\mathcal{D}_{\rho_T}$. This corresponds to (7) of [Li5]; (8) of [Li5] also drops out by direct calculation.

CASE 2. T is a partial isometry (i.e., an isometry on the orthogonal complement of its kernel) with

$$\dim\{(I - T^*T)\mathcal{H}\} = \dim\{(I - TT^*)\mathcal{H}\} = m < \infty .$$

DISCUSSION. For ease of exposition we shall assume that there exists a point $\mu \in \mathbb{T}$ such that $\mu I - T$ is invertible. Then by formula (9.19) for $\Omega_+ = \mathbb{D}$,

$$\Theta_\mu(\lambda) = I_m - (1 - \lambda\mu^*)\Gamma^*(I - \lambda T^*)^{-1}(I - \mu^*T)^{-1}\Gamma J .$$

Therefore, since

$$\Gamma^*\Gamma = I_m \quad \text{and} \quad \Gamma J\Gamma^* = I - TT^* \geq 0 ,$$

it follows easily that $J = I_m$, and hence that

$$\begin{aligned}\Theta_\mu(\lambda) &= \Gamma^*\left\{I_m - (1 - \lambda\mu^*)(I - \lambda T^*)^{-1}(I - \mu^*T)^{-1}\right\}\Gamma \\ &= \Gamma^*(I - \lambda T^*)^{-1}\left\{\mu^*(\lambda I - T) - \lambda T^*(I - \mu^*T)\right\}(I - \mu^*T)^{-1}\Gamma .\end{aligned}$$

But now as

$$\Gamma : \mathbb{C}^m \text{ onto } (I - TT^*)\mathcal{H} = \ker T^* ,$$

it follows that

$$T^*\Gamma = 0 .$$

Therefore

$$\Theta_\mu(\lambda) = \mu^*\Gamma^*(I - \lambda T^*)^{-1}(\lambda I - T)(I - \mu^*T)^{-1}\Gamma ,$$

and

$$\begin{aligned}\Gamma^*(I - \lambda T^*)^{-1}T &= \Gamma^*\{(I - \lambda T^*)^{-1} - I\}T \\ &= \Gamma^*(I - \lambda T^*)^{-1}\lambda T^*T ,\end{aligned}$$

which combine to yield

$$\Theta_\mu(\lambda) = \mu^*\lambda\Gamma^*(I - \lambda T^*)^{-1}(I - T^*T)(I - \mu^*T)^{-1}\Gamma \ .$$

This last formula can be rewritten as

$$\Theta_\mu(\lambda) = \Psi(\lambda)\Psi(\mu)^* \tag{9.22}$$

with

$$\Psi(\lambda) = \lambda\Gamma^*(I - \lambda T^*)^{-1}\Gamma_1 \ , \tag{9.23}$$

where Γ_1 is a linear mapping from $\mathbb{C}^m$ onto $(I - T^*T)\mathcal{H}$ such that

$$\Gamma_1^*\Gamma_1 = I_m \quad \text{and} \quad \Gamma_1\Gamma_1^* = I - T^*T \ .$$

(The existence of such a Γ_1 follows from Theorem 9.2 with $X = I - T^*T$.) Moreover, since $\Theta_\mu(\mu) = I_m$, it follows that $\Psi(\mu)$ is unitary. Formula (9.23) agrees with formula (20) of [Li4] up to a sign, and so $\Theta_\mu(\lambda)$ is unitarily equivalent to the Livšic characteristic function in this case also.

For the role of the Livšic characteristic function in the study of the indicated classes of operators, the reader is referred to the original papers of Livšic: [Li1]–[Li5] and [LiP]. The recent survey article [BC] of Ball and Cohen is also recommended.

10. REFERENCES

[AG] N.I. Akhiezer and I.M. Glazman, *Theory of linear operators in Hilbert space*, Pitman Advanced Publishing Program, London, 1980.

[A] D. Alpay, *Reproducing kernel Krein spaces of analytic functions and inverse scattering*, Ph.D. Thesis, Dept. of Theoretical Mathematics, The Weizmann Institute of Science, Rehovot, Israel (submitted October, 1985).

[AD1] D. Alpay and H. Dym, *Hilbert spaces of analytic functions, inverse scattering, and operator models I*, Integral Equations Operator Theory **7** (1984) 589-641.

[AD2] ———, *On applications of reproducing kernel spaces to the Schur algorithm and rational J unitary factorization*, in: I. Schur Methods in Operator Theory and Signal Processing (I. Gohberg, ed.), Oper. Theory: Adv. Appl. **OT18**, Birkhauser Verlag, Basel, 1986, pp. 89-159.

[AD3] ———, *Structured invariant spaces of vector valued rational functions, Hermitian matrices, and a generalization of the Iohvidov laws*, Linear Algebra Appl. **137/138** (1990), 137-181.

[AD4] ———, *On reproducing kernel spaces, the Schur algorithm and interpolation in a general class of domains*, in: Operator Theory and Complex Analysis (T. Ando and I. Gohberg, eds.), Oper. Theory: Adv. Appl. **OT59**, Birkhäuser Verlag, Basel, 1992, pp. 30-77.

[AD5] ———, *On a new class of reproducing kernel spaces and a new generalization of the Iohvidov laws*, Linear Algebra Appl. **178** (1993), 109-183.

[AD6] ———, *On a new class of structured reproducing kernel spaces*, J. Funct. Anal. **111**, 1993, 1-28.

[AGo] D. Alpay and I. Gohberg, *Unitary rational matrix functions*, in: Topics in Interpolation Theory of Rational Matrix-Valued Functions, Oper. Theory: Adv. Appl. **OT33**, Birkhäuser Verlag, Basel, 1988, pp. 175-222.

[Arn] N. Aronszajn, *Theory of reproducing kernels*, Trans. Amer. Math. Soc. **68** (1950), 337-404.

[Ba] J.A. Ball, *Models for non contractions*, J. Math. Anal. Appl. **52** (1975), 235-254.

[BC] J.A. Ball and N. Cohen, *de Branges Rovnyak operator models and systems: a survey*, in Topics in Matrix and Operator Theory (H. Bart, I. Gohberg and M. A. Kaashoek, eds.), Oper. Theory: Adv. Appl. **OT50**, 1991, Birkhäuser Verlag, Basel, pp. 93-136.

[BGR] J.A. Ball, I. Gohberg and L. Rodman, *Interpolation of Rational Matrix Functions*, Birkhäuser Verlag, Basel, 1990.

[BR] J.A. Ball and A. Ran, *Local inverse spectral problems for rational matrix functions*, Integral Equations Operator Theory **10** (1987), 350-415.

[dB] L. de Branges, *Some Hilbert spaces of analytic functions I*, Trans. Amer. Math. Soc. **106** (1963), 445-468.

[dBR] L. de Branges and J. Rovnyak, *Canonical models in quantum scattering theory*, in: Perturbation Theory and its Applications in Quantum Mechanics (C. Wilcox, ed.), Wiley, New York, 1966, pp. 295-392.

[D1] H. Dym, *Hermitian block Toeplitz matrices, orthogonal polynomials, reproducing kernel Pontryagin spaces, interpolation and extension*, in: Orthogonal Matrix-Valued Polynomials and Applications (I. Gohberg, ed.), Oper. Theory: Adv. Appl. **OT34**, Birkhäuser Verlag, Basel, 1988, pp. 79-135.

[D2] ———, *J Contractive Matrix Functions, Reproducing Kernel Hilbert Spaces and Interpolation*, CBMS Regional Conference Series in Mathematics, No. 71, Amer. Math. Soc., Providence, R.I., 1989.

[D3] ———, *On reproducing kernel spaces, J unitary matrix functions, interpolation and displacement rank*, in: The Gohberg Anniversary Collection II (H. Dym, S. Goldberg, M.A. Kaashoek and P. Lancaster, eds.), Oper. Theory: Adv. Appl. **OT41**, Birkhäuser Verlag, Basel, 1989, pp. 173-239.

[D4] ———, *On Hermitian block Hankel matrices, matrix polynomials, the Hamburger moment problem, interpolation and maximum entropy*, Integral Equations Operator Theory **12** 1989, 757-812.

[D5] ———, *RKHS, LIS and Interpolation*, in: Signal Processing, Scattering and Operator Theory, and Numerical Methods (M.A. Kaashoek, J.H. van Schuppen and A.C.M. Ran, eds.), Birkhäuser, Boston, 1990, pp. 65-78.

[D6] ———, *A Hermite theorem for matrix polynomials*, in: Topics in Matrix and Operator Theory (H. Bart, I Gohberg and M.A. Kaashoek, eds.), Oper. Theory: Adv. Appl. **OT50**, 1991, Birkhäuser Verlag, Basel, pp. 191–214.

[D7] ———, *On the zeros of some continuous analogues of matrix and a related extension problem with negative squares*, Comm. Pure Appl. Math., in press.

[D8] ———, *Methods and Examples in Interpolation Theory*, (tentative title), in preparation.

[G–M] Y. Genin, P. Van Dooren, T. Kailath, J. M. Delosme and M. Morf, *On Σ-lossless transfer functions and related questions*, Linear Algebra Appl. **50** (1983), 251-275.

[G] K. Glover, *All optimal Hankel-norm approximations of linear multivariable systems and their L^∞ error bounds*, Int. J. Control **39** (1984), 1115-1193.

[K] T. Kailath, *Linear Systems*, Prentice Hall, Englewood Cliffs, Jew Jersey, 1980.

[Ki] H. Kimura, *Directional interpolational approach to H^∞-optimization and robust stabilization*, IEEE Trans. on Automatic Control **32** (1987), 1085-1093.

[LAK] H. Lev-Ari and T. Kailath, *Triangular factorization of structured Hermitian matrices*, in: I. Schur Methods in Operator Theory and Signal Processing (I. Gohberg, ed.), Oper. Theory: Adv. Appl. **OT18**, Birkhäuser Verlag, Basel, 1986, pp. 301-324.

[LA] D.J.N. Limebeer and B.D.O. Anderson, *An interpolation theory approach to H^∞ controller degree bounds*, Linear Alg. Appl. **98** (1988), 347-386.

[Li1] M.S. Livšic, *On a class of linear operators in Hilbert space*, Mat. Sbornik N.S. **19** (1946), 239-262; English transl. Amer. Math. Soc. Transl. (2) **13** (1960), 61-83.

[Li2] ———, *On the theory of isometric operators with equal deficiency indices*, Dokl. Akad. Nauk SSSR **58** (1947), 13-15.

[Li3] ———, *On the theory of elementary divisors of nonhermitian operators*, Dokl. Akad. Nauk SSSR **60** (1948), 17-20.

[Li4] ———, *Isometric operators with equal deficiency indices, quasi-unitary operators*, Mat. Sbornik N.S. **26** (1950), 247-264; English transl. Amer. Math. Soc. Transl. (2) **13** (1960), 85-103.

[Li5] ———, *On the spectral decomposition of linear non-selfadjoint operators*, Mat. Sbornik N.S. **34** (1954), 145-199; English transl. Amer. Math. Soc. Transl. (2) **5** (1957), 67-114.

[LiP] M.S. Livšic and V.P. Potapov *A theorem on the multiplication of characteristic matrix functions*, Dokl. Akad. Nauk SSSR **72** (1950), 625-628.

[Nu1] A. A. Nudelman, *On a new problem of moment type*, Soviet Math. Dokl. **18** (1977), 507-510.

[Nu2] ———, *A generalization of classical interpolation problems*, Soviet Math. Dokl. **23** (1981), 125-128.

[Nu3] ———, *Lecture at Workshop on Operator Theory and Complex Analysis*, Sapporo, Japan, June 1991.

[P] V.P. Potapov, *The multiplicative structure of J-contractive matrix functions*, Trudy Moskov. Mat. Obšč. **4** (1955), 125-163; English trans. Amer. Math. Soc. Transl. (2) **15** (1960), 131–243.

[Ros] M. Rosenblum, *On the operator equation* $BX - XA = Q$, Duke Math. J. **23** (1956), 263-269.

[Rov] J. Rovnyak, *Characterization of spaces* $\mathcal{K}(M)$, unpublished manuscript.

[S] L. A. Sakhnovich, *Factorization problems and operator identities*, Russian Math. Surveys **41** (1986), 1-64.

[SK] A. H. Sayed and T. Kailath, *Recursive solutions to rational interpolation problems*, ISCAS, 1992.

Department of Theoretical Mathematics
The Weizmann Institute of Science
Rehovot 76100, Israel

AMS Classification Numbers: 30E05, 47A57, 47B38, 93B28.

Operator Theory:
Advances and Applications, Vol. 73

Arveson's Distance Formulae and Robust Stabilization for Linear Time-Varying Systems

Avraham Feintuch

To my dear friend and colleague Prof. Moshe Livsic on the occasion of his retirement.

1 Introduction

Robustness properties for time invariant systems were first studied in the context of Youla parameterization by Vidyasagar and Kimura ([13], [12]). There, plants are given by co-prime factorizations and neighbourhoods of the plant are characterized in terms of perturbations of the numerator and denominator of a given co-prime factorization. In this paper we give a generalization of this result to the time-varying setting. Because the natural notions of numerator, denominator, poles, zeroes for time-invariant systems do not have an obvious (or in some cases any) meaning in the time-varying case, another formulation is useful. We use the framework described in [2], which allows us to give a purely operator-theoretic formulation of this problem, strongly related to that given for time-invariant systems in the fundamental paper of Georgiou and Smith ([1]). Our main result (Theorem 4.3) is strongly related to that of Shamma ([10]) on the necessary condition for the Small-Gain Theorem for time-varying systems and our proof can be adapted to give what seems to be a more transparent proof of that result.

As a consequence of this theorem we obtain upper and lower bounds for the maximal radius of neighbourhoods of stability. In the time-invariant case these numbers are the same and reduce to the formulae of Glover and McFarlane. It is of interest to point out that both endpoints of the obtained interval have natural interpretations as the distance from a given operator to an operator algebra (see [1]). This situation arises by relating the robust optimization problem to a 2-block uniform optimization problem ([4]).

The approach in this paper is strongly related to the fundamental paper of Georgiou-Smith [8] and relates to issues raised in [7]. My thanks to T. Georgiou for a valuable discussion related to ([7], Theorem 2) as well as to my colleague A. Markus for our discussions dealing with the proof of Theorem 4.3.

2 Preliminaries

Let h^2 denote the Hilbert sequence space

$$\{\langle x_0, x_1, x_2, \dots \rangle : \sum_{i=0}^{\infty} |x_i|^2 < \infty\}.$$

The x_i's can be scalars or vectors in $\mathbb{C}^n$. The dimension (as long as it is finite) doesn't matter. The truncation projections on h^2 are denote by P_n and defined by:

$$P_n\langle x_0, x_1, \dots, x_n, x_{n+1} \dots \rangle = \langle x_0, x_1, \dots, x_n, 0 \dots \rangle.$$

The following definition is a special case of Definition 3.1 of [7]. The motivation behind this definition is given there.

Definition 2.1 *A linear system on h^2 is a lower triangular infinite complex matrix A which defines a linear transformation on h^2 by matrix multiplication.*

Of course, in general, for $x \in h^2$, Ax may not be in h^2. We associate with A the linear manifold

$$\mathcal{D} = \mathcal{D}(A) = \{x \in h^2 : Ax \in h^2\}.$$

It follows from [7] that A is a closed operator; i.e. its graph $\mathcal{J}(A) = \{\langle x, Ax\rangle : x \in \mathcal{D}\}$, is a closed subspace of $h^2 \oplus h^2$. (It is of course possible that $\mathcal{D}(A) = \{0\}$. See [8] for such an example.

The set $\mathcal{L}$ of linear systems, with the standard operations of addition and matrix multiplication forms an algebra. The invertible elements of this algebra are those with non-singular entries on their diagonal.

We are concerned here with linear systems that have the additional property of stability.

Definition 2.2 *$A \in \mathcal{L}$ is stable if $Ah^2 \subset h^2$.*

Since A is closed, it follows from the Closed Graph Theorem that A is stable if and only if A defines a bounded operator on h^2. The stable systems can therefore be identified with the lower triangular matrices which define bounded operators on h^2. These also form an algebra which we will denote by $\mathcal{C}$. $\mathcal{C}$ is, of course, a nest algebra and is determined by its complete nest of invariant subspace $\{(I - P_n)h^2 : n \geq 0\}$. The following formula is a special case (for this particular nest) of a distance formula due to Arveson ([1]):

Theorem 2.3 *If T is a bounded linear operator on h^2, the $d(T, \mathcal{C}) = \sup_{n\geq 0}\{\|P_nT(I - P_n)\|\}$.*

Let K denote the ideal of compact operators on h^2. Then $\mathcal{C} + \mathcal{K}$ is a norm-closed algebra of operators and for T as above

$$d(T, \mathcal{C} + \mathcal{K}) = \limsup_n\{\|P_nT(I - P_n)\|\}.$$

The range of an operator T will be denoted by $R(T)$ and its null space by $N(T)$.

3 Stabilization and Proper Representations:

Most of the material in this section is standard and brought for later reference. We adopt the neat point of view of [10], [2].

Consider $L \in \mathcal{L}$ with graph $\mathcal{J}(L) \subset h^2 \oplus h^2$ whch we will relate to as the range of the linear tranformation

$$\begin{bmatrix} I \\ L \end{bmatrix} : \mathcal{D}(L) \to h^2 \oplus h^2.$$

The inverse graph $\mathcal{J}^{-1}(L)$ is the range of

$$\begin{bmatrix} L \\ I \end{bmatrix} : \mathcal{D}(L) \to h^2 \oplus h^2.$$

Definition 3.1 *For $L, C \in \mathcal{L}$, the operator*

$$\begin{bmatrix} I & -C \\ -L & I \end{bmatrix} : \mathcal{D}(C) \oplus \mathcal{D}(L) \to h^2 \oplus h^2.$$

is the feedback system with plant P and compensator C.

This 2×2 operator matrix arises from the system of equations

$$\begin{bmatrix} u_1 \\ u_2 \end{bmatrix} = \begin{bmatrix} I & -C \\ -L & I \end{bmatrix} \begin{bmatrix} e_1 \\ e_2 \end{bmatrix}$$

where the feedback property is expressed by the fact that u_1, u_2 are external inputs and e_1, e_2 are internal ones. In the standard block diagram form this can be represented as:

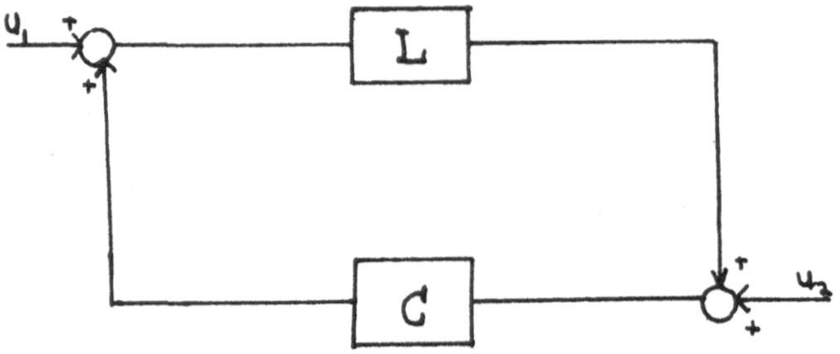

Figure 1: Standard Feedback Configuration

The configuration, denoted by $\{L, C\}$ is said to be ***stable*** if the operators $u_i \to e_j$, for $i, j = 1, 2$, are bounded. This is equivalent to the operator $\begin{bmatrix} I & -C \\ -L & I \end{bmatrix}$ having a bounded inverse, or, geometrically to $\mathcal{J}(-L) \cap \mathcal{J}^{-1}(-C) = \{0\}$, $\mathcal{J}(-L) + \mathcal{J}^{-1}(-C) = h^2 \oplus h^2$.

When this is the case the inverse matrix is given by

$$\begin{bmatrix} (I - CL)^{-1} & C(I - LC)^{-1} \\ L(I - CL)^{-1} & (I - LC)^{-1} \end{bmatrix}.$$

Definition 3.2 *$L \in \mathcal{L}$ is stabilizeable if there exists $C \in \mathcal{L}$ such that the feedback configuration $\{L, C\}$ is stable.*

In this case we say that C stabilizes L. This notion is symmetric in L and C.

It is classical ([16]) that finite-dimensional time-invariant linear systems are stabilizeable. However, it is not hard to construct linear systems (as defined above) which are not. A trivial way to do this is to take $L \in \mathcal{L}$ with $\mathcal{D}(L) = \{0\}$. The following example is much more interesting.

Example 3.3:

Let L be the lower triangular infinte matrix with all entries zero except on the first subdiagonal and with $[L]_{i,i-1} = i$, $i = 1, 2, \ldots$. We show there is no $C \in \mathcal{L}$ such that $L(I - CL)^{-1}$ is a bounded operator. Since L is strictly lower triangular, so is CL and thus $(I - CL)$ is invertible in $\mathcal{L}$ and $(I - CL)^{-1}$ has all its diagonal elements equal to one. Thus $L(I - CL)^{-1}$ is strictly lower triangular, with the same first sub-diagonal as L. It therefore can't be bounded.

A similar argument can be given for any $L \in \mathcal{L}$ with an unbounded subdiagonal.

Stabilizeable systems turn out to be those whose graphs have nice representations.

Definition 3.4 *$L \in \mathcal{L}$ has a right representation $\begin{bmatrix} M \\ N \end{bmatrix}$ if $M, N \in \mathcal{C}$ and $\mathcal{R}\left(\begin{bmatrix} M \\ N \end{bmatrix}\right) = \mathcal{J}(L)$. L has a left representation $[-\hat{N}, \hat{M}]$ if $\hat{M}, \hat{N} \in \mathcal{C}$ and $\mathcal{N}(-\hat{N}, \hat{M}]) = \mathcal{J}(L)$.*

If $\{L, C\}$ is stable then L has right representation $\begin{bmatrix} (I - CL)^{-1} \\ L(I - CL)^{-1} \end{bmatrix}$ and left representation $[-(I - LC)^{-1}L, (I - CL)^{-1}]$. However non-stabilizeable systems may also have such representations.

Example 3.5: Let L be as in Example 3.3. Take

$$M = \operatorname{diag}\{1, \frac{1}{2}, \frac{1}{3}, \ldots\}$$

and let N be the unilateral shift. Take

$$\hat{M} = \operatorname{diag}\{1, 1, \frac{1}{2}, \frac{1}{3}, \ldots\}$$

$$\hat{N} = N.$$

Noting that

$$\begin{aligned} \mathcal{D}(L) &= \mathcal{D}(M^{-1}) \\ &= \{ \langle x_0, x_1, \ldots \rangle \; : \; \textstyle\sum_{i=0}^{\infty} |(i+1)x_i|^2 < \infty \}. \end{aligned}$$

it follows that for $x \in \mathcal{D}(L)$, $M^{-1}x \in h^2$, and

$$\begin{bmatrix} I \\ L \end{bmatrix} x = \begin{bmatrix} I \\ NM^{-1} \end{bmatrix} x = \begin{bmatrix} M \\ N \end{bmatrix} M^{-1}x.$$

Thus $\mathcal{J}(L) \subset R\left(\begin{bmatrix} M \\ N \end{bmatrix}\right)$. Also, for $y \in h^2$,

$$\begin{bmatrix} M \\ N \end{bmatrix} u = \begin{bmatrix} I \\ L \end{bmatrix} My \text{ where } My \in \mathcal{D}(M^{-1}) = \mathcal{D}(L).$$

This gives the opposite inclusion.

To see that $\mathcal{J}(L) = \mathcal{N}([-\hat{N}, \hat{M}])$, note that $L = \hat{M}^{-1}\hat{N}$. Therefore

$$[-\hat{N}, \hat{M}] \begin{bmatrix} I \\ L \end{bmatrix} x = -\hat{N}x + \hat{M}\hat{M}^{-1}\hat{N}x = 0.$$

On the other hand, if $x, y \in h^2$ and $[-\hat{N}, \hat{M}] \begin{bmatrix} x \\ y \end{bmatrix} - 0$, then $\hat{N}x = \hat{M}y$. Thus $\hat{N}x \in R(\hat{M}) = \mathcal{D}(\hat{M}^{-1})$, and $y = \hat{M}^{-1}\hat{N}x$. Thus

$$\begin{bmatrix} x \\ y \end{bmatrix} = \begin{bmatrix} I \\ L \end{bmatrix} x$$

and $\mathcal{N}([-\hat{N}, \hat{M}]) = \mathcal{J}(L)$.

We need a stronger notion.

Definition 3.6 *A right (left) representation* $\begin{bmatrix} M \\ N \end{bmatrix}$ *of* $L \in \mathcal{L}$ *is* proper *if it has a left (right) inverse whose entries are in* $\mathcal{C}$*; equivalently, if there exist* $X, Y (\hat{X}, \hat{Y}) \in \mathcal{C}$ *such that* $YM + XN = I$ $(\hat{M}\hat{Y} + \hat{N}\hat{X} = I)$.

The main result of [2] is that stabilizeabilty is equivalent to the existance of right and left proper representations.

Theorem 3.7 *(i) If* $\{L, C\}$ *is stable then there exist stable operators* $M, N, X, Y, \hat{M}, \hat{N}, \hat{X}, \hat{Y}$ *such that*

(1) $\begin{bmatrix} M \\ N \end{bmatrix}$ *and* $[-\hat{N}, \hat{M}]$ *are, respectively, right and left proper represenations for* L.

(2) $\begin{bmatrix} M & \hat{X} \\ N & \hat{Y} \end{bmatrix} \begin{bmatrix} Y & -X \\ -\hat{N} & \hat{M} \end{bmatrix} = \begin{bmatrix} Y & -X \\ -\hat{N} & \hat{M} \end{bmatrix} \begin{bmatrix} M & \hat{X} \\ N & \hat{Y} \end{bmatrix} = \begin{bmatrix} I & 0 \\ 0 & I \end{bmatrix}$.

(ii) IF $L \in \mathcal{L}$ *is given with proper left and right representations then*

(a) *L is stabilizeable.*

(b) *The representations can be chosen so that they satisfy (2).*

(c) $C \in \mathcal{L}$ *stabilizes* L *if and only if it has a proper right representation* $\begin{bmatrix} \hat{Y} + NQ \\ \hat{X} + MQ \end{bmatrix}$ *and a proper left representation* $[X + Q\hat{M}, Y + Q\hat{N}]$*, for some* $Q \in \mathcal{C}$.

Remark 3.8

(1) When is a right representation $\begin{bmatrix} M \\ N \end{bmatrix}$ of L proper? This has been answered in [1]; if and only if there exists $\varepsilon > 0$ such that for all truncation projections P_n and for all $x \in h^2$,

$$\left\| \begin{bmatrix} P_n M \\ P_n N \end{bmatrix} x \right\| \geq \varepsilon \|P_n x\|.$$

(2) Proper representations are not unique. If $\begin{bmatrix} M \\ N \end{bmatrix}$ is a proper right representation for L, and T is invertible in $\mathcal{C}$, then $\begin{bmatrix} MT \\ NT \end{bmatrix}$ is also a proper right representation. In fact, all proper right represenations for L are given in this form. A similar statement holds for left representations.

We will need a particular proper representation introduced by Vidyasagar ([16]).

Definition 3.9 *The proper right representation* $\begin{bmatrix} M \\ N \end{bmatrix}$ *of* L *is normalized if* $\begin{bmatrix} M \\ N \end{bmatrix}$ *is an isometry* $(M^*M + N^*N = I)$.

Such representations are constructed as follows. If $\begin{bmatrix} M \\ N \end{bmatrix}$ is a given proper right representation then since it is left invertible,

$$M^*M + N^*N \geq cI > 0.$$

By spectral factorization ([4]), there exists T invertible in $\mathcal{C}$ with

$$M^*M + N^*N = T^*T.$$

Then $\begin{bmatrix} M \\ N \end{bmatrix} T^{-1}$ is an isometry.

Theorem 3.10 *([4]): Suppose* $L \in \mathcal{L}$ *with right representation* $\begin{bmatrix} M \\ N \end{bmatrix}$ *and* $C \in \mathcal{L}$ *has proper left representation* $[-A, B]$. *Then* C *stabilizes* L *if and only if* $[B, -A]\begin{bmatrix} M \\ N \end{bmatrix}$ *is invertible.*

4 Robust Stabilization: Proper Representation Uncertainty

Up to this point we have considered the problem of stabilizing a single given plant. However, the essential uncertainty in modelling procedures suggests that a more realistic problem would be to find a controller C which stabilizes a family of plants related to the given model L. What family to consider is an issue of considerable controversy. Here we will take the approach of Vidyasagar and Kimura ([15]) and consider neighborhoods of a plant obtained by taking perturbations of its proper representations. More precisely, for $L \in \mathcal{L}$ stabilizable, with right proper representation $\begin{bmatrix} M \\ N \end{bmatrix}$, consider the ball

$$B(L,R) = \{\begin{bmatrix} M+\Delta M \\ N+\Delta N \end{bmatrix} : \ \Delta M, \Delta N \in C, \quad \left\| \begin{bmatrix} \Delta M \\ \Delta N \end{bmatrix} \right\| < r\}.$$

Note that not all elements of $B(L,R)$ will represent linear systems; i.e. they won't have as their range the graph of some $S \in \mathcal{L}$. It is of interest to characterize those that do represent linear systems.

Theorem 4.1 *Given $M, N \in C$, the operator matrix $\begin{bmatrix} M \\ N \end{bmatrix}$ is a right representation of some $S \in \mathcal{L}$ if and only if*

(1) $R\left(\begin{bmatrix} M \\ N \end{bmatrix}\right)$ *is closed.*

(2) *For each n, $Ker P_n M P_n \subseteq Ker P_n N P_n$.*

$\begin{bmatrix} M \\ N \end{bmatrix}$ *is a right proper representation of some $S \in \mathcal{L}$ if and only if*

(1') *there exist $X, Y \in C$ such that $XN + YM = I$.*

(2') *M is invertible in $\mathcal{L}$.*

Suppose $L \in \mathcal{L}$ is stabilizeable with proper right representation $\begin{bmatrix} M \\ N \end{bmatrix}$ and that $C \in \mathcal{L}$ with proper left represenattion $[-A, B]$ stabilizes L. Consider the function

$$f\left(\begin{bmatrix} M_1 \\ N_1 \end{bmatrix}\right) = \left\|(BM_1 + AN_1)^{-1}\right\|$$

defined at the points of the neighborhood $B(L,r)$ where the inverse exists. Here $M_1 = M + \Delta M, N + \Delta N$ with $\left\| \begin{bmatrix} \Delta M \\ \Delta N \end{bmatrix} \right\| < r$ and $\Delta M, \Delta N \in C$. Note that by Theorem 3.7, $[-A, B]$ must be of the form $[-(X + Q\hat{M}), (Y + Q\hat{N})]$ and that

$$\begin{aligned} [B, -A] \begin{bmatrix} M \\ N \end{bmatrix} &= [Y + Q\hat{N}, -X - Q\hat{M}] \begin{bmatrix} M \\ N \end{bmatrix} \\ &= YM - XN + Q(\hat{N}M - \hat{M}N) \\ &= I. \end{aligned}$$

Also, if $|||[B,-A]||| \leq \frac{1}{r}$,

$$[B,-A]\begin{bmatrix} M_1 \\ N_1 \end{bmatrix} = [B,-A]\begin{bmatrix} M+\Delta M \\ N+\Delta N \end{bmatrix} = I + [B,-A]\begin{bmatrix} \Delta M \\ \Delta N \end{bmatrix}$$

Since $\left\|[B,-A]\begin{bmatrix} \Delta M \\ \Delta N \end{bmatrix}\right\| < 1$, it follows that $[B,-A]\begin{bmatrix} M_1 \\ N_1 \end{bmatrix}$ is invertible. In particular, if $\begin{bmatrix} M_1 \\ N_1 \end{bmatrix}$ is the right representation of a linear system S, it follows from Theorem 3.10 that S is stabilized by C. Also, $f\left(\begin{bmatrix} M_1 \\ N_1 \end{bmatrix}\right)$ is defined on $\mathcal{B}(h,r)$ and it is easily seen that it is uniformly bounded there. We state this formally.

Theorem 4.2 *Suppose $L \in \mathcal{L}$ is stabilizeable with proper right representation $\begin{bmatrix} M \\ N \end{bmatrix}$ chosen as in Theorem 3.7. Suppose C stabilizes L and has proper left representation $[-A,B]$ such that $|||[-A,B]||| \leq \frac{1}{r}$. Then C stabilizes all linear systems with right representations in $B(L,r)$ and f is uniformly bounded there.*

This is the sufficiency part of the Vidyasagar-Kimura Theorem for finite dimensional time-invariant systems. In that case the condition $|||[-A,B]||| \leq \frac{1}{r}$ is also necessary. If C staiblizes all linear systems in $\mathcal{B}(L,R)$ then $|||[-A,B]||| \leq \frac{1}{r}$. This leads to the fact that r_{opt} the radius of the larges ball $\mathcal{B}(L,r_{opt})$ stabilized by a fixed $C \in \mathcal{L}$ is given as

$$\frac{1}{r_{opt}} = \inf_{Q \in \mathcal{C}} |||[Y+Q\hat{N}, -X-Q\hat{N}]|||$$

and this can be transformed to a standard 2-block problem. For time-varying systems this is not the case.

Example 4.3: Take $r=1$. Consider the operator in $\mathcal{C}$ whose matrix representation is

$$A = \begin{bmatrix} 0 & 0 & \dots \\ a_{10} & 0 & \dots \\ a_{20} & 0 & \dots \\ \vdots & \vdots & \end{bmatrix}$$

where $2 > \|A\| > 1$. Let $B = \frac{1}{2}I$, and C have proper left representation $[-A,B]$. Take $M = -A + 2I$, $N = \frac{1}{2}I$. Then $[B,-A]\begin{bmatrix} M \\ N \end{bmatrix} = \frac{1}{2}(A+2I) - \frac{1}{2}A = I$, so that $L = NM^{-1}$ with right representation $\begin{bmatrix} M \\ N \end{bmatrix}$ is stabilized by $C = AB^{-1}$. Also $|||[-A,B]||| > \|A\| > 1$.

Suppose $\begin{bmatrix} \Delta M \\ \Delta N \end{bmatrix}$ is given with $\Delta M, \Delta N \in \mathcal{C}$ and $\left\|\begin{bmatrix} \Delta M \\ \Delta N \end{bmatrix}\right\| < 1$. We claim that

$$[B,-A]\begin{bmatrix} M+\Delta M \\ N+\Delta N \end{bmatrix} = I + \frac{1}{2}\Delta M - A\Delta N$$

is invertible. Since $\|\frac{1}{2}\Delta M\| < 1$, $I + \frac{1}{2}\Delta M$ is invertible. Now $A\Delta N$ has a matrix representation of the form

$$\begin{bmatrix} 0 & 0 & \dots \\ & 0 & \dots \\ & 0 & \dots \\ & 0 & \dots \\ \vdots & \vdots & \end{bmatrix}$$

Since $T \in C$ is invertible if and only if for some $k \geq 0$, $P_k T P_k$ and $(I - P_k)T(I - P_k)$ are invertible, respectively, on $P_k h^2$ and $(I - P_k)h^2$, it suffices to produce such a k. Note that $P_k(I + \frac{1}{2}\Delta M - A\Delta N)P_k$ is invertible for all k. Choose k sufficiently large so that

$$\|(\frac{1}{2}\Delta M - A\Delta N)(I - P_k)\| < 1.$$

This is possible since $\|\frac{1}{2}\Delta M\| < 1$, and $\|A\Delta N(I - P_k)\| = \|A(I - P_k)\Delta N(I - P_k)\| \leq \|A(I - P_k)\| = \left(\sum_{j\geq k} |a_{j0}|\right)^{\frac{1}{2}} \to 0$ as $k \to \infty$. ∎

The necessary condition must take into consideration the time-variance of the system.

Theorem 4.4 *Suppose $L \in \mathcal{L}$ is stabilizable with proper right represenattion $\begin{bmatrix} M \\ N \end{bmatrix}$. Suppose $C \in \mathcal{L}$ stabilizes L and has proper left representation $[-A, B]$. Then:*

(1) *if for all $\begin{bmatrix} M_1 \\ N_1 \end{bmatrix} \in \mathcal{B}(L,r)$, $[B, -A]\begin{bmatrix} M \\ N \end{bmatrix}$ is invertible then $b = \inf_{n\geq 0} \|[B, -A](I - Q_n)\| \leq \frac{1}{r}$ where $Q_n = P_n \oplus P_n$ acting on $h^2 \oplus h^2$.*

(2) *if $br > 1$, then either there exists a linear system $S \in \mathcal{L}$ with right representation in $\mathcal{B}(L,R)$ which is not stabilized by C or there exists a sequence $\{S_n\} \in \mathcal{L}$ with right representations $\begin{bmatrix} M_n \\ N_n \end{bmatrix}$ such that $f\left(\begin{bmatrix} M_n \\ N_n \end{bmatrix}\right) \to \infty$ as $n \to \infty$.*

Proof: See [8].

If the situation described in (2) of Theorem 4.4 occurs we will say that C doesn't stabilize $\mathcal{B}(L,r)$. Thus if we wish to compute r_{opt}, the supremum over all r such that there exists a fixed compensator C which stabilizes $\mathcal{B}(L,r)$ we can compute upper and lower boundes for r_{opt} as follows: write $[B, -A]$ as $[Y + Q\hat{N}, -X - Q\hat{M}]$. By the above

$$\inf_{n\geq 0}\{\inf_{Q\in C} \|[Y + Q\hat{N}, -X - Q\hat{M}](I - Q_n)\|\} \leq \frac{1}{r_{opt}} \leq \inf_{Q\in C} \|[Y + Q\hat{N}, -X - Q\hat{M}]\|$$

Note that for time-invariant systems these two numbers are equal.

If we assume that the given representations $\begin{bmatrix} M \\ N \end{bmatrix}$ and $[-\hat{N}, \hat{M}]^*$ of L are normalized, then

$$Z = \begin{bmatrix} \hat{N}^* & M \\ -\hat{M}^* & N \end{bmatrix}$$

is a co-isometry. Thus

$$\|[Y+Q\hat{N},-X-Q\hat{M}]\| = \left\|\{[Y+Q\hat{N},-X-Q\hat{M}]\}Z\right\|$$
$$= \|[Y\hat{N}^* + X\hat{M}^* + Q, I]\|.$$

We compute the right and left sides of the above inequality using Arveson's distance formulas and obtain, denoting $Y\hat{N}^* + X\hat{M}^*$ by R, that

$$[1 + d(R, \mathcal{C} + \mathcal{K})^2]^{\frac{1}{2}} \leq \frac{1}{r_{opt}} \leq [1 + d(R, \mathcal{C})^2]^{\frac{1}{2}}.$$

In particular $r_{opt} \leq 1$.

5 Gap Metric Robustness

Suppose $L_1, L_2 \in \mathcal{L}$ with normalized proper right representations $\begin{bmatrix} M_1 \\ N_1 \end{bmatrix}$ and $\begin{bmatrix} M_2 \\ N_2 \end{bmatrix}$ respectively. Denote the projections on their ranges by π_1 and π_2 respectively. The gap between L_1 and L_2 is defined as

$$\delta(L_1, L_2) = \|\pi_1 - \pi_2\|$$

and the directed gap from L_1 to L_2 is

$$\vec{\delta}(L_1, L_2) = \|[I - \pi_2]\pi_1\|$$

It is well known that

$$\delta(L_1, L_2) = \max\{\vec{\delta}(L_1, L_2), \vec{\delta}(L_2, L_1)\}$$

and that if $\delta(L_1, L_2) < 1$ then all these numbers are equal. For these and other basic properties of the gap which defines a metric on $\mathcal{L}$ see [14].

By our assumptions on $\begin{bmatrix} M_1 \\ N_1 \end{bmatrix}$ and $\begin{bmatrix} M_2 \\ N_2 \end{bmatrix}$, it follows that for all n and $i = 1, 2$, $\begin{bmatrix} M_i \\ N_i \end{bmatrix}(I - P_n)$ are isometries on $(I - P_n)h^2$ with range in $(I - P_n)h^2 \oplus (I - P_n)h^2$. Let π_{in} denote the orthogonal projection on the range of $\begin{bmatrix} M_i \\ N_i \end{bmatrix}(I - P_n)$ and define

$$\vec{\delta}(L_1, L_2) = \left\| \left\{ \begin{bmatrix} I - P_n & 0 \\ 0 & I - P_n \end{bmatrix} - \pi_{2n} \right\} \pi_{1n} \right\|$$

and

$$\begin{aligned} \delta_n(L_1, L_2) &= \|\pi_{1n} - \pi_{2n}\| \\ &= \max\{\vec{\delta}_n(L_1, L_2), \vec{\delta}(L_2, L_1)\}. \end{aligned}$$

Definition 5.1 $\vec{\alpha}(L_1, L_2) = \sup_{n \geq 0} \vec{\delta}_n(L_1, L_2)$ *is the directed time-varying gap from* L_1 *to* L_2 *and* $\alpha(L_1, L_2) = \max\{\vec{\alpha}(L_1, L_2), \vec{\alpha}(L_2, L_1)\}$ *is the time-varying gap between* L_1 *and* L_2.

Note that if L_1 and L_2 are time-invariant systems these gaps just reduce to the standard gaps. The importance of the time-varying gap is due to the following.

Theorem 5.2 $\vec{\alpha}(L_1, L_2) = \inf_{Q \in C} \left\| \begin{bmatrix} M_1 \\ N_1 \end{bmatrix} - \begin{bmatrix} M_2 \\ N_2 \end{bmatrix} Q \right\|$.

Proof: See [9].

For time-invariant systems this just reduces to the formula of Georgiou.

This theorem is the main tool used to show the connection between gap metric robustness and robustness discussed in the previous section.

Let

$$\vec{s}(L,r) = \{\, L_1 \text{ stabilizable} \; : \; \vec{\alpha}(L, L_1) < r\}$$

$$s(L,r) = \{\, L_1 \text{ stabilizable} \; : \; \vec{\alpha}(L, L_1) < r\}$$

These are called the directed gap ball and gap ball with centre L and radius r. The gap metric robust stabilization problem is: find the maximal r such that $\alpha(L, L_1)$ is stabilzied by a fixed controller C.

Let

$$r_1 = \frac{1}{[1 + d(R,C)^2]^{\frac{1}{2}}}, \quad r_2 = \frac{1}{[1 + d(R, C + K)^2]^{\frac{1}{2}}}$$

for $R = Y\hat{N}^* + X\hat{M}^*$ as defined in the previous section.

Theorem 5.3 *Consider a stabilizable system L with normalized proper right representation $\begin{bmatrix} M \\ N \end{bmatrix}$, and consider a controller C which stabilizes L. For $r \leq r_2$ the following statements are equivalent:*

(a) *$\{L_1, C\}$ is stable for all $L_1 \in B(L,r)$.*

(b) *$\{L_1, C\}$ is stable for all $L_1 \in \vec{C}(L,r)$.*

(c) *$\{L_1, C\}$ is stable for all $L_1 \in C(L,r)$.*

For $r > r_1$ there is no C which stabilizes $C(L,r)$.

Proof: See [9].

6 Bibliography

[1] W. Arveson, "Interpolation in Nest Algebras", J. Funct. Anal. 20, 208-233, 1975.

[2] W. Dale, M. Smith, "Stabilizability and Existence of System Representations for Discrete-Time, Time Varying Systems", preprint.

[3] K. Davidson, "Nest Algebras", Pitman Research Notes in Mathematics, 191, Longman Scientifc and Technical, U.K., 1988.

[4] A. Feintuch and R. Saeks, "System Theory: A Hilbert Space Approach", Academic Press, New York, 1982.

[5] A. Feintuch and B. A. Francis, "Uniformly Optimal Control of Linear Feedback Systems", Automatica 21, 563-574, 1985.

[6] A. Feintuch and B. A. Francis, "Distance Formulas for Operator Algebras Arising in Optimal Control Problems", Operator Theory: Advances and Applications, 29, 151-170, 1988.

[7] A. Feintuch, "Graphs of Time-Varying Linear Systems and Co-prime Factorizations", J. Math. Anal. and Applic., 163, 1, 79-85, 1992.

[8] A. Feintuch, "Robustness for Time-Varying Systems", preprint.

[9] A. Feintuch, "Gap-Metric Robustness for Linear Time-Varying Systems", preprint.

[10] C. Foias, T. Georgiou, M. Smith, "Geometric Techniques for Robust Stabilization of Linear Time-Varying Systems", preprint.

[11] T. Georgiou, "On the Computation of the Gap Metric", Syst. Contr. Lett. 11, 253-257, 1988.

[12] T. Georgiou and M. Smith, "Optimal Robustness in the Gap Metric", IEEE Trans. Autom. Contr. 35, 6, 673-686, 1990.

[13] K. Glover and D. McFarlane, "Robust Stabilization of normalized Coprime Factor Plant Descriptions with H^∞-bounded Uncertainties", IEEE Trans. Autom. Contr. 34, 821-830, 1989.

[14] T. Kato, "Perturbation Theory for Linear Operators", Springer-Verlag, New York, 1966.

[15] M. Vidyasagar and H. Kimura, "Robust Controllers for Uncertain Linear Multivariable Systems", Automatica, 22, 85-94, 1986.

[16] M. Vidyasagar, "Control System Synthesis: A Factorization Approach, M.I.T. Press, Cambridge, 1986.

[17] G. Zames and A.K. El-Sakkary, "Unstable Systems and Feedback: the Gap Metric", Proc. Allerton Conf. 1980, 380-385.

[18] G. Zames, "Feedback and Optimal Sensitivity: Model reference transformations, multiplicative seminorms and approximate inverses", IEEE Trans. Autom. Contr. Ac-26, 301-320, 1981.

Dept. of Mathematics
Ben-Gurion University of the Negev
Beer Sheva, Israel

AMS Classification 47C05 93C50

Operator Theory:
Advances and Applications, Vol. 73

ENTIRE CYCLIC COHOMOLOGY OF BANACH ALGEBRAS

Peter Fillmore and Masoud Khalkhali*

Dedicated to Moshe Livsic

Connes has introduced the notion of cyclic cohomology as a replacement for de Rham cohomology in the non-commutative setting. Entire cyclic cohomology is an infinite dimensional version of cyclic cohomology. The technical complications of this subject are severe and consequently some of the fundamental properties that one would expect have until now been conjectural. We report on some recent results in this area, notably Morita and homotopy invariance. The details will appear elsewhere [10].

§1. BACKGROUND

The subject has its roots in index theory, which began with Fredholm's results on integral operators (1903) and reached a more or less definitive form in the work of Atkinson and Gohberg (c. 1951). Then came the K-theory of Grothendieck (1958) and Atiyah and Hirzebruch (1959), the Atiyah-Singer Index Theorem (1963) and the theorem of Atiyah and Jänich (1964). The latter, which asserts that $K^0(X) \cong [X, \mathcal{F}]$, where $\mathcal{F}$ denotes the Fredholm operators on a Hilbert space, put in evidence in a striking way the connection between index theory and K-theory.

It was realized almost immediately that the natural domain of K-theory is Banach algebras, the case of topological spaces X being the restriction to the commutative algebra $C(X)$. This is perhaps the origin of the idea of "non-commutative topology". In a parallel development, the theory of von Neumann algebras can be viewed as "non-commutative measure theory", since the commutative von Neumann algebras are the algebras $L^\infty(X,\mathcal{S},\mu)$.

About ten years ago Alain Connes began to point out the need to develop "non-commutative differential geometry". Such methods would be

---

* The work described here, except for 3.4, forms part of Khalkhali's Ph.D. dissertation (Dalhousie University, 1991), written under the supervision of Fillmore.

useful in working with spaces more general than manifolds:

- leaf spaces of foliations
- dual spaces of certain non-abelian groups (e.g. Lie groups and finitely generated discrete groups)
- orbit spaces of such groups.

The "topology" and "measure theory" of these spaces has been successfully studied:

- C*- and W*-algebras of a foliation
- group C*-algebras
- C*- and W*-crossed products.

One goal of non-commutative differential geometry would be the construction of a general de Rham theory and its application to index theory for families of elliptic operators. This is precisely the cyclic cohomology introduced in 1983 independently by Connes [4] and Tsygan [13].

One way to think of cyclic cohomology is by analogy with K-homology. This exists abstractly as the dual of K-theory, so for a space X one has a pairing

$$K_*(X) \times K^*(X) \to \mathbb{Z}.$$

Concrete cycles were constructed in the 1970's by Atiyah [1], Brown, Douglas and Fillmore [2] and Kasparov [8]. Thus the elements of $K_0(X)$ arise from elliptic operators D on X and the above pairing for $* = 0$ is

$$\langle [D], [E] \rangle = \text{index } D_E$$

for any vector bundle E, where D_E is the operator D with coefficients in E. For $* = 1$, elements of $K_1(X)$ come from homomorphisms τ of $C(X)$ into the Calkin algebra and the pairing is given by

$$\langle [\tau], [f] \rangle = \text{index } \tau_n(f)$$

for any $f : X \to GL_n(\mathbb{C})$, where τ_n is τ applied entry-wise.

Cyclic cohomology is the non-commutative replacement for this. If A is any algebra over $\mathbb{C}$, the (abstract) cyclic cocycles are the multilinear functionals $\phi : A^{n+1} \to \mathbb{C}$ that are appropriately related to the action of $\mathbb{Z}/(n+1)$, namely

$$\phi(a^0, a^1, \ldots, a_n) = (-1)^n \phi(a^n, a^0, \ldots, a^{n-1}),$$

and are cocycles for the Hochschild coboundary

$$b\phi(a^0, \ldots, a^{n+1}) = \phi(a^0 a^1, \ldots, a^{n+1}) - \phi(a^0, a^1 a^2, \ldots, a^{n+1}) + \ldots + (-1)^{n+1} \phi(a^{n+1} a^0, \ldots, a^n).$$

In this situation elements of $K_0(A)$ come from idempotents e in

$M_k(A) = M_k(\mathbb{C}) \otimes A$ and the pairing is given by

$$\langle[\phi], [e]\rangle = \phi_k(e, e,\dots, e),$$

where $\phi_k(\mu_0 \otimes a^0, \dots, \mu_n \otimes a^n) = \mathrm{tr}(\mu_0 \mu_1 \dots \mu_n)\, \phi(a^0, a^1,\dots, a^n)$.
Concrete cocycles are obtained from Fredholm modules $(\mathcal{H}, D)$ over A. Here $\mathcal{H}$ is a Hilbert space and an A-module and D is a self-adjoint operator on $\mathcal{H}$ such that the commutator $[D, a]$ is bounded for all $a \in A$. One takes

$$\phi(a^0,\dots, a^n) = \mathrm{tr}\, P[P, a^0] \dots [P, a^n],$$

where $P = D|D|^{-1}$, and says that $(\mathcal{H}, D)$ is n-summable if this exists.

As examples we mention:

(i) $A = C(\mathbb{T}^p)$, $\mathcal{H} = L^2(\mathbb{T}^p)$, $D = \Sigma \frac{\partial^2}{\partial x_i^2}$

This is n-summable for $2n > p$.
(ii) $A = C^*_{red}(\Gamma)$, where Γ is a finitely-generated discrete group, $\mathcal{H} = \ell^2(\Gamma)$, and D = multiplication by the word-length function L. This is finitely summable if and only if

$$\mathrm{card}\{\, \gamma \in \Gamma \mid L(\gamma) \le n \,\} = O(n^k)$$

for some k (i.e. Γ is of polynomial growth).

These examples suggest that in some cases of interest there will be no finitely-summable Fredholm modules. It may be, however, that there still exist Fredholm modules that yield cocycles in an appropriate extension of cyclic cohomology. This is in fact the case for the θ-summable modules introduced by Connes [5]. Moreover such modules, which have the defining property that $\exp(-tD^2)$ is trace class, have arisen in the work of Jaffe, Lesniewski and Osterwalder on quantum field theory [7]. The relevant extension is the entire cyclic theory [5], which we now describe.

§2. DEFINITIONS.

Both theories, the cyclic and the entire cyclic, are defined via the Connes (b,B) bicomplex $\mathcal{B}(A)$:

$$\begin{array}{ccccccc}
 & & & & C^0(A) & \xrightarrow{b} & \cdots \\
 & & & & B\uparrow & & \\
 & & C^0(A) & \xrightarrow{b} & C^1(A) & \xrightarrow{b} & \cdots \\
 & & B\uparrow & & B\uparrow & & \\
C^0(A) & \xrightarrow{b} & C^1(A) & \xrightarrow{b} & C^2(A) & \xrightarrow{b} & \cdots \\
B\uparrow & & B\uparrow & & B\uparrow & & \\
\vdots & & \vdots & & \vdots & &
\end{array}$$

where $C^n(A)$ = (n+1)-linear maps $\phi : A^{n+1} \to \mathbb{C}$
b = Hochschild coboundary (so each row is the Hochschild complex)
$B = (1 + \lambda + \ldots + \lambda^n)\, S(1-\lambda)$
$(\lambda\phi)(a^0,\ldots, a^n) = (-1)^n\, \phi(a^n, a^0,\ldots, a^{n-1})$
$(S\phi)(a^0,\ldots, a^{n-1}) = \phi(1,a^0,\ldots, a^{n-1})$.

Note in particular that the definition of the map S requires that A be unital, and that $\mathcal{B}(A)$ is functorial for unital maps. The **periodic cyclic cohomology** groups of the unital algebra A are then defined by

$$HC^*_{per}(A) = H^*(\mathrm{Tot}\,\mathcal{B}(A))\ ;$$

that is, as the homology of the complex Tot $\mathcal{B}(A)$

$$\cdots\longrightarrow \sum_{n\geq 0} C^{2n} \xrightarrow{b+B} \sum_{n\geq 0} C^{2n+1} \xrightarrow{b+B} \sum_{n\geq 0} C^{2n} \longrightarrow\cdots$$

Connes has observed [5] that the complex $\mathrm{Tot}_\infty\,\mathcal{B}(A)$

$$\cdots\longrightarrow \prod_{n\geq 0} C^{2n} \xrightarrow{b+B} \prod_{n\geq 0} C^{2n+1} \xrightarrow{b+B} \prod_{n\geq 0} C^{2n} \longrightarrow\cdots$$

of cochains with infinite support has trivial homology (i.e. the sequence is exact), but that this need not be so if one restricts to cochains in which the growth of $\|\phi_n\|$ is suitably controlled. In order for this to be meaningful we restrict to unital Banach algebras A and to continuous functionals.

2.1 Definition. The **entire cyclic cohomology** of the unital Banach algebra A is defined by

$$HC^*_\varepsilon(A) = H^*(\mathrm{Tot}_\varepsilon\,\mathcal{B}(A))\ ,$$

where $\mathrm{Tot}_\varepsilon\,\mathcal{B}(A)$ is the subcomplex of $\mathrm{Tot}_\infty\,\mathcal{B}(A)$ consisting of cochains (ϕ_{2n}) and (ϕ_{2n+1}) for which the power series

$$\sum \frac{(2n)!}{n!} \|\phi_{2n}\|\, z^n \quad , \quad \sum \frac{(2n)!}{n!} \|\phi_{2n+1}\|\, z^n$$

converge everywhere in $\mathbb{C}$.

Note that there is an obvious map
$HC^*_{per,\ conts}(A) \to HC^*_\varepsilon(A)$.

2.2 Example. $A = \mathbb{C}$. Here $C^n(A)$ can be identified with $\mathbb{C}$ since any $\phi \in C^n$ is of the form

$$\phi(a^0,\ldots, a^n) = \mu a^0 \ldots a^n \quad \text{for some } \mu \in \mathbb{C}\ .$$

Hence Tot $\mathcal{B}(A)$, $\mathrm{Tot}_\varepsilon\,\mathcal{B}(A)$ and $\mathrm{Tot}_\infty\,\mathcal{B}(A)$ are identified with the polynomials, the formal power series and the entire functions. Moreover

$$b\mu_n = \begin{cases} 0, & n \text{ even} \\ \mu_n, & n \text{ odd} \end{cases} \qquad B\mu_n = \begin{cases} 0, & n \text{ even} \\ 2n\mu_n, & n \text{ odd} \end{cases}$$

so that $b+B$ is 0 in even dimensions and multiplication by $2(z+1)$ in odd dimensions. It follows that

$$HC^{odd}_{per}(\mathbb{C}) = HC^{odd}_{\varepsilon}(\mathbb{C}) = 0$$

$$HC^{ev}_{per}(\mathbb{C}) \xrightarrow[\text{id}]{} HC^{ev}_{\varepsilon}(\mathbb{C}) \equiv \mathbb{C} .$$

§3. RESULTS.

The main results are concerned with derivations and automorphisms, stability, additivity and homotopy invariance. For the cyclic theory, such results are obtained using the Connes long exact sequence [4] (relating the cyclic and Hochschild cohomology), facts about Hochschild cohomology and spectral sequences. These methods are not available for the entire theory. Instead new proofs are found for the cyclic case that can be extended to the entire case with the help of a "comparison theorem" for $\mathcal{B}(A)$ and the cyclic bicomplex $\mathcal{C}(A)$ defined by Loday-Quillen-Tsygan [12,13] as follows:

$$\begin{array}{ccccccccc}
 & & \vdots & & \vdots & & \vdots & & \\
 & & -b'\uparrow & & b\uparrow & & -b'\uparrow & & \\
\cdots & \xrightarrow{1-\lambda} & C^2(A) & \xrightarrow{N} & C^2(A) & \xrightarrow{1-\lambda} & C^2(A) & \xrightarrow{N} & \cdots \\
 & & -b'\uparrow & & b\uparrow & & -b'\uparrow & & \\
\cdots & \xrightarrow{1-\lambda} & C^1(A) & \xrightarrow{N} & C^1(A) & \xrightarrow{1-\lambda} & C^1(A) & \xrightarrow{N} & \cdots \\
 & & -b'\uparrow & & b\uparrow & & -b'\uparrow & & \\
\cdots & \xrightarrow{1-\lambda} & C^0(A) & \xrightarrow{N} & C^0(A) & \xrightarrow{1-\lambda} & C^0(A) & \xrightarrow{N} & \cdots
\end{array}$$

where A = any algebra over $\mathbb{C}$ (not necessarily unital)
$b' = b$ with the "wrap-around" term deleted
$N = 1 + \lambda + \ldots + \lambda^n$ (on C^n)
so that the columns are alternately the Hochschild and cobar complexes.

3.1 Comparison. Loday and Quillen define a chain map
$I : \text{Tot}\, \mathcal{C}(A) \to \text{Tot}\, \mathcal{B}(A)$
and show that it induces an isomorphism of cohomology groups. Their proof

does not extend to the entire case. However it can be shown that the left homotopy inverse for I given by Kassel [9] is also a right homotopy inverse, and that the homotopy operators preserve the growth condition. Hence

$$H^*(\mathrm{Tot}_\varepsilon\, \mathscr{C}(A)) \cong H^*(\mathrm{Tot}_\varepsilon\, \mathscr{B}(A)) = HC^*_\varepsilon(A)\,.$$

It should be noted that this can be used to define the functor HC^*_ε in the non-unital case.

3.2 Derivations. For any derivation δ of A, it can be checked that the "Lie derivative" $L\delta$, defined by

$$(L\delta\phi)(a^0,\ldots,a^n) = \sum_j \phi(a^0,\ldots,\delta(a^j),\ldots,a^n)$$

induces endomorphisms L^*_δ of $HC^*_{per}(A)$ and (in the continuous case) of $HC^*_\varepsilon(A)$. For A unital and δ inner it is known that $L^*_\delta = 0$ in both the cyclic [4] and (up to simplicial normalization) entire [6] cases. With the help of the Comparison Theorem one can get rid of the technicalities about the use of simplicially normalized cochains in the entire theory. Among the consequences of this are:

(i) Any inner automorphism of A induces the identity map on $HC^*_\varepsilon(A)$.

(ii) (Stability). The map $a \to \begin{pmatrix} a & 0 \\ 0 & 0 \end{pmatrix} : A \to M_2(A)$ induces an isomorphism on HC^*_ε. This was already known for ordinary cyclic homology of Banach algebras with a bounded approximate unit [14].

(iii) (Additivity). $HC^*_\varepsilon(A \oplus B) \cong HC^*_\varepsilon(A) \oplus HC^*_\varepsilon(B)$.

3.3 The Vanishing Theorem. Let $f : A \to B$ be a continuous homomorphism of Banach algebras and $\delta : A \to B$ a continuous linear map that is a derivation in the sense that

$$\delta(a_1 a_2) = \delta(a_1)f(a_2) + f(a_1)\delta(a_2) \quad \text{for all} \quad a_1, a_2 \in A\,.$$

Then $L^*_\delta : HC^*_\varepsilon(B) \to HC^*_\varepsilon(A)$ is 0.

An immediate consequence of this result is homotopy invariance: if $f_t : A \to B$ is a uniformly bounded family of homomorphisms such that for all $a \in A$ the map $t \to f_t(a)$ is C^1 with uniformly bounded derivative, then $f_0^* = f_1^*$.

3.4 Amenable algebras. If A is an amenable Banach algebra, then the map

$$HC^*_{per,\, conts}(A) \to HC^*_{\varepsilon}(A)$$

described in §2 is an isomorphism [11]. For amenable C*-algebras the latter is known to be the space of bounded traces on A or 0, according as * is even or odd [3]. Thus for the algebra $\mathcal{K}$ of compact operators we have, since every compact operator is a sum of commutators, $HC^{ev}_{\varepsilon}(\mathcal{K}) = 0$. It follows (see the example at the end of section 2) that entire cyclic cohomology, unlike K-theory, is not stable under tensoring with the algebra of compact operators.

REFERENCES

1. M. Atiyah, Global theory of elliptic operators, Proc. Internat. Conf. on Functional Analysis and Related Topics, Univ. of Tokyo Press, Tokyo, 1970.

2. L. Brown, R. Douglas and P. Fillmore, Extensions of C*-algebras and K-homology, Ann. Math. 105 (1977) 265-324.

3. E. Christensen and A. Sinclair, On the vanishing of $H^n(A,A^*)$ for certain C*-algebras, Pac. J. Math. 137 (1989) 55-63.

4. A. Connes, Non-commutative differential geometry, Publ. Math. I.H.E.S. 62 (1986) 257-360.

5. A. Connes, Entire cyclic cohomology of Banach algebras and characters of θ-summable Fredholm modules, K-Theory 1 (1988) 519-548.

6. E. Getzler and A. Szenes, On the Chern character of a theta-summable Fredholm module, J. Funct. Anal. 84 (1989) 343-357.

7. A. Jaffe, A. Lesniewski and K. Osterwalder, Quantum K-theory I: The Chern character, Commun. Math. Phys. 118 (1988) 1-14.

8. G. Kasparov, Topological invariants of elliptic operators I: K-homology, Math. USSR Izvestija 9 (1975) 751-792.

9. C. Kassel, Homologie cyclique, charactere de Chern et lemme de perturbation, J. Reine Angew. Math. 408 (1990) 159-180.

10. M. Khalkhali, On the entire cyclic cohomology of Banach algebras: I. Morita invariance; II. Homotopy invariance, preprints, October 1992, University of Heidelberg.

11. M. Khalkhali and J. Phillips, Separability and entire cyclic cohomology, in preparation.

12. J.-L. Loday and D. Quillen, Cyclic homology and the Lie algebra homology of matrices, Comment. Math. Helvetici 59 (1984) 565-591.

13. B. Tsygan, Homology of matrix Lie algebras over rings and Hochschild homology, Uspekhi Math. Nauk. 38 (1983) 217-218.

14. M. Wodzicki, Excision in cyclic homology and in rational algebraic K-theory, Ann. Math. 129 (1989) 591-639.

Peter Fillmore
Department of Mathematics, Statistics and C.S.
Dalhousie University
Halifax N.S.
Canada B3H 3J5

Masoud Khalkhali
Mathematics Institute
University of Heidelberg
W-6900 Heidelberg
Germany

Operator Theory:
Advances and Applications, Vol. 73

THE BOUNDED REAL CHARACTERISTIC FUNCTION AND NEHARI EXTENSIONS

P. A. Fuhrmann*†

Dedicated to Professor M.S. Livsic

For the class of bounded real functions in the open right half plane we define a map, we call it the B-characteristic, which associates with each bounded real unction a stable function. The definition is based on normalized coprime factorizations of bounded real functions, the normalization being with respect to an indefinite metric. The construction bears great similarity to the characteristic functions studied in Fuhrmann and Ober [1993] for other classes of functions. We study the properties of the B-characteristic, the inverse map and the connections to balancing. We apply the construction to shed some new light on the problem of suboptimal Nehari complementation.

1 Introduction

In Fuhrmann and Ober [1993] two types of characteristic functions, the S-characteristic for the class of antistable transfer functions bounded in the left half plane and the L-charactristic for the class of minimal systems, were introduced. These characteristic functions played a central role in the development of a theory clarifying the duality between problems of model reduction on the one hand and robust control on the other. The L-characteristic, though not under this name, has initially been used by Glover and McFarlane [1989], see also Georgiou and Smith [1990], for characterizing the optimally robust stability margin for normalized coprime factor uncertainty. The S-characteristic has been introduced on the model of the L-characteristic by replacing the standard metric by a degenerate. Other than that the procedure remains very much the same.

In the present paper we take the natural step of generalizing the construction to the case of an indefinite metric, a metric induced by the matrix

$$J_B = \begin{pmatrix} I & 0 \\ 0 & -I \end{pmatrix}.$$

*Earl Katz Family Chair in Algebraic System Theory

†Partially supported by the Israeli Academy of Sciences under Grant No. 249/90 and by GIF under Grant No. I 184

It turns out that this metric is the right one for the study of the class of bounded real functions, namely the analytic functions in the open right half plane which have norm bound of less than one.

Our object in this paper is to study the class of bounded real functions via their normalized coprime factorizations, the normalization being with respect to the indefinite metric. Once we have the normalized coprime factors we can, using the Youla-Kucera parametrization, proceed with the construction of the B-characteristic. The B-characteristic can be used in the analysis of a special balancing which uses the positive definite solutions to the bounded real Riccati equations rather than the standard Lyapunov equations used by Moore [1981] in the case of stable transfer functions or the standard Riccati equations used by Jonckheere and Silverman [1983] for LQG balancing.

However our main object will be to focus on the application of the B-characteristic to the problem of parametrizing all suboptimal solutions to the Nehari complementation problem. This problem, solved originally by Nehari [1957], characterized the class of bounded Hankel operators.
It has long been known that some classical interpolation problems, connected with the work of Schur, Nevanlinna and Pick, Fejer and Caratheodory etc, could be cast in the framework of Nehari's theorem. In a seemingly independent line of research, starting with Beurling's theorem, Beurling [1949] and Lax [1959], and the work of M.S. Livsic [1954] on characteristic functions, a deep connection was made between problems of interpolation and lifting of operators. This seems to have been initialized in the work of Sz.-Nagy and Koranyi [1958]. However the big advance was made by Sarason [1967] who proved the first version of the commutant lifting theorem. This was quickly generalized by Sz.-Nagy and Foias [1970] who proved the version of the commutant lifting theorem that is used widely today. Almost simultaneously with the availability of this theorem came the observation, due to Page [1970], that Nehari's theorem could be derived from the commutant lifting theorem.

Parallel to this, operator theoretic oriented, development there was a direct study of Hankel operators and their relevance to various interpolation and approximation problems. This development initialized by a series of seminal papers by Adamjan, Arov and Krein [1968a,1968b,1971,1978], is currently being referred to as AAK theory. Though it seems a bit less general, AAK theory can be easily used to derive the most useful versions of the commutant lifting theorem. This is based on a very close connection between certain classes of Hankel operators and the model operators based on compressions of shifts, see Sz.-Nagy and Foias [1970], Rota [1960], Fuhrmann [1981] and Nikolskii [1985]. In the early eighties, in the wake of the influential work of Zames [1981] and its change of emphasis in control problems, came the recognition of the great relevance of AAK theory to control and this triggered a vast amount of research. This came to be known as H^∞-control. Probably the most influential paper in this connection is Glover [1984] which brought AAK theory into the realm of standard state space theory. While, for the finite dimensional situation, it became possible to develop AAK theory almost solely from the matrix theoretic point of view, this tendency has the disadvantage that it loses many of the geometric insights that the operator theoretic approach has to offer.

Mostly in the model matching problem, see Francis [1987], and in other problems that can be reduced to it, the role of Nehari complementation is central. Over the years several distinct methods have been put forward to solve the Nehari extension problem. These include the one step extension methods of Adamjan, Arov and Krein [1978], Young [1988], the method of solving a set of equations that have are induced by the Hankel operator and its adjoint, a method that originated with Krein and Melik-Adamyan [1984] see also Dym [1989], methods based on geometric considerations and J-spectral factorizations as in Ball and Ran [1983] and Francis [1987], state space methods based on balanced realizations and the solution of pairs of Lyapunov equations as in Glover [1984]. Another method used by Adamjan, Arov and Krein [1971] is based on Schmidt pair analysis. This has been put, for the rational case, into algebraic form using polynomial based methods in Fuhrmann [1991,1993a].

It is a proof of the richness of this circle of ideas that so many seemingly different methods converge on the same problem. However not much attention has been paid to elucidating the connections between the various methods. In this paper we will propose yet another approach to this problem, based on the analysis and use of the B-characteristic and embedding results. We will not stop with that but study the connection of this method with the Krein and Melik-Adamyan method as well as the method based on J-spectral factorization. The main advantage of the approach presented in this paper is not only its relative simplicity but the way it unifies several other approaches to Nehari extensions. For a problem area to be fully understood one needs to be able to view the results from several perspectives. It is strongly felt by the author that the perspective presented in this paper, with its emphasis on coprime factorizations, model operators, intertwining maps and Hankel operators, all leading directly to state space formulas based on the shift realization, is a particularly advantegeous perspective. Moreover, the material presented in section 3 opens up interesting possibilities of extending AAK theory to other contexts.

Some of the work presented in this paper was done during a visit to the Department of Mathematics at the University of Kaiserslautern. The hospitality of the department and in particular of my host D. Prätzel-Wolters are gratefully acknowledged. I am also indebted to the DFG for its support. Finally I want to thank Raimund Ober for the numerous discussions during the work on this paper.

2 Bounded real functions

In this section we quote several results on the special normalized coprime factorizations associated with bounded real functions that lead quite naturally to the construction of the bounded real characteristic. All the material in this section is based on Fuhrmann and Ober [1993b] and Ober and Fuhrmann [1993]. We arrange the results in a series of lemmas. For proofs we refer to the above mentioned papers.

Definition 2.1 *1. A rational transfer function G is called* bounded real *if it is analytic and strictly contractive in the closed right half plane i.e.*

$$I - G(s)^*G(s) > 0$$

We assume this inequality holds also at infinity.

2. Let

$$J_B = \begin{pmatrix} I & 0 \\ 0 & -I \end{pmatrix}.$$

A right coprime factorization $G = NM^{-1}$ of G is called a J_B normalized right coprime factorization, or J_B-$NRCF$, of G if

$$\begin{pmatrix} M^* & N^* \end{pmatrix} \begin{pmatrix} I & 0 \\ 0 & -I \end{pmatrix} \begin{pmatrix} M \\ N \end{pmatrix} = M^*M - N^*N = I.$$

Similarly a left coprime factorization $G = \overline{M}^{-1}\overline{N}$ of the transfer function G is called a J_B-$NLCF$ of G if

$$\overline{M}\overline{M}^* - \overline{N}\overline{N}^* = I.$$

We have the following useful lemmas.

Lemma 2.1 *Let $G = NM^{-1}$ be a J_B-$NRCF$ and $G = \overline{M}^{-1}\overline{N}$ a J_B-$NLCF$ of the transfer function G. Then*

1. $$\begin{pmatrix} M^* & N^* \\ \overline{N} & \overline{M} \end{pmatrix} \begin{pmatrix} I & 0 \\ 0 & -I \end{pmatrix} \begin{pmatrix} M & \overline{N}^* \\ N & \overline{M}^* \end{pmatrix} = \begin{pmatrix} I & 0 \\ 0 & -I \end{pmatrix}. \quad (1)$$

2. $$\begin{pmatrix} M & \overline{N}^* \\ N & \overline{M}^* \end{pmatrix} \begin{pmatrix} I & 0 \\ 0 & -I \end{pmatrix} \begin{pmatrix} M^* & N^* \\ \overline{N} & \overline{M} \end{pmatrix} = \begin{pmatrix} I & 0 \\ 0 & -I \end{pmatrix} \quad (2)$$

3. $$\begin{pmatrix} M^* & N^* \\ \overline{N} & \overline{M} \end{pmatrix} \begin{pmatrix} M & -\overline{N}^* \\ -N & \overline{M}^* \end{pmatrix} = \begin{pmatrix} I & 0 \\ 0 & I \end{pmatrix} \quad (3)$$

Definition 2.2 *Let $G = NM^{-1}$ be a J_B-$NRCF$ and $G = \overline{M}^{-1}\overline{N}$ a J_B-$NLCF$ of the transfer function G and let $K = UV^{-1}$ be a RCF and $K = \overline{V}^{-1}\overline{U}$ a LCF of the transfer function K. If*

$$\begin{pmatrix} \overline{V} & -\overline{U} \\ -\overline{N} & \overline{M} \end{pmatrix} \begin{pmatrix} M & U \\ N & V \end{pmatrix} = \begin{pmatrix} I & 0 \\ 0 & I \end{pmatrix}$$

*holds, then the coprime factorizations are said to satisfy the J_B-*doubly coprime factorization *of G and K.*

Of course doubly coprime factorizations are related to the Youla-Kucera parametrization of stabilizing controllers, see Francis [1987] for the details. The next lemma relates coprime factorizations of the plant and controller to doubly coprime factorizations.

Lemma 2.2 *Let $G = NM^{-1}$ be a $J_B - NRCF$ and $G = \overline{M}^{-1}\overline{N}$ a $J_B - NLCF$ of the transfer function G and let $K = UV^{-1}$ be a RCF and $K = \overline{V}^{-1}\overline{U}$ a LCF of the function K. These coprime factorizations satisfy the J_B-doubly coprime factorization if and only if*

$$\overline{V}M - \overline{U}N = I$$

and

$$\overline{M}V - \overline{N}U = I.$$

Lemma 2.3 *Let $G = NM^{-1}$ be a $J_B - NRCF$ and $G = \overline{M}^{-1}\overline{N}$ a $J_B - NLCF$ of the transfer function G.*

1. *Consider the Bezout equation $\overline{M}V - \overline{N}U = I$. Then,*
   (a) *there exists a unique solution $(U_B, V_B) \in H_+^\infty$ to the Bezout equation such that*
   $$R_B^* := M^*U_B - N^*V_B \in H_-^\infty,$$
   and is strictly proper.
   (b) *let $(U, V) \in H_+^\infty$ be an arbitrary solution to the Bezout equation then R_B^* is the strictly proper antistable part of $M^*U - N^*V$.*
   (c) *the Hankel operator*
   $$H_{M^*U-N^*V}$$
   is independent of the solution to the Bezout equation.
2. *Consider the Bezout equation $\overline{V}M - \overline{U}N = I$. Then,*
   (a) *there exists a unique solution $(\overline{U}_B, \overline{V}_B) \in H_+^\infty$ to the Bezout equation such that*
   $$\overline{R}_B^* := \overline{U}_B\overline{M}^* - \overline{V}_B\overline{N}^* \in H_-^\infty,$$
   and is strictly proper.
   (b) *let $(\overline{U}, \overline{V}) \in H_+^\infty$ be an arbitrary solution to the Bezout equation then $\overline{R}_B^*$ is the strictly proper antistable part of $\overline{U}\overline{M}^* - \overline{V}\overline{N}^*$.*
   (c) *the Hankel operator*
   $$H_{\overline{U}\overline{M}^*-\overline{V}\overline{N}^*}$$
   is independent of the solution to the Bezout equation.

Proposition 2.1 *Let $G = NM^{-1}$ be a $J_B - NRCF$ and $G = \overline{M}^{-1}\overline{N}$ a $J_B - NLCF$ of the transfer function G and let $K_1 = U_1V_1^{-1}$ be a RCF of the transfer function K and $K_2 = \overline{V}_2^{-1}\overline{U}_2$ a LCF of the transfer function K_2 such that*

$$\overline{M}V_1 - \overline{N}U_1 = I$$

and

$$\overline{V}_2M - \overline{U}_2N = I.$$

Then the following are equivalent:

1. $K_1 = K_2$.

2. $-N^*V_1 + M^*U_1 = -\overline{V}_2\overline{N}^* + \overline{U}_2\overline{M}^*$

3. $\begin{pmatrix} -\overline{U}_2 & \overline{V}_2 \\ \overline{M} & -\overline{N} \end{pmatrix}\begin{pmatrix} N & V_1 \\ M & U_1 \end{pmatrix} = \begin{pmatrix} I & 0 \\ 0 & I \end{pmatrix}$

Proof: Using Lemma 2.1 we have that

$$\begin{pmatrix} N & \overline{M}^* \\ M & \overline{N}^* \end{pmatrix}\begin{pmatrix} I & 0 \\ 0 & -I \end{pmatrix}\begin{pmatrix} N^* & M^* \\ \overline{M} & \overline{N} \end{pmatrix}\begin{pmatrix} -I & 0 \\ 0 & I \end{pmatrix} = \begin{pmatrix} I & 0 \\ 0 & I \end{pmatrix}.$$

The proposition follows from the following equality.

$$\begin{pmatrix} I & \overline{V}_2U_1 - \overline{U}_2V_1 \\ 0 & I \end{pmatrix} = \begin{pmatrix} -\overline{U}_2 & \overline{V}_2 \\ \overline{M} & -\overline{N} \end{pmatrix}\begin{pmatrix} N & V_1 \\ M & U_1 \end{pmatrix}$$

$$= \begin{pmatrix} -\overline{U}_2 & \overline{V}_2 \\ \overline{M} & -\overline{N} \end{pmatrix}\begin{pmatrix} N & \overline{M}^* \\ M & \overline{N}^* \end{pmatrix}\begin{pmatrix} I & 0 \\ 0 & -I \end{pmatrix}\begin{pmatrix} N^* & M^* \\ \overline{M} & \overline{N} \end{pmatrix}\begin{pmatrix} -I & 0 \\ 0 & I \end{pmatrix}\begin{pmatrix} N & V_1 \\ M & U_1 \end{pmatrix}$$

$$= \begin{pmatrix} I & \overline{V}_2\overline{N}^* - \overline{U}_2\overline{M}^* \\ 0 & I \end{pmatrix}\begin{pmatrix} I & 0 \\ 0 & -I \end{pmatrix}\begin{pmatrix} I & -N^*V_1 + M^*U_1 \\ 0 & -I \end{pmatrix}$$

$$= \begin{pmatrix} I & [-N^*V_1 + M^*U_1] - [-\overline{V}_2\overline{N}^* + \overline{U}_2\overline{M}^*] \\ 0 & I \end{pmatrix}.$$

∎

Corollary 2.1 *We have*

1. $R_B^* = \overline{R}_B^*$ (4)

2. $U_BV_B^{-1} = \overline{V}_B^{-1}\overline{U}_B$ (5)

3. $\begin{pmatrix} \overline{V}_B & -\overline{U}_B \\ -\overline{N} & \overline{M} \end{pmatrix}\begin{pmatrix} M & U_B \\ N & V_B \end{pmatrix} = \begin{pmatrix} I & 0 \\ 0 & I \end{pmatrix}$ (6)

4. $\begin{pmatrix} M & U_B \\ N & V_B \end{pmatrix}\begin{pmatrix} \overline{V}_B & -\overline{U}_B \\ -\overline{N} & \overline{M} \end{pmatrix} = \begin{pmatrix} I & 0 \\ 0 & I \end{pmatrix}$ (7)

Proof: From the proof of the proposition we have that

$$\begin{pmatrix} I & \overline{V}_BU_B - \overline{U}_BV_B \\ 0 & I \end{pmatrix} = \begin{pmatrix} I & [-N^*V_1 + M^*U_1] - [-\overline{V}_2\overline{N}^* + \overline{U}_2\overline{M}^*] \\ 0 & I \end{pmatrix}$$

$$= \begin{pmatrix} I & R_B^* - \overline{R}_B^* \\ 0 & I \end{pmatrix}$$

which implies that

$$\overline{V}_BU_B - \overline{U}_BV_B = R_B^* - \overline{R}_B^*.$$

Since the right hand side is in H_-^∞ and is strictly proper and the left hand side is in H_+^∞, we have that both sides are zero which implies the result. ∎

We will refer to $R_B = \overline{R}_B$ as the **bounded real characteristic** of G, or equivalently, as the **B-characteristic** of G.

We note that equation (7) can be written also as

$$\begin{pmatrix} M \\ N \end{pmatrix}\begin{pmatrix} \overline{V}_B & -\overline{U}_B \end{pmatrix} + \begin{pmatrix} U_B \\ V_B \end{pmatrix}\begin{pmatrix} -\overline{N} & \overline{M} \end{pmatrix} = \begin{pmatrix} I & 0 \\ 0 & I \end{pmatrix} \tag{8}$$

Lemma 2.4 *Let*

$$R_B^* = \begin{pmatrix} M^* & N^* \end{pmatrix}\begin{pmatrix} I & 0 \\ 0 & -I \end{pmatrix}\begin{pmatrix} U_B \\ V_B \end{pmatrix}$$

and

$$\overline{R}_B^* = \begin{pmatrix} \overline{V}_B & \overline{U}_B \end{pmatrix}\begin{pmatrix} I & 0 \\ 0 & -I \end{pmatrix}\begin{pmatrix} \overline{M}^* \\ \overline{N}^* \end{pmatrix}.$$

Then

1. $$\begin{pmatrix} \overline{V}_B & \overline{U}_B \end{pmatrix} = R_B^*\begin{pmatrix} \overline{N} & \overline{M} \end{pmatrix} + \begin{pmatrix} M^* & N^* \end{pmatrix} \tag{9}$$

   and

2. $$\begin{pmatrix} U_B \\ V_B \end{pmatrix} = \begin{pmatrix} M \\ N \end{pmatrix}\overline{R}_B^* + \begin{pmatrix} \overline{N}^* \\ \overline{M}^* \end{pmatrix} \tag{10}$$

Proof:

1. We multiply the identity, which is derived from (3),

$$\begin{pmatrix} M & -\overline{N}^* \\ -N & \overline{M}^* \end{pmatrix}\begin{pmatrix} M^* & N^* \\ \overline{N} & \overline{M} \end{pmatrix} = \begin{pmatrix} I & 0 \\ 0 & I \end{pmatrix}$$

on the right by $\begin{pmatrix} U_B \\ -V_B \end{pmatrix}$, to get

$$\begin{aligned}\begin{pmatrix} U_B \\ -V_B \end{pmatrix} &= \begin{pmatrix} M & -\overline{N}^* \\ -N & \overline{M}^* \end{pmatrix}\begin{pmatrix} M^* & N^* \\ \overline{N} & \overline{M} \end{pmatrix}\begin{pmatrix} U_B \\ -V_B \end{pmatrix} = \begin{pmatrix} M & -\overline{N}^* \\ -N & \overline{M}^* \end{pmatrix}\begin{pmatrix} R_B^* \\ -I \end{pmatrix} \\ &= \begin{pmatrix} M \\ -N \end{pmatrix} R_B^* - \begin{pmatrix} -\overline{N}^* \\ \overline{M}^* \end{pmatrix}\end{aligned}$$

or

$$\begin{pmatrix} U_B \\ V_B \end{pmatrix} = \begin{pmatrix} M \\ N \end{pmatrix} R_B^* + \begin{pmatrix} \overline{N}^* \\ \overline{M}^* \end{pmatrix}$$

2. Similarly, multiplying the same identity on the left by $\begin{pmatrix} \overline{V}_B & \overline{U}_B \end{pmatrix}$, we get

$$\begin{aligned}
\begin{pmatrix} \overline{V}_B & \overline{U}_B \end{pmatrix} &= \begin{pmatrix} \overline{V}_B & \overline{U}_B \end{pmatrix} \begin{pmatrix} M & -\overline{N}^* \\ -N & \overline{M}^* \end{pmatrix} \begin{pmatrix} M^* & N^* \\ \overline{N} & \overline{M} \end{pmatrix} \\
&= \begin{pmatrix} I & R_B^* \end{pmatrix} \begin{pmatrix} M^* & N^* \\ \overline{N} & \overline{M} \end{pmatrix} \\
&= \begin{pmatrix} M^* & N^* \end{pmatrix} + R_B^* \begin{pmatrix} \overline{N} & \overline{M} \end{pmatrix}
\end{aligned}$$

∎

Lemma 2.5 *Let G be a bounded real, proper, rational function and let*

$$G = NM^{-1} = \overline{M}^{-1}\overline{N}$$

be its normalized J_B coprime factorizations. Then

1.
$$\begin{cases} MM^* - \overline{N}^*\overline{N} = I \\ MN^* = \overline{N}^*\overline{M} \\ NM^* = \overline{M}^*\overline{N} \\ \overline{M}^*\overline{M} - NN^* = I. \end{cases} \tag{11}$$

2. The function $M^{-1}\overline{N}^*$ is in the unit ball of L^∞.

3. We have the following equality

$$-M^{-1}\overline{N}^* = R_B^* - M^{-1}U_B \tag{12}$$

4. The singular values of $H_{R_B^*}$ are of absolute value less than 1.

Proof:

1. Follows from the identity, which is equivalent to (3),

$$\begin{pmatrix} M & -\overline{N}^* \\ -N & \overline{M}^* \end{pmatrix} \begin{pmatrix} M^* & N^* \\ \overline{N} & \overline{M} \end{pmatrix} = \begin{pmatrix} I & 0 \\ 0 & I \end{pmatrix}$$

2. From the equality $MM^* - \overline{N}^*\overline{N} = I$ we get

$$I - (M^{-1}\overline{N}^*)(\overline{N}M^{-*}) = M^{-1}M^{-*} > 0.$$

So $M^{-1}\overline{N}^*$ is contractive on the imaginary axis, i.e. it is in the unit ball of L^∞.

3. From the equality

$$\begin{pmatrix} U_B \\ V_B \end{pmatrix} = \begin{pmatrix} M \\ N \end{pmatrix} \overline{R_B^*} + \begin{pmatrix} \overline{N}^* \\ \overline{M}^* \end{pmatrix}$$

we get

$$M^{-1}U_B = R_B^* + M^{-1}\overline{N}^*$$

or (12) follows.

4. Since $M^{-1}U_B$ is stable we have $H_{R_B^*} = -H_{M^{-1}\overline{N}^*}$ and so

$$\mu_1 = \|H_{R_B^*}\| = \|H_{M^{-1}\overline{N}^*}\| \leq \|M^{-1}\overline{N}^*\|_\infty < 1.$$

This, of course, implies that all the singular values of $H_{R_B^*}$ are less than 1.

∎

We want to point out that equation (12) i.e.

$$-M^{-1}\overline{N}^* = R_B^* - M^{-1}U_B$$

shows that the function R_B^* is the strictly proper, antistable part of the contractive L^∞ function $-M^{-1}\overline{N}^*$. In the language of Nehari's theorem this shows that $-M^{-1}\overline{N}^*$ is a suboptimal Nehari extension of R_B^*. In fact this simple equality points out the way to yet another aapproach to the Nehari extension problem. We will return to this subject in section 5.

3 Hankel operators

In this section we study several Hankel operators associated with the normalized coprime factorizations of a bounded real functions. Many of the results are based on a similar analysis carried out in Fuhrmann and Ober [1993a].

As usual, given a matrix function $\Phi \in L^\infty$ we define the **Hankel operator** $H_\Phi : H_+^2 \longrightarrow H_-^2$ by

$$H_\Phi f = P_-\Phi f.$$

Similarly, the **involuted Hankel operator** $\hat{H}_\Phi : H_-^2 \longrightarrow H_+^2$ is defined by

$$\hat{H}_\Phi h = P_+\Phi h.$$

Clearly we have $H_\Phi^* = \hat{H}_{\Phi^*}$.

Let now R_B^* be the B-characteristic of the bounded real function G. As R_B^* is rational and in H_-^∞ it has coprime Douglas, Shapiro and Shields [1971], or DSS, factorizations of the form

$$R_B^* = \Phi_K S_K^* = S_I^* \Phi_I \tag{13}$$

with S_K and S_I inner functions in H_+^∞. For more on these factorizations we refer to Fuhrmann [1975,1981] and Fuhrmann and Ober [1993b]. We know that these factorizations imply

$$\begin{cases} KerH_{R^*} &= S_K H_+^2 \\ ImH_{R^*} &= \{S_I^* H_-^2\}^\perp = H_-^2 \ominus S_I^* H_-^2 \end{cases}$$

and the induced map from $\{S_K H_+^2\}^\perp$ to $\{S_I^* H_-^2\}^\perp$ is invertible. Moreover, by the rationality of R_B^*, these are finite dimensional spaces.

Given an antistable function A^* we will say that an inner function S is a **minimal left inner function** of A^* if the following conditions are satisfied

1. $SA^* \in H_+^\infty$
2. For any inner function T for which $TA^* \in H_+^\infty$ we have $T = QS$.

A **minimal right inner function** is defined similarly.

Lemma 3.1 *Let G be a bounded real function with J_B normalized coprime factorizations*

$$G = NM^{-1} = \overline{M}^{-1}\overline{N} \tag{14}$$

and let R_B^* be its BR-characteristic. Assume (13) are the DSS factorizations of R^*. Then

1. *The minimal right inner function of* $\begin{pmatrix} \overline{N}^* \\ \overline{M}^* \end{pmatrix}$ *is* S_K.
2. *The minimal left inner function of* $\begin{pmatrix} N^* & M^* \end{pmatrix}$ *is* S_I.

Proof:

1. Since $KerH_{R^*} = S_K H_+^2$ we have

$$P_- R_B^* S_K H_+^2 = P_- \begin{pmatrix} \overline{V}_B & -\overline{U}_B \end{pmatrix} \begin{pmatrix} \overline{N}^* \\ \overline{M}^* \end{pmatrix} S_K H_+^2 = 0.$$

However it is clear that also

$$P_- \begin{pmatrix} -\overline{N} & \overline{M} \end{pmatrix} \begin{pmatrix} \overline{N}^* \\ \overline{M}^* \end{pmatrix} S_K H_+^2 = 0.$$

Since $\begin{pmatrix} \overline{V}_B & -\overline{U}_B \\ -\overline{N} & \overline{M} \end{pmatrix}$ is invertible in H_+^∞ we have, for all $f \in H_+^2$,

$$\begin{aligned} 0 &= P_- \begin{pmatrix} \overline{V}_B & -\overline{U}_B \\ -\overline{N} & \overline{M} \end{pmatrix} \begin{pmatrix} \overline{N}^* \\ \overline{M}^* \end{pmatrix} S_K f \\ &= P_- \begin{pmatrix} \overline{N}^* \\ \overline{M}^* \end{pmatrix} S_K f \end{aligned}$$

This implies $\begin{pmatrix} \overline{N}^* \\ \overline{M}^* \end{pmatrix} S_K \in H_+^\infty$. We set

$$\begin{pmatrix} J_1 \\ J_2 \end{pmatrix} = \begin{pmatrix} \overline{N}^* \\ \overline{M}^* \end{pmatrix} S_K. \tag{15}$$

Let now Q be any inner function for which $\begin{pmatrix} \overline{N}^* \\ \overline{M}^* \end{pmatrix} Q \in H_+^\infty$. This implies that for an arbitrary $f \in H_+^2$ we have $P_- \begin{pmatrix} \overline{V}_B & -\overline{U}_B \end{pmatrix} \begin{pmatrix} \overline{N}^* \\ \overline{M}^* \end{pmatrix} Qf = 0$. But we also have for an arbitrary $f \in H_+^2$ that $P_- \begin{pmatrix} -\overline{N} & \overline{M} \end{pmatrix} \begin{pmatrix} \overline{N}^* \\ \overline{M}^* \end{pmatrix} Qf = P_- Qf = 0$. It follows as before that $H_{R^*} f = P_- \begin{pmatrix} \overline{N}^* \\ \overline{M}^* \end{pmatrix} Qf = 0$, i.e. $Qf \in Ker H_{R^*} = S_K H_+^2$. Thus we have the inclusion $QH_+^2 \subset S_K H_+^2$. This implies the factorization $Q = S_K T$ for some inner function T.

2. The proof is similar and omitted. We set

$$\begin{pmatrix} K_1 & K_2 \end{pmatrix} = S_I \begin{pmatrix} N^* & M^* \end{pmatrix} \tag{16}$$

∎

Corollary 3.1 *Let S_K be the minimal right inner function of $\begin{pmatrix} \overline{N}^* \\ \overline{M}^* \end{pmatrix}$, then*

$$Ker H_{\begin{pmatrix} \overline{N}^* \\ \overline{M}^* \end{pmatrix}} = S_K H_+^2. \tag{17}$$

Proof: Clearly we have $S_K H_+^2 \subset Ker H_{\begin{pmatrix} \overline{N}^* \\ \overline{M}^* \end{pmatrix}}$. To see the converse inclusion we note that, by Beurling's theorem, $Ker H_{\begin{pmatrix} \overline{N}^* \\ \overline{M}^* \end{pmatrix}} = QH_+^2$ for some inner function Q. Therefore, for an arbitrary $f \in H_+^2$, we have $P_- \begin{pmatrix} \overline{N}^* \\ \overline{M}^* \end{pmatrix} Qf = 0$. This implies that $\begin{pmatrix} \overline{N}^* \\ \overline{M}^* \end{pmatrix} Q \in H_+^\infty$. So, by the minimality of S_K we have the factorization $Q = S_K T$, i.e.

$$Ker H_{\begin{pmatrix} \overline{N}^* \\ \overline{M}^* \end{pmatrix}} = QH_+^2 \subset S_K H_+^2.$$

The two inclusions, taken together, imply equality (17).

∎

Given the $p \times m$ bounded real function G with the J_B-normalized coprime factorizations (14), and with J_i defined by (15), we define the $(p+m)\times(p+m)$ function Ω_K by

$$\Omega_K = \begin{pmatrix} M & J_1 \\ N & J_2 \end{pmatrix} \tag{18}$$

Next we consider the H_+^2 space of $\mathbf{C}^{p+m}$ valued functions. On this space we define a new, indefinite, inner product by letting

$$[\begin{pmatrix} f_1 \\ f_2 \end{pmatrix}, \begin{pmatrix} g_1 \\ g_2 \end{pmatrix}] = (J_B \begin{pmatrix} f_1 \\ f_2 \end{pmatrix}, \begin{pmatrix} g_1 \\ g_2 \end{pmatrix}) = (f_1, g_1) - (f_2, g_2) \tag{19}$$

We denote this space, supressing in the notation the dependence on p and m, by $H_+^{2,J}$. Thus $H_+^{2,J}$ is a Krein space, see Bognar [1974] for the general theory of these spaces.

In $H_+^{2,J}$ we define a map, based on the matrix function Ω_K. Thus we set, for $f \in H_+^{2,J}$,

$$P_{\{\Omega_K H_+^{2,J}\}^{[\perp]}} f = \Omega_K J_B P_- \Omega_K^* J_B f \tag{20}$$

Then we have

Proposition 3.1 *Let $P_{\{\Omega_K H_+^{2,J}\}^{[\perp]}}$ be defined by (20).*

1. *The following identities hold.*

$$\begin{cases} \Omega_K^* J_B \Omega_K &= J_B \\ \Omega_K^* J_B \Omega_K J_B &= I \\ \Omega_K J_B \Omega_K^* J_B &= I \\ \Omega_K J_B \Omega_K^* &= J_B \end{cases} \tag{21}$$

2. *The map $P_{\{\Omega_K H_+^{2,J}\}^{[\perp]}}$ is a projection map in $H_+^{2,J}$.*

3. *We have*

$$Ker P_{\{\Omega_K H_+^{2,J}\}^{[\perp]}} = \Omega_K H_+^{2,J} \tag{22}$$

4. *We have*

$$\{\Omega_K H_+^{2,J}\}^{[\perp]} = \{f \in H_+^{2,J} | \Omega_K^* J_B f \in H_-^2\} \tag{23}$$

5. *We have*

$$Im P_{\{\Omega_K H_+^{2,J}\}^{[\perp]}} = \{\Omega_K H_+^{2,J}\}^{[\perp]} \tag{24}$$

6. *We have*

$$H_+^{2,J} = \{\Omega_K H_+^{2,J}\}^{[\perp]} [\oplus] \Omega_K H_+^{2,J} \tag{25}$$

Proof:

1. These identities follow by direct computation.

2. We use the identity $\Omega_K^* J_B \Omega_K = J_B$ to compute

$$\begin{aligned}(P_{\{\Omega_K H_+^{2,J}\}^{[\perp]}})^2 f &= \Omega_K J_B P_- \Omega_K^* J_B \Omega_K J_B P_- \Omega_K^* J_B f \\ &= \Omega_K J_B P_- J_B^2 P_- \Omega_K^* J_B f \\ &= \Omega_K J_B P_- \Omega_K^* J_B f = P_{\{\Omega_K H_+^{2,J}\}^{[\perp]}} f\end{aligned}$$

3. Let $f \in \Omega_K H_+^{2,J}$, with $f = \Omega_K g$, then

$$\begin{aligned}P_{\{\Omega_K H_+^{2,J}\}^{[\perp]}} f &= P_{\{\Omega_K H_+^{2,J}\}^{[\perp]}} \Omega_K g \\ &= \Omega_K J_B P_- \Omega_K^* J_B \Omega_K g \\ &= \Omega_K J_B P_- J_B g = 0\end{aligned}$$

as $J_B g \in H_+^2$. This shows that $\Omega_K H_+^{2,J} \subset Ker P_{\{\Omega_K H_+^{2,J}\}^{[\perp]}}$.

Conversely, assume $f \in Ker P_{\{\Omega_K H_+^{2,J}\}^{[\perp]}}$. Then

$$0 = P_{\{\Omega_K H_+^{2,J}\}^{[\perp]}} f = \Omega_K J_B P_- \Omega_K^* J_B f$$

This implies $g = \Omega_K^* J_B f \in H_+^2$. But then $\Omega_K J_B g = \Omega_K J_B \Omega_K^* J_B f = f$, i.e. $f \in \Omega_K H_+^{2,J}$.

4. Assume $f \in \{\Omega_K H_+^{2,J}\}^{[\perp]}$ i.e. $[f, \Omega_K g] = 0$ for all $g \in H_+^{2,J}$. Then

$$0 = [f, \Omega_K g] = (J_B f, \Omega_K g) = (\Omega_K^* J_B f, g)$$

i.e. $\Omega_K^* J_B f \in H_-^2$.

5. We have $f \in Im P_{\{\Omega_K H_+^{2,J}\}^{[\perp]}}$ if and only if $f = P_{\{\Omega_K H_+^{2,J}\}^{[\perp]}} f = \Omega_K J_B P_- \Omega_K^* J_B f$. However this equality holds if and only if $\Omega_K^* J_B f \in H_-^2$.

6. Follows from the characterization of the Kernel and Image of the projection $P_{\{\Omega_K H_+^{2,J}\}^{[\perp]}}$.

∎

In a similar way, given the $p \times m$ bounded real function G with the J_B-normalized coprime factorizations (14). Let us define

$$\begin{pmatrix} K_1 & K_2 \end{pmatrix} = S_I \begin{pmatrix} N^* & M^* \end{pmatrix} \tag{26}$$

We define the $(p+m) \times (p+m)$ function Ω_I by

$$\Omega_I = \begin{pmatrix} \overline{M} & \overline{N} \\ K_1 & K_2 \end{pmatrix} \tag{27}$$

Also, we consider the H_-^2 space of $\mathbf{C}^{p+m}$ valued functions. On this space we define a new, indefinite, inner product by letting

$$[\begin{pmatrix} f_1 \\ f_2 \end{pmatrix}, \begin{pmatrix} g_1 \\ g_2 \end{pmatrix}] = (J_B \begin{pmatrix} f_1 \\ f_2 \end{pmatrix}, \begin{pmatrix} g_1 \\ g_2 \end{pmatrix}) = (f_1, g_1) - (f_2, g_2) \tag{28}$$

We denote this space, supressing in the notation the dependence on p and m, by $H_-^{2,J}$. Thus $H_-^{2,J}$ is also a Krein space.

In $H_-^{2,J}$ we define a map, based on the matrix function Ω_I. Thus we set, for $h \in H_-^{2,J}$,

$$P_{\{\Omega_I^* H_-^{2,J}\}^{[\perp]}} h = \Omega_I^* J_B P_+ \Omega_I J_B h \tag{29}$$

Now we can prove the counterpart to Proposition 3.1.

Proposition 3.2 *Let $P_{\{\Omega_I^* H_-^{2,J}\}^{[\perp]}}$ be defined by (29).*

1. *The following identities hold.*

$$\begin{cases} \Omega_I J_B \Omega_I^* &= J_B \\ \Omega_I^* J_B \Omega_I J_B &= I \\ \Omega_I J_B \Omega_I^* J_B &= I \\ \Omega_I^* J_B \Omega_I &= J_B \end{cases} \tag{30}$$

2. *The map $P_{\{\Omega_I^* H_-^{2,J}\}^{[\perp]}}$ is a projection map in $H_-^{2,J}$.*

3. *We have*

$$Ker P_{\{\Omega_I^* H_-^{2,J}\}^{[\perp]}} = \Omega_I^* H_-^{2,J} \tag{31}$$

4. *We have*

$$\{\Omega_I^* H_-^{2,J}\}^{[\perp]} = \{h \in H_-^{2,J} | \Omega_I J_B h \in H_+^2\} \tag{32}$$

5. *We have*

$$Im P_{\{\Omega_I^* H_-^{2,J}\}^{[\perp]}} = \{\Omega_I^* H_-^{2,J}\}^{[\perp]} \tag{33}$$

6. *We have*

$$H_+^{2,J} = \{\Omega_I^* H_-^{2,J}\}^{[\perp]} [\oplus] \Omega_I^* H_-^{2,J} \tag{34}$$

Proof:

1. The identities follow by direct computation.

2. We use the identity $\Omega_I J_B \Omega_I^* = J_B$ to compute

$$\begin{aligned}(P_{\{\Omega_I^* H_-^{2,J}\}^{[\perp]}})^2 f &= \Omega_I^* J_B P_+ \Omega_I J_B \Omega_I^* J_B P_+ \Omega_I J_B f\\ &= \Omega_I^* J_B P_+ J_B^2 P_+ \Omega_I J_B f\\ &= \Omega_I^* J_B P_+ \Omega_I J_B f = P_{\{\Omega_I^* H_-^{2,J}\}^{[\perp]}} f\end{aligned}$$

3. Let $f \in \Omega_I^* H_-^{2,J}$, with $f = \Omega_I^* g$, then

$$\begin{aligned}P_{\{\Omega_I^* H_-^{2,J}\}^{[\perp]}} f &= P_{\{\Omega_I^* H_-^{2,J}\}^{[\perp]}} \Omega_I^* g\\ &= \Omega_I^* J_B P_+ \Omega_I J_B \Omega_I^* g\\ &= \Omega_I^* J_B P_+ J_B g = 0\end{aligned}$$

as $J_B g \in H_-^2$. This shows that $\Omega_I^* H_-^{2,J} \subset Ker P_{\{\Omega_I^* H_-^{2,J}\}^{[\perp]}}$.

4. Assume $f \in \{\Omega_I^* H_-^{2,J}\}^{[\perp]}$ i.e. $[f, \Omega_I^* g] = 0$ for all $g \in H_-^{2,J}$. Then

$$0 = [f, \Omega_I^* g] = (J_B f, \Omega_I^* g) = (\Omega_I J_B f, g)$$

i.e. $\Omega_I J_B f \in H_+^2$.

5. We have $f \in Im P_{\{\Omega_I^* H_-^{2,J}\}^{[\perp]}}$ if and only if $f = P_{\{\Omega_I^* H_-^{2,J}\}^{[\perp]}} f = \Omega_I^* J_B P_+ \Omega_I J_B f$. However this equality holds if and only if $\Omega_I J_B f \in H_+^2$.

6. Follows from the characterization of the Kernel and Image of the projection $P_{\{\Omega_I^* H_-^{2,J}\}^{[\perp]}}$.

∎

Given a linear transformation T in a Krein space we define its adjoint with respect to the indefinite metric as the unique transformation $T^\sharp$ satisfying, for all x, y,

$$[Tx, y] = [x, T^\sharp y] \tag{35}$$

Proposition 3.3 *For*

$$\Omega_K = \begin{pmatrix} M & J_1 \\ N & J_2 \end{pmatrix} \tag{36}$$

we define the map $Z_K : \{S_K H_+^2\}^\perp \longrightarrow \{\Omega_K H_+^{2,J}\}^{[\perp]}$ *by*

$$Z_K f = P_{\{\Omega_K H_+^{2,J}\}^{[\perp]}} \begin{pmatrix} U_B \\ V_B \end{pmatrix} f \qquad \text{for } f \in \{S_K H_+^2\}^\perp \tag{37}$$

and the map $Y_K : \{\Omega_K H_+^{2,J}\}^{[\perp]} \longrightarrow \{S_K H_+^2\}^\perp$ *by*

$$Y_K \begin{pmatrix} g_1 \\ g_2 \end{pmatrix} = -P_{\{S_K H_+^2\}^\perp} \begin{pmatrix} \overline{N} & \overline{M} \end{pmatrix} J_B \begin{pmatrix} g_1 \\ g_2 \end{pmatrix} \qquad \text{for } \begin{pmatrix} g_1 \\ g_2 \end{pmatrix} \in \{S_K H_+^2\}^\perp \tag{38}$$

Then

1. *The map Z is invertible and $Z_K^{-1} = Y_K$.*
2. *The indefinite adjoint of Z_K is given by*

$$Z_K^{\sharp}\begin{pmatrix} g_1 \\ g_2 \end{pmatrix} = P_{\{S_K H_+^2\}^{\perp}} P_+ \begin{pmatrix} U_B^* & V_B^* \end{pmatrix} J_B \begin{pmatrix} g_1 \\ g_2 \end{pmatrix} \quad \textit{for} \begin{pmatrix} g_1 \\ g_2 \end{pmatrix} \in \{\Omega_K H_+^{2,J}\}^{[\perp]} \tag{39}$$

3. *The indefinite adjoint of Y_K is given by*

$$Y_K^{\sharp} f = -J_B P_{\{\Omega_K H_+^{2,J}\}^{[\perp]}} P_+ \begin{pmatrix} \overline{N}^* \\ \overline{M}^* \end{pmatrix} f \quad \textit{for } f \in \{S_K H_+^2\}^{\perp} \tag{40}$$

Proof:

1. For $f \in \{S_K H_+^2\}^{\perp}$ we have

$$\begin{aligned}
Y_K Z_K f &= -P_{\{S_K H_+^2\}^{\perp}} \begin{pmatrix} \overline{N} & \overline{M} \end{pmatrix} J_B P_{\{\Omega_K H_+^{2,J}\}^{[\perp]}} \begin{pmatrix} U_B \\ V_B \end{pmatrix} f \\
&= -P_{\{S_K H_+^2\}^{\perp}} \begin{pmatrix} \overline{N} & \overline{M} \end{pmatrix} J_B \Omega_K J_B P_- \Omega_K^* J_B \begin{pmatrix} U_B \\ V_B \end{pmatrix} f \\
&= P_{\{S_K H_+^2\}^{\perp}} \begin{pmatrix} -\overline{N} & \overline{M} \end{pmatrix} \begin{pmatrix} M & J_1 \\ N & J_2 \end{pmatrix} J_B P_- \begin{pmatrix} M^* & N^* \\ J_1^* & J_2^* \end{pmatrix} J_B \begin{pmatrix} U_B \\ V_B \end{pmatrix} f \\
&= S_K P_- S_K^* \begin{pmatrix} 0 & S_K \end{pmatrix} J_B P_- \begin{pmatrix} R_B^* \\ -S_K^* \end{pmatrix} f \\
&= S_K P_- \begin{pmatrix} 0 & -I \end{pmatrix} P_- \begin{pmatrix} R_B^* \\ -S_K^* \end{pmatrix} f \\
&= S_K P_- S_K^* f = P_{\{S_K H_+^2\}^{\perp}} f = f
\end{aligned}$$

Similarly, for $\begin{pmatrix} g_1 \\ g_2 \end{pmatrix} \{\Omega_K H_+^{2,J}\}^{[\perp]}$, and using equality (8) we have

$$\begin{aligned}
Z_K Y_K \begin{pmatrix} g_1 \\ g_2 \end{pmatrix} &= P_{\{\Omega_K H_+^{2,J}\}^{[\perp]}} \begin{pmatrix} U_B \\ V_B \end{pmatrix} (-1) P_{\{S_K H_+^2\}^\perp} \begin{pmatrix} \overline{N} & \overline{M} \end{pmatrix} J_B \begin{pmatrix} g_1 \\ g_2 \end{pmatrix} \\
&= \Omega_K J_B P_- \Omega_K^* J_B \begin{pmatrix} U_B \\ V_B \end{pmatrix} S_K P_- S_K^* \begin{pmatrix} -\overline{N} & \overline{M} \end{pmatrix} \begin{pmatrix} g_1 \\ g_2 \end{pmatrix} \\
&= \Omega_K J_B P_- \begin{pmatrix} M^* & N^* \\ J_1^* & J_2^* \end{pmatrix} \begin{pmatrix} U_B \\ -V_B \end{pmatrix} S_K P_- S_K^* \begin{pmatrix} -\overline{N} & \overline{M} \end{pmatrix} \begin{pmatrix} g_1 \\ g_2 \end{pmatrix} \\
&= \Omega_K J_B P_- \begin{pmatrix} R_B^* \\ -S_K^* \end{pmatrix} S_K P_- S_K^* \begin{pmatrix} -\overline{N} & \overline{M} \end{pmatrix} \begin{pmatrix} g_1 \\ g_2 \end{pmatrix} \\
&= \Omega_K J_B P_- J_B \begin{pmatrix} R_B^* \\ S_K^* \end{pmatrix} \begin{pmatrix} -\overline{N} & \overline{M} \end{pmatrix} \begin{pmatrix} g_1 \\ g_2 \end{pmatrix} \\
&= \Omega_K J_B P_- \Omega_K^* J_B \Omega_K \begin{pmatrix} R_B^* \\ S_K^* \end{pmatrix} \begin{pmatrix} -\overline{N} & \overline{M} \end{pmatrix} \begin{pmatrix} g_1 \\ g_2 \end{pmatrix} \\
&= \Omega_K J_B P_- \Omega_K^* J_B \left[\begin{pmatrix} M \\ N \end{pmatrix} R_B^* + \begin{pmatrix} \overline{N}^* \\ \overline{M}^* \end{pmatrix} \right] \begin{pmatrix} -\overline{N} & \overline{M} \end{pmatrix} \begin{pmatrix} g_1 \\ g_2 \end{pmatrix} \\
&= P_{\{\Omega_K H_+^{2,J}\}^{[\perp]}} \begin{pmatrix} U_B \\ V_B \end{pmatrix} \begin{pmatrix} -\overline{N} & \overline{M} \end{pmatrix} \begin{pmatrix} g_1 \\ g_2 \end{pmatrix} \\
&= P_{\{\Omega_K H_+^{2,J}\}^{[\perp]}} \left\{ \begin{pmatrix} I & 0 \\ 0 & I \end{pmatrix} - \begin{pmatrix} M \\ N \end{pmatrix} \begin{pmatrix} \overline{V} & -\overline{U} \end{pmatrix} \right\} \begin{pmatrix} g_1 \\ g_2 \end{pmatrix} \\
&= \begin{pmatrix} g_1 \\ g_2 \end{pmatrix}
\end{aligned}$$

2. Let $f \in \{S_K H_+^2\}^\perp$, $\begin{pmatrix} g_1 \\ g_2 \end{pmatrix} \in \{\Omega_K H_+^{2,J}\}^{[\perp]}$. Then

$$\begin{aligned}
[Zf, \begin{pmatrix} g_1 \\ g_2 \end{pmatrix}] &= (J_B \Omega_K J_B P_- \Omega_K^* J_B \begin{pmatrix} U_B \\ V_B \end{pmatrix} f, \begin{pmatrix} g_1 \\ g_2 \end{pmatrix}) \\
&= (f, P_+ \begin{pmatrix} U_B^* & V_B^* \end{pmatrix} J_B \Omega_K P_- J_B \Omega_K^* J_B \begin{pmatrix} g_1 \\ g_2 \end{pmatrix}) \\
&= (f, P_{\{\Omega_K H_+^{2,J}\}^{[\perp]}} P_+ \begin{pmatrix} U_B^* & V_B^* \end{pmatrix} J_B \Omega_K J_B P_- \Omega_K^* J_B \begin{pmatrix} g_1 \\ g_2 \end{pmatrix})
\end{aligned}$$

Thus (39) follows.

3. Let $\begin{pmatrix} g_1 \\ g_2 \end{pmatrix} \in \{\Omega_K H_+^{2,J}\}^{[\perp]}$, $f \in \{S_K H_+^2\}^\perp$. Then

$$\begin{aligned}
(Y_K \begin{pmatrix} g_1 \\ g_2 \end{pmatrix}, f) &= -(P_{\{S_K H_+^2\}^\perp} \begin{pmatrix} \overline{N} & \overline{M} \end{pmatrix} \begin{pmatrix} g_1 \\ g_2 \end{pmatrix}, f) \\
&= -(\begin{pmatrix} g_1 \\ g_2 \end{pmatrix}, P_+ \begin{pmatrix} \overline{N}^* \\ \overline{M}^* \end{pmatrix} f) \\
&= -(\begin{pmatrix} g_1 \\ g_2 \end{pmatrix}, P_{\{\Omega_K H_+^{2,J}\}^{[\perp]}} P_+ \begin{pmatrix} \overline{N}^* \\ \overline{M}^* \end{pmatrix} f) \\
&= -[\begin{pmatrix} g_1 \\ g_2 \end{pmatrix}, J_B P_{\{\Omega_K H_+^{2,J}\}^{[\perp]}} P_+ \begin{pmatrix} \overline{N}^* \\ \overline{M}^* \end{pmatrix} f]
\end{aligned}$$

which proves (40).

∎

Proposition 3.4 *For*

$$\Omega_I = \begin{pmatrix} \overline{M} & \overline{N} \\ K_1 & K_2 \end{pmatrix} \tag{41}$$

we define the map $Z_I : \{S_I^* H_-^2\}^\perp \longrightarrow \{\Omega_I^* H_-^{2,J}\}^{[\perp]}$ *by*

$$Z_I h = P_{\{\Omega_I^* H_-^{2,J}\}^{[\perp]}} \begin{pmatrix} \overline{U}_B^* \\ \overline{V}_B^* \end{pmatrix} h \quad \textit{for } h \in \{S_I^* H_-^2\}^\perp \tag{42}$$

and the map $Y_I : \{\Omega_I^* H_-^{2,J}\}^{[\perp]} \longrightarrow \{S_I^* H_-^2\}^\perp$ *by*

$$Y_I \begin{pmatrix} h_1 \\ h_2 \end{pmatrix} = -P_{\{S_I^* H_+^2\}^\perp} \begin{pmatrix} N^* & M^* \end{pmatrix} J_B \begin{pmatrix} h_1 \\ h_2 \end{pmatrix} \quad \textit{for } \begin{pmatrix} h_1 \\ h_2 \end{pmatrix} \in \{\Omega_I^* H_-^{2,J}\}^{[\perp]} \tag{43}$$

Then

1. *The maps Z_I and Y_I are invertible and $Z_I^{-1} = Y_I$.*
2. *The indefinite adjoint of Z_I is given by*

$$Z_I^\sharp \begin{pmatrix} h_1 \\ h_2 \end{pmatrix} = P_{\{S_I^* H_-^2\}^\perp} P_- \begin{pmatrix} \overline{U} & \overline{V} \end{pmatrix} J_B \begin{pmatrix} h_1 \\ h_2 \end{pmatrix} \quad \textit{for } \begin{pmatrix} h_1 \\ h_2 \end{pmatrix} \in \{\Omega_I^* H_-^{2,J}\}^{[\perp]} \tag{44}$$

3. *The indefinite adjoint of Y_I is given by*

$$Y_I^\sharp h = -P_{\{\Omega_I^* H_-^{2,J}\}^{[\perp]}} P_- \begin{pmatrix} N \\ M \end{pmatrix} h \quad \textit{for } h \in \{S_I^* H_-^2\}^\perp \tag{45}$$

Proof:

1. For $h \in \{S_I^* H_-^2\}^\perp$ we have

$$\begin{aligned}
Y_I Z_I &= -P_{\{S_I^* H_+^2\}^\perp} \begin{pmatrix} N^* & M^* \end{pmatrix} J_B P_{\{\Omega_I^* H_-^{2,J}\}^{[\perp]}} \begin{pmatrix} \overline{U}_B^* \\ \overline{V}_B^* \end{pmatrix} h \\
&= P_{\{S_I^* H_+^2\}^\perp} \begin{pmatrix} -N^* & M^* \end{pmatrix} P_{\{\Omega_I^* H_-^{2,J}\}^{[\perp]}} \begin{pmatrix} \overline{U}_B^* \\ \overline{V}_B^* \end{pmatrix} h \\
&= S_I^* P_+ S_I \begin{pmatrix} -N^* & M^* \end{pmatrix} \begin{pmatrix} \overline{M}^* & K_1^* \\ \overline{N}^* & K_2^* \end{pmatrix} J_B P_+ \begin{pmatrix} \overline{M} & \overline{N} \\ K_1 & K_2 \end{pmatrix} J_B \begin{pmatrix} \overline{U}_B^* \\ \overline{V}_B^* \end{pmatrix} h \\
&= S_I^* P_+ S_I \begin{pmatrix} 0 & -S_I^* \end{pmatrix} P_+ \begin{pmatrix} R_B \\ -S_I \end{pmatrix} h = h.
\end{aligned}$$

On the other hand, for $\begin{pmatrix} h_1 \\ h_2 \end{pmatrix} \in \{\Omega_I^* H_-^{2,J}\}^{[\perp]}$, we compute, using equality (9) as well as the following equality

$$\begin{pmatrix} \overline{U}_B^* \\ \overline{V}_B^* \end{pmatrix} \begin{pmatrix} -N^* & M^* \end{pmatrix} + \begin{pmatrix} \overline{M}^* \\ \overline{N}^* \end{pmatrix} \begin{pmatrix} V^* & -U^* \end{pmatrix} = \begin{pmatrix} I & 0 \\ 0 & I \end{pmatrix}$$

$$\begin{aligned}
Z_I Y_I \begin{pmatrix} h_1 \\ h_2 \end{pmatrix} &= -P_{\{\Omega_I^* H_-^{2,J}\}^{[\perp]}} \begin{pmatrix} \overline{U}_B^* \\ \overline{V}_B^* \end{pmatrix} P_{\{S_I^* H_+^2\}^\perp} \begin{pmatrix} N^* & M^* \end{pmatrix} J_B \begin{pmatrix} h_1 \\ h_2 \end{pmatrix} \\
&= -\Omega_I^* J_B P_+ \Omega_I J_B \begin{pmatrix} \overline{U}_B^* \\ \overline{V}_B^* \end{pmatrix} = S_I^* P_+ S_I \begin{pmatrix} N^* & M^* \end{pmatrix} J_B \begin{pmatrix} h_1 \\ h_2 \end{pmatrix} \\
&= -\Omega_I^* J_B P_+ \begin{pmatrix} R \\ -S_I \end{pmatrix} \begin{pmatrix} N^* & M^* \end{pmatrix} J_B \begin{pmatrix} h_1 \\ h_2 \end{pmatrix} \\
&= -\Omega_I^* J_B P_+ \Omega_I J_B \Omega_I^* \begin{pmatrix} R \\ S_I \end{pmatrix} \begin{pmatrix} N^* & M^* \end{pmatrix} J_B \begin{pmatrix} h_1 \\ h_2 \end{pmatrix} \\
&= -P_{\{\Omega_I^* H_-^{2,J}\}^{[\perp]}} \left[\begin{pmatrix} \overline{M}^* \\ \overline{N}^* \end{pmatrix} R_B + \begin{pmatrix} N \\ M \end{pmatrix} \right] \begin{pmatrix} N^* & M^* \end{pmatrix} J_B \begin{pmatrix} h_1 \\ h_2 \end{pmatrix} \\
&= -P_{\{\Omega_I^* H_-^{2,J}\}^{[\perp]}} \begin{pmatrix} \overline{U}^* \\ \overline{V}^* \end{pmatrix} \begin{pmatrix} N^* & M^* \end{pmatrix} J_B \begin{pmatrix} h_1 \\ h_2 \end{pmatrix} \\
&= P_{\{\Omega_I^* H_-^{2,J}\}^{[\perp]}} \begin{pmatrix} \overline{U}^* \\ \overline{V}^* \end{pmatrix} \begin{pmatrix} -N^* & M^* \end{pmatrix} \begin{pmatrix} h_1 \\ h_2 \end{pmatrix} \\
&= P_{\{\Omega_I^* H_-^{2,J}\}^{[\perp]}} \left\{ \begin{pmatrix} I & 0 \\ 0 & I \end{pmatrix} - \begin{pmatrix} \overline{M}^* \\ \overline{N}^* \end{pmatrix} \begin{pmatrix} V^* & -U^* \end{pmatrix} \right\} \begin{pmatrix} h_1 \\ h_2 \end{pmatrix} \\
&= \begin{pmatrix} h_1 \\ h_2 \end{pmatrix}
\end{aligned}$$

2. Let $h \in \{S_I^* H_+^2\}^\perp$ and $\begin{pmatrix} h_1 \\ h_2 \end{pmatrix} \in \{\Omega_I^* H_-^{2,J}\}^{[\perp]}$. Then

$$\begin{aligned}
[Z_I h, \begin{pmatrix} h_1 \\ h_2 \end{pmatrix}] &= (J_B P_{\{\Omega_I^* H_-^{2,J}\}^{[\perp]}} \begin{pmatrix} \overline{U}_B^* \\ \overline{V}_B^* \end{pmatrix} h, \begin{pmatrix} h_1 \\ h_2 \end{pmatrix}) \\
&= (\Omega_I^* J_B P_+ \Omega_I J_B \begin{pmatrix} \overline{U}_B^* \\ \overline{V}_B^* \end{pmatrix} h, J_B \begin{pmatrix} h_1 \\ h_2 \end{pmatrix}) \\
&= (h, \begin{pmatrix} \overline{U}_B & \overline{V}_B \end{pmatrix} J_B \Omega_I^* P_+ J_B \Omega_I J_B \begin{pmatrix} h_1 \\ h_2 \end{pmatrix}) \\
&= (h, P_{\{S_I^* H_+^2\}^\perp} \begin{pmatrix} \overline{U}_B & \overline{V}_B \end{pmatrix} J_B \Omega_I^* P_+ J_B \Omega_I J_B \begin{pmatrix} h_1 \\ h_2 \end{pmatrix})
\end{aligned}$$

This proves (44).

3. Let $h \in \{S_I^* H_+^2\}^\perp$ and $\begin{pmatrix} h_1 \\ h_2 \end{pmatrix} \in \{\Omega_I^* H_-^{2,J}\}^{[\perp]}$. Then

$$\begin{aligned}
\left(Y_I \begin{pmatrix} h_1 \\ h_2 \end{pmatrix}, h\right) &= \left(-P_{\{S_I^* H_+^2\}^\perp} \begin{pmatrix} N^* & M^* \end{pmatrix} J_B \begin{pmatrix} h_1 \\ h_2 \end{pmatrix}, h\right) \\
&= \left(\begin{pmatrix} h_1 \\ h_2 \end{pmatrix}, \begin{pmatrix} -N \\ M \end{pmatrix} h\right) = \left(P_- \begin{pmatrix} h_1 \\ h_2 \end{pmatrix}, \begin{pmatrix} -N \\ M \end{pmatrix}, h\right) \\
&= \left(\begin{pmatrix} h_1 \\ h_2 \end{pmatrix}, P_- \begin{pmatrix} -N \\ M \end{pmatrix}, h\right) \\
&= \left(P_{\{\Omega_I^* H_-^{2,J}\}^{[\perp]}} \begin{pmatrix} h_1 \\ h_2 \end{pmatrix}, P_- \begin{pmatrix} -N \\ M \end{pmatrix}, h\right) \\
&= \left(\Omega_I^* J_B P_+ \Omega_I J_B \begin{pmatrix} h_1 \\ h_2 \end{pmatrix}, P_- \begin{pmatrix} -N \\ M \end{pmatrix}, h\right) \\
&= \left(\begin{pmatrix} h_1 \\ h_2 \end{pmatrix}, J_B \Omega_I^* P_+ J_B \Omega_I J_B^2 P_- \begin{pmatrix} -N \\ M \end{pmatrix}, h\right) \\
&= \left[\begin{pmatrix} h_1 \\ h_2 \end{pmatrix}, P_{\{\Omega_I^* H_-^{2,J}\}^{[\perp]}} J_B P_- \begin{pmatrix} -N \\ M \end{pmatrix}, h\right] \\
&= -\left[\begin{pmatrix} h_1 \\ h_2 \end{pmatrix}, P_{\{\Omega_I^* H_-^{2,J}\}^{[\perp]}} P_- \begin{pmatrix} N \\ M \end{pmatrix}, h\right]
\end{aligned}$$

This proves (45).

∎

In Proposition 3.3 we studied the matrix function $\Omega_K = \begin{pmatrix} M & J_1 \\ N & J_2 \end{pmatrix}$ in great detail. The reason for this is that it provides the representation of the kernel of an important Hankel operator, namely the operator $H_{\begin{pmatrix} M^* & N^* \end{pmatrix} J_B} = H_{\begin{pmatrix} M^* & -N^* \end{pmatrix}}$. We make this more specific in the following.

Proposition 3.5 *1. With the operator Z_K defined by (37), we have*

$$H_{R_B^*} = H_{\begin{pmatrix} M^* & N^* \end{pmatrix} J_B} Z_K \tag{46}$$

2. We have

$$Ker H_{\begin{pmatrix} M^* & N^* \end{pmatrix} J_B} = \Omega_K H_+^{2,J} \tag{47}$$

Proof:

1. We compute

$$\begin{aligned} H_{\begin{pmatrix} M^* & N^* \end{pmatrix} J_B} Z_K f &= P_- \begin{pmatrix} M^* & N^* \end{pmatrix} J_B P_{\{\Omega_K H_+^{2,J}\}^{[\perp]}} \begin{pmatrix} U_B \\ V_B \end{pmatrix} f \\ &= P_- \begin{pmatrix} M^* & N^* \end{pmatrix} J_B \Omega_K J_B P_- \Omega_K^* J_B \begin{pmatrix} U_B \\ V_B \end{pmatrix} f \\ &= P_- \begin{pmatrix} M^* & N^* \end{pmatrix} J_B \begin{pmatrix} M & J_1 \\ N & J_2 \end{pmatrix} J_B P_- \Omega_K^* J_B \begin{pmatrix} U_B \\ V_B \end{pmatrix} f \\ &= P_- \begin{pmatrix} I & 0 \end{pmatrix} J_B P_- \begin{pmatrix} M^* & N^* \\ J_1^* & J_2^* \end{pmatrix} \begin{pmatrix} U_B \\ -V_B \end{pmatrix} f \\ &= P_- \begin{pmatrix} I & 0 \end{pmatrix} P_- \begin{pmatrix} R_B^* \\ -S_K^* \end{pmatrix} f \\ &= H_{R_B^*} f \end{aligned}$$

2. Clearly we have the inclusion $\begin{pmatrix} M & J_1 \\ N & J_2 \end{pmatrix} H_+^{2,J} \subset Ker H_{\begin{pmatrix} M^* & N^* \end{pmatrix} J_B}$.

Conversely, assume $\begin{pmatrix} g_1 \\ g_2 \end{pmatrix} \in Ker H_{\begin{pmatrix} M^* & N^* \end{pmatrix} J_B}$ and $f \in \{\Omega_K H_+^{2,J}\}^{[\perp]}$. Then $\begin{pmatrix} g_1 \\ g_2 \end{pmatrix} = Zf$ for some $f \in \{S_K H_+^2\}^\perp$. Now $H_{R_B^*} = H_{\begin{pmatrix} M^* & N^* \end{pmatrix} J_B} Z_K$ and the invertibility of Z_K imply $H_{R_B^*} f = 0$. But on $\{S_K H_+^2\}^\perp$ the Hankel operator $H_{R_B^*}$ is injective. So it follows that $f = 0$ and hence also $\begin{pmatrix} g_1 \\ g_2 \end{pmatrix} = 0$.

∎

In the same way we can state the following. We omit the proof.

Proposition 3.6 *With the operators Z_I and Y_I defined by (42) and (43) respectively, we have*

1. $$\hat{H}_{R_B} = \hat{H}_{\begin{pmatrix} \overline{M} & \overline{N} \end{pmatrix} J_B} Z_I \tag{48}$$

2. $$H_{\begin{pmatrix} \overline{M}^* \\ \overline{N}^* \end{pmatrix}} = Y_I^\sharp H_{R_B^*} \tag{49}$$

3. *We have*

$$Ker \hat{H}_{\begin{pmatrix} \overline{M} & \overline{N} \end{pmatrix} J_B} = \Omega_I^* H_-^{2,J} \tag{50}$$

In analogy with the key diagram studied in great detail in Fuhrmann and Ober [1993] and in Fuhrmann [1993b] we have another equivalent diagram connecting the various Hankel and intertwining operators derived from the bounded real function G. Thus we have.

Proposition 3.7 *The operator* $T : \{\Omega_K H^2_+\}^\perp \longrightarrow \{\Omega_I^* H^2_-\}^\perp$

$$\begin{pmatrix} f_1 \\ f_2 \end{pmatrix} \mapsto P_- \begin{pmatrix} \overline{M}^* \\ \overline{N}^* \end{pmatrix} (-\overline{N} \ \ \overline{M}) \begin{pmatrix} f_1 \\ f_2 \end{pmatrix} \tag{51}$$

is such that the following diagram commutes.

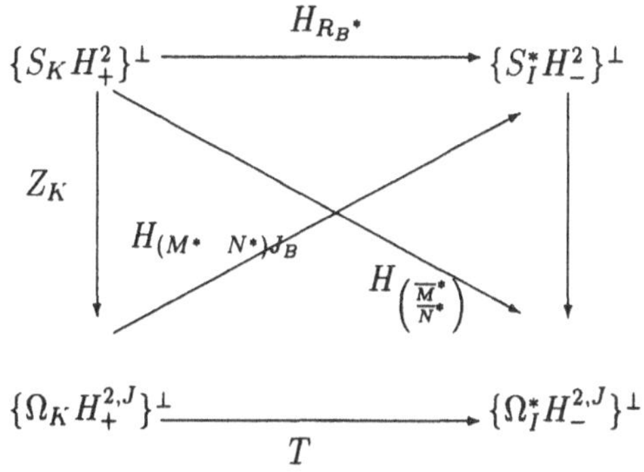

Proof: In view of Propositions 3.5 and 3.6 it suffices to prove that

$$H_{\begin{pmatrix} \overline{M}^* \\ \overline{N}^* \end{pmatrix}} Y_K = Y_I^\sharp \hat{H}_{\begin{pmatrix} \overline{M} & \overline{N} \end{pmatrix} J_B}. \tag{52}$$

We begin by computing

$$\begin{aligned} H_{\begin{pmatrix} \overline{M}^* \\ \overline{N}^* \end{pmatrix}} Y_K \begin{pmatrix} g_1 \\ g_2 \end{pmatrix} &= (Y_I^\sharp H_{R_B^*}) Y_K \\ &= Y_I^\sharp (H_{R_B^*} Y_K) \\ &= Y_I^\sharp H_{\begin{pmatrix} M^* & N^* \end{pmatrix} J_B} \end{aligned}$$

Next we compute

$$Y_I^\sharp \hat{H}_{\begin{pmatrix} M^* & N^* \end{pmatrix} J_B} = P_{\{\Omega_I^* H^{2,J}_-\}[\perp]} J_B P_- \begin{pmatrix} M \\ N \end{pmatrix} P_- \begin{pmatrix} M^* & N^* \end{pmatrix} J_B$$

∎

As a corollary to the results in this section we have the following statement about the dimensions of the spaces with which we are dealing.

Corollary 3.2 *If any of the spaces* $\{S_K H^2_+\}^\perp$, $\{S_I^* H^2_-\}^\perp$, $\{\Omega_K H^{2,J}_+\}^\perp$, $\{\Omega_I^* H^{2,J}_-\}^\perp$ *is finite dimensional, so are all the other and*

$$dim(\{S_K H^2_+\}^\perp) = dim(\{S_I^* H^2_-\}^\perp) = dim(\{\Omega_K H^{2,J}_+\}^\perp) = dim(\{\Omega_I^* H^{2,J}_-\}^\perp).$$

If the McMillan degree of G *is* n, *then* n *is the dimension of these spaces.*

Corollary 3.3 *Let G be a rational, bounded real function of McMillan degree n. Then its B-characteristic R_B^* has also McMillan degree n.*

Proof: The McMillan degree of R_B^* is equal to the dimension of any of the spaces $\{S_K H_+^2\}^\perp$ or $\{S_I^* H_-^2\}^\perp$. These are equal, by the previous corollary, to the dimensions of any of the spaces $\{\Omega_K H_+^{2,J}\}^{[\perp]}$ or $\{\Omega_I^* H_-^{2,J}\}^{[\perp]}$. These on the other hand are equal to the McMillan degrees of $\begin{pmatrix} M^* & N^* \end{pmatrix}$ or $\begin{pmatrix} \overline{N}^* \\ \overline{M}^* \end{pmatrix}$. However, by the results of Fuhrmann and Ober [1993b], these have the same McMillan degree as G. ∎

Proposition 3.8 *Let the operators Z_K and Y_K be defined by (37) and (38) respectively. Then the following identities are satisfied.*

1. $$Z_K^\sharp Z_K = H_{R_B^*}^* H_{R_B^*} - I|_{\{S_K H_+^2\}^\perp} \tag{53}$$

2. $$Y_K Y_K^\sharp = -I + H^*_{\begin{pmatrix} \overline{N}^* \\ \overline{M}^* \end{pmatrix}} J_B H_{\begin{pmatrix} \overline{N}^* \\ \overline{M}^* \end{pmatrix}} |_{\{S_K H_+^2\}^\perp} \tag{54}$$

Proof:

1. We compute

$$\begin{aligned}
Z_K^\sharp Z_K &= P_{\{S_K H_+^2\}^\perp} P_+ \begin{pmatrix} U_B^* & V_B^* \end{pmatrix} J_B P_{\{\Omega_K H_+^{2,J}\}^{[\perp]}} \begin{pmatrix} U_B \\ V_B \end{pmatrix} f \\
&= S_K P_- S_K^* P_+ \begin{pmatrix} U_B^* & V_B^* \end{pmatrix} J_B \Omega_K J_B P_- \Omega_K^* J_B \begin{pmatrix} U_B \\ V_B \end{pmatrix} f \\
&= S_K P_- S_K^* P_+ \begin{pmatrix} U_B^* & V_B^* \end{pmatrix} J_B \begin{pmatrix} M & J_1 \\ N & J_2 \end{pmatrix} J_B P_- \begin{pmatrix} M^* & N^* \\ J_1^* & J_2^* \end{pmatrix} J_B \begin{pmatrix} U_B \\ V_B \end{pmatrix} f \\
&= S_K P_- S_K^* P_+ \begin{pmatrix} R & -S_K \end{pmatrix} J_B P_- \begin{pmatrix} R^* \\ -S^* \end{pmatrix} f \\
&= P_{\{S_K H_+^2\}^\perp} H_{R_B^*}^* H_{R_B^*} f - S_K P_- S_K^* P_+ S_K P_- S_K^* f
\end{aligned}$$

2. We compute

$$\begin{aligned}
Y_K Y_K^\sharp &= P_{\{S_K H_+^2\}^\perp} \begin{pmatrix} \overline{N} & \overline{M} \end{pmatrix} J_B P_{\{\Omega_K H_+^{2,J}\}^{[\perp]}} P_+ \begin{pmatrix} \overline{N}^* \\ \overline{M}^* \end{pmatrix} f \\
&= P_{\{S_K H_+^2\}^\perp} \begin{pmatrix} \overline{N} & \overline{M} \end{pmatrix} J_B \Omega_K J_B P_- \Omega_K^* J_B (I - P_-) \begin{pmatrix} \overline{N}^* \\ \overline{M}^* \end{pmatrix} f \\
&= P_{\{S_K H_+^2\}^\perp} \begin{pmatrix} \overline{N} & \overline{M} \end{pmatrix} J_B \Omega_K J_B P_- \Omega_K^* J_B \begin{pmatrix} \overline{N}^* \\ \overline{M}^* \end{pmatrix} f \\
&\quad - P_{\{S_K H_+^2\}^\perp} \begin{pmatrix} \overline{N} & \overline{M} \end{pmatrix} J_B \Omega_K J_B P_- \Omega_K^* J_B P_- \begin{pmatrix} \overline{N}^* \\ \overline{M}^* \end{pmatrix} f
\end{aligned}$$

We compute each term separately.

$$\begin{aligned}
&P_{\{S_K H_+^2\}^\perp} \begin{pmatrix} \overline{N} & \overline{M} \end{pmatrix} J_B \Omega_K J_B P_- \Omega_K^* J_B \begin{pmatrix} \overline{N}^* \\ \overline{M}^* \end{pmatrix} f \\
&= P_{\{S_K H_+^2\}^\perp} \begin{pmatrix} \overline{N} & \overline{M} \end{pmatrix} J_B \Omega_K J_B (I - P_+) \Omega_K^* J_B \begin{pmatrix} \overline{N}^* \\ \overline{M}^* \end{pmatrix} f \\
&= P_{\{S_K H_+^2\}^\perp} \begin{pmatrix} \overline{N} & \overline{M} \end{pmatrix} J_B \Omega_K J_B \Omega_K^* J_B \begin{pmatrix} \overline{N}^* \\ \overline{M}^* \end{pmatrix} f \\
&\quad - P_{\{S_K H_+^2\}^\perp} \begin{pmatrix} \overline{N} & \overline{M} \end{pmatrix} J_B \Omega_K J_B P_+ \Omega_K^* J_B \begin{pmatrix} \overline{N}^* \\ \overline{M}^* \end{pmatrix} f \\
&= P_{\{S_K H_+^2\}^\perp} \begin{pmatrix} \overline{N} & \overline{M} \end{pmatrix} J_B \begin{pmatrix} \overline{N}^* \\ \overline{M}^* \end{pmatrix} f - P_{\{S_K H_+^2\}^\perp} \begin{pmatrix} 0 & S_K \end{pmatrix} P_+ \Omega_K^* J_B \begin{pmatrix} \overline{N}^* \\ \overline{M}^* \end{pmatrix} f \\
&= -f
\end{aligned}$$

On the other hand

$$\begin{aligned}
&P_{\{S_K H_+^2\}^\perp} \begin{pmatrix} \overline{N} & \overline{M} \end{pmatrix} J_B \Omega_K J_B P_- \Omega_K^* J_B P_- \begin{pmatrix} \overline{N}^* \\ \overline{M}^* \end{pmatrix} f \\
&= S_K P_- S_K^* \begin{pmatrix} \overline{N} & \overline{M} \end{pmatrix} J_B \begin{pmatrix} M & J_1 \\ N & J_2 \end{pmatrix} J_B P_- \begin{pmatrix} M^* & N^* \\ J_1^* & J_2^* \end{pmatrix} J_B P_- \begin{pmatrix} \overline{N}^* \\ \overline{M}^* \end{pmatrix} f \\
&= S_K P_- S_K^* \begin{pmatrix} 0 & S_K \end{pmatrix} J_B P_- \begin{pmatrix} M^* & N^* \\ J_1^* & J_2^* \end{pmatrix} J_B P_- \begin{pmatrix} \overline{N}^* \\ \overline{M}^* \end{pmatrix} f \\
&= S_K P_- \begin{pmatrix} 0 & -I \end{pmatrix} P_- \begin{pmatrix} M^* & N^* \\ J_1^* & J_2^* \end{pmatrix} J_B P_- \begin{pmatrix} \overline{N}^* \\ \overline{M}^* \end{pmatrix} f \\
&= S_K P_- \begin{pmatrix} 0 & -I \end{pmatrix} \begin{pmatrix} M^* & N^* \\ J_1^* & J_2^* \end{pmatrix} J_B P_- \begin{pmatrix} \overline{N}^* \\ \overline{M}^* \end{pmatrix} f \\
&= -S_K P_- S_K^* \begin{pmatrix} \overline{N} & \overline{M} \end{pmatrix} J_B P_- \begin{pmatrix} \overline{N}^* \\ \overline{M}^* \end{pmatrix} f \\
&= -S_K P_- S_K^* P_+ \begin{pmatrix} \overline{N} & \overline{M} \end{pmatrix} J_B P_- \begin{pmatrix} \overline{N}^* \\ \overline{M}^* \end{pmatrix} f
\end{aligned}$$

Taken together, these computations imply (54).

∎

The next proposition is proved analogously.

Proposition 3.9 *Let the operators Z_I and Y_I be defined by (42) and (43) respectively. Then the following identities are satisfied.*

1. $$Z_I^\sharp Z_I = H_{R_B^*} H_{R_B^*}^* - I|_{\{S_I^* H_-^2\}^\perp} \tag{55}$$

2. $$Y_I Y_I^\sharp = -I + H_{\begin{pmatrix} N^* & M^* \end{pmatrix}} J_B \hat{H}_{\begin{pmatrix} N \\ M \end{pmatrix}} |_{\{S_I^* H_-^2\}^\perp} \tag{56}$$

The next theorem follows easily from the relations proved in the previous propositions. We omit the simple proof. This theorem has implications as far as balanced realizations are concerned, however we will not pursue this theme in the present paper.

Theorem 3.1 *Let G be a rational, proper bounded real function and let R_B be its B-characteristic. Let $1 > \mu_1 \geq \cdots \geq \mu_n > 0$ be the singular values of the Hankel operator $H_{R_B^*}$ and $\{f_i, h_i\}$ the corresponding Schmidt pairs. We set*

$$g_i = (1-\mu_i^2)^{-\frac{1}{2}} Z_K f_i \in \{\Omega_K H_+^{2,J}\}^{[\perp]} \tag{57}$$

and

$$k_i = (1-\mu_i^2)^{-\frac{1}{2}} Z_I h_i \in \{\Omega_I^* H_-^{2,J}\}^{[\perp]}. \tag{58}$$

Then

1. $$\begin{cases} Z_K f_i &= (1-\mu_i^2)^{\frac{1}{2}} g_i \\ Z_K^\sharp g_i &= -(1-\mu_i^2)^{\frac{1}{2}} f_i \end{cases} \tag{59}$$

2. $$\begin{cases} Y_K g_i &= (1-\mu_i^2)^{-\frac{1}{2}} f_i \\ Y_K^\sharp f_i &= -(1-\mu_i^2)^{-\frac{1}{2}} g_i \end{cases} \tag{60}$$

3. $$\begin{cases} Z_I h_i &= (1-\mu_i^2)^{\frac{1}{2}} k_i \\ Z_I^\sharp k_i &= -(1-\mu_i^2)^{\frac{1}{2}} h_i \end{cases} \tag{61}$$

4. $$\begin{cases} Y_I k_i &= (1-\mu_i^2)^{-\frac{1}{2}} h_i \\ Y_I^\sharp h_i &= -(1-\mu_i^2)^{-\frac{1}{2}} k_i \end{cases} \tag{62}$$

5. $$\begin{cases} H_{\begin{pmatrix} M^* & N^* \end{pmatrix}} J_B g_i &= \mu_i(1-\mu_i^2)^{-\frac{1}{2}} h_i \\ H_{\begin{pmatrix} \overline{M^*} \\ \overline{N^*} \end{pmatrix}} f_i &= -\mu_i(1-\mu_i^2)^{-\frac{1}{2}} k_i \end{cases} \tag{63}$$

4 State space realizations

We devote this section to the presentation of state space formulas of the various objects under study. We quote the results obtained previously in Fuhrmann and Ober [1993b] and Ober and Fuhrmann [1993] on the state space representations for the J_B normalized coprime factor of a bounded real function. We proceed to the formulas for the doubly coprime factors using the BR-controller. This leads to the state space representation for the B-characteristic R_B. The constructions of these state space realizations are based on the use of the stabilizing solutions of two Riccati equations. As the B-characteristic R_B of the bounded real function G is stable we have two Lyapunov equations associated with the derived state space realization for R_B. The solutions of these equations are necessary for reconstructing the bounded real function G from its B-characteristic R_B. However these solutions, naturally positivedefinite, are expressible in terms of the stabilizing solutions of the two Riccati equations. With the solutions to the Lyapunov equations at hand we can reconstruct not only G but actually also the normalized coprime factors. This is of importance as these yields state space formulas for the fractional linear transformation that parametrizes the set of all suboptimal Nehari extensions. This theme is picked up in Section 5.

As in previous studies, see Fuhrmann and Ober [1993a] and Fuhrmann [1993], i.e. in the case of stable systems and minimal systems the normalized coprime factorizations, the normalization being with respect to an appropriate metric, play a major role in the construction of the characteristic functions. The same is true here and we proceed next with the presentation of state space formulas for the J_B-normalized coprime factors of a bounded real function. For the purpose of this paper, focusing on the Nehari problem, we do not need the general case and may, without loss of generality, restrict ourselves to strictly proper bounded real functions. For this class of functions the state space formulas simplify and are given by the following theorem. This is a specialization of the general result in Fuhrmann and Ober [1993b]. In turn these formulas are extensions of previous similar results obtained by Meyer and Franklin [1987] and Vidyasagar [1988]. Throughout this section the proofs, mostly computational, are omitted.

Theorem 4.1 *Let G be a rational, bounded real, strictly proper, transfer function, and let*

$$G = NM^{-1} = \overline{M}^{-1}\overline{N} \tag{64}$$

be its normalized right and left J_B-coprime factorization. Let

$$G = \left(\begin{array}{c|c} A & B \\ \hline C & 0 \end{array}\right)$$

be a minimal state space realization. Then

1. *A state space realization for $\begin{pmatrix} M \\ N \end{pmatrix}$ is given by*

$$\begin{pmatrix} M \\ N \end{pmatrix} = \left(\begin{array}{c|c} A + BB^*X & B \\ \hline B^*X & I \\ C & 0 \end{array}\right) \tag{65}$$

where X is the stabilizing solution of the Bounded Real Algebraic Riccati Equation (BRRE),

$$A^*X + XA + C^*C + XBB^*X = 0 \tag{66}$$

2. *A state space realization for $\begin{pmatrix} -\overline{N} & \overline{M} \end{pmatrix}$ is given by*

$$\begin{pmatrix} -\overline{N} & \overline{M} \end{pmatrix} = \left(\begin{array}{c|cc} A + ZC^*C & -B & ZC^* \\ \hline C & 0 & I \end{array}\right), \tag{67}$$

where Z is the stabilizing solution to the Bounded Real Algebraic Riccati Equation (BRRE),

$$ZA^* + AZ + BB^* + ZC^*CZ = 0 \tag{68}$$

We point out that the minimal positive definite solutions can be shown to be actually the stabilizing solutions, i.e. the solutions of the Riccati equations for which $A + BB^*Y$ and $A + ZC^*C$ are stable.

Next we state a theorem giving a state space description of the coprime factors of the B-controller. These were characterized in Lemma 2.3.

Theorem 4.2 *1. Let* $\begin{pmatrix} U \\ V \end{pmatrix}$ *be any solution of the Bezout equation* $\overline{M}V - \overline{N}U = I$ *for which* $\begin{pmatrix} M & U \\ N & V \end{pmatrix}$ *has McMillan degree n. Then we have the following state space realization for it*

$$\begin{pmatrix} U \\ V \end{pmatrix} = \left(\begin{array}{c|c} A - BB^*X & BD' + ZC^* \\ \hline B^*X & D' \\ C & I \end{array}\right) \tag{69}$$

where D' is arbitrary and X is the stabilizing solution of the Bounded Real Algebraic Riccati Equation (BRRE),

$$A^*X + XA + C^*C + XBB^*X = 0 \tag{70}$$

For the bounded real controller's coprime factors we have

$$\begin{pmatrix} U_B \\ V_B \end{pmatrix} = \left(\begin{array}{c|c} A + BB^*X & -ZC^* \\ \hline B^*X & 0 \\ C & I \end{array}\right) \tag{71}$$

2. Let $\begin{pmatrix} \overline{V} & -\overline{U} \end{pmatrix}$ *be any solution of the Bezout equation* $\overline{V}M - \overline{U}N = I$ *for which* $\begin{pmatrix} \overline{V} & -\overline{U} \\ -\overline{N} & \overline{M} \end{pmatrix}$ *has McMillan degree n. Then we have the following state space realization for it*

$$\begin{pmatrix} \overline{V} & -\overline{U} \end{pmatrix} = \left(\begin{array}{c|cc} A + ZC^*C & -B & ZC^* \\ \hline \overline{D}'C + B^*X & I & \overline{D}' \end{array}\right) \tag{72}$$

where $\overline{D}'$ is arbitrary and Z is the stabilizing solution of the Bounded Real Algebraic Riccati Equation (BRRE),

$$ZA^* + AZ + ZC^*CZ + BB^* = 0 \tag{73}$$

For the bounded real controller's coprime factors we have

$$\begin{pmatrix} \overline{V}_B & -\overline{U}_B \end{pmatrix} = \left(\begin{array}{c|cc} A + ZC^*C & -B & ZC^* \\ \hline B^*X & I & 0 \end{array}\right) \tag{74}$$

The following theorem is a special case of a result in Ober and Fuhrmann [1993] and it gives a state space representations of the B-characteristic of G.

Theorem 4.3 *Let G be a bounded real function, vanishing at ∞, with a minimal state space realization*

$$G=\left(\begin{array}{c|c}A&B\\\hline C&0\end{array}\right)\tag{75}$$

Let R be its B-characteristic. Then minimal state space realizations of R are given by

$$R=\left(\begin{array}{c|c}A+BB^*X&B\\\hline -C(I-ZX)&0\end{array}\right)\tag{76}$$

and

$$R=\left(\begin{array}{c|c}A+ZC^*C&(I-ZX)B\\\hline -C&0\end{array}\right)\tag{77}$$

where X and Z are the stabilizing solutions of the Riccati equations

$$\left\{\begin{array}{rcl}A^*X+XA+XBB^*X+C^*C&=&0\\AZ+ZA^*+ZC^*CZ+BB^*&=&0\end{array}\right.\tag{78}$$

That the two realizations of R are isomorphic follows from the Bucy relation

$$(I-ZX)(A+BB^*X)=(A+ZC^*C)(I-ZX).\tag{79}$$

The next lemma gives a one to one relation between the solutions to the Lyapunov equations associated with the B-characteristic and the stabilizing solutions of the two Riccati equations associated with the bounded real function G. In particular the inverse relations (82) are instrumental in the proof of Theorem 5.3.

Lemma 4.1 *Let G be a bounded real function, vanishing at ∞, with a minimal state space realization*

$$G=\left(\begin{array}{c|c}A&B\\\hline C&0\end{array}\right)\tag{80}$$

and let X, Z be the stabilizing, or alternatively the minimal positive definite solutions of the Riccati equations (66) and (68). Then

1. *For the realization*

$$R=\left(\begin{array}{c|c}A+BB^*X&B\\\hline -C(I-ZX)&0\end{array}\right)=\left(\begin{array}{c|c}\mathcal{A}&\mathcal{B}\\\hline \mathcal{C}&0\end{array}\right)$$

of the B-characteristic, the Lyapunov equations

$$\left\{\begin{array}{rcl}\mathcal{A}P+P\mathcal{A}^*&=&-\mathcal{B}\mathcal{B}^*\\\mathcal{A}^*Q+Q\mathcal{A}&=&-\mathcal{C}^*\mathcal{C}\end{array}\right.\tag{81}$$

have solutions given by

$$\begin{cases} P &= (I - ZX)^{-1}Z = Z(I - XZ)^{-1} \\ Q &= X(I - ZX) = (I - XZ)X \end{cases}$$

where X, Z are the stabilizing solutions to the Riccati equations (66) and (68). The stabilizing solutions of these Riccati equations are given by

$$\begin{cases} X &= Q(I - PQ)^{-1} = (I - QP)^{1}Q \\ Z &= (I - PQ)P = P(I - QP) \end{cases} \tag{82}$$

2. *For the realization*

$$R = \left(\begin{array}{c|c} A + ZC^*C & (I - ZX)B \\ \hline -C & 0 \end{array}\right) = \left(\begin{array}{c|c} \mathcal{A} & \mathcal{B} \\ \hline \mathcal{C} & 0 \end{array}\right)$$

of the B-characteristic, the Lyapunov equations

$$\begin{cases} \mathcal{A}P + P\mathcal{A}^* &= -\mathcal{B}\mathcal{B}^* \\ \mathcal{A}^*Q + Q\mathcal{A} &= -\mathcal{C}^*\mathcal{C} \end{cases} \tag{83}$$

have solutions given by

$$\begin{cases} P &= Z(I - XZ) = (I - ZX)Z \\ Q &= (I - XZ)^{-1}X = X(I - ZX)^{-1} \end{cases}$$

The stabilizing solutions of these Riccati equations are given by

$$\begin{cases} X &= Q(I - PQ) = (I - QP)Q \\ Z &= (I - PQ)^{-1}P = P(I - QP)^{-1} \end{cases} \tag{84}$$

Our next step is the inversion, via state space formulas, of the characteristic map. This is based on Ober and Fuhrmann [1993].

Theorem 4.4 *Let G be a bounded real function, vanishing at ∞, Let R be its B-characteristic have a state space realization*

$$R = \left(\begin{array}{c|c} \mathcal{A} & \mathcal{B} \\ \hline \mathcal{C} & 0 \end{array}\right) \tag{85}$$

Let P and Q be the solutions of the Lyapunov equations

$$\begin{cases} \mathcal{A}P + P\mathcal{A}^* &= -\mathcal{B}\mathcal{B}^* \\ \mathcal{A}^*Q + Q\mathcal{A} &= -\mathcal{C}^*\mathcal{C} \end{cases} \tag{86}$$

Then state space realization of G are given by

$$G = \left(\begin{array}{c|c} A & B \\ \hline C & 0 \end{array}\right) = \left(\begin{array}{c|c} \mathcal{A} - \mathcal{B}\mathcal{B}^*Q(I-PQ)^{-1} & \mathcal{B} \\ \hline -\mathcal{C}(I-PQ)^{-1} & 0 \end{array}\right) \tag{87}$$

and

$$G = \left(\begin{array}{c|c} A & B \\ \hline C & 0 \end{array}\right) = \left(\begin{array}{c|c} \mathcal{A} - P(I-QP)^{-1}\mathcal{C}^*\mathcal{C} & -(I-PQ)^{-1}\mathcal{B} \\ \hline \mathcal{C} & 0 \end{array}\right) \tag{88}$$

We conclude this section by giving state space representations for the J_B-normalized coprime factors of G in terms of a minimal realization of the B-characteristic of G.

Theorem 4.5 *Let G be a bounded real function, vanishing at ∞, with a minimal state space realization of its B-characteristic given by*

$$R = \left(\begin{array}{c|c} \mathcal{A} & \mathcal{B} \\ \hline \mathcal{C} & 0 \end{array}\right)$$

Let P and Q be the solutions of the Lyapunov equations

$$\left\{\begin{array}{rcl} \mathcal{A}P + P\mathcal{A}^* & = & -\mathcal{B}\mathcal{B}^* \\ \mathcal{A}^*Q + Q\mathcal{A} & = & -\mathcal{C}^*\mathcal{C} \end{array}\right. \tag{89}$$

Let

$$G = NM^{-1} = \overline{M}^{-1}\overline{N} \tag{90}$$

be the normalized J_B-coprime factorizations of G. Then

1. *A state space realization of the right normalized coprime factors is given by*

$$\begin{pmatrix} M \\ N \end{pmatrix} = \left(\begin{array}{c|c} \mathcal{A} & \mathcal{B} \\ \hline \mathcal{B}^*Q(I-PQ)^{-1} & I \\ -\mathcal{C}(I-PQ)^{-1} & 0 \end{array}\right) \tag{91}$$

2. *A state space realization of the left normalized coprime factors is given by.*

$$\begin{pmatrix} -\overline{N} & \overline{M} \end{pmatrix} = \left(\begin{array}{c|cc} \mathcal{A} & (I-PQ)^{-1}\mathcal{B} & (I-PQ)^{-1}P\mathcal{C}^* \\ \hline \mathcal{C} & 0 & I \end{array}\right) \tag{92}$$

5 Suboptimal Nehari extensions

There is an extensive literature on various aspects of the Nehari extension problem, its applications and its connections to other basic mathematical problems. Nehari's theorem is very closely associated to the commutant lifting theorem and hence to a whole class of interpolation problems. Another way to look at it is as an optimization problem where, given a function $G \in L^\infty(i\mathbf{R})$, we look for a best approximation for it in H_+^∞ (considered as a subspace of $L^\infty(i\mathbf{R})$.

There is an immediate lower bound on the best approximation given in terms of the norm of an induced Hankel operator. In fact if the Hankel operator $H_G : H_+^2 \longrightarrow H_-^2$ is defined by

$$H_G f = P_- G f, \qquad f \in H_+^2$$

then, noting that for every $Q \in H_+^\infty$

$$\|H_G\| = \|H_{G-Q}\| \leq \|G-Q\|_\infty$$

we have

$$\|H_G\| \leq inf_{Q \in H_+^\infty} \|G-Q\|_\infty.$$

In fact, the content of Nehari's theorem is that this lower bound is achievable. In particular, if $\|H_G\| < 1$, then there exists a $Q \in H_+^\infty$ for which $\|H_G\| = \|G-Q\|_\infty$. The function $G-Q$ is referred to as an optimal Nehari extension of G. If however, with $\|H_G\| < 1$, we look for a $Q \in H_+^\infty$ such that $\|G-Q\|_\infty \leq 1$ then we refer to $G-Q$ as a suboptimal Nehari extension of G. If our starting point is a rational, strictly proper function $G \in H_-^\infty$ then such a function Q is called a suboptimal Nehari complement of G.

Our approach to the solution of the suboptimal Nehari extension problem will be based on the observation made after the proof of Lemma 2.5. There we saw that in terms of the NJ_BCF of a bounded real function G we immediately can write down a suboptimal Nehari extension of R_B^*, the B-characteristic of G. This allows us to reverse the reasoning. If we start with an arbitrary, rational, strictly proper antistable function $R^* \in H_-^\infty$, our first step will be to identify it as the B-characteristic of some unique, strictly proper, bounded real function. This yields at least one solution of the suboptimal Nehari extension problem, given by equation (12). From there it is a short road to the parametrization of all suboptimal solutions.

5.1 Feedback and scattering formalism

Before proceeding with the analysis of the suboptimal Nehari extension problem we find it of use to collect the basic facts concerning the Julia transform which is the tool connecting the feedback and scattering formalisms.

The standard feedback configuration is given by the following diagram.

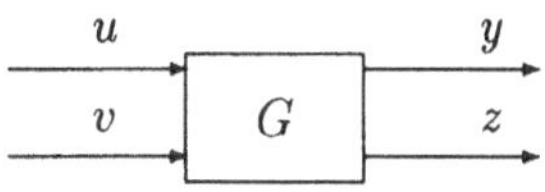

The corresponding closed loop system, given the dynamic feedback $v = Kz$ is given by

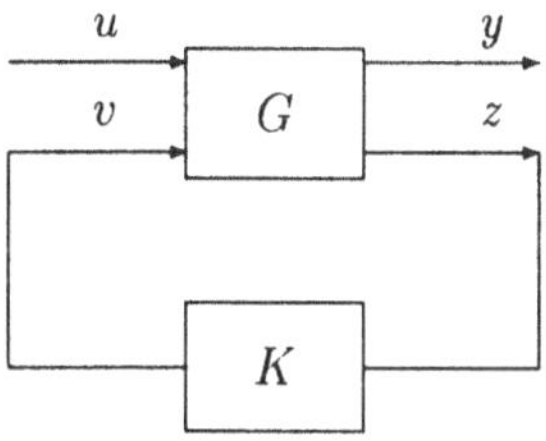

This is equivalent to the equations

$$\begin{pmatrix} y \\ z \end{pmatrix} = \begin{pmatrix} G_{11} & G_{12} \\ G_{21} & G_{22} \end{pmatrix} \begin{pmatrix} u \\ v \end{pmatrix} \tag{93}$$

and

$$v = Kz \tag{94}$$

The closed loop transfer function is easily computed to be $G_{11} + G_{12}K(I - G_{22}K)^{-1}G_{21}$

The scattering formalism is obtained by changing our view on the inputs and outputs. This is summed by the following diagram.

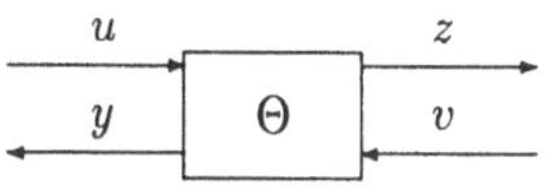

In turn this is equivalent to

$$\begin{pmatrix} u \\ y \end{pmatrix} = \begin{pmatrix} \Theta_{11} & \Theta_{12} \\ \Theta_{21} & \Theta_{22} \end{pmatrix} \begin{pmatrix} z \\ v \end{pmatrix} \tag{95}$$

Closing the loop by the same feedback as before, namely by $v = Kz$ leads to

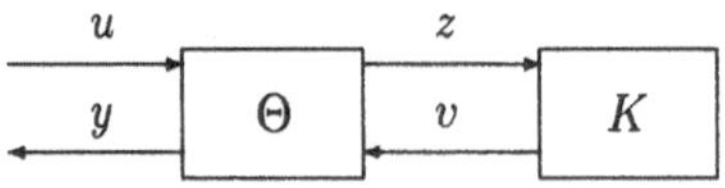

In terms of the scattering function Θ the closed loop transfer function is

$$(\Theta_{21}+\Theta_{22}K)(\Theta_{11}+\Theta_{12}K)^{-1}$$

.

The passage from the feedback to scattering formalism, that is from G to Θ, is given by the **Julia transform**, namely

$$\Theta = \begin{pmatrix} G_{21}^{-1} & -G_{21}^{-1}G_{22} \\ G_{11}G_{21}^{-1} & G_{12}-G_{11}G_{21}^{-1}G_{22} \end{pmatrix} \tag{96}$$

The inverse transformation, that is from the scattering to the feedback formalism is given by

$$G = \begin{pmatrix} \Theta_{21}\Theta_{11}^{-1} & \Theta_{22}-\Theta_{21}\Theta_{11}^{-1}\Theta_{12} \\ \Theta_{11}^{-1} & -\Theta_{11}^{-1}\Theta_{12} \end{pmatrix} \tag{97}$$

Clearly the Julia transform is dependent on our choice of variables as well as their order. Thus if we would write

$$\begin{pmatrix} z \\ y \end{pmatrix} = \begin{pmatrix} G_{11} & G_{12} \\ G_{21} & G_{22} \end{pmatrix} \begin{pmatrix} v \\ u \end{pmatrix} \tag{98}$$

and

$$\begin{pmatrix} u \\ y \end{pmatrix} = \begin{pmatrix} \Theta_{11} & \Theta_{12} \\ \Theta_{21} & \Theta_{22} \end{pmatrix} \begin{pmatrix} v \\ z \end{pmatrix} \tag{99}$$

then the Julia transform would be

$$\Theta = \begin{pmatrix} -G_{12}^{-1}G_{11} & G_{12}^{-1} \\ G_{21}-G_{22}G_{12}^{-1}G_{11} & G_{22}G_{12}^{-1} \end{pmatrix}. \tag{100}$$

The Julia transform and its inverse preserve McMillan degree. State space formulas for the Julia transform can be found in Gohberg and Rubinstein [1988].

The following is well known, see Potapov [1955].

Theorem 5.1 *Assume* $G = \begin{pmatrix} G_{11} & G_{12} \\ G_{21} & G_{22} \end{pmatrix}$ *is inner and that* G_{21} *is invertible. Then, with* $J = \begin{pmatrix} I & 0 \\ 0 & -I \end{pmatrix}$ *the matrix function* $\Theta = \begin{pmatrix} G_{21}^{-1} & -G_{21}^{-1}G_{22} \\ G_{11}G_{21}^{-1} & G_{12}-G_{11}G_{21}^{-1}G_{22} \end{pmatrix}$ *is J-inner.*

5.2 Parametrization

We begin this section with the study of a special all-pass function, defined in the following theorem.

Theorem 5.2 *Let G be a bounded real function of McMillan degree n and let*

$$G = NM^{-1} = \overline{M}^{-1}\overline{N}$$

be NJ_BCF. Then

1. $\begin{pmatrix} NM^{-1} & \overline{M}^{-1} \\ M^{-1} & -N^*\overline{M}^{-1} \end{pmatrix}$ *is an all-pass function of McMillan degree $2n$.*

2. *The Julia transform of* $\begin{pmatrix} NM^{-1} & \overline{M}^{-1} \\ M^{-1} & -N^*\overline{M}^{-1} \end{pmatrix}$ *is* $\begin{pmatrix} M & \overline{N}^* \\ N & \overline{M}^* \end{pmatrix}$.

3. *We have the following additive decomposition*

$$\begin{pmatrix} NM^{-1} & \overline{M}^{-1} \\ M^{-1} & -N^*\overline{M}^{-1} \end{pmatrix} = \begin{pmatrix} NM^{-1} & \overline{M}^{-1} \\ M^{-1} & -M^{-1}U_B \end{pmatrix} + \begin{pmatrix} 0 & 0 \\ 0 & R_B^* \end{pmatrix}$$

4. *The McMillan degrees of* $\begin{pmatrix} NM^{-1} & \overline{M}^{-1} \\ M^{-1} & -M^{-1}U_B \end{pmatrix}$ *and* $\begin{pmatrix} 0 & 0 \\ 0 & R_B^* \end{pmatrix}$ *are both equal to n.*

5. *The singular values of the involuted Hankel operator* $\hat{H}_{\begin{pmatrix} NM^{-1} & \overline{M}^{-1} \\ M^{-1} & -M^{-1}U_B \end{pmatrix}}$ *are equal to the singular values of $H_{R_B^*}$.*

6. *The Julia transform of* $\begin{pmatrix} NM^{-1} & \overline{M}^{-1} \\ M^{-1} & -M^{-1}U_B \end{pmatrix}$ *is* $\begin{pmatrix} M & U_B \\ N & V_B \end{pmatrix}$.

Proof:

1. We compute, using the equalities $\overline{M}^*\overline{M} - NN^* = I$ and $M^*M - N^*N = I$,

$$\begin{pmatrix} M^{-*}N^* & M^{-*} \\ \overline{M}^{-*} & \overline{M}^{-*}N \end{pmatrix}\begin{pmatrix} NM^{-1} & \overline{M}^{-1} \\ M^{-1} & -N^*\overline{M}^{-1} \end{pmatrix}$$
$$= \begin{pmatrix} M^{-*}(N^*N+I)M^{-1} & 0 \\ 0 & \overline{M}^{-*}(NN^*+I)\overline{M}^{-1} \end{pmatrix}\begin{pmatrix} I & 0 \\ 0 & I \end{pmatrix}$$

2. By direct computation, using equation (96).

3. Follows from equation (12).

4. We know that the McMillan degrees of both $\begin{pmatrix} M \\ N \end{pmatrix}$ and $\begin{pmatrix} \overline{N}^* \\ \overline{M}^* \end{pmatrix}$ are equal to n, hence the McMillan degree of $\begin{pmatrix} M & \overline{N}^* \\ N & \overline{M}^* \end{pmatrix}$ is equal to $2n$. Now the Julia transform preserves McMillan degree so the McMillan degree of $\begin{pmatrix} NM^{-1} & \overline{M}^{-1} \\ M^{-1} & -N^*\overline{M}^{-1} \end{pmatrix}$ is equal to $2n$. Since the B-characteristic map preserves McMillan degree, we must have that the McMillan degree of R_B^* is equal to n, and the result follows.
5. Follows from Theorem 1.6 in Appendix 5 of Nikolskii [1985].
6. By direct computation, using the Julia transform equation (96) and the Bezout equation $\overline{M}V - \overline{N}U = I$.

∎

We pass on now to the parametrization of all suboptimal Nehari extensions of a rational, strictly proper, antistable function $R^* \in H_-^\infty$.

Theorem 5.3 *Let $R^* \in H_-^\infty$ be a rational, strictly proper, antistable function with $\|H_{R^*}\| < 1$. Let G be the bounded real function obtained as the inverse B-characteristic of R^* and let*

$$G = NM^{-1} = \overline{M}^{-1}\overline{N}$$

be NJ_BCF of G. Then the set

$$\{-(N^* + M^*Q)(\overline{M} + \overline{N}Q)^{-1} | Q \in BH_+^\infty\} \tag{101}$$

is a parametrization of the set of all suboptimal Nehari extensions of R^*.

Proof: Since all the Hankel singular values of R^* are less than 1 we can consider it to be the B-characteristic function of some, uniquely determined, strictly proper H_+^∞ contraction G.
We consider now the NJ_BCF of G, i.e.

$$G = NM^{-1} = \overline{M}^{-1}\overline{N} \tag{102}$$

and go back to the all-pass function

$$\begin{pmatrix} NM^{-1} & \overline{M}^{-1} \\ M^{-1} & -N^*\overline{M}^{-1} \end{pmatrix} \tag{103}$$

We consider it as the transfer function of the two port

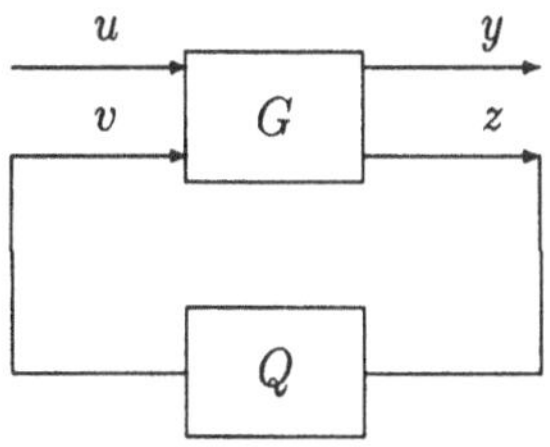

and write this time

$$\begin{pmatrix} z \\ y \end{pmatrix} = \begin{pmatrix} G_{11} & G_{12} \\ G_{21} & G_{22} \end{pmatrix} \begin{pmatrix} v \\ u \end{pmatrix} \tag{104}$$

which we terminate by

$$v = -Qz \tag{105}$$

The closed loop transfer function is easily computed to be $G_{22} - G_{21}Q(I + G_{11}Q)^{-1}G_{12}$. For the particular choice $Q = 0$, we get $-N^*\overline{M}^{-1}$. As $\begin{pmatrix} NM^{-1} & \overline{M}^{-1} \\ M^{-1} & -N^*\overline{M}^{-1} \end{pmatrix}$ is an all-pass function, i.e. it is pointwise unitary on the imaginary axis, then for every Q of norm ≤ 1 the transfer function $G_{22} - G_{21}Q(I + G_{11}Q)^{-1}G_{12}$ is contractive.
This we compute to be

$$\begin{aligned}
&-N^*\overline{M}^{-1} - M^{-1}Q(I + NM^{-1}Q)^{-1}\overline{M}^{-1} \\
&\quad = -M^{-1}\overline{N}^* - M^{-1}Q(I + \overline{M}^{-1}\overline{N}Q)^{-1}\overline{M}^{-1} \\
&\quad = -M^{-1}\overline{N}^* - M^{-1}Q(\overline{M} + \overline{N}Q)^{-1} \\
&\quad = -M^{-1}(\overline{N}^*(\overline{M} + \overline{N}Q) + Q)(\overline{M} + \overline{N}Q)^{-1} \\
&\quad = -M^{-1}(\overline{N}^*\overline{M} + \overline{N}^*\overline{N}Q + Q)(\overline{M} + \overline{N}Q)^{-1} \\
&\quad = -M^{-1}(\overline{N}^*\overline{M} + MM^*Q)(\overline{M} + \overline{N}Q)^{-1} \\
&\quad = -M^{-1}(MN^* + MM^*Q)(\overline{M} + \overline{N}Q)^{-1} \\
&\quad = -(N^* + M^*Q)(\overline{M} + \overline{N}Q)^{-1}
\end{aligned} \tag{106}$$

or

$$N^*\overline{M}^{-1} + M^{-1}Q(I + NM^{-1}Q)^{-1}\overline{M}^{-1} = (N^* + M^*Q)(\overline{M} + \overline{N}Q)^{-1}. \tag{107}$$

Therefore

$$N^*\overline{M}^{-1} - (N^* + M^*Q)(\overline{M} + \overline{N}Q)^{-1} = -M^{-1}Q(I + NM^{-1}Q)^{-1}\overline{M}^{-1} \tag{108}$$

which is clearly in H_+^∞ whenever $Q \in H_+^\infty$ is contractive. Thus for each contraction $Q \in H_+^\infty$ the function $(N^* + M^*Q)(\overline{M} + \overline{N}Q)^{-1}$ is a Nehari extension of R^*.

Note that the previous computation is basically the computation of the Julia transform, i.e. we write

$$\begin{pmatrix} u \\ y \end{pmatrix} = \begin{pmatrix} \Theta_{11} & \Theta_{12} \\ \Theta_{21} & \Theta_{22} \end{pmatrix} \begin{pmatrix} v \\ z \end{pmatrix} \tag{109}$$

terminated by $v = -Qz$. Again a standard computation leads to

$$\Theta = \begin{pmatrix} -G_{12}^{-1}G_{11} & G_{12}^{-1} \\ G_{21} - G_{22}G_{12}^{-1}G_{11} & G_{22}G_{12}^{-1} \end{pmatrix} = \begin{pmatrix} -\overline{N} & \overline{M} \\ M^* & -N^* \end{pmatrix} \tag{110}$$

Here we used the fact that $M^{-1} + N^*\overline{M}^{-1}\overline{M}NM^{-1} = (I + N^*N)M^{-1} = M^*$. So with the termination $z = -Qv$, we have

$$y = -(N^* + M^*Q)(\overline{M} + \overline{N}Q)^{-1}u$$

and of course with $Q = 0$ we get the transfer function $-N^*\overline{M}^{-1}$. This particular Nehari extension will be referred to as the *central Nehari extension.*

Next we show that

$$\{-(N^* + M^*Q)(\overline{M} + \overline{N}Q)^{-1} | Q \in BH_+^\infty\} \tag{111}$$

exhausts the set of all Nehari extensions.
To see this let $K \in L^\infty$ be any suboptimal Nehari extension of R^*. We set

$$Q = -(M^* + K\overline{N})^{-1}(N^* + K\overline{M}) = -(\overline{N}^* + MK)(\overline{M} + NK)^{-1}$$

Since $\|K\|_\infty \le 1$ it follows, by the fact that $\begin{pmatrix} N^* & \overline{M} \\ M^* & \overline{N} \end{pmatrix}$ is J-unitary, that Q is also in the unit ball of L^∞. As the previous fractional linear transformation can be inverted, we can write

$$K = -(N^* + M^*Q)(\overline{M} + \overline{N}Q)^{-1} = -(M + QN)^{-1}(\overline{N}^* + Q\overline{M}^*).$$

So it remains to show that $Q \in H_+^\infty$.
As both $-N^*\overline{M}^{-1} = -M^{-1}\overline{N}^*$ and K are Nehari extensions of R^*, we have

$$\begin{aligned}
-N^*\overline{M}^{-1} - K &= -N^*\overline{M}^{-1} + (N^* + M^*Q)(\overline{M} + \overline{N}Q)^{-1} \\
&= -M^{-1}\overline{N}^* + (N^* + M^*Q)(\overline{M} + \overline{N}Q)^{-1} \\
&= -M^{-1}\left[\overline{N}^*(\overline{M} + \overline{N}Q) - M(N^* + M^*Q)\right](\overline{M} + \overline{N}Q)^{-1} \\
&= -M^{-1}\left[(\overline{N}^*\overline{M} - MN^*) + (\overline{N}^*\overline{N} - MM^*)Q\right](\overline{M} + \overline{N}Q)^{-1} \\
&= M^{-1}Q(\overline{M} + \overline{N}Q)^{-1} \\
&= M^{-1}Q(I + \overline{M}^{-1}\overline{N}Q)^{-1}\overline{M}^{-1} \\
&= M^{-1}Q(I + GQ)^{-1}\overline{M}^{-1} \in H_+^\infty.
\end{aligned}$$

Since both M and $\overline{M}$ are invertible in H_+^∞ it follows that $Q(I+GQ)^{-1} \in H_+^\infty$. The fact that $G \in H_+^\infty$ implies $GQ(I+GQ)^{-1} \in H_+^\infty$, $(I+GQ)^{-1} = I - GQ(I+GQ)^{-1} \in H_+^\infty$ and $(I-GQ)(I+GQ)^{-1} \in H_+^\infty$. Now, as GQ is contractive $(I-GQ)(I+GQ)^{-1}$ is positive real. So $I + (I-GQ)(I+GQ)^{-1} = 2(I+GQ)^{-1}$ is strictly positive real and hence invertible in H_+^∞. Thus we conclude that $Q = Q(I+GQ)^{-1}(I+GQ) \in H_+^\infty$. This completes the proof.

∎

Thus the solvability of the problem of parametrizing all suboptimal Nehari extensions is seen to be through a special type of Nehari extension, namely the extension of $\begin{pmatrix} 0 & 0 \\ 0 & R_B^* \end{pmatrix}$ to an all-pass function. In a sense this seems to be the dual to the standard four block problem in H^∞-control theory. The further clarification of this connection is worthy of further study.

Theorem 5.3 is of interest only if, starting with a rational, strictly proper, antistable function $R^* \in H_-^\infty$ with $\|H_{R^*}\| < 1$, we can find the J_B-normalized coprime factors of a bounded real function G obtained as the inverse B-characteristic of R^*. This problem we solved in Theorem 4.4, inverting the B-characteristic function using state space methods. In fact we can do better than that, for we have derived, in Theorem 4.5, state space formulas for the normalized coprime factors of G directly in terms of a minimal realization of R. Thus we can state.

Corollary 5.1 *Let $R^* \in H_-^\infty$ be a rational, strictly proper, antistable function with $\|H_{R^*}\| < 1$. Let*

$$R = \left(\begin{array}{c|c} \mathcal{A} & \mathcal{B} \\ \hline \mathcal{C} & 0 \end{array}\right) \tag{112}$$

be a minimal realization.
Let P and Q be the solutions of the Lyapunov equations

$$\begin{cases} \mathcal{A}P + P\mathcal{A}^* &= -\mathcal{B}\mathcal{B}^* \\ \mathcal{A}^*Q + Q\mathcal{A} &= -\mathcal{C}^*\mathcal{C} \end{cases} \tag{113}$$

Define

$$\begin{pmatrix} M \\ N \end{pmatrix} = \left(\begin{array}{c|c} \mathcal{A} & \mathcal{B} \\ \hline \mathcal{B}^*Q(I-PQ)^{-1} & I \\ -\mathcal{C}(I-PQ)^{-1} & 0 \end{array}\right) \tag{114}$$

and

$$\begin{pmatrix} -\overline{N} & \overline{M} \end{pmatrix} = \left(\begin{array}{c|cc} \mathcal{A} & (I-PQ)^{-1}\mathcal{B} & (I-PQ)^{-1}P\mathcal{C}^* \\ \hline \mathcal{C} & 0 & I \end{array}\right) \tag{115}$$

Then the set

$$\{-(N^* + M^*Q)(\overline{M} + \overline{N}Q)^{-1} | Q \in BH_+^\infty\} \tag{116}$$

is a parametrization of the set of all suboptimal Nehari extensions of R^*.

These formulas can be found in Francis [1987]. That exposition is based in turn on Ball and Helton [1983] and Ball and Ran [1986].

In the next two sections we present two more methods for this inversion, one based on J_B-spectral factorizations, the other a method of solving a Hankel operator based system of equations.

5.3 Connection to J spectral factorizations

We explore now the connections to the J-spectral factorization approach to the problem of parametrizing all suboptimal Nehari extensions.

We observed already the importance of the additive decomposition

$$\begin{pmatrix} NM^{-1} & \overline{M}^{-1} \\ M^{-1} & -N^*\overline{M}^{-1} \end{pmatrix} = \begin{pmatrix} NM^{-1} & \overline{M}^{-1} \\ M^{-1} & -\overline{U}_B\overline{M}^{-1} \end{pmatrix} + \begin{pmatrix} 0 & 0 \\ 0 & R_B^* \end{pmatrix}. \tag{117}$$

The Julia transform of $\begin{pmatrix} NM^{-1} & \overline{M}^{-1} \\ M^{-1} & -N^*\overline{M}^{-1} \end{pmatrix}$ was computed to be $\begin{pmatrix} M & \overline{N}^* \\ N & \overline{M}^* \end{pmatrix}$ whereas that of $\begin{pmatrix} NM^{-1} & \overline{M}^{-1} \\ M^{-1} & -\overline{U}_B\overline{M}^{-1} \end{pmatrix}$ is $\begin{pmatrix} M & U_B \\ N & V_B \end{pmatrix}$. Here U, V correspond to the BR controller. It is of interest to compute the difference of the two Julia transforms. Recall that we have the relation

$$\begin{pmatrix} \overline{N}^* \\ \overline{M}^* \end{pmatrix} = \begin{pmatrix} U_B \\ V_B \end{pmatrix} - \begin{pmatrix} M \\ N \end{pmatrix} R_B^* \tag{118}$$

This leads to

$$\begin{pmatrix} M & \overline{N}^* \\ N & \overline{M}^* \end{pmatrix} - \begin{pmatrix} M & U_B \\ N & V_B \end{pmatrix} = -\begin{pmatrix} 0 & MR^* \\ 0 & NR^* \end{pmatrix} \tag{119}$$

or

$$\begin{pmatrix} M & \overline{N}^* \\ N & \overline{M}^* \end{pmatrix} = \begin{pmatrix} M & U_B \\ N & V_B \end{pmatrix} \begin{pmatrix} I & -R^* \\ 0 & I \end{pmatrix}. \tag{120}$$

Equivalently,

$$\begin{pmatrix} M & U_B \\ N & V_B \end{pmatrix} = \begin{pmatrix} M & \overline{N}^* \\ N & \overline{M}^* \end{pmatrix} \begin{pmatrix} I & R^* \\ 0 & I \end{pmatrix}. \tag{121}$$

Note that $\begin{pmatrix} M & \overline{N}^* \\ N & \overline{M}^* \end{pmatrix}$ is J-unitary whereas $\begin{pmatrix} I & R^* \\ 0 & I \end{pmatrix}$. is a unit in H_-^∞, therefore this provides an analogue of inner-outer factorizations.

Equation (121) can be rewritten, using the equality

$$\begin{pmatrix} M^* & N^* \\ \overline{N} & \overline{M} \end{pmatrix} \begin{pmatrix} I & 0 \\ 0 & -I \end{pmatrix} \begin{pmatrix} M & \overline{N}^* \\ N & \overline{M}^* \end{pmatrix} = \begin{pmatrix} I & 0 \\ 0 & -I \end{pmatrix}$$

or an equivalent one derived from it, namely

$$\begin{pmatrix} M & \overline{N}^* \\ N & \overline{M}^* \end{pmatrix}\begin{pmatrix} I & 0 \\ 0 & -I \end{pmatrix}\begin{pmatrix} M^* & N^* \\ \overline{N} & \overline{M} \end{pmatrix} = \begin{pmatrix} I & 0 \\ 0 & -I \end{pmatrix}$$

as

$$\begin{pmatrix} I & -R^* \\ 0 & I \end{pmatrix} = \begin{pmatrix} \overline{V}_B & -\overline{U}_B \\ -\overline{N} & \overline{M} \end{pmatrix}\begin{pmatrix} M & \overline{N}^* \\ N & \overline{M}^* \end{pmatrix}. \tag{122}$$

This leads to

$$\begin{aligned} &\begin{pmatrix} I & -R^* \\ 0 & I \end{pmatrix}\begin{pmatrix} I & 0 \\ 0 & -I \end{pmatrix}\begin{pmatrix} I & 0 \\ -R & I \end{pmatrix} \\ &= \begin{pmatrix} \overline{V}_B & -\overline{U}_B \\ -\overline{N} & \overline{M} \end{pmatrix}\begin{pmatrix} M & \overline{N}^* \\ N & \overline{M}^* \end{pmatrix}\begin{pmatrix} I & 0 \\ 0 & -I \end{pmatrix}\begin{pmatrix} M^* & N^* \\ \overline{N} & \overline{M} \end{pmatrix}\begin{pmatrix} \overline{V}_B^* & -\overline{N}^* \\ -\overline{U}_B^* & \overline{M}^* \end{pmatrix} \\ &= \begin{pmatrix} \overline{V}_B & -\overline{U}_B \\ -\overline{N} & \overline{M} \end{pmatrix}\begin{pmatrix} I & 0 \\ 0 & -I \end{pmatrix}\begin{pmatrix} \overline{V}_B^* & -\overline{N}^* \\ -\overline{U}_B^* & \overline{M}^* \end{pmatrix} \end{aligned} \tag{123}$$

So $\begin{pmatrix} \overline{V}_B & -\overline{U}_B \\ -\overline{N} & \overline{M} \end{pmatrix} \in H_+^\infty$ can be obtained from $\begin{pmatrix} I & -R^* \\ 0 & I \end{pmatrix}$ by a process of spectral factorization.

Thus we can state.

Theorem 5.4 *Let $R^* \in H_-^\infty$ be a rational, strictly proper, antistable function with $\|H_{R^*}\| < 1$. Let $\Gamma = \begin{pmatrix} I & -R^* \\ 0 & I \end{pmatrix} \in H_-^\infty$ and let*

$$\Gamma_+ = \begin{pmatrix} \overline{V}_B & -\overline{U}_B \\ -\overline{N} & \overline{M} \end{pmatrix} \in H_+^\infty \tag{124}$$

be the solution to the J_B-spectral factorization problem

$$\begin{cases} \Gamma_+ J_B \Gamma_+^* &= \Gamma J_B \Gamma^* \\ \Gamma_+(\infty) &= \begin{pmatrix} I & 0 \\ 0 & I \end{pmatrix}. \end{cases} \tag{125}$$

Defining the H_+^∞ functions M, N, U, V by

$$\Gamma_+^{-1} = \begin{pmatrix} M & U \\ N & V \end{pmatrix} \tag{126}$$

then the set

$$\{-(N^* + M^*Q)(\overline{M} + \overline{N}Q)^{-1} | Q \in BH_+^\infty\} \tag{127}$$

is a parametrization of the set of all suboptimal Nehari extensions of R^*.

Proof: We have to show that if $\Gamma_+ = \begin{pmatrix} \overline{V}_B & -\overline{U}_B \\ -\overline{N} & \overline{M} \end{pmatrix}$ is the solution to the J_B-spectral factorization problem there exists abounded real function G whose normalized coprime factors are $\overline{M}, \overline{N}$, whose B-controller coprime factors are $\overline{U}, \overline{V}$ and whose B-characteristic is R^*. So assume now that $\Gamma_+ = \begin{pmatrix} \overline{V}_B & -\overline{U}_B \\ -\overline{N} & \overline{M} \end{pmatrix}$ is the unique solution to the J_B spectral factorization problem (125). Obviously, in view of (126), we have the doubly coprime factorization

$$\begin{pmatrix} \overline{V}_B & -\overline{U}_B \\ -\overline{N} & \overline{M} \end{pmatrix}\begin{pmatrix} M & U \\ N & V \end{pmatrix} = \begin{pmatrix} I & 0 \\ 0 & I \end{pmatrix}$$

Now, with $L = \Gamma_+^{-1}\Gamma$, we have

$$\begin{aligned} \begin{pmatrix} L_{11} & L_{12} \\ L_{21} & L_{22} \end{pmatrix} &= \begin{pmatrix} M & U \\ N & V \end{pmatrix}\begin{pmatrix} I & -R^* \\ 0 & I \end{pmatrix} \\ &= \begin{pmatrix} M & U - MR^* \\ N & V - NR^* \end{pmatrix}. \end{aligned} \tag{128}$$

This implies $L_{11} = M$ and $L_{21} = N$. Similarly, using the equalities $L = \Gamma_+^{-1}\Gamma$ and $L^*J_BL = LJ_BL^* = J_B$, we have $L^* = \Gamma^*\Gamma_+^{-*}$ so

$$J_B = LJ_BL^* = LJ_B\Gamma^*\Gamma_+^{-*}$$

This implies

$$\begin{aligned} L &= J_B\Gamma_+^*\Gamma^{-*}J_B \\ &= \begin{pmatrix} I & 0 \\ 0 & -I \end{pmatrix}\begin{pmatrix} \overline{V}^* & -\overline{N}^* \\ -\overline{U}^* & \overline{M}^* \end{pmatrix}\begin{pmatrix} I & 0 \\ R & I \end{pmatrix}\begin{pmatrix} I & 0 \\ 0 & -I \end{pmatrix} \\ &= \begin{pmatrix} \overline{V}^* & -\overline{N}^* \\ \overline{U}^* & -\overline{M}^* \end{pmatrix}\begin{pmatrix} I & 0 \\ R & -I \end{pmatrix} \end{aligned}$$

which implies $L_{12} = \overline{N}^*$ and $L_{22} = \overline{M}^*$. Thus $J_B = LJ_BL^* = L^*J_BL$ implies the identities

$$\begin{pmatrix} M & \overline{N}^* \\ N & \overline{M}^* \end{pmatrix}\begin{pmatrix} I & 0 \\ 0 & -I \end{pmatrix}\begin{pmatrix} M^* & N^* \\ \overline{N} & \overline{M} \end{pmatrix} = \begin{pmatrix} I & 0 \\ 0 & -I \end{pmatrix} \tag{129}$$

and

$$\begin{pmatrix} M^* & N^* \\ \overline{N} & \overline{M} \end{pmatrix}\begin{pmatrix} I & 0 \\ 0 & -I \end{pmatrix}\begin{pmatrix} M & \overline{N}^* \\ N & \overline{M}^* \end{pmatrix} = \begin{pmatrix} I & 0 \\ 0 & -I \end{pmatrix} \tag{130}$$

We define now $G = NM^{-1} = \overline{M}^{-1}\overline{N}$. Obviously G is bounded real and $UV^{-1} = \overline{V}^{-1}\overline{U}$ is a stabilizing controller. Since from (128) we have

$$\begin{pmatrix} U \\ V \end{pmatrix} = \begin{pmatrix} M \\ N \end{pmatrix}\overline{R}^* + \begin{pmatrix} \overline{N}^* \\ \overline{M}^* \end{pmatrix}$$

we get

$$M^*U - N^*V = (M^*M - N^*N)R^* + (M^*\overline{N}^* - N^*\overline{M}^*) = R^* \in H_-^\infty,$$

and R^* strictly proper. So U, V and $\overline{U}, \overline{V}$ are the right and left coprime factors respectively of the BR controller of G. In particular R^* is the B-characteristic of G.
We can apply now Theorem 5.3 to complete the proof. ∎

5.4 Connection to Krein and Melik-Adamyan Method

We outline now the Krein and Melik-Adamyan [1984a,b] approach, another close exposition is Dym [1989], to the problem of parametrization of all suboptimal Nehari complements. We will do so by showing that soving the previous J_B-spectral factorization problem is equivalent to solving a special, Hankel based, system of equations.

Theorem 5.5 *Let $R^* \in H_-^\infty$ be rational and strictly proper and satisfy $\|H_{R^*}\| < 1$. Let*

$$\Gamma = \begin{pmatrix} I & -R^* \\ 0 & I \end{pmatrix}. \tag{131}$$

Then $\Gamma_+ \in H_+^\infty$ is a solution to the J_B-spectral factorization problem

$$\begin{cases} \Gamma J_B \Gamma^* &= \Gamma_+ J_B \Gamma_+^* \\ \Gamma_+(\infty) &= \begin{pmatrix} I & 0 \\ 0 & I \end{pmatrix} \end{cases} \tag{132}$$

if and only if $\Gamma = \Gamma_+ L$ with

$$L = \begin{pmatrix} L_{11} & L_{12} \\ L_{21} & L_{22} \end{pmatrix} = \begin{pmatrix} I + \Phi_+ & \Phi_- \\ \overline{\Phi}_+ & I + \overline{\Phi}_- \end{pmatrix} = \begin{pmatrix} M & \overline{N}^* \\ N & \overline{M}^* \end{pmatrix} \tag{133}$$

and Φ_+, Φ_- and $\overline{\Phi}_+, \overline{\Phi}_-$ are the, unique, solutions to the systems of equations

$$\begin{cases} \Phi_-^* + H_{R^*}^* \Phi_+^* &= -R \\ \Phi_+^* + H_{R^*} \Phi_-^* &= 0 \end{cases} \tag{134}$$

and

$$\begin{cases} \overline{\Phi}_-^* + H_{R^*}^* \overline{\Phi}_+^* &= 0 \\ \overline{\Phi}_+^* + H_{R^*} \overline{\Phi}_-^* &= -R^* \end{cases} \tag{135}$$

Proof: Assume first $R^* \in H_-^\infty$ and $\|H_{R^*}\| < 1$, and Γ is defined by (131). We proceed with solving the J_B-spectral factorization problem (132). Since

$$(\Gamma_+^{-1}\Gamma)J_B(\Gamma^*\Gamma_+^{-*}) = J_B \tag{136}$$

it follows that, with L defined by

$$L = \Gamma_+^{-1}\Gamma, \tag{137}$$

that

$$LJ_BL^* = J_B. \tag{138}$$

By our normalization of Γ_+ we have $L(\infty) = \begin{pmatrix} I & 0 \\ 0 & I \end{pmatrix}$.

We set now

$$\Gamma_+ = \begin{pmatrix} K_{11} & K_{12} \\ K_{21} & K_{22} \end{pmatrix} \tag{139}$$

with $K_{ij} \in H_+^\infty$. We also set

$$\Gamma_+^{-1} = \begin{pmatrix} H_{11} & H_{12} \\ H_{21} & H_{22} \end{pmatrix} \tag{140}$$

and again $H_{ij} \in H_+^\infty$.

Since $L = \Gamma_+^{-1}\Gamma$ we have

$$\begin{pmatrix} H_{11} & H_{12} \\ H_{21} & H_{22} \end{pmatrix}\begin{pmatrix} I & -R^* \\ 0 & I \end{pmatrix} = \begin{pmatrix} H_{11} & H_{12} - H_{11}R_B^* \\ H_{21} & H_{22} - H_{21}R_B^* \end{pmatrix} = \begin{pmatrix} L_{11} & L_{12} \\ L_{21} & L_{22} \end{pmatrix}$$

This implies the inclusions

$$L_{11}, L_{21} \in H_+^\infty. \tag{141}$$

Next note that as, $LJ_BL^* = J_B$, it follows that

$$L^{-*} = J_BLJ_B. \tag{142}$$

Also $\Gamma = \Gamma_+L$ implies $\Gamma^* = L^*\Gamma_+^*$ and hence $\Gamma_+^* = L^{-*}\Gamma^* = (J_BLJ_B)\Gamma^*$ and hence

$$\begin{aligned}\begin{pmatrix} K_{11}^* & K_{21}^* \\ K_{12}^* & K_{22}^* \end{pmatrix} &= \begin{pmatrix} I & 0 \\ 0 & -I \end{pmatrix}\begin{pmatrix} L_{11} & L_{12} \\ L_{21} & L_{22} \end{pmatrix}\begin{pmatrix} I & 0 \\ 0 & -I \end{pmatrix}\begin{pmatrix} I & 0 \\ -R & I \end{pmatrix} \\ &= \begin{pmatrix} L_{11} & -L_{12} \\ -L_{21} & L_{22} \end{pmatrix}\begin{pmatrix} I & 0 \\ -R & I \end{pmatrix}\end{aligned}$$

Thus we get

$$L_{12}, L_{22} \in H_-^\infty \tag{143}$$

Similarly

$$\begin{pmatrix} H_{11} & H_{12} \\ H_{21} & H_{22} \end{pmatrix} \begin{pmatrix} I & -R^* \\ 0 & I \end{pmatrix} = \begin{pmatrix} L_{11} & L_{12} \\ L_{21} & L_{22} \end{pmatrix} \tag{144}$$

we have

$$\begin{pmatrix} I & 0 \\ -R & I \end{pmatrix} \begin{pmatrix} H_{11}^* & H_{21}^* \\ H_{12}^* & H_{22}^* \end{pmatrix} = \begin{pmatrix} L_{11}^* & L_{21}^* \\ L_{12}^* & L_{22}^* \end{pmatrix} \tag{145}$$

or

$$\begin{pmatrix} H_{11}^* & H_{21}^* \\ H_{12}^* & H_{22}^* \end{pmatrix} = \begin{pmatrix} I & 0 \\ R & I \end{pmatrix} \begin{pmatrix} L_{11}^* & L_{21}^* \\ L_{12}^* & L_{22}^* \end{pmatrix} \tag{146}$$

In particular we have the following equalities

$$\begin{cases} H_{12}^* &= RL_{11}^* + L_{12}^* \\ H_{22}^* &= RL_{21}^* + L_{22}^* \end{cases} \tag{147}$$

We apply the orthogonal projection P_+ to get

$$\begin{aligned} P_+H_{12}^* = P_+R(L_{11}^* - L_{11}(\infty)^*) + L_{12}^* + P_+RL_{11}(\infty)^* = 0 \\ P_+(H_{22}^* - H_{22}(\infty)^*) = P_+RL_{21}^* + P_+(L_{22}^* - L_{22}(\infty)^*) \end{aligned} \tag{148}$$

On the other hand

$$\begin{pmatrix} K_{11} & K_{12} \\ K_{21} & K_{22} \end{pmatrix} = \begin{pmatrix} I & -R^* \\ 0 & I \end{pmatrix} \begin{pmatrix} L_{11}^* & -L_{21}^* \\ -L_{12}^* & L_{22}^* \end{pmatrix} \tag{149}$$

we obtain

$$\begin{cases} K_{11} &= L_{11}^* + R^*L_{12}^* \\ K_{12} &= -L_{21}^* + R^*L_{22}^* \end{cases} \tag{150}$$

and applying the orthogonal projection P_- we get

$$\begin{cases} P_-(L_{11}^* - I) + H_{R^*}L_{12}^* &= 0 \\ -L_{21}^* - P_-R^*(L_{22}^* - I) - R^* &= 0 \end{cases} \tag{151}$$

We set now

$$\begin{aligned} \Phi_+ &= L_{11} - I, & \Phi_- &= L_{12} \\ \overline{\Phi}_+ &= L_{21}, & \overline{\Phi}_- &= L_{22} - I \end{aligned} \tag{152}$$

Obviously, $\Phi_+, \overline{\Phi}_+ \in H_+^\infty \cap H_+^2$ and $\Phi_-, \overline{\Phi}_- \in H_-^\infty \cap H_-^2$.

In this notation the equations (148) and (151) can be rewritten together as

$$\begin{cases} \Phi_-^* + H_{R^*}^* \Phi_+^* &= -R \\ \overline{\Phi}_-^* + H_{R^*}^* \overline{\Phi}_+^* &= 0 \\ & \\ \overline{\Phi}_+^* + H_{R^*} \overline{\Phi}_-^* &= -R^* \\ \Phi_+^* + H_{R^*} \Phi_-^* &= 0 \end{cases} \tag{153}$$

We split equations (153) into the two systems of equations (134) and (135).

Obviously, by our assumption that $\|H_{R^*}\| < 1$, these equations have a unique solution. We define now

$$\begin{cases} M = I + \Phi_+, & N = \overline{\Phi}_+ \\ & \\ \overline{M} = I + \overline{\Phi}_-^*, & \overline{N} = \Phi_-^* \end{cases} \tag{154}$$

Therefore

$$L = \begin{pmatrix} L_{11} & L_{12} \\ L_{21} & L_{22} \end{pmatrix} = \begin{pmatrix} I + \Phi_+ & \Phi_- \\ & \\ \overline{\Phi}_+ & I + \overline{\Phi}_- \end{pmatrix} = \begin{pmatrix} M & \overline{N}^* \\ N & \overline{M}^* \end{pmatrix} \tag{155}$$

Since $LJ_BL^* = J_B$, we have

$$\begin{pmatrix} M & \overline{N}^* \\ N & \overline{M}^* \end{pmatrix} \begin{pmatrix} I & 0 \\ 0 & -I \end{pmatrix} \begin{pmatrix} M^* & N^* \\ \overline{N} & \overline{M} \end{pmatrix} = \begin{pmatrix} I & 0 \\ 0 & -I \end{pmatrix} \tag{156}$$

and hence also

$$\begin{pmatrix} M^* & N^* \\ \overline{N} & \overline{M} \end{pmatrix} \begin{pmatrix} I & 0 \\ 0 & -I \end{pmatrix} \begin{pmatrix} M & \overline{N}^* \\ N & \overline{M}^* \end{pmatrix} = \begin{pmatrix} I & 0 \\ 0 & -I \end{pmatrix} \tag{157}$$

This implies the identities

$$\begin{cases} M^*M - N^*N &= I \\ \overline{M}N - \overline{N}M &= 0 \\ M^*\overline{N}^* - N^*\overline{M}^* &= 0 \\ \overline{M}\,\overline{M}^* - \overline{N}\,\overline{N}^* &= I \end{cases} \tag{158}$$

and

$$\begin{cases} \overline{M}^*\overline{M} - NN^* &= I \\ \overline{M}^*\overline{N} - NM^* &= 0 \\ \overline{N}^*\overline{M} - MN^* &= 0 \\ MM^* - \overline{N}^*\overline{N} &= I \end{cases} \tag{159}$$

From this we deduce that for G defined by

$$G = NM^{-1} = \overline{M}^{-1}\overline{N} \tag{160}$$

is a bounded real function and these are J_B normalized coprime factorization of it.

$$T = N^*\overline{M}^{-1} = M^{-1}\overline{N}^* \tag{161}$$

is an L^∞ contraction.

We will show now that $\chi_B(G) = R^*$. To this end we define matrix functions U, V and $\overline{U}, \overline{V}$ by

$$\begin{pmatrix} U \\ V \end{pmatrix} = \begin{pmatrix} M \\ N \end{pmatrix}\overline{R}^* + \begin{pmatrix} \overline{N}^* \\ \overline{M}^* \end{pmatrix} \tag{162}$$

and

$$\begin{pmatrix} \overline{V} & \overline{U} \end{pmatrix} = R^*\begin{pmatrix} \overline{N} & \overline{M} \end{pmatrix} + \begin{pmatrix} M^* & N^* \end{pmatrix} \tag{163}$$

respectively. Equations (134) and (135) imply that all these functions are in H_+^∞. Next we compute

$$\begin{pmatrix} \overline{V} & \overline{U} \end{pmatrix}\begin{pmatrix} M \\ -N \end{pmatrix} = [R^*\begin{pmatrix} \overline{N} & \overline{M} \end{pmatrix} + \begin{pmatrix} M^* & N^* \end{pmatrix}]\begin{pmatrix} M \\ -N \end{pmatrix} = I,$$

and

$$\begin{pmatrix} -\overline{N} & \overline{M} \end{pmatrix}\begin{pmatrix} U \\ V \end{pmatrix} = \begin{pmatrix} -\overline{N} & \overline{M} \end{pmatrix}\left[\begin{pmatrix} M \\ N \end{pmatrix}\overline{R}^* + \begin{pmatrix} \overline{N}^* \\ \overline{M}^* \end{pmatrix}\right] = I.$$

Moreover

$$\begin{pmatrix} \overline{V} & -\overline{U} \end{pmatrix}\begin{pmatrix} U \\ V \end{pmatrix} = [R^*\begin{pmatrix} \overline{N} & -\overline{M} \end{pmatrix} + \begin{pmatrix} M^* & -N^* \end{pmatrix}]\left[\begin{pmatrix} M \\ N \end{pmatrix}\overline{R}^* + \begin{pmatrix} \overline{N}^* \\ \overline{M}^* \end{pmatrix}\right] = 0.$$

Since

$$M^*U - N^*V = \begin{pmatrix} M^* & -N^* \end{pmatrix}\left[\begin{pmatrix} M \\ N \end{pmatrix}\overline{R}^* + \begin{pmatrix} \overline{N}^* \\ \overline{M}^* \end{pmatrix}\right] = \overline{R}^* \in H_-^\infty$$

and

$$\overline{U}\overline{M}^* - \overline{V}\overline{N}^* = [R^*\begin{pmatrix} \overline{N} & \overline{M} \end{pmatrix} + \begin{pmatrix} M^* & N^* \end{pmatrix}]\begin{pmatrix} -\overline{N}^* \\ \overline{M}^* \end{pmatrix} = R^*,$$

it follows that R^* is the B-charcteristic of G.

Conversely, assume that $\Phi_+, \Phi_-, \overline{\Phi}_+, \overline{\Phi}_-$ are the solutions to the system of equations (134) and (134) . We define the H_+^∞ functions $N, M, \overline{N}, \overline{M}$ via (154). Obviously $N, \overline{N}$ are strictly proper, whereas $M, \overline{M}$ are proper and satisfy $M(\infty) = I$ and $\overline{M}(\infty) = I$. We show now that

$$L = \begin{pmatrix} M & \overline{N}^* \\ N & \overline{M}^* \end{pmatrix}$$

satisfies $L^*J_BL = J_B$.

To this end we rewrite the equation $\Phi_-^* + H_{R^*}^*\Phi_+^* = -R$ as $\overline{N} + P_+R(M^* - I) + R = P_+(\overline{N} + RM^*) = 0$, i.e. $\overline{N} + RM^* \in H_-^\infty$. Hence, going through similar computations, we get that the functions defined by

$$\begin{cases} U &= MR^* + \overline{N}^* \\ V &= NR^* + \overline{M}^* \\ \overline{U} &= R^*\overline{M} + N^* \\ \overline{V} &= R^*\overline{N} + M^* \end{cases} \tag{164}$$

are all in H_+^∞.

Note that these equations can be written also as

$$\begin{pmatrix} U \\ V \end{pmatrix} = \begin{pmatrix} M \\ N \end{pmatrix} R^* + \begin{pmatrix} \overline{N}^* \\ \overline{M}^* \end{pmatrix} \tag{165}$$

and

$$\begin{pmatrix} \overline{V} & \overline{U} \end{pmatrix} = R^* \begin{pmatrix} \overline{N} & \overline{M} \end{pmatrix} + \begin{pmatrix} M^* & N^* \end{pmatrix} \tag{166}$$

These equations imply also the following

$$\begin{pmatrix} M & U_B \\ N & V_B \end{pmatrix} = \begin{pmatrix} M & \overline{N}^* \\ N & \overline{M}^* \end{pmatrix} \begin{pmatrix} I & R^* \\ 0 & I \end{pmatrix}. \tag{167}$$

and

$$\begin{pmatrix} \overline{V}_B & -\overline{U} \\ -\overline{N} & \overline{M} \end{pmatrix} = \begin{pmatrix} I & -R^* \\ 0 & I \end{pmatrix} \begin{pmatrix} M^* & -N^* \\ -\overline{N} & \overline{M} \end{pmatrix}. \tag{168}$$

Now we can write

$$MM^* - \overline{N}^*\overline{N} = (M + \overline{N}^*R)M^* - \overline{N}^*(\overline{N} + RM^*) \in H_-^\infty$$

as all factors are in H_-^∞. Since $MM^* - \overline{N}^*\overline{N}$ is self adjoint we have also $MM^* - \overline{N}^*\overline{N} \in H_+^\infty$, and hence it must be a constant. Using the strict properness of $\overline{N}$ and the fact that $M(\infty) = I$ we get

$$MM^* - \overline{N}^*\overline{N} = I.$$

Similarly

$$\overline{M}^*\overline{M} - NN^* = \overline{M}^*(\overline{M} + RN^*) - (N + \overline{M}^*R)N^* \in H_-^\infty.$$

Again this implies

$$\overline{M}^*\overline{M} - NN^* = I.$$

Next we write

$$\overline{M}^*\overline{N} - NM^* = (V - NR^*)\overline{N} - N(\overline{V} - R^*\overline{N}) = V\overline{N} - N\overline{V} \in H_+^\infty.$$

On the other hand

$$\overline{M}^*\overline{N} - NM^* = \overline{M}^*(\overline{N} + RM^*) - (N + \overline{M}^*R)M^* = \overline{M}^*U^*) - \overline{U}^*M^* \in H_-^\infty.$$

These two inclusions imply that $\overline{M}^*\overline{N} - NM^*$ is a constant, and evaluating at ∞ we get

$$\overline{M}^*\overline{N} - NM^* = 0,$$

and also

$$\overline{N}^*\overline{M} - MN^* = 0.$$

Thus we have derived the identities (159). This is equivalent to $L^*J_BL = J_B$ and therefore also to $LJ_BL^* = J_B$.

We set now

$$\Gamma_+ = \Gamma L^{-1}.$$

Clearly we have Γ_+ is proper and $\Gamma_+(\infty) = I$. Moreover we compute

$$\begin{aligned} \Gamma J_B \Gamma^* &= \Gamma_+ L J_B L^* \Gamma_+^* \\ &= \Gamma_+ J_B \Gamma_+^* \end{aligned}$$

So Γ_+ as defined is the solution to the J_B spectral factorization problem

$$\begin{aligned} \Gamma_+ J_B \Gamma_+^* &= \Gamma J_B \Gamma^* \\ \Gamma_+(\infty) &= \begin{pmatrix} I & 0 \\ 0 & I \end{pmatrix}. \end{aligned} \tag{169}$$

Note that, using the identities (158) and (159), we have the doubly coprime factorization

$$\begin{pmatrix} \overline{V}_B & -\overline{U} \\ -\overline{N} & \overline{M} \end{pmatrix} \begin{pmatrix} M & U_B \\ N & V_B \end{pmatrix} = \begin{pmatrix} I & 0 \\ 0 & I \end{pmatrix}.$$

∎

In view of the previous theorem, we can apply Theorem 5.4 to state.

Theorem 5.6 *Let $R^* \in H_-^\infty$ be a rational, strictly proper, antistable function with $\|H_{R^*}\| < 1$. Let $\Phi_+, \Phi_-, \overline{\Phi}_+, \overline{\Phi}_-$ be the solutions of the systems of equations (134) and (135). Let $M, N, \overline{M}, \overline{N}$ be defined by equation (154). Then the set*

$$\{-(N^* + M^*Q)(\overline{M} + \overline{N}Q)^{-1} | Q \in BH_+^\infty\} \tag{170}$$

is a parametrization of the set of all suboptimal Nehari extensions of R^*.

References

[1968a] V. M. Adamjan, D. Z. Arov and M. G. Krein, "Infinite Hankel matrices and generalized problems of Caratheodory-Fejer and F. Riesz", *Funct. Anal. Appl.* 2, 1-18.

[1968b] V. M. Adamjan, D. Z. Arov and M. G. Krein, "Infinite Hankel matrices and generalized problems of Caratheodory-Fejer and I. Schur", *Funct. Anal. Appl.* 2, 269-281.

[1971] V. M. Adamjan, D. Z. Arov and M. G. Krein, "Analytic properties of Schmidt pairs for a Hankel operator and the generalized Schur-Takagi problem", *Math. USSR Sbornik* 15 (1971), 31-73.

[1978] V. M. Adamjan, D. Z. Arov and M. G. Krein, "Infinite Hankel block matrices and related extension problems", Amer Math. Soc. Transl., series 2, Vol. 111, 133-156.

[1983] J.A. Ball and J. W. Helton, "A Beurling-Lax theorem for the Lie group $U(m,n)$ which contains most classical interpolation theory", *J. Op. Theory*, 9, 107-142.

[1986] J. A. Ball and A.C.M. Ran, "Optimal Hankel norm model reduction and Wiener-Hopf factorizations I: the canonical case", *SIAM J. Contr. and Opt.*,.

[1949] A. Beurling, "On two problems concerning linear transformations in Hilbert space", *Acta Math.*, 81, pp. 239-255.

[1974] J. Bognar, *Indefinite Inner Product Spaces*, Springer Verlag, Berlin and New York.

[1989] H. Dym, "J Contractive matrix Functions, Reproducing Kernel Hilbert Spaces and Interpolation", *Regional conference series in mathematics*, 71, Amer. Math. Soc., Providence, R.I.

[1987] B. Francis, *A course in H_∞ Control Theory*, Springer-Verlag.

[1975] P. A. Fuhrmann, "On Hankel operator ranges, meromorphic pseudo-continuation and factorization of operator valued analytic functions", *J. Lon. Math. Soc.* (2) 13, 323-327.

[1981] P. A. Fuhrmann, *Linear Systems and Operators in Hilbert Space*, McGraw-Hill, New York.

[1991] P. A. Fuhrmann, "A polynomial approach to Hankel norm and balanced approximations", *Lin. Alg. Appl.*, 146, 133-220.

[1993] P. A. Fuhrmann, "An algebraic approach to Hankel norm approximation problems", *L. Markus Festschrift, Lecture Notes in Pure and Applied Mathematics*, M. Dekker.

[1994] P. A. Fuhrmann, "A duality theory for robust stabilization and model reduction", to appear in *Lin. Alg. Appl.*.

[1993a] P. A. Fuhrmann and R. Ober, " A functional approach to LQG balancing", *Int. J. Contr.* 57, 627-741.

[1993b] P. A. Fuhrmann and R. Ober, "State space formulas for coprime factorizations", *T. Ando Festschrift*, Birkhauser Verlag.

[1990] T. T. Georgiou and M. C. Smith, "Optimal robustness in the gap metric", *IEEE Trans. on Auto. Contr.*, 35, 673-686.

[1984] K. Glover, "All optimal Hankel-norm approximations and their L^∞-error bounds", *Int. J. Contr.* 39, 1115-1193.

[1989] K. Glover and D. McFarlane, "Robust stabilization of normalized coprime factor plant descriptions with H_∞-bounded uncertainty", *IEEE Trans. on Auto. Contr.* 34, 821-830.

[1988] I. Gohberg and S. Rubinstein, "Proper contractions and their unitary minimal completions", *Operator Theory: Advances and Applications*, Birkhauser Verlag, Basel, Vol. 33, 223-247.

[1983] E. Jonckheere and L. Silverman, "A new set of invariants for linear systems - application to reduced order compensator design", *IEEE Trans. on Auto. Contr.* 28, 953-964.

[1984a] M.G. Krein and F.E. Melik-Adamyan, "Integral Hankel operators and related continuation problems", *Izvest. Akad. Nauk Armyanskoi SSR*, 19, 311-332.

[1984b] M.G. Krein and F.E. Melik-Adamyan, "Integral Hankel operators and related continuation problems", *Izvest. Akad. Nauk Armyanskoi SSR*, 19, 339-360.

[1959] P.D. Lax, "Translation invariant subspaces", *Acta Math.*, 101, 163-178.

[1954] M.S. Livsic, "On the spectral decomposition of linear non-selfadjoint operators", *Math. Sb.*, 34(76), 145-199. English translation, *Amer. Math. Soc. Transl.* (2) 5 (1957), 67-114.

[1990] D. McFarlane and K. Glover, *Robust Controller Design using Normalized Coprime Factor Plant Description*, Springer-Verlag.

[1987] D. Meyer and G. Franklin, "A connection between normalized oprime factorizations and linear quadratic regulator theory", *IEEE Trans. on Auto. Contr.* 32, 227-228.

[1981] B. C. Moore, "Principal component analysis in linear systems: controllability, observability and model reduction", *IEEE Trans. on Auto. Contr.* 26, 17-32.

[1957] Z. Nehari, "On bounded bilinear forms", *Ann. of Math.*, 65, 153-162.

[1985] N. K. Nikolskii, *Treatise on the Shift Operator*, Springer Verlag, Berlin.

[1993] R. Ober and P.A. Fuhrmann, "Diffeomorphisms between sets of linear systems", *The IMA Volumes in Mathematics and its Applications, P. Van Dooren Editor*, Springer-Verlag.

[1970] L.B. Page, "Applications of the Sz.-Nagy and Foias lifting theorem", *Indiana Univ. Math. J.*, 20, 135-145.

[1955] V.P. Potapov, "The multiplicative structure of J-contractive matrix-functions", *Trudy Moskov. Mat. Obs.*, 4, 125-236, English Traslation: *Amer. Math. Soc Transl.*, (2), 15, (1960), 131-244.

[1960] G.C. Rota, "On models for linear operators", *Comm. Pure and Appl. Math.*, 13, 469-472.

[1967] D. Sarason, "Generalized interpolation in H^∞", *Trans. Amer. Math. Soc.* 127, 179-203.

[1970] B. Sz.-Nagy and C. Foias, *Harmonic Analysis of Operators on Hilbert Space*, North Holland, Amsterdam.

[1958] B. Sz.-Nagy and A. Koranyi, "Operatortheoretische Behandlung und Verallgemeinierung eines Problemkreis in der Komplexen Funktionentheorie", *Acta Math.*, 171-202.

[1988] M. Vidyasagar. "Normalized coprime factorizations for non strictly proper systems". Automatica, 85-94.

[1988] N. Young, *An Introduction to Hilbert Space*, Cambridge University Press, Cambridge.

[1981] G. Zames, Feedback and optimal sensitivity: model reference transformations, multiplicative seminorms and approximate inverses", *IEEE Trans. on Auto. Contr.*, 28, 1030-1035.

Department of Mathematics
Ben-Gurion University of the Negev
84105 Beer Sheva, Israel

MSC: 47A15 47A68 47B35 47B50 47N70 93B36

Operator Theory:
Advances and Applications, Vol. 73

ON ISOMETRIC ISOMORPHISM BETWEEN THE SECOND DUAL TO THE "SMALL" LIPSCHITZ SPACE AND THE "BIG" LIPSCHITZ SPACE

Leonid Hanin [1)]

Dedicated to Professor Moshe Livsic on his seventy fifth birthday

Described are all compact metric spaces (K,ρ) for which the natural mapping from the second dual to the "small" Lipschitz space $\mathrm{lip}(K,\rho)$ to the "big" Lipschitz space $\mathrm{Lip}(K,\rho)$ is an isometric isomorphism or equivalently for which the completion of the space of finite Borel measures on K in the Kantorovich-Rubinstein norm is isometrically isomorphic to $(\mathrm{lip}(K,\rho))^*$. A theorem of the Stone-Weierstrass type for algebras $\mathrm{lip}(K,\rho)$ is obtained.

PRELIMINARY OBSERVATIONS AND NOTATION

For a metric space (K,ρ), we denote by $\mathrm{Lip}(K,\rho)$ the Banach space of all real functions f on K with the finite norm

$$\|f\|_{K,\rho} := \max\{\|f\|_K , |f|_{K,\rho}\} ,$$

where

$$\|f\|_K := \sup\{|f(x)| : x \in K\}$$

and

$$|f|_{K,\rho} := \sup\{(f(x) - f(y))/\rho(x,y) : x,y \in K , x \neq y\} .$$

Functions $f \in \mathrm{Lip}(K,\rho)$ satisfying the condition

$$\lim_{\rho(x,y)\to 0} (f(x) - f(y))/\rho(x,y) = 0$$

constitute a closed linear subspace $\mathrm{lip}(K,\rho)$ in $\mathrm{Lip}(K,\rho)$.

With every functional $\psi \in (\mathrm{lip}(K,\rho))^{**}$, we associate a function $j(\psi) \in \mathrm{Lip}(K,\rho)$ defined by means of the formula

1) This work was partly supported by the grant No. 3763292 from Israel Ministry of Science and by the Maagara, a special program for absorption of new immigrants in the Department of Mathematics of the Technion.

$$j(\psi)(x) := \psi(\delta_x) \quad , \ x \in K \ ,$$

where δ_x is the Dirac measure at a point x (viewed as a functional on $lip(K,\rho)$). Obviously, $j : (lip(K,\rho))^{**} \to Lip(K,\rho)$ is a linear mapping with the norm ≤ 1. This mapping is canonical in the sense that if π is the natural imbedding of $lip(K,\rho)$ into $(lip(K,\rho))^{**}$, then $j\circ\pi$ is the identical inclusion of $lip(K,\rho)$ into $Lip(K,\rho)$.

In general, mapping j is neither surjective nor injective, as it is shown by the following two examples.

1. Let (K,ρ) be an interval in $\mathbb{R}$ with the Euclidean metric. Then the space $lip(K,\rho)$ contains only constants, hence j is not onto.

2. Let $K = \mathbb{N}$ and let $\rho(x,y) = 2$ for all $x,y \in \mathbb{N}$, $x \ne y$. In this case $lip(K,\rho) = Lip(K,\rho) = l^{\infty}$, and we observe that the natural projection $j : (l^{\infty})^{**} \to l^{\infty}$ is not injective, because the space l^{∞} is not reflexive.

In the present paper we give a complete description of all compact metric spaces (K,ρ) such that $j : (lip(K,\rho))^{**} \to Lip(K,\rho)$ is an isometric isomorphism (see Theorems 1 and 2 below). These results improve essentially the corresponding assertions in [4,5]. By slight modification of the argument all results of the paper can be carried over to noncompact metric spaces with compact closed balls. In this case the space $lip(K,\rho)$ should be replaced by the subspace $lip_0(K,\rho)$ of functions vanishing at "infinity".

The study of the problem of existence of a canonical isometric isomorphism between the spaces $((lip(K,\rho))^{**}$ and $Lip(K,\rho)$ was initiated by K. deLeeuw [10] who proved that this is the case for $K = [0,1]$, $\rho(x,y) = |x - y|^{\alpha}$, $0 < \alpha < 1$. For compact metric spaces (K,d) (or for those with compact close balls), isometric isomorphism $(lip(K,d^{\alpha}))^{**} \cong Lip(K,d^{\alpha})$, $0 < \alpha < 1$, was established in [2,3], see also [1].

However the methods used in the above mentioned works seem to be insufficient for obtaining the required general result. In this present work we develop a quite different approach. The main idea is to find a normed space M for which the following natural isometric isomorphisms hold,

$$M^{*} \cong Lip(K,\rho) \quad , \ M^{c} \cong (lip(K,\rho))^{*} \ ,$$

where M^{c} stands for the completion of M. As shown by L.V. Kantorovich and G.Sh. Rubinstein [7-9], such space M can be identified with the space of finite Borel measures on K supplied with a special norm which definition and properties are discussed in the next section.

THE KANTOROVICH - RUBINSTEIN NORM

Let (K,ρ) be a metric space and $M(K)$ be the set of all finite real Borel measures on K. For a measure $\mu \in M(K)$, we denote by μ_+, μ_-, and $|\mu|$ its corresponding variations and set $\mathrm{Var}\mu = |\mu|(K)$. Let $M_0(K)$ be the subspace of measures μ in $M(K)$ such that $\mu(K) = 0$. We correspond to every $\mu \in M_0(K)$ the family Γ_μ of all measures $\phi \in M_+(K\times K)$ such that

$$\phi(K\times e) - \phi(e\times K) = \mu(e) \qquad \text{for all Borel sets } e \subset K . \tag{1}$$

The value $\phi(e_1\times e_2)$ may be thought of as the mass carried from a set e_1 to a set e_2. Then ϕ can be interpreted as a mass transfer on K which converts the available mass distribution μ_- into the required one μ_+ (note that $\mu_-(K) = \mu_+(K)$) and produces the "mechanical work" $\int_{K\times K} \rho d\phi$, while (1) becomes the natural balance condition.

The Monge-Kantorovich mass transfer problem [6-9] consists of finding the value

$$\|\mu\|_\rho^0 := \inf_{\phi \in \Gamma_\mu} \int_{K\times K} \rho d\phi , \qquad \mu \in M_0(K) ,$$

and the corresponding set of optimal mass transfers.

It is easy to see that $\|\delta_x - \delta_y\|_\rho^0 = \rho(x,y)$ for all $x,y \in K$. Also, if ρ is a bounded metric on K, then the functional $\|\ \|_\rho^0$ is a *norm* on $M_0(K)$ which is called the Kantorovich-Rubinstein (KR) norm.

We will now define an analog of the KR norm on the *whole* space $M(K)$

$$\|\mu\|_\rho := \inf\{\|\nu\|_\rho^0 + \mathrm{Var}(\mu - \nu) : \nu \in M_0(K)\} , \qquad \mu \in M(K) ,$$

see [5]. This norm is equivalent on $M_0(K)$ to the original one,

$$\|\mu\|_\rho \le \|\mu\|_\rho^0 \le \frac{1}{2} \max\{\mathrm{diam}(K,\rho),2\}\|\mu\|_\rho , \quad \mu \in M_0(K) .$$

In particular, if $\mathrm{diam}(K,\rho) \le 2$, then $\|\mu\|_\rho = \|\mu\|_\rho^0$ for each $\mu \in M_0(K)$. Note also that

$$\|\delta_x\|_\rho = 1 , \ \|\delta_x - \delta_y\|_\rho = \min\{\rho(x,y),2\} , \quad x,y \in K .$$

REMARK. For metric $\tilde{\rho} := \min\{\rho,2\}$, we have $\mathrm{Lip}(K,\tilde{\rho}) \cong \mathrm{Lip}(K,\rho)$ and $\mathrm{lip}(K,\tilde{\rho}) \cong \mathrm{lip}(K,\rho)$. Hence, as far as the problem of isometric isomorphism between the spaces $(\mathrm{lip}(K,\rho))^{**}$ and $\mathrm{Lip}(K,\rho)$ is concerned, we may assume without loss of generality that $\mathrm{diam}(K,\rho) \le 2$.

The following two important properties of the norm $\|\ \|_\rho$ can be established in essentially the same way as it was done in the case of compact metric spaces for its classical counterpart $\| \|_\rho^0$ [7-9].

LEMMA 0. *If* (K,ρ) *is a separable bounded metric space, then measures with finite support are dense in* $(M(K), \| \|_\rho)$.

THEOREM 0. *Suppose that* (K,ρ) *is a separable bounded metric space. Then* $(M(K), \| \|_\rho)^* \cong \mathrm{Lip}(K,\rho)$ *with respect to the duality*

$$<f,\mu> = \int_K f d\mu \quad , \quad \mu \in M(K) \ , \ f \in \mathrm{Lip}(K,\rho) \ .$$

REMARK. In the work [9], a norm on $M(K)$ was defined in such a way that it is always (even in the case when $\mathrm{diam}(K,\rho) > 2$) an extension of $\| \|_\rho^0$. Accordingly, the isomorphism between the conjugate space to $M(K)$ with this norm and $\mathrm{Lip}(K,\rho)$ established in [9] is not isometric.

Comparing for two distinct points $x , y \in K$ the equalities $\|\delta_x - \delta_y\|_\rho = \min\{\rho(x,y),2\}$ and $\mathrm{Var}(\delta_x - \delta_y) = 2$, we see readily that, for infinite set K, the normed space $(M(K), \| \|_\rho)$ is incomplete. In what follows we identify its completion and thus establish the required isomorphism between the spaces $(\mathrm{lip}(K,\rho))^{**}$ and $\mathrm{Lip}(K,\rho)$.

COMPLETION OF THE SPACE OF MEASURES IN THE KR NORM

The formula

$$i(\mu)(f) := \int_K f d\mu \quad , \quad f \in \mathrm{lip}(K,\rho) \ , \quad \mu \in M(K) \ ,$$

defines a linear mapping $i : (M(K), \| \|_\rho) \to (\mathrm{lip}(K,\rho))^*$. According to Theorem 0 the norm of i does not exceed 1. We denote hereafter by $\| \|$ the usual norm on $(\mathrm{lip}(K,\rho))^*$.

LEMMA 1. *Suppose that* (K,ρ) *is a compact metric space. Then* $i(M(K))$ *is dense in* $(\mathrm{lip}(K,\rho))^*$ *in the norm topology.*

PROOF. Let Δ_K be the "diagonal" in $K \times K$, $\Delta_K = \{(x,x) : x \in K \}$. Clearly, the disjoint union $S = K \cup ((K \times K) \setminus \Delta_K)$ is a locally compact metric space. To every function $f \in \mathrm{lip}(K,\rho)$, we correspond a function $\hat{f}$ on S in the following way,

$$\hat{f}(x) := f(x) \ , \ x \in K \ ;$$
$$\hat{f}(x,y) := (f(x) - f(y))/\rho(x,y) \ , \ (x,y) \in (K \times K) \setminus \Delta_K \ .$$

We denote by $C_0(S)$ the space of all continuous functions on S vanishing at "infinity" and supply it with the supremum norm. The correspondence $f \to \hat{f}$ gives rise to a linear isometry $\tau : \mathrm{lip}(K,\rho) \to C_0(S)$ whose image will be denoted by L.

Now take $\lambda \in (\mathrm{lip}(K,\rho))^*$. Observe that $\lambda \circ \tau^{-1}$ is a bounded linear functional on L. Extending it by the Hahn-Banach theorem to a bounded linear functional on $C_0(S)$ and applying the Riesz representation theorem, we obtain measures $\nu \in M(K)$ and $m \in M((K \times K)\setminus\Delta_K)$ such that

$$\lambda(f) = \int_K f d\nu + \int_{(K\times K)\setminus\Delta_K} \frac{f(x) - f(y)}{\rho(x,y)} dm(x,y) , \quad f \in \mathrm{lip}(K,\rho) .$$

For $n \in \mathbb{N}$, denote

$$E_n := \{(x,y) \in K \times K : \rho(x,y) \geq 1/n\} ,$$

$$F_n := \{(x,y) \in K \times K : 0 < \rho(x,y) < 1/n\} ,$$

and set

$$\lambda_n(f) := \int_K f d\nu + \int_{E_n} \frac{f(x) - f(y)}{\rho(x,y)} dm(x,y) , \quad f \in \mathrm{lip}(K,\rho) .$$

We have

$$|\lambda_n(f)| \leq \|f\|_K (\mathrm{Var}\nu + 2n\, \mathrm{Var}m) , \quad f \in \mathrm{lip}(K,\rho) .$$

Applying again the Hahn-Banach theorem, we extend the functional λ_n on $\mathrm{lip}(K,\rho)$ to $C(K)$ and thus obtain a measure $\mu_n \in M(K)$ such that

$$\lambda_n(f) = \int_K f d\mu_n , \quad f \in \mathrm{lip}(K,\rho) .$$

Hence for every function $f \in \mathrm{lip}(K,\rho)$ with $\|f\|_{K,\rho} \leq 1$

$$\lambda(f) - (i(\mu_n))(f) = \int_{F_n} \frac{f(x) - f(y)}{\rho(x,y)} dm(x,y) \leq m(F_n) .$$

Therefore,

$$\|\lambda - i(\mu_n)\| \leq m(F_n) \to 0 \text{ as } n \to \infty ,$$

which completes the proof.

REMARK. For Hölder classes $\mathrm{lip}\,\alpha$ $(0 < \alpha < 1)$ on an interval, Lemma 1 was proved in [10]. In the more general case of metric spaces (K,d^α), $0 < \alpha < 1$, where d is a compact metric on K, Lemma 1 can be found in [1].

The mapping $i : (M(K), \|\ \|_\rho) \to (\mathrm{lip}(K,\rho))^*$ can be extended to the mapping $(M(K), \|\ \|_\rho)^c \to (\mathrm{lip}(K,\rho))^*$ having the same norm ≤ 1 (and notation). We denote by B and b closed unit balls in $\mathrm{Lip}(K,\rho)$ and $\mathrm{lip}(K,\rho)$, respectively.

THEOREM 1. *Suppose that* (K,ρ) *is a compact metric space. The mapping* $i : (M(K), \| \ \|_\rho)^c \to (\mathrm{lip}(K,\rho))^*$ *is an isometric isomorphism if and only if the following condition is satisfied:*

(A) *for every two distinct points* $x, y \in K$ *and for each* $\varepsilon > 0$ *there exists a function* $g \in \mathrm{lip}(K,\rho)$ *such that* $g(x) = 1$, $g(y) = 0$, *and* $|g|_{K,\rho} \le (1+\varepsilon)/\min\{\rho(x,y),2\}$.

PROOF. Sufficiency. Suppose condition (A) is fulfilled. We claim that then (K,ρ) has the following more general property.

($\tilde{A}$) For every finite set $F \subset K$, for every function f on F , and for each $\varepsilon > 0$ there exists a function $g \in \mathrm{lip}(K,\rho)$ such that $g_{|F} = f$ and $\|g\|_{K,\rho} \le (1+\varepsilon)\|f\|_{F,\rho}$.

To show this, take a finite subset F in K , a function f on F , and $\varepsilon > 0$. For any two points $x, y \in K$, define the function

$$g_{x,y}(t) := f(y) + [f(x) - f(y)]g(t) \quad , \ t \in K ,$$

where g is a function provided by condition (A). Obviously, $g \in \mathrm{lip}(K,\rho)$, $g_{x,y}(x) = f(x)$, $g_{x,y}(y) = f(y)$, and

$$|g_{x,y}|_{K,\rho} = |f(x) - f(y)||g|_{K,\rho} \le (1+\varepsilon)\frac{|f(x) - f(y)|}{\min\{\rho(x,y),2\}} \le (1+\varepsilon)\|f\|_{F,\rho} .$$

Now set

$$g_0 := \max_{x\in F}\min_{y\in F} g_{x,y} .$$

For every $z \in F$, $g_0(z) \ge \min_{y\in F} g_{z,y}(z) = f(z)$. Conversely, there is a point $u \in F$ depending on z such that $g_0(z) = \min_{y\in F} g_{u,y}$, hence $g_0(z) \le g_{u,z}(z) = f(z)$. Resuming we conclude that $g_{0|F} = f$.

Observe that if every function from a finite collection $\{g_\alpha\}$ belongs to the set $\{g \in \mathrm{lip}(K,\rho) : |g|_{K,\rho} \le C\}$, then the same is also true for $\max_\alpha g_\alpha$ and for $\min_\alpha g_\alpha$. Setting $C = (1+\varepsilon)\|f\|_{F,\rho}$ and applying this argument to the function g_0 we see that that $|g_0| \le C$. Then the function

$$g(x) := \begin{cases} C , & \text{if} \quad g_0(x) > C \\ g\ (x) , & \text{if} \quad |g_0(x)| \le C \\ - C , & \text{if} \quad g_0(x) < - C \end{cases}$$

meets condition ($\tilde{A}$).

In view of Lemma 1 it suffices to show that $\|i(\mu)\| = \|\mu\|_\rho$ for all $\mu \in M(K)$. Take any such μ and fix $\varepsilon > 0$. By Lemma 0 there exists a measure ν with finite support F such that $\|\nu - \mu\|_\rho \le \varepsilon$. Let g be a function given by condition ($\tilde{A}$). In view of Theorem 0 we have for all $f \in B$

$$\int_K f d\mu = \int_K f d(\mu-\nu) + \int_K g d(\nu-\mu) + \int_K g d\mu \le \varepsilon + \varepsilon(1+\varepsilon) + \int_K g d\mu .$$

Hence again by Theorem 0

$$\|\mu\|_\rho = \sup\{\int_K f d\mu : f \in B\} = \sup\{\int_K g d\mu : g \in b\} = \|i(\mu)\| .$$

Necessity. Suppose that mapping i is an isometry. Let x, y be a pair of distinct points in K and let $\varepsilon > 0$. Then $\sup\{h(x)-h(y) : h \in b\} = \|i(\delta_x-\delta_y)\| = \|\delta_x-\delta_y\|_\rho = \min\{\rho(x,y),2\}$. Hence there is a function $h \in b$ such that $h(x)-h(y) \ge \min\{\rho(x,y),2\}/(1+\varepsilon)$. Set $g(t) = (h(t)-h(y))/(h(x)-h(y))$, $t \in K$. We have readily $g \in lip(K,\rho)$, $g(x) = 1$, $g(y) = 0$, and $|g|_{K,\rho} \le (1+\varepsilon)/\min\{\rho(x,y),2\}$. Thus function g satisfies condition (A).

Combining Theorem 0 and Theorem 1 we obtain immediately

THEOREM 2 . *Let* (K,ρ) *be a compact metric space. The mapping* $j : (lip(K,\rho))^{**} \to Lip(K,\rho)$ *is an isometric isomorphism if and only if* (K,ρ) *satisfies condition (A).*

Implication (A) $\Rightarrow$ ($\tilde{A}$) is an adjusted to the case of Lipschitz spaces simplified version of the classical Stone's proof of the Stone-Weierstrass theorem . The idea to use this type of argument belongs to S.V. Kislyakov. Together with Theorem 3 from [5] this leads us to the following theorem of the Stone-Weierstrass type for algebras $lip(K,\rho)$.

THEOREM 3. *Let* (K,ρ) *be a compact metric space, and* L *be a unital subalgebra (or sublattice) in* $lip(K,\rho)$ *satisfying the following condition:*

there is a constant $C \ge 1$ *such that for every two distinct points* $x,y \in K$ *there exists a function* $g \in L$ *with the properties* $g(x) = 1$, $g(y) = 0$, *and* $|g|_{K,\rho} \le C/\min\{\rho(x,y),2\}$.

Then L *is dense in* $lip(K,\rho)$.

In the sequel, metric spaces for which condition (A) is fulfilled will be called *noncritical* while those failing to satisfy condition (A) will be referred to as *critical* metric spaces.

CRITICAL AND NONCRITICAL METRIC SPACES

Let Ω be the set of all nondecreasing functions $\omega : \mathbb{R}_+ \to \mathbb{R}_+$ such that $\omega(0) = \lim_{t\to 0} \omega(t) = 0$, $\lim_{t\to 0} \omega(t)/t = +\infty$, and the function $\omega(t)/t$ is nonincreasing for $t > 0$. It can be easily seen that if (K,ρ) is a (compact) metric space and $\omega \in \Omega$, then the same is $(K,\omega(\rho))$.

PROPOSITION 1 [5]. *Suppose* (K,d) *is any compact metric space and* $\omega \in \Omega$. *Then* $(K,\omega(d))$ *is a noncritical metric space.*

In particular, for all compact metric spaces of the form $(K,\omega(d))$ with $\omega \in \Omega$ we have $(\mathrm{lip}(K,\omega(d)))^{**} \cong \mathrm{Lip}(K,\omega(d))$. For $\omega(t) = t^\alpha$, $0 < \alpha < 1$, this was proved in [10,2,3] and [1].

As it was mentioned earlier, an interval with the Euclidean metric is a critical metric space. Indeed, this is also true for any metric space containing it as a subspace and for every metric space which is isometrically isomorphic to it, in particular, for a rectifiable curve in $\mathbb{R}^n$ with a metric defined as the arc length. A more general class of critical metric spaces is defined below.

Given a metric space (K,ρ) and $\delta > 0$, we set for $x,y \in K$

$$\rho_\delta(x,y) := \inf \sum_{i=1}^{N} \rho(t_{i-1},t_i) ,$$

where infimum is taken over all sequences $t_0 ,\ldots, t_N$ of points in K such that $t_0 = x$, $t_N = y$ and $\rho(t_{i-1},t_i) \le \delta$, $i = 1,\ldots,N$ (if there are no such sequences, we set $\rho_\delta(x,y) := +\infty$). Observe that the family of functions ρ_δ , $\delta > 0$, is nonincreasing with the growth of δ , and define $\rho_0(x,y) := \lim_{\delta\to 0} \rho_\delta(x,y)$, $x,y \in K$. Clearly, functions ρ_δ , $\delta \ge 0$, possess all properties of a metric except possibly finiteness of their values.

PROPOSITION 2. *Let* (K,ρ) *be a metric space. Suppose that there exists a pair of points* $x,y \in K$, $x \ne y$, *with* $\rho_0(x,y) < +\infty$. *Then* (K,ρ) *is a critical metric space.*

PROOF. Take any function $g \in \mathrm{lip}(K,\rho)$ and fix $\varepsilon > 0$. There is a number $\delta > 0$ such that $|g(x) - g(y)| \le \varepsilon\, \rho(x,y)$ for all $x,y \in K$ with $\rho(x,y) \le \delta$. For every sequence of points $t_0 ,\ldots, t_N \in K$ such that $t_0 = x$, $t_N = y$ and $\rho(t_{i-1},t_i) \le \delta$, $i = 1,\ldots,N$, we have

$$|g(x) - g(y)| \le \sum_{i=1}^{N} |g(t_{i-1}) - g(t_i)| \le \varepsilon \sum_{i=1}^{N} \rho(t_{i-1},t_i) ,$$

hence $|g(x) - g(y)| \le \varepsilon\, \rho_\delta(x,y) \le \varepsilon\, \rho_0(x,y)$. Since $\rho_0(x,y) < +\infty$, this implies via arbitrariness of ε that $g(x) = g(y)$. Therefore condition (A) is not satisfied for points x, y, and we conclude that (K,ρ) is a critical metric space.

REFERENCES

1. Bade, W.G., Curtis, P.C.,Jr., and Dales, H.G.: Amenability and weak amenability for Beurling and Lipschitz algebras, Proc. London Math. Soc. 55(1987), 359-377.

2. Jenkins, T.M.: Banach spaces of Lipschitz functions on an abstract metric space, Thesis, Yale University, New Haven, CT, 1967.

3. Johnson, J.A.: Banach spaces of Lipschitz functions and vector-valued Lipschitz functions, Trans. Amer. Math. Soc. 148(1970), 147-169.

4. Hanin, L.G.: Kantorovich-Rubinstein duality for Lipschitz spaces defined by differences of arbitrary order, Soviet Math. Dokl. 42 (1991), 220-224.

5. Hanin, L.G.: Kantorovich-Rubinstein norm and its application in the theory of Lipschitz spaces, Proc. Amer. Math. Soc. 115(1992), 345-352.

6. Kantorovich, L.V.: On mass transfer, Dokl. Akad. Nauk SSSR 37(1942), 227-229 (Russian).

7. Kantorovich, L.V. and Akilov, G.P.: Functional Analysis, 2nd ed., Pergamon Press, New York, 1982.

8. Kantorovich, L.V. and Rubinstein, G.Sh.: On functional space and certain extremal problems, Dokl. Akad. Nauk SSSR 115(1957), 1058-1061 (Russian).

9. Kantorovich, L.V. and Rubinstein, G.Sh.: On a space of completely additive functions, Vestnik Leningrad Univ. Math. 13(1958), 52-59 (Russian).

10. deLeeuw, K.: Banach spaces of Lipschitz functions, Studia Math. 21 (1961), 55-66.

Department of Mathematics,
Technion - Israel Institute of Technology,
Haifa 32000,
Israel

MSC: Primary 46B04, 46E15, 46E27; Secondary 28A33, 54E35.

Operator Theory:
Advances and Applications, Vol. 73

RULES FOR COMPUTER SIMPLIFICATION OF THE FORMULAS IN OPERATOR MODEL THEORY AND LINEAR SYSTEMS

J. William Helton * and John J. Wavrik

Dedicated to Moshe Livsic on the occasion of his 70th birthday

This article formulates and treats questions in operator theory arising from computer simplification of formulas commonly found in the study of operator models. Operator model theory originated with Moshe Livsic and subsequently became one of the main branches of operator theory. In studying a particular operator model polynomials in certain expressions occur repeatedly. This makes it a natural area for exploring computer algebra simplification.

The purpose of a simplification theory is to provide a means for replacing complex expressions by expressions which are "simpler" in some sense. The main task is to obtain a list of rules each of which replaces a "complicated" monomial which occurs in an expression by a sum of "simpler" monomials. To simplify such an expression one applies the rules to the expression until no further reduction is possible. The result is called an N-Form (normal form) for the original expression. The reduction of an expression to an N-Form can be easily implemented on a computer. This article provides a collection of reduction rules for expressions which arise in the Nagy-Foias operator model. Our simplifying rules were obtained by applying an algorithm for computing a Grobner basis for an ideal in a polynomial ring. It is applied to the ideal generated by a set of fundamental relations which obviously hold for NF calculations. We conjecture that all appropriate relations are in this ideal. The algorithm produces an infinite set of rules in the NF case. The traditional operator theorist's functional calculus is used to produce a nice formulation of the rules as a finite set.

If a set of generators is a Grobner Basis, then the reduction to an N-form has very nice properties; the N-form is independent of the order in which rules are applied and equality of N-forms can be used as a test of equivalence of expressions mod the ideal. We have established that our rules form a Grobner Basis for several situations. A proof outline and computer tests provide strong evidence that this is true for the NF case as well.

The results here have potential applications to engineering systems theory since this algebraic structure occurs in formulas arising in H^∞ control. Also since the entries of a unitary 2x2 block matrix have algebraic structure similar to the NF model, this applies to 'all pass' functions from engineering.

* Research on this paper was supported in part by the Air Force Office of Scientific Research and the National Science Foundation.

I. INTRODUCTION

Our goal is to find Gröbner bases for polynomials in four different sets of expressions:

(RESOL)	$x,\quad x^{-1},\quad (1-x)^{-1}$
(EB)	$x,\quad x^{-1},\quad (1-xy)^{-1}$
	$y,\quad y^{-1},\quad (1-yx)^{-1}$
(preNF)	(EB) plus $(1-x)^{-1}$ and $(1-y)^{-1}$
(NF)	(preNF) plus $(1-xy)^{1/2}$ and $(1-yx)^{1/2}$

Most formulas in the theory of the Nagy-Foias operator model [NF] are polynomials in these expressions where $x = T$ and $y = T^*$. Complicated polynomials can often be simplified by applying "replacement rules". For example, the polynomial

$$(1-xy)^{-2} - 2xy(1-xy)^{-2} + xy^2\,(1-xy)^{-2} - 1$$

simplifies to 0. This can be seen by three applications of the replacement rule $(1-xy)^{-1}xy \longrightarrow (1-xy)^{-1} - 1$ which is true because of the definition of $(1-xy)^{-1}$.

A replacement rule consists of a left hand side (LHS) and a right hand side (RHS). The LHS will always be a monomial. The RHS will be a polynomial whose terms are "simpler" (in a sense to be made precise) than the LHS. An expression is reduced by repeatedly replacing any occurrence of a LHS by the corresponding RHS. The monomials will be well-ordered, so the reduction procedure will terminate after finitely many steps. Our aim is to provide a list of substitution rules for the classes of expressions above. These rules, when implemented on a computer, provide an efficient automatic simplification process.

We discuss and define the ordering on monomials later. Such an ordering typically is chosen to reflect the intuitive idea that complicated expressions are to be replaced by simpler ones.

We find that for cases (RESOL) and (EB) the ideal of relations (considered as a polynomial ring in the non-commuting atomic expressions above) is finitely generated. A finite Gröbner basis is obtained using a modification of the Buchberger algorithm due to F. Mora [Mora]. The ideal of relations in the other two cases can be shown not to be finitely generated—and the Mora algorithm does not terminate. We find, however, that the set of relations generated by the algorithm ultimately consists of a collection of classes of general relations which, with some particular relations, forms an infinite Gröbner Basis.

We introduce a formulation for our results in terms of an algebra which contains the operator expressions of interest. This formulation is very intuitive to a functional analyst in that it incorporates the operator functional calculus. An advantage here is that the infinite set of replacement rules obtained by applying the Basis Algorithm can be expressed by a finite set in the new algebra.

It will be interesting to see if other areas of operator theory behave the same way since it is natural to try what we have done on many of them. We expect that a major challenge for operator theorists will be to to take the infinite lists of rules which are appropriate for a particular field and find parameterizations which make them a natural looking finite list.

Much of the technique used here is algebra of the kind common in operator theory, while the non-commutative Gröbner Basis theory is not familiar in the operator community. Since this article is intended to be accessible to someone who is not familiar with this theory we shall outline basics. This article, however, is not intended to be a complete survey of the field.

The authors wish to thank Mark Stankus for helpful comments.

II. THE REDUCTION AND BASIS ALGORITHMS

A. Commuting Variables

1. The Reduction Process.

We impose an ordering (a "term order") on the monomials in the ring of polynomials in several commuting variables $(x_1, \ldots, x_n)$ over a field. Once such an ordering is imposed, we obtain a reduction process:

Let $F = \{f_1, \ldots, f_k\}$ be a set of polynomials in commuting variables. Using the given ordering, each can be written with the monomials in descending order. Now let f be any polynomial. We say that f is reducible to g with respect to F if $g = f - cx^\alpha f_i$ where c and α are chosen so that the leading term of $cx^\alpha f_i$ coincides with one of the terms of f. The effect is to replace a term of f with terms of lower order. A given f can be "simplified" by iterating this process until a polynomial is obtained which can no longer be reduced (none of its terms are divisible by any of the leading terms of the f_i). We call this a normal form (N-form) for f with respect to the process. Notice that any polynomial obtained from f by reduction is congruent to f mod the ideal generated by F.

2. The Gröbner Basis Algorithm.

Buchberger [Buch] and Hironaka [Hiron] independently showed that every ideal in a polynomial ring has a basis for which the reduction process is well-behaved. Buchberger provided a constructive procedure for obtaining such a basis from an arbitrary system of generators for the ideal.

> **Definition.** A set, F, of polynomials is called a Gröbner Basis for the ideal, I, it generates if it has the property that a polynomial f is in I if and only if it reduces to 0.

Buchberger shows that the order in which reductions are applied is immaterial in the case of a Gröbner Basis. He also supplies some equivalent characterizations of a Gröbner Basis among which is a simple algorithmic test for a basis to have this property. The Gröbner Basis Algorithm is important because it provides an algorithmic method for determining ideal membership, and for determining a canonical simple representative for congruence classes mod an ideal.

B. Operator Expressions.

We regard x, y, x^{-1}, y^{-1}, $(1-x)^{-1}$, etc. which arise in matrix expressions as atomic symbols—the variables in a ring of polynomials in non-commuting variables over the rational numbers. We let F be a set of relations known to hold among these symbols—the relations are expressed as polynomials in the atomic variables which are known to be zero. We have, for example, the relations $xx^{-1}-1$ and $x^{-1}x-1$ which are "defining relations" (they define what is meant by the symbol x^{-1} to be the inverse of the symbol x). Any member of the ideal generated by F is clearly a relation on the atomic symbols. The reduction process described above can be carried over to this situation (provided that care is taken to respect the non-commutative nature of multiplication).

1. The Reduction Process.

Let $F = \{f_1, \ldots, f_k\}$ be a set of polynomials. Using the given ordering, each can be written with the monomials in descending order. Now let f be any polynomial. We say that f is reducible to g with respect to F if $g = f - cuf_iv$ where c is a constant; u, v are monomials chosen so that the leading term of cuf_iv coincides with one of the terms of f. The effect is to replace a term of f with terms of lower order.

A reduction step can be conceived of as a replacement LHS$\longrightarrow$RHS of a term in f containing LHS as a factor by a term or sum of terms of lower order. If, for example, a term of f contains $x\ x^{-1}$, then it can be replaced by 1. It should be noted that this process is like the method that analysts use when simplifying operator expressions "by hand"; complicated subexpressions are repeatedly replaced by simpler ones.

The description of the reduction procedure in terms of replacement rules corresponds to the way it is commonly implemented. The "handedness" of replacement rules is determined by the term ordering. The LHS is always taken to be the leading term of a polynomial relation, while RHS is the negative of the sum of the remaining terms (which are of lower order).

2. The Basis Algorithm.

The reduction process is simpler than and independent of the Basis Algorithm which generates a set of relations "once and for all" for each set of atomic expressions. The generation of relations can be quite time-consuming. Once the relations are at hand, reduction of expressions is relatively fast. (The Basis Algorithm algorithm is part of a Forth-based system designed by Wavrik for the study of reduction processes of this type. The reduction process implementing some of the (NF) relations is available as part of a Mathematica linear algebra package [NFA].)

As in the case of commuting variables, it cannot be expected that an arbitrary set of relations provides a "good" set of replacement rules. Our aim is to apply a non-commutative version of the Gröbner Basis Algorithm to obtain a set of replacement rules which allow the simplification process to be applied mechanically.

As in the commutative case, a Gröbner Basis, G, for an ideal, I, is a set of polynomials in I having the property that a polynomial f is in I if and only if N-Form$(f, G) = 0$. In the commutative case a finite Gröbner Basis exists for any ideal and can be obtained by the Buchberger Algorithm applied to any system of generators for the idea. For ideals in a non-commutative polynomial ring it need not be true that an ideal has a finite Gröbner Basis. The following algorithm, due to F. Mora [Mora], is an adaptation of Buchberger's Algorithm to the case of non-commutative polynomial rings.

Let S be the free semigroup generated by a finite alphabet A. Let (m_1, m_2) be an ordered pair of elements of S. By a match of (m_1, m_2) we mean a 4-tuple (l_1, r_1, l_2, r_2) of elements of S which satisfy one of the following conditions:

(1) $l_1 = r_1 = 1,\ m_1 = l_2 m_2 r_2$.

(2) $l_2 = r_2 = 1,\ m_2 = l_1 m_1 r_1$.

(3) $l_1 = r_2 = 1,\ l_2 \neq 1,\ r_1 \neq 1$, there is a $w \neq 1$ with $m_1 = l_2 w,\ m_2 = w r_1$.

(4) $l_2 = r_1 = 1,\ l_1 \neq 1,\ r_2 \neq 1$, there is a $w \neq 1$ with $m_1 = w r_2,\ m_2 = l_1 w$.

These conditions make $l_1 m_1 r_1 = l_2 m_2 r_2$ a common multiple of m_1 and m_2 which is minimal in some sense.

If f is a polynomial, let $lt(f)$ denote the leading term of f (with respect to the given ordering). $lt(f) = lc(f)\ lm(f)$ where $lc(f)$ is the leading coefficient and $lm(f)$ the leading monomial. In the commutative case, the Basis Algorithm makes use of a kind of resolvent of two polynomials called the S-Polynomial. S-Pol$(f, g) = c_2 u_1 f_1 - c_1 u_2 f_2$ where $c_i = lc(f_i)$ and where u_i is chosen so that $u_i lm(f_i)$ is the least common multiple of $lm(f_1)$ and $lm(f_2)$. In the non-commutative case, there are several such resolvents—one for each match. We denote by M a 6-tuple $(f_1, f_2, l_1, r_1, l_2, r_2)$ where (l_1, r_1, l_2, r_2) is a match for $(lm(f_1), lm(f_2))$ and $c_i = lc(f_i)$. Let

$$\text{S-Pol}(M) = c_2 l_1 f_1 r_1 - c_1 l_2 f_2 r_2 .$$

Example:

Consider the polynomials

$$f_1 = aaba + ab$$

$$f_2 = abba + ba$$

There are four matches for (f_1, f_2):

1. $(aba, 1, 1, aba)$. In this case the S-Polynomial is

$$(aba)(f_1) - (f_2)(aba) = -baaba + abaab .$$

2. $(ab, 1, 1, ba)$ In this case the S-polynomial is

$$(ab)(f_1) - (f_2)(ba) = -baba + abab\ .$$

3. $(1, baa, aab, 1)$ In this case the S-Polynomial is

$$(f_1)(baa) - (aab)(f_2) = abbaa - aabba\ .$$

4. $(1, a, a, 1)$ In this case the S-Polynomial is

$$(f_1)(a) - (a)(f_2) = 0\ .$$

The algorithm takes the given basis and adds to it all nonzero reductions of S-Polynomials for all pairs in the original basis. The process repeats as long as there are S-Polynomials with nonzero reduction.

MORA'S BASIS ALGORITHM (MBA):

Input: A set F of polynomials
Output: A Gröbner basis, G, for the ideal generated by F
(if the algorithm terminates)

```
n := 1, H_1 := F, G_1 := F
While not-empty(H_n) do
        set-empty(B_n), set-empty(H_{n+1})
        For every match M = (f_1, f_2, l_1, r_1, l_2, r_2)
            with f_1 in G_n, f_2 in H_n do
            adjoin(B_n, M)
        While not-empty(B_n) do
                extract M from B_n
                f := S-Pol(M)
                f := N-Form(f, union(G_n, H_{n+1}))
                If f ≠ 0 then adjoin(H_{n+1}, f)
        G_{n+1} := union(G_n, H_{n+1})
        n := n + 1
G := G_n
```

In the current work, a final reduction step is used:

```
Repeat
    unchanged := true
    set-empty(G_1)
    While not-empty(G) do
        extract f from G
        f_1 :=N-Form(f, union(G, G_1))
        If f_1 ≠ f then unchanged := false
        If f_1 ≠ 0 then adjoin(G_1, f_1)
    G := G_1
Until unchanged
```

Mora shows that, if the process terminates, G is a finite Gröbner Basis for the ideal generated by F.

C. Term Ordering

The term ordering used in this paper is the graded lexicographic order. Let M_1 and M_2 be monomials in the atomic symbols. We define $M_1 < M_2$ if either

$$\text{length}\,(M_1) < \text{length}\,(M_2)$$

or length (M_1) = length (M_2) and M_1 comes before M_2 in lexicographic order. This ordering depends on the choice of ordering (say lexicographic) on the atomic symbols.

Example. If $a < b < c$ then the monomials of length 3 are ordered

$$aaa < aab < aac < aba < abb < abc < aca < acb < acc < baa < bab <$$
$$bac < bba < bbb < bbc < bca < bcb < bcc < caa < cab < cac < cba <$$
$$cbb < cbc < cca < ccb < ccc$$

The concept of simplification by use of replacement rules involves an underlying ordering on the terms of expressions. For the cases we study there are many natural ways to put an ordering on terms. We select one which reflects our notion of going from complicated to less complicated. While other orderings are possible, there are advantages to the user if a theoretically "simplified" expression looks simple. Here is the one we use for EB: The intuition we capture in our ordering is:

square roots of complicated polynomials = most complicated.

square roots of simple polynomials = second most complicated.

inv of complicated polynomials = third complicated.

inv of simple polynomials = next most complicated.

complicated monomials

simple monomials

commuting elements and expressions in them.

This is merely a partial order rather than a total order so in addition we specify lexicographic order on the underlying variables.

In other notation the ordering on EB is

$$\begin{matrix} x \\ y \end{matrix} < \begin{matrix} x^{-1} \\ y^{-1} \end{matrix} < \begin{matrix} (1-xy)^{-1} \\ (1-yx)^{-1} \end{matrix}$$

It is a partial order, but both in computing the basis and in applying the reduction process it is made a total ordering by imposing the lexicographic order $x < y$.

In this ordering we will always write rules so that they move complicated expressions to the right. The effect is that any replacement rule will replace expressions of high order by those of lower order. (A set of replacement rules which is not consistent in this way can produce infinite loops of replacements.)

For example, the identity $(1-xy)^{-1}\,x = x(1-yx)^{-1}$ can be converted to a replacement rule in two ways. The one consistent with the above order is:

$$(1-xy)^{-1}x \longrightarrow x(1-yx)^{-1}$$

in that the complicated inv moves right. The wrong rule is

$$x(1-yx)^{-1} \longrightarrow (1-xy)^{-1}x \,.$$

For the NF case we use

$$x < y < x^{-1} < y^{-1} < (1-x)^{-1} < (1-y)^{-1} < (1-xy)^{-1}$$
$$< (1-yx)^{-1} < (1-xy)^{1/2} < (1-yx)^{1/2} \,.$$

It should be noted that expressions involving higher powers of $(1-yx)^{1/2}$ and $(1-xy)^{1/2}$ are ordered in the standard way. However, among the replacement rules are

$$((1-yx)^{1/2})^2 \longrightarrow 1-yx$$

$$((1-xy)^{1/2})^2 \longrightarrow 1-xy$$

so any expression involving higher powers of $(\ )^{1/2}$ is reduced. All higher powers will be replaced by lower order expressions.

III. OPERATOR RELATIONS WITH FINITE BASIS FOR RULES

A. The (EB) and (RESOL) Relations

The (RESOL) case uses graded lexicographic ordering induced by the following order of the variables:

$$x < x^{-1} < (1-x)^{-1}$$

while for the (EB) case we use:

$$x < y < x^{-1} < y^{-1} < (1-x)^{-1} < (1-y)^{-1} .$$

Input to the algorithm are the defining relations for the inverses which are involved.

Theorem III.1. *The Mora Basis algorithm applied to* (EB) *and* (RESOL) *with the ordering above stops in finitely many steps and produces the relations below. The resulting set of relations is a finite Gröbner Basis.*

(RESOL)

Relations 0-3 are the starting relations, 4-5 were deduced by the algorithm.

$$\mathrm{Res}_0 = x^{-1}x - 1$$

$$\mathrm{Res}_1 = xx^{-1} - 1$$

$$\mathrm{Res}_2 = (1-x)^{-1}\, x - (1-x)^{-1} + 1$$

$$\mathrm{Res}_3 = x\,(1-x)^{-1} - (1-x)^{-1} + 1$$

$$\mathrm{Res}_4 = (1-x)^{-1}\, x^{-1} - (1-x)^{-1} - x^{-1}$$

$$\mathrm{Res}_5 = x^{-1}\,(1-x)^{-1} - (1-x)^{-1} - x^{-1} .$$

(EB)

The starting relations for the process are the defining relations for the inverse function:

$$\mathrm{EB}_0 = x^{-1}\, x - 1$$

$$\mathrm{EB}_1 = x x^{-1} - 1$$

$$\mathrm{EB}_2 = y^{-1}\, y - 1$$

$$\mathrm{EB}_3 = y y^{-1} - 1$$

$$\mathrm{EB}_4 = xy\,(1 - xy)^{-1} - (1 - xy)^{-1} + 1$$

$$\mathrm{EB}_5 = yx\,(1 - yx)^{-1} - (1 - yx)^{-1} + 1$$

with

$$\mathrm{EB}_6 = (1 - xy)^{-1} xy - (1 - xy)^{-1} + 1$$

$$\mathrm{EB}_7 = (1 - yx)^{-1} yx - (1 - yx)^{-1} + 1$$

The Mora Basis Algorithm terminates and produces precisely the following additional relations

$$\mathrm{EB}_8 = (1 - yx)^{-1}\, x^{-1} - y(1 - xy)^{-1} - x^{-1}$$

$$\mathrm{EB}_9 = (1 - xy)^{-1}\, y^{-1} - x\,(1 - yx)^{-1} - y^{-1}$$

$$\mathrm{EB}_{10} = x^{-1}\,(1 - xy)^{-1} - y\,(1 - xy)^{-1} - x^{-1}$$

$$\mathrm{EB}_{11} = y^{-1}\,(1 - yx)^{-1} - x\,(1 - yx)^{-1} - y^{-1}$$

$$\mathrm{EB}_{12} = (1 - yx)^{-1}\, y - y\,(1 - xy)^{-1}$$

$$\mathrm{EB}_{13} = (1 - xy)^{-1}\, x - x\,(1 - yx)^{-1}$$

The termination of the algorithm implies that these form a Gröbner Basis for the ideal generated by the initial relations. When used as replacement rules they provide a canonical simplification for any expression that only involves these variables.

IV. OPERATOR RELATIONS WITH INFINITE BASIS FOR RULES

A. (preNF) and (NF) Relations

The ordering used is:

(preNF) $x < y < x^{-1} < y^{-1} < (1 - x)^{-1} < (1 - y)^{-1} < (1 - xy)^{-1} < (1 - yx)^{-1}$

(NF) *above* $< (1-xy)^{1/2} < (1-yx)^{1/2}$.

Notice that the (preNF) expressions equal (RESOL) and (EB) together with a particular ordering. In the preNF, NF and NF+ situations we find that the Mora Basis Algorithm does not terminate. It produces, instead, a small collection of special relations followed by an unending set of sequences of relations. These can be regarded as general relations. A typical relation of this type is

$$x\,(1-yx)^{-n}\,(1-x)^{-1}-(1-xy)^{-n}\,\{(1-x)^{-1}-1\}$$

expanded out. The relations listed below are valid for $n=1,2,\ldots$. Relations with a * are also valid for $n=0$. The relations for each of these cases are based on the following set of 16 general relations.

*I**. $x(1-yx)^{-n}(1-x)^{-1}-(1-xy)^{-n}\{(1-x)^{-1}-1\}$

II. $x(1-yx)^{-n}(1-y)^{-1}-(1-xy)^{-n}(1-y)^{-1}-x(1-yx)^{-n}+(1-xy)^{-(n-1)}(1-y)^{-1}$

*III**. $x(1-yx)^{-n}(1-yx)^{1/2}(1-x)^{-1}-(1-xy)^{-n}(1-xy)^{1/2}\{(1-x)^{-1}-1\}$

IV. $x(1-yx)^{-n}(1-yx)^{1/2}(1-y)^{-1}-\{(1-xy)^{-n}-(1-xy)^{-(n-1)}\}(1-xy)^{1/2}(1-y)^{-1}-x(1-yx)^{-n}(1-yx)^{1/2}$

V. $y(1-xy)^{-n}(1-x)^{-1}-\{(1-yx)^{-n}-(1-yx)^{-(n-1)}\}(1-x)^{-1}-y(1-xy)^{-n}$

*VI**. $y(1-xy)^{-n}(1-y)^{-1}-(1-yx)^{-n}\{(1-y)^{-1}-1\}$

*VII**. $y(1-xy)^{-n}(1-xy)^{1/2}(1-x)^{-1}-\{(1-yx)^{-n}-(1-yx)^{-(n-1)}\}(1-yx)^{1/2}(1-x)^{-1}-y(1-xy)^{-n}(1-xy)^{1/2}$

*VIII**. $y(1-xy)^{-n}(1-xy)^{1/2}(1-y)^{-1}-(1-yx)^{-n}(1-yx)^{1/2}\{(1-y)^{-1}-1\}$

*IX**. $(1-x)^{-1}(1-yx)^{-n}(1-x)^{-1}-(1-x)^{-1}(1-xy)^{-n}(1-x)^{-1}-(1-yx)^{-n}(1-x)^{-1}+(1-x)^{-1}(1-xy)^{-n}$

X. $(1-x)^{-1}(1-yx)^{-n}(1-y)^{-1}-(1-x)^{-1}\{(1-xy)^{-n}-(1-xy)^{-(n-1)}\}(1-y)^{-1}-(1-yx)^{-n}\{(1-y)^{-1}-1\}-(1-x)^{-1}(1-yx)^{-n}$

XI. $(1-x)^{-1}(1-yx)^{-n}(1-yx)^{1/2}(1-x)^{-1}-(1-x)^{-1}(1-xy)^{-n}(1-xy)^{1/2}\{(1-x)^{-1}-1\}-(1-yx)^{-n}(1-yx)^{1/2}(1-x)^{-1}$

XII. $(1-x)^{-1}(1-yx)^{-n}(1-yx)^{1/2}(1-y)^{-1}-(1-x)^{-1}\{(1-xy)^{-n}-(1-xy)^{-(n-1)}\}$ $(1-xy)^{1/2}(1-y)^{-1}-(1-yx)^{-n}(1-yx)^{1/2}\{(1-y)^{-1}-1\}-(1-x)^{-1}(1-yx)^{-n}(1-yx)^{1/2}$

XIII. $(1-y)^{-1}(1-yx)^{-n}(1-x)^{-1}-(1-y)^{-1}\{(1-xy)^{-n}+\ldots+1\}(1-x)^{-1}+$ $\{(1-xy)^{-n}+\ldots+(1-xy)^{-1}\}(1-x)^{-1}+(1-y)^{-1}\{(1-xy)^{-n}+\ldots+(1-xy)^{-1}\}-$ $\{(1-xy)^{-n}+\ldots+(1-xy)^{-1}\}$

*XIV**. $(1-y)^{-1}(1-yx)^{-n}(1-y)^{-1}-(1-y)^{-1}(1-xy)^{-n}(1-y)^{-1}+(1-xy)^{-n}(1-y)^{-1}-$ $(1-y)^{-1}(1-yx)^{-n}$

XV. $(1-y)^{-1}(1-yx)^{-n}(1-yx)^{1/2}(1-x)^{-1}-(1-y)^{-1}\{(1-xy)^{-n}+\ldots+(1-xy)^{-1}\}$ $(1-xy)^{1/2}(1-x)^{-1}+\{1-xy)^{-n}+\ldots+(1-xy)^{-1}\}1-xy)^{1/2}(1-x)^{-1}+(1-y)^{-1}$ $\{1-xy)^{-n}+\ldots+(1-xy)^{-1}\}(1-xy)^{1/2}-\{(1-xy)^{-n}+\ldots+(1-xy)^{-1}\}(1-xy)^{1/2}-$ $(1-y)^{-1}(1-yx)^{1/2}(1-x)^{-1}$

*XVI**. $(1-y)^{-1}(1-yx)^{-n}(1-yx)^{1/2}(1-y)^{-1}-(1-y)^{-1}(1-xy)^{-n}(1-xy)^{1/2}(1-y)^{-1}+$ $(1-xy)^{-n}(1-xy)^{1/2}(1-y)^{-1}-(1-y)^{-1}(1-yx)^{-n}(1-yx)^{1/2}$

The rules can be paired. In the left column below are the relations for the preNF case which do not involve $[\ ]^{1/2}$. The corresponding right member is obtained from the left by inserting a $[\ ]^{1/2}$ factor.

I	*III*
II	*IV*
V	*VII*
VI	*VIII*
IX	*XI*
X	*XII*
XIII	*XV*
XIV	*XVI*

The transformation of the relations on the left to those on the right is (with the exception of $XIII \longrightarrow XV$) to replace

$$(1-xy)^{-n} \quad \text{by} \quad (1-xy)^{-n}(1-xy)^{1/2}$$
$$(1-yx)^{-n} \quad \text{by} \quad (1-yx)^{-n}(1-yx)^{1/2}$$

While the set of relations produced by the Gröbner Basis Algorithm is infinite in the polynomial setting, in Section V we will discuss below a general algebra, $A_{\mathcal{F}}$, in which the infinite sequences can be considered as members of a finite set.

1. (preNF)

The starting relations used in the preNF case are the defining relations for z^{-1} where z is x, y, $1-x$, $1-y$, $1-xy$, and $1-yx$. Note that these are the same as the starting relations for (EB) and (RESOL) on x and on y.

Theorem IV.1. *The Mora Basis algorithm for* (preNF) *does not terminate but ultimately generates relations contained in the eight classes listed in the left column above (viz* I, II, V, VI, IX, X, XIII, XIV*). In addition to the starting relations, the algorithm produces the relations from* (EB) *and* (RESOL) *listed earlier. It also produces the following particular relations which do not fit into the eight general classes.*

$$\text{rel[Pre1]} = (1-x)^{-1}y(1-xy)^{-1} - (1-x)^{-1}(1-xy)^{-1} - y(1-xy)^{-1} + (1-x)^{-1}$$

$$\text{rel[Pre2]} = (1-y)^{-1}x(1-yx)^{-1} - (1-y)^{-1}(1-yx)^{-1} - x(1-yx)^{-1} + (1-y)^{-1}$$

All of these relations are algebraic consequences of the starting relations (which are just the defining relations for inverse). They are obtained from the starting relations strictly by the application of polynomial arithmetic.

2. (NF)

We add the symbols $(1-xy)^{1/2}$ and $(1-yx)^{1/2}$ to the (preNF) setting and add the relations

$$\text{NFDEF}_1 = ((1-xy)^{1/2})^2 + xy - 1$$

$$\text{NFDEF}_2 = ((1-yx)^{1/2})^2 + yx - 1$$

(defining relations for $(\ )^{1/2}$) to the starting relations. We have seen that the Mora process for (preNF) produces

$$(1-xy)^{-1}x - x(1-yx)^{-1}$$

$$(1-yx)^{-1}y - y(1-xy)^{-1}$$

These yield important simplifications but appear relatively late in the process. They are, in the interest of efficiency, added to the starting relations for (NF).

Companion relations for $(\ )^{1/2}$

$$(1-xy)^{1/2}x - x(1-yx)^{1/2}$$

$$(1-yx)^{1/2}y - y(1-xy)^{1/2}$$

These relations will not arise from applying the Basis Algorithm to the defining relations. We have therefore added them to the starting relations in the (NF) case. These can be proved by an analytic argument. These relations are extremely important to those working with the Nagy-Foias model.

Theorem IV.2. *The Mora Basis algorithm for* (NF) *does not terminate but ultimately generates the relations from* (preNF) *together with the remaining eight classes from the* 16 *above (viz* III, IV, VII, VIII, XI, XII, XV, XVI*). It also produces the following particular relations which do not fit into the* 16 *general classes.*

$$\text{rel[NF1]} = (1-xy)^{1/2}y^{-1} - y^{-1}(1-yx)^{1/2}$$

$$\text{rel[NF2]} = (1-xy)^{1/2}(1-xy)^{-1} - (1-xy)^{-1}(1-xy)^{1/2}$$

$$\text{rel[NF3]} = (1-yx)^{1/2}x^{-1} - x^{-1}(1-xy)^{1/2}$$

$$\text{rel[NF4]} = (1-yx)^{1/2}(1-yx)^{-1} - (1-yx)^{-1}(1-yx)^{1/2}$$

$$\text{rel[NF5]} = x^{-1}(1-xy)^{1/2}(1-x)^{-1} - (1-yx)^{1/2}(1-x)^{-1} - x^{-1}(1-xy)^{1/2}$$

$$\text{rel[NF6]} = y^{-1}(1-yx)^{1/2}(1-y)^{-1} - (1-xy)^{1/2}(1-y)^{-1} - y^{-1}(1-yx)^{1/2}$$

B. A Note on Symmetry

In terms of operator theory, nothing distinguishes x from y. The Simplification Process, however, requires a term ordering—and we have imposed an ordering in which y has a higher precedence than x. We examine the extent to which the reduction rules are symmetric in x and y. The starting relations for these classes are symmetric if one interchanges x and y, but the Basis Algorithm depends on the term ordering and reduction rules which are not entirely symmetric. For reference below we denote by $xchg(r)$ the result of making the exchange of x with y in relation r. The change also takes place within atomic symbols—thus $xchg((1-x)^{-1}) = (1-y)^{-1}$ and $xchg((1-xy)^{-1}) = (1-yx)^{-1}$.

Here are the rules related by $xchg$:

$$
\begin{array}{llll}
\text{I} & = xchg(\text{VI}) & \text{III} & = xchg(\text{VIII}) \\
\text{II} & = xchg(\text{V}) & \text{IV} & = xchg(\text{VII}) \\
\text{IX} & = xchg(\text{XIV}) & \text{XI} & = xchg(\text{XVI})
\end{array}
$$

The classes X and XIII (also XII and XV) are not directly related by $xchg$ but satisfy the following inductive formulas:

$$
\begin{aligned}
\text{XIII}(n) &= -xchg(\text{X}(n)) + \text{XIII}(n-1) \\
\text{XV}(n) &= -xchg(\text{XII}(n)) + \text{XV}(n-1)
\end{aligned}
$$

V. A NEW ALGEBRA CONTAINING THE FUNCTIONAL CALCULUS OF OPERATOR THEORY

A. The General Relations

To avoid dazzling with abstraction we begin by stating some explicit rules which contain one half of the rules in section IV as special cases. These seem like basic simplification rules for much of operator theory. Next we show a setting natural for treating such rules and obtaining Gröbner basis properties.

(GENR)

GR0 $\quad h(xy)x \longrightarrow xh(yx)$.

GR1 $\quad h(yx)x^{-1} \longrightarrow x^{-1}h(xy)$.

GR2 $\quad xh(yx)(1-x)^{-1} \longrightarrow -h(xy) + h(xy)(1-x)^{-1}$.

GR3 $\quad x^{-1}h(xy)(1-x)^{-1} \longrightarrow x^{-1}h(xy) + h(yx)(1-x)^{-1}$.

GR4 $\quad (1-x)^{-1}h(yx)(1-x)^{-1} - (1-x)^{-1}h(xy)(1-x)^{-1}$
$\longrightarrow h(yx)(1-x)^{-1} - (1-x)^{-1}h(xy)$.

for all operators x, y on a Hilbert space H with x, $1-x$ invertible and functions h analytic on the spectrum of xy and yx.

Operators of the form $h(xy)$ are of course what is called the *functional calculus* of the operator xy. Consequently we are preparing to treat general rules which are true for general functional calculi; one of the most important constructions in operator theory.

There are several observations about (GENR). First while (GR4) is not in the same form as the reduction rules considered previously, it transforms a sum rather than a monomial. To build a conventional rule from this one must use lexicographic order. The second observation is that (GENR) are easy to prove directly without reference to any algorithm for their construction.

Proof. (GENR) The proof is easy. A typical example is

$$\text{(GR3)}\ x^{-1}h(xy)(1-x)^{-1} = h(yx)x^{-1}(1-x)^{-1} =$$
$$= h(yx)(x^{-1}+(1-x)^{-1}) = h(yx)x^{-1} + h(yx)(1-x)^{-1}$$

which yields (GR3). The other cases are similar.

The main point about these rules is that they contain many of the preNF and NF rules as a special case. Indeed

THEOREM V.1. *Rules* I, III, VI, VIII, IX, XI, XIV, XVI *of section* 4 *follow from* (GENR) *provided xy and yx have spectrum inside the open unit disk.*

Proof. The proof consists of comparing the list in section IV against (GENR). For example rule III is obtained from (GR2) by setting $h[s] = (1-s)^{-n}(1-s)^{1/2}$ for s in the unit disk in the complex plane $\mathbf{C}$. Since the spectrum of xy is in the unit disk and h is analytic there (GR2) applies to give rule III of section IV.

B. The Functional Calculus Algebra

What is missing in our study so far is a unified way of dealing with infinite sets of rules. We now introduce a setting which does this for the situation here and may be good for doing lots of the algebra which occurs in functional analysis.

Let $\mathcal{F}$ denote the set of analytic functions

$$f : \mathcal{D} \subset \mathbf{C} \longrightarrow \mathbf{C},$$

on an open domain $\mathcal{D}$. Given A the algebra of all bounded operators on H define the *functional calculus algebra* $A_{\mathcal{F}}$ for A to be the set of A valued functions on $\mathcal{F}$. It is an algebra with the multiplication on

$$A_{\mathcal{F}} = \{g : \mathcal{F} \longrightarrow A\}$$

defined by $uv[h] = u[h]v[h]$. Note the constant functions in $A_{\mathcal{F}}$ can be identified with A as follows: Denote by c_v the element c in $A_{\mathcal{F}}$ defined by $c[h] = v$ where v is a constant function. Then $c_a c_b = c_{ab}$.

We organize our approach differently than before to take advantage of the power of this more general framework.

We now define certain elements in the algebra $A_{\mathcal{F}}$ and give the defining relations for them so that when we apply the (MBA) algorithm we get (GENR).

C. RESOL **Plus Generalized** EB

Fix x, y that is the constant functions c_x c_y in the algebra $A_{\mathcal{F}}$. Then x^{-1}, y^{-1}, $(1-x)^{-1}$, $(1-y)^{-1}$ correspond to $c_{x^{-1}}$, $c_{y^{-1}}$, $c_{(1-x)^{-1}}$, $c_{(1-y)^{-1}}$ in the algebra $A_{\mathcal{F}}$ and indeed satisfy the same relations, namely, the (RESOL) relations.

Fix two functions S, T in $A_{\mathcal{F}}$ defined by

$$S(h) = h(xy) \quad T(h) = h(yx)$$

These satisfy

$$\begin{aligned} S\,c_x = c_x\,T \quad &\text{since} \quad h(xy)x = xh(yx) \\ c_y\,S = T\,c_y \quad &\text{since} \quad yh(xy) = h(yx)y \end{aligned}$$

This somewhat complicated setup has the following rather simple algebraic structure which is what we use when implementing (MBA) on a computer.

We denote by $\mathcal{R}_1^{\mathcal{F}}$ the elements c_x, c_y, $c_{x^{-1}}$, $c_{y^{-1}}$, $c_{(1-x)^{-1}}$, $c_{(1-y)^{-1}}$, S and T of $A_{\mathcal{F}}$ together with the ordering on formal monomials in them based on degree of the monomial plus

$$\begin{matrix} c_x \\ c_y \end{matrix} < \begin{matrix} c_{x^{-1}} \\ c_{y^{-1}} \end{matrix} < \begin{matrix} c_{(1-x)^{-1}} \\ c_{(1-y)^{-1}} \end{matrix} < \begin{matrix} S \\ T \end{matrix}$$

Instead of working directly with $\mathcal{R}_1^{\mathcal{F}}$, a subset of $A_{\mathcal{F}}$, we introduce an abstraction of the algebra involved.

DEFINITION: Let $\mathcal{R}_1$ denote the polynomial ring generated by the atomic symbols used in (RESOL)

$$\begin{matrix} x,\ (1-x)^{-1},\ x^{-1} \\ y,\ (1-y)^{-1},\ y^{-1} \end{matrix}$$

plus two additional elements S and T which satisfy (5.1)

$$\begin{aligned} Sx &= xT \\ yS &= Ty \end{aligned}$$

The ordering on monomials is the usual plus

$$\begin{matrix} x \\ y \end{matrix} < \begin{matrix} x^{-1} \\ y^{-1} \end{matrix} < \begin{matrix} (1-x)^{-1} \\ (1-y)^{-1} \end{matrix} < \begin{matrix} S \\ T \end{matrix}$$

The relations we use in $\mathcal{R}_1$ are satisfied by the corresponding elements of $\mathcal{R}_1^{\mathcal{F}}$ considered as elements of $A_{\mathcal{F}}$. Let $\mathcal{I}_1$ be the ideal in $\mathcal{R}_1$ generated by the starting relations. $\mathcal{I}_1$ is invariant under the substitution sending S to S^n and T to T^n. This is true of the starting generators hence of any element in the ideal. Our main result shows that (GENR) is a type of Gröbner basis produced for $\mathcal{R}_1^{\mathcal{F}}$. We now state it formally, but with the obvious abbreviated notation c_v replaced by v.

THEOREM V.2. *The* (RESOL) *relations and the list of relations:*

(GA0) $$Sx - xT \qquad yS - Ty$$

plus

(GA1) $$x^{-1}S - Tx^{-1} \quad Sy^{-1} - y^{-1}T$$

(GA2) $$\begin{aligned} &xT(1-x)^{-1} - S(1-x)^{-1} - S \\ &yS(1-y)^{-1} - T(1-y)^{-1} + T \end{aligned}$$

(GA3) $$\begin{aligned} &x^{-1}S(1-x)^{-1} - T(1-x)^{-1} - x^{-1}S \\ &y^{-1}T(1-y)^{-1} - S(1-y)^{-1} - y^{-1}T \end{aligned}$$

(GA4) $$\begin{aligned} &(1-x)^{-1}T(1-x)^{-1} - (1-x)^{-1}S(1-x)^{-1} - T(1-x)^{-1} + (1-x)^{-1}S \\ &(1-y)^{-1}T(1-y)^{-1} - (1-y)^{-1}S(1-y)^{-1} + S(1-y)^{-1} - (1-y)^{-1}T \end{aligned}$$

form a Gröbner basis for the ideal they generate (inside $\mathcal{R}_1$) provided we use the property imposed by $\mathcal{R}_1^{\mathcal{F}}$ that if a relation contains S, T it is also true for S, T replaced by S^n, T^n. Note these relations are produced by the (MBA).

The proof is given in section VI where we also give some basics about Gröbner bases.

D. The preNF Relations

The previous section concerned relations involving a general analytic function h. If one restricts h to be in a special class of functions, then it is likely that more rules are true and that one gets a set of rules which contain (GR0–4). For example if $h(s) = (1-s)^{-n}$, then (with the previous choice of ordering) one gets the preNF rules and these contain rules in addition to those arising from (RES) alone, (EB) alone, and (GR0–4) In this section we show how treating addition of the special structure $h(s) = (1-s)^{-n}$ and $h(s) = (1-s)^{-n}(1-s)^{-1/2}$ can be put in the context of the functional calculus algebra. Thereby we can treat problems of varying levels of generality.

We start with

DEFINITION: $\mathcal{R}_2$ is the algebra $\mathcal{R}_1$ plus U, V defined by

$$U = (1-xy)S$$
$$V = (1-yx)T$$

$Ux = xV \quad yU = Vy$.

Now we set the ordering so that

$$\text{all the rest } < \begin{matrix} U \\ V \end{matrix} < \begin{matrix} S \\ T \end{matrix}.$$

This is counter intuitive because $U = (1-xy)S$ seems formally to have higher degree than S.

We motivate this construction by pointing out that it naturally sits in $A_{\mathcal{F}}$ where one defines U and V by

$$U(h) = (1-xy)S(h)$$

$$V(h) = (1-yx)T(h)$$

and to motivate the ordering think of h as having an expansion

$$h(s) = \sum_{n=0}^{\infty} a_n(1-s)^{-n}.$$

Then $U(h) = (1-xy)S(h)$ because of cancellation has lower degree than $S(h) = h(xy)$.

What we find upon running Mora's algorithm are relations which can be summarized by:

Those gotten from (RESOL) and(EB) plus the two special relations rel[Pre1] and rel[Pre2]. Plus

$\text{GENREL}[1] = xT^nV^m(1-x)^{-1} - S^nU^m(1-x)^{-1} + S^nU^m$

$\text{GENREL}[2]$
$= xT^nV^m(1-y)^{-1} + S^{n-1}U^{m+1}(1-y)^{-1} - S^nU^m(1-y)^{-1} - xT^nV^m$

$\text{GENREL}[3] = yS^nU^m(1-x)^{-1} + T^{n-1}V^{m+1}(1-x)^{-1} - T^nV^m(1-x)^{-1} - yS^nU^m$

$\text{GENREL}[4] = yS^nU^m(1-y)^{-1} - T^nV^m(1-y)^{-1} + T^nV^m$

$\text{GENREL}[5] = x^{-1}S^n(1-x)^{-1} - T^n(1-x)^{-1} - x^{-1}S^n$

$\text{GENREL}[6] = x^{-1}S^nU - x^{-1}S^{n+1} + yS^{n+1}$

$\text{GENREL}[7] = y^{-1}T^n(1-y)^{-1} - S^n(1-y)^{-1} - y^{-1}T^n$

$\text{GENREL}[8] = y^{-1}T^nV - y^{-1}T^{n+1} + xT^{n+1}$

$\text{GENREL}[9] = (1-x)^{-1}T^nV^m(1-x)^{-1} - (1-x)^{-1}S^nU^m(1-x)^{-1} -$
$T^nV^m(1-x)^{-1} + (1-x)^{-1}S^nU^m$

$\text{GENREL}[10] =$
$(1-x)^{-1}T^nV^m(1-y)^{-1} + (1-x)^{-1}S^{n-1}U^{m+1}(1-y)^{-1} -$
$(1-x)^{-1}S^nU^m(1-y)^{-1} - T^nV^m(1-y)^{-1} - (1-x)^{-1}T^nV^m + T^nV^m$

$\text{GENREL}[11] =$
$(1-x)^{-1}U^n(1-y)^{-1} + (1-x)^{-1}TV^{n-1}(1-y)^{-1} -$
$(1-x)^{-1}SU^{n-1}(1-y)^{-1} - TV^{n-1}(1-y)^{-1} - (1-x)^{-1}TV^{n-1} + TV^{n-1}$

$\text{GENREL}[12] =$
$(1-y)^{-1}T^nV^m(1-x)^{-1} - (1-y)^{-1}T^{n+m}(1-x)^{-1} + (1-y)^{-1}\sigma(1-x)^{-1}$
$-\sigma(1-x)^{-1} - (1-y)^{-1}\sigma + \sigma$

where $\sigma = \sum_{\imath=0}^{m-1} S^{m+n-\imath}U^\imath$ for $\imath = 0 \ldots m-1$.

$\text{GENREL}[13] =$
$(1-y)^{-1}T^nV^m(1-y)^{-1} - (1-y)^{-1}S^nU^m(1-y)^{-1} + S^nU^m(1-y)^{-1} -$
$(1-y)^{-1}T^nV^m$

We may use these to deduce some of the general relations in section IV. To see this observe that the elements S, T, U, V of $A_{\mathcal{F}}$ satisfy GENREL[1–13] whenever they are applied to functions h of the form $h(s) = (1-s)^{-k}$. Now we use that

$$S^m(h) = (1-xy)^{-mk} \qquad T^m(h) = (1-yx)^{-mk}$$
$$U^m(h) = (1-xy)^{-(k-1)m} \quad V^m(h) = (1-yx)^{-(k-1)m}$$

Thus the common expressions T^nV^m and S^nU^m become

$$T^nV^m \longrightarrow (1-yx)^{-nk-(k-1)m}$$
$$S^nU^m \longrightarrow (1-xy)^{-nk-(k-1)m}$$

Now we substitute these into GENREL[1–13] to reach the conclusion. As an example we treat GENREL[2]. When applied to $h(s) = (1-s)^{-k}$ it becomes

$$(1-xy)^{-r}(1-y)^{-1} - (1-xy)^{-(r-1)}(1-y)^{-1} + (1-xy)^{-r} - (1-yx)^{-r} \qquad \text{(IV.1)}$$

where we set $r = nk + (k-1)n$ and observe that $(n-1)k + (k-1)(m+l) = r-1$. Now (IV.1) equals II of section IV. Similar calculations yield preNF relations according to the table:

GENREL	preNF
1	I
2	II
3	V
4	VI
9	IX
10	X
13	XIV

This produces seven out of eight preNF relations; only XIII is missing. GENREL 5-8 and 12 after substituting $(1-xy)^{-1}$ become rules which are reduced to 0 by the rules already produced. GENREL 12 gives a relation which when subjected to induction gives XIII.

Consequently, we have an alternative way of producing the preNF relations to that given in section IV. Clearly it amounts to feeding different starting relations and ordering into the (MBA).

E. The NF Relations in the $A_{\mathcal{F}}$ Setting

We simply give the starting relations and ordering which seem appropriate to NF in the $A_{\mathcal{F}}$ setting.

DEFINITION: $\mathcal{R}_3$ is the algebra obtained by adjoining two elements Q, R to $\mathcal{R}_2$.

$$Q^2 - (1-xy) = 0$$
$$R^2 - (1-yx) = 0$$

and

$$Qx - xR = 0$$
$$yQ - Ry = 0\,.$$

(Intuitively Q and R behave like $(1-xy)^{1/2}$ and $(1-yx)^{1/2}$.) The ordering reflects

$$x < x^{-1} < (1-x)^{-1} < (1-xy)^{-1} < (1-xy)^{1/2} < (1-xy)^{-1/2}$$

and is

$$x < x^{-1} < (1-x)^{-1} < U < S < Q < Q^{-1}$$

with a similar ordering on T, V, R.

We have not tested this thoroughly but expect it will imply the NF

VI. Gröbner BASIS PROPERTY

Let I be a two-sided ideal in a polynomial ring, let I^* denote the invertible elements in I, and let G be a (possibly infinite) set of elements in I^*.

DEFINITION We say that a polynomial f has a finite d-representation in terms of G if there is a finite set of a_i in K, l_i, r_i monomials, g_i in G for $i = 1, \ldots, k$ so that $f = a_1 l_1 g_1 r_1 + \cdots + a_k l_k g_k r_k$ and where $lm(f) = lm(l_1 g_1 r_1) > \cdots > lm(l_k g_k r_k)$.

THEOREM V.3. *The following two conditions are equivalent:*

(1) *f is in I^* if and only if f reduces to 0 with respect to G.*

(2) *For every match $M = (g_1, g_2, l_1, r_1, l_2, r_2)$ with g_1, g_2 in G, S-Pol(M) has a finite d-representation in terms of G.*

Proof: See [Mora] Theorem 3.3

G is a Gröbner Basis if these equivalent conditions hold. For Gröbner bases the reduction process is well-behaved. Notice that condition (1) tells us that N-form(f, G) gives a canonical representative for the equivalence class of f mod I (equivalence can be tested by complete reduction).

A. Finite Gröbner Bases

If the Basis Algorithm terminates, the resulting set of relations is a Gröbner Basis—for, by the nature of the algorithm, condition (2) is satisfied. Any match for which S-Pol(M)

does not have a finite d-representation will produce a new basis element. The algorithm terminates when this no longer happens.

We have seen that the MBA terminates for (RESOL) and (EB). Thus we may assert that the relations we obtained in those cases form a Gröbner Basis.

It is of interest to note that, since rings of polyomials in commuting variables are Noetherian rings, the Basis Algorithm always terminates in the classical case. Our application to Operator Theory has produced examples in which the Basis Algorithm does not terminate—but which do yield Gröbner Bases.

B. An Infinite Gröbner Basis

Many of the defining relations in this paper do not give finite Gröbner Bases. However, they may give infinite ones.

We shall illustrate this by proving Theorem V.2.

Proof of Theorem V.2. The relations arising from $\mathcal{R}_1^{\mathcal{F}}$ equal those produced by

$$x < x^{-1} < (1-x)^{-1} < S < T \tag{5.1x}$$

union those arising from

$$y < y^{-1} < (1-y)^{-1} < S < T\,. \tag{5.1y}$$

We shall analyze only one of them (5.1x), and since the other has a similar structure a similar argument will prevail. The core of the proof is

Proposition VI.1. *The Mora algorithm applied to elements* (5.1x) *with starting relations the defining relations for* $(\ )^{-1}$ *as well as the relation* $Sx - xT$, *yield*

$$\begin{aligned}
\mathrm{Res}_0 &:= x^{-1}x - 1 \\
\mathrm{Res}_1 &:= xx^{-1} - 1 \\
\mathrm{Res}_2 &:= (1-x)^{-1}x - (1-x)^{-1} + 1 \\
\mathrm{Res}_4 &:= (1-x)^{-1}x^{-1} - (1-x)^{-1} - x^{-1} \\
\mathrm{rel}[\mathcal{V}_1] &:= Sx - xT \\
\mathrm{rel}[\mathcal{V}_2] &:= Tx^{-1} - x^{-1}S
\end{aligned} \tag{5.2}$$

together with the three general relations

$$
\begin{aligned}
A[n] &:= xT^n(1-x)^{-1} - S^n(1-x)^{-1} + S^n \\
C[n] &:= x^{-1}S^n(1-x)^{-1} - T^n(1-x)^{-1} - x^{-1}S^n \\
E[n] &:= (1-x)^{-1}T^n(1-x)^{-1} - (1-x)^{-1}S^n(1-x)^{-1} \\
&\quad -T^n(1-x)^{-1} + (1-x)^{-1}S^n .
\end{aligned} \tag{5.3}
$$

These form a Gröbner basis for the ideal of relations.

The general relations GA0 and GA1 are obtained from rel[$\mathcal{V}_1$] and rel[$\mathcal{V}_2$]. The general relations GA2–GA4 are obtained from A, C and E. For convenience we make A, B, and C into functions

$$
\begin{aligned}
A(h) &:= xT(h)(1-x)^{-1} - S(h)(1-x)^{-1} + S(h) \\
C(h) &:= x^{-1}S(h)(1-x)^{-1} - T(h)(1-x)^{-1} - x^{-1}S(h) \\
E(h) &:= (1-x)^{-1}T(h)(1-x)^{-1} - (1-x)^{-1}S(h)(1-x)^{-1} \\
&\quad - T(h)(1-x)^{-1} + (1-x)^{-1}S(h)
\end{aligned}
$$

We have $A[n](h) = A(h^n)$, etc.

For convenience, we also introduce

$$
\begin{aligned}
TC[n] &:= T^n x^{-1} - x^{-1}S^n \\
SA[n] &:= S^n x - xT^n
\end{aligned}
$$

Notice that $TC[n]$ reduces to

$$\sum_{i+j=n-1} T^i \mathrm{rel}[\mathcal{V}_2] S^j$$

and $SA[n]$ reduces to

$$\sum_{i+j=n-1} S^i \mathrm{rel}[\mathcal{V}_1] T^j .$$

Proof of Proposition VI.1. The (MBA) has been described in terms of "matches". We denote by M a 6-tuple

$$(f_1, f_2, l_1, r_1, l_2, r_2)$$

where (l_1, r_1, l_2, r_2) is a match for $(\mathrm{lm}(f_1), \mathrm{lm}(f_2))$. Mora [Mora] (Theorem 3.3) shows that a set of polynomials is a Gröbner Basis for the ideal they generate if

$$S\text{-}Pol(M) := l_1 f_1 r_1 - l_2 f_2 r_2$$

has a finite d-representation for every M formed from pairs in the set. We will show that the set of relations described above satisfies this criterion.

The only matches among the general relations involve $(E[n], E[m])$, $(E[n], C[m])$, and $(E[n], A[m])$. We could view these in the present notation using powers or describe these matches as

$$(E(h), E(k)),\ (E(h), C(k)),\quad \text{and}\quad (E(h), A(k))\,.$$

Either would do and in this proof we shall stick with powers. We now list each match, M, and a finite d-representation (see definition sec. VI) for $S\text{-}Pol(M)$.

$$\begin{aligned} M &= (E[n], E[m], (1-x)^{-1}T^m, 1, 1, T^n(1-x)^{-1}) \\ S\text{-}Pol(M) &= -E[m]S^n(1-x)^{-1} + (1-x)^{-1}S^m E[n] \\ &\quad +T^m E[n] - E[n+m] + E[m]S^n \end{aligned}$$

$$\begin{aligned} M &= (E[n], C[m], x^{-1}S^m, 1, 1, T^n(1-x)^{-1}) \\ S\text{-}Pol(M) &= -C[m]S^n(1-x)^{-1} + T^m E[n] \\ &\quad -C[n+m] + C[m]S^n \end{aligned}$$

$$\begin{aligned} M &= (E[n], A[m], xT^m, 1, 1, T^n(1-x)^{-1}) \\ S\text{-}Pol(M) &= -A[m](1-x)^{-1} + S^m E[n] \\ &\quad -A[n+m] + A[m]S^n \end{aligned}$$

The six special relations (5.2) interact with each other to give, by the nature of the Basis Algorithm, relations in the final set. Thus we need only describe interactions between the special relations (5.2) and the general ones (5.3).

$$M = (C[n], \text{Res}_0, x, 1, 1, S^n(1-x)^{-1})$$
$$S\text{-}Pol(M) = -A[n] - \text{Res}_0 S^n$$

$$M = (C[n], \text{Res}_3, (1-x)^{-1}, 1, 1, S^n(1-x)^{-1})$$
$$S\text{-}Pol(M) = -E[n] - \text{Res}_3 S^n + C[n]$$

$$M = (C[n], \text{rel}[\mathcal{V}_2], T, 1, 1, S^n(1-x)^{-1})$$
$$S\text{-}Pol(M) = C[n+1] - \text{rel}[\mathcal{V}_2]S^n$$

$$M = (A[n], \text{Res}_1, x^{-1}, 1, 1, T^n)$$

$$S\text{-}Pol(M) = -C[n]$$

$$M = (A[n], \text{Res}_2, (1-x)^{-1}, 1, 1, T^n(1-x)^{-1})$$
$$S\text{-}Pol(M) = E[n]$$

$$M = (A[n], \text{rel}[\mathcal{V}_1], S, 1, 1, T^n)$$
$$S\text{-}Pol(M) = A[n+1]$$

$$M = (\text{Res}_3, A[n], xT^n, 1, 1, x^{-1})$$
$$S\text{-}Pol(M) = S^n\text{Res}_3 - A[n] - xTC[n] - \text{Res}_0 S^n$$

$$M = (\text{Res}_3, C[n], x^{-1}S^n, 1, 1, x^{-1})$$
$$S\text{-}Pol(M) = T^n\text{Res}_3 - C[n] - TC[n]$$

$$M = (\text{Res}_3, E[n], (1-x)^{-1}T(c), 1, 1, x^{-1})$$
$$S\text{-}Pol(M) = (1-x)^{-1}S^n\text{Res}_3 + T^n\text{Res}_3 - E[n] - (1-x)^{-1}TC[n] - \text{Res}_3 S^n + TC[n]$$

$$M = (\text{Res}_2, A[n], xT^n, 1, 1, x)$$
$$S\text{-}Pol(M) = S^n\text{Res}_2 - A[n]$$

$$M = (\text{Res}_2, C[n], x^{-1}S^n, 1, 1, x^{-1})$$
$$S\text{-}Pol(M) = T^n\text{Res}_2 - C[n] + x^{-1}SA[n] + \text{Res}_1 T^n$$

$$M = (\text{Res}_2, E[n], (1-x)^{-1}T^n, 1, 1, x^{-1})$$
$$S\text{-}Pol(M) = (1-x)^{-1}S^n\text{Res}_2 + T^n\text{Res}_2 - E[n] - (1-x)^{-1}\text{SA}[n] - \text{Res}_2 T^n$$

No other pairs give matches, so condition (2) of the theorem is satisfied.

The authors strongly believe that the Grobner Basis Property holds for the other sets of relations which occur in this paper. We now list evidence for this. In the (NF) case, condition (2) of the Theorem has been verified by machine for all pairs containing basis elements arising from the special rules and all instances of the general rules up to n=12. A general proof like that given above could be given but the greater number of combinations of rules in the (preNF) and (NF) case would make it cumbersome to write.

VII. SUMMARY OF PRACTICAL RULES YOU MIGHT USE

Here we list in the most condensed form rules that we have generated. Some of these might be worth teaching a functional analysis class. Note that we have replaced resolvents such as $(\lambda - a)^{-1}$ with $(1-x)^{-1}$. This is a normalization easily implemented by dividing a by λ.

A. Rules corresponding to the classical resolvent identity:

(RES) $\qquad x \qquad x^{-1} \qquad (1-x)^{-1}$

$$xx^{-1} \longrightarrow 1 \qquad x^{-1}x \longrightarrow 1$$

$$(1-x)^{-1}x \longrightarrow (1-x)^{-1} - 1$$

$$x(1-x)^{-1} \longrightarrow (1-x)^{-1} - 1$$

$$x^{-1}(1-x)^{-1} \longrightarrow (1-x)^{-1} - x^{-1}$$

$$(1-x)^{-1}x^{-1} \longrightarrow (1-x)^{-1} - x^{-1}$$

B. Rules for general analytic functions h of xy

$$\text{(GENR)} \qquad = \begin{cases} x \quad x^{-1} \quad (1-x)^{-1} \quad \text{(RES)} \\ h(xy) \quad h(yx) \end{cases}$$

here h is analytic on $\sigma(XY)$, $\sigma(YX)$ yields

$$h(xy)x \longrightarrow xh(yx)$$

$$h(yx)x^{-1} \longrightarrow x^{-1}h(xy)$$

$$xh(yx)(1-x)^{-1} \longrightarrow -h(xy) + h(xy)(1-x)^{-1}$$

$$x^{-1}h(xy)(1-x)^{-1} \longrightarrow x^{-1}h(xy) + h(yx)(1-x)^{-1}$$

$$(1-x)^{-1}h(yx)(1-x)^{-1} - (1-x)^{-1}h(xy)(1-x)^{-1}$$

$$\longrightarrow (1-x)^{-1}h(xy) - h(yx)(1-x)^{-1}$$

Note these rules hold for any operators a, b replacing x y. The rules in A and B have broad application; enough so that teaching them in a general functional analysis class might be an interesting experiment.

C. Rules special to $h(s) = (1-s)^{-n}$

These apply to polynomials in

$$x \quad x^{-1} \quad (1-x)^{-1} \quad y \quad y^{-1} \quad (1-y)^{-1} \quad (1-xy)^{-1} \quad (1-yx)^{-1}$$

which is the same as RES + GENR with h specialized to $K(s) = (1-s)^{-n}$ and yields in addition to the rules in A and C.

$$(1-xy)^{-1}xy \longrightarrow (1-xy)^{-1} - 1$$

$$xy(1-xy)^{-1} \longrightarrow (1-xy)^{-1} - 1$$

$$(1-xy)^{-1}y^{-1} \longrightarrow (1-yx)^{-1}x + x^{-1}$$

$$(1-x)^{-1}y(1-xy)^{-1} \longrightarrow (1-x)^{-1}(1-xy)^{-1} + (1-xy)^{-1} - (1-x)^{-1}$$

together with an infinite class of relations which holds for functions $K(s) = (1-s)^{-n}$ for $n \geq 0$:

$$xK(yx)(1-y)^{-1} \longrightarrow$$
$$K(xy)(1-y)^{-1} + xK(yx) - (1-xy)K(xy)(1-y)^{-1}$$

Note that the following rules depend on alphabetical order.

$$(1-x)^{-1}K(xy)(1-y)^{-1} - (1-x)^{-1}K(xy)(1-y)^{-1}$$
$$\longrightarrow (1-x)^{-1}(1-xy)K(xy)(1-y)^{-1} + K(yx)\{(1-y)^{-1} - 1\} - (1-x)^{-1}K(yx)$$

$$(1-y)^{-1}K(yx)(1-x)^{-1} - (1-y)^{-1}K(xy)(1-x)^{-1} \longrightarrow$$
$$+ (1-y)^{-1}\Sigma(n-1)(1-x)^{-1} + \Sigma(n)(1-x)^{-1} + (1-y)^{-1}\Sigma(n) - \Sigma(n)$$

where $\Sigma(n) = (1-xy)^{-n} + (1-xy)^{-n+1} + \cdots + 1$ AND $n \geq 1$.

D. Rules special to $K(xy) = (1-xy)^{-n}(1-xy)^{1/2}$

Add $(1-xy)^{1/2}$ and $(1-xy)^{1/2}$ to the list of generators in C.

This yields the same rules as C with

$$K(s) = (1-s)^{-n}(1-s)^{1/2} \qquad \hat{\Sigma}(n) = \Sigma(n)(1-xy)^{1/2}$$

In other words adjoining the square root does nothing fundementally new.

REFERENCES

[Buch] B. Buchberger "Grobner bases: an algorithmic method in polynomial ideal theory" Recent Trends in multidimensional system theory, Reidel (1985) pp 184-232.

[Hiron] H. Hironaka, "Resolution of singularities of an algebraic variety over a field of characteristic zero" Ann of Math 79 (1964) pp 109-326.

[Mora] F. Mora, "Groebner Bases for Non-commutative Polynomial Rings" Lecture Notes in Computer Science, number 229 (1986) pp 353-362.

[NCA] J.W. Helton and R.L. Miller "NCAlgebra: A Mathematica Package for Doing Non Commuting Algebra" available from ncalg@ucsd.edu

[NF] B. Sz-Nagy and C. Foias, Harmonic Analysis of Operators on Hilbert Space North Holland 1970

Department of Mathematics
University of California, San Diego
La Jolla, California 92093-0112

AMS subject classification: 47,93,15-04,16-04

Operator Theory:
Advances and Applications, Vol. 73

Some Global Properties of Fractional-Linear Transformations

Victor Khatskevich[1)]

Dedicated to Professor M. Livsic

Preliminaries

We consider some properties of fractional-linear transformations (such as focusedness and compactness of the image) which may be applied to the study behavior of solution of evolution problems (see, for example, [1], [2], [3]). Consider a solution $x(t) = U(t)x_0$, $U(0) = I$, of an evolution problem in a Hilbert space $\mathfrak{H}$ with the scalar product $(\cdot,\cdot)$. The space $\mathfrak{H}$ is endowed with the corresponding indefinite metric $V(t)$, the operator $U(t)$ is a plus-operator (see below) with respect to $V(t)$: $(V(t)U(t)x, U(t)x) \geq 0$ for all $x \in \mathfrak{H}$ such that $(V(t)x, x) \geq 0$. The study of the behavior of $x(t)$ is naturally divided into the following three cases: 1) $U(t)$ is invertible, i.e. $(U(t))^{-1}$ exists, is defined on the whole of $\mathfrak{H}$ and is bounded (hyperbolic); 2) $U(t)$ is bounded (parabolic); 3) $U(t)$ is unbounded (elliptic).

Our considerations are connected with the first two cases: hyperbolic and parabolic.

Let $\mathfrak{H}$ be a complex or real Hilbert space with scalar product $\langle x,y\rangle$ and regular indefinite metric $[x,y] = \langle Vx,y\rangle$, where $x,y \in \mathfrak{H}$ and $V(= V^*)$ is a linear bounded invertible operator in $\mathfrak{H}$ with bounded inverse, i.e., 0 is a regular point of V.

Let us recall some notions of Krein space theory. We will use as a rule the terminology from the book [4].

We put $\mathfrak{P}_+ = \{x \in \mathfrak{H} : [x,x] \geq 0\}$ and $\mathfrak{P}_- = \{x \in \mathfrak{H} : [x,x] \leq 0\}$. Let $\mathfrak{M}_+$ ($\mathfrak{M}_-$) denote the set of all maximal non-negative (maximal non-positive) subspaces of $\mathfrak{H}$ and let $\mathfrak{M}_+^0$ ($\mathfrak{M}_-^0$) denote the set of all maximal uniformly positive (maximal uniformly negative) subspaces of $\mathfrak{H}$. It is perhaps well to recall that if $L \in \mathfrak{M}_+^0$ ($\mathfrak{M}_-^0$), then $[x,x] \geq c(L) \parallel x \parallel^2$, $c(L) > 0$ ($[x,x] \leq -d(L) \parallel x \parallel^2$, $d(L) > 0$) for all $x \in L$. If $\mathfrak{H}_+ \in \mathfrak{M}_+^0$, then $\mathfrak{H}_- = \mathfrak{H}_+^{[\perp]}$ (the orthogonal complement of $\mathfrak{H}_+$ with respect to the indefinite metric $[\cdot,\cdot]$) belongs to $\mathfrak{M}_-^0$ and

$$\mathfrak{H} = \mathfrak{H}_+[\dot{+}]\mathfrak{H}_-. \tag{1}$$

If $P_\pm$ are the projections corresponding to the decomposition (1) ($P_+ + P_- = I$), then with respect to the new scalar product $(x,y) = [Jx,y]$, $J = P_+ - P_-$, $x,y \in \mathfrak{H}$ we get

$$[x,y] = (Jx,y), \quad x,y \in \mathfrak{H}. \tag{2}$$

---

1) Supported in part by the Ministry of Absorption and the Rashi Foundation

The scalar products $(\cdot,\cdot)$ and $\langle\cdot,\cdot\rangle$ generate the equivalent norms $\| x \| = \sqrt{(x,x)}$ and $|x| = \sqrt{\langle x,x\rangle}$. The decomposition (1) with corresponding formula (2) is called canonical. With respect to this decomposition, any $L_+ \in \mathfrak{M}_+$ is of the form $L_+ = \{x_+ + K_+x_+ : x_+$ runs over the whole of $\mathfrak{H}_+\}$, where K_+ is the angular operator of the subspace L_+, $\| K_+ \| \leq 1$. The correspondence between L_+ and K_+ is one-to-one. If $L_+ \in \mathfrak{M}_+^0$, then $\| K_+ \| < 1$. Let $\mathfrak{K}$ denote the family of all angular operators of the subspaces $L_+ \in \mathfrak{M}_+$, and let $\mathfrak{K}_+^0$ be the interior of $\mathfrak{K}_+$. Evidently $\mathfrak{K}_+$ is the closed unit ball of the space $L(\mathfrak{H}_+, \mathfrak{H}_-)$, and there is a one-to-one correspondence between $\mathfrak{M}_-$ and $\mathfrak{K}_- = \mathfrak{K}_+^*(= \{K_+^* : K_+ \in \mathfrak{K}_+\})$.

Let V_1 and V_2 be a pair of bounded linear invertible operators with bounded inverses which induce indefinite metrics with the same signature on $\mathfrak{H}$: $\nu_+^1 = \nu_+^2 = \nu_+$, $\nu_-^1 = \nu_-^2 = \nu_-$ ($\nu_\pm = \dim \mathfrak{H}_\pm$).

A bounded linear operator U defined on the whole of $\mathfrak{H}$ is called a *plus-operator* if $U\mathfrak{P}_+^1 \subseteq \mathfrak{P}_+^2$. It is known that for a plus-operator U the following inequality is true:

$$[Ux, Ux] \geq \mu(U)[x,x], \quad x \in \mathfrak{H}, \tag{3}$$

where $\mu(U) \geq 0$. If in (3) $\mu(U) > 0$, then the plus-operator U is called *strict*. If together with U the adjoint operator U^* is a strict plus-operator too, then U is called a *bistrict* plus-operator. A plus-operator U is called *focused* if

$$[Ux, Ux] \geq c \| x \|^2, \quad c = c(U) > 0, \quad x \in \mathfrak{P}_+^1.$$

There are the similar definitions for a minus-operator.

Let us recall some basic facts from the theory of plus-operators [5], [6] (for minus-operators the facts are similar).

(I) If U is a non-strict plus-operator ($\mu(U) = 0$), then $\mathfrak{R}(U) \subseteq \mathfrak{P}_+^2$.

(II) A strict plus-operator U is bistrict iff its adjoint U^* is a plus-operator too.

(III) A strict plus-operator U is bisttrict iff $UL_+ \in \mathfrak{M}_+^2$ for some $L_+ \in \mathfrak{M}_+^1$.

(IV) A bistrict plus-operator U generates a fractional-linear transformation F_U : $\mathfrak{K}_+^1 \to \mathfrak{K}_+^2$ via the formula:

$$F_U(K_+) = (U_{21} + U_{22}K_+)(U_{11} + U_{12}K_+)^{-1},$$

where $U_{11} = P_+^2UP_+^1$, $U_{12} = P_+^2UP_-^1$, $U_{21} = P_-^2UP_+^1$, $U_{22} = P_-^2UP_-^1$.

1 The Case of Invertible Plus-Operators

In this section we consider an invertible operator U: U^{-1} exists and is defined on the whole of $\mathfrak{H}$ (consequently, U^{-1} is bounded).

1.1 The Case of a Space $\Pi_\kappa : \min\{\nu_+, \nu_-\} = \kappa < \infty$

Proposition 1. *An invertible plus-operator U is a bistrict plus-operator and U^{-1} is a bistrict minus-operator.*

Proof. By virtue of (1) and the equality $\mathfrak{R}(U) = \mathfrak{H}$ the plus-operator U is strict. As a consequence of $U\mathfrak{P}^1_+ \subseteq \mathfrak{P}^2_+$ we obtain $U^{-1}(\mathfrak{H} \setminus \mathfrak{P}^2_+) \subseteq (\mathfrak{H} \setminus \mathfrak{P}^1_+)$. Evidently $\mathfrak{H}\setminus\mathfrak{P}^i_+ = \mathfrak{P}^i_-\setminus\mathfrak{P}_0, i = 1, 2$. Hence by virtue of the continuity of U^{-1} we have $U^{-1}\mathfrak{P}^2_- \subseteq \mathfrak{P}^1_-$, i.e., U^{-1} is a minus-operator. As above from $\mathfrak{R}(U^{-1}) = \mathfrak{H}$ it follows that U^{-1} is strict. As the space $\mathfrak{H}$ is Π_κ, by virtue of (III) at least one of the operators U, U^{-1} is bistrict. Let, for example, U be a bistrict plus-operator. Then U^* is a plus-operator. As above hence we obtain that $U^{*-1}(= U^{-1*})$ is a minus-operator and by virtue of (II) U^{-1} is a bistrict minus-operator. The proof for U^{-1} is similar. □

Theorem 1. *If U is an invertible plus-operator, then the fractional-linear transformations $F_U : \mathfrak{K}^1_+ \to \mathfrak{K}^2_+$ and $F_{U^{-1}} : \mathfrak{K}^2_- \to \mathfrak{K}^1_-$ are defined and the sets $F_U(\mathfrak{K}^1_+)$, $F_{U^{-1}}(\mathfrak{K}^2_-)$ are compact in the weak operator topology (w.o.t.).*

Proof. By Proposition 1 U is a bistrict plus-operator, U^{-1} is a bistrict minus-operator. By virtue of (IV) the fractional-linear transformations F_U, $F_{U^{-1}}$ are defined and, as at least one of the numbers ν_+, ν_- is finite, they are continuous in the w.o.t. [7]. Hence by virtue of the compactness of the balls $\mathfrak{K}^1_+$, $\mathfrak{K}^2_-$ in this topology we get the conclusion of the theorem. □

1.2 The Case of an Arbitrary Signature (ν_+, ν_-)

In the case that both of the numbers ν_+ and ν_- are infinite, an invertible strict plus-operator U may not be bistrict, so the transformations F_U, $F_{U^{-1}}$ may not be defined.

Example 1. Let $\mathfrak{H}_- = CLin\{e_{-n} : n \in \mathbb{N}\}$, $\mathfrak{H}_+ = CLin\{e_n : n \in (\{0\} \cup \mathbb{N})\}$, $\mathfrak{H} = \mathfrak{H}_+ \oplus \mathfrak{H}_-$, $(e_i, e_j) = \delta_{ij}$, $i, j = 0, \pm1, \pm2, \ldots$, $[x, y] = (x_+, y_+) - (x_-, y_-)$, where $x_\pm, y_\pm \in \mathfrak{H}_\pm$. We put $Ue_i = e_{i+1}$, $i = 0, \pm1, \pm2, \ldots$. Then U is a unitary operator in $\mathfrak{H}$, U is a plus-operator, but $U^*(= U^{-1}) : e_i \to e_{i-1}$ is not a plus-operator (it is a minus-operator). As neither U, nor U^{-1} is bistrict, the transformations F_U, $F_{U^{-1}}$ are not defined.

Suppose in addition a plus-operator U be bistrict (see (II)–(III)).

Theorem 2. *Let U be an invertible bistrict plus-operator. Then both of the transformations F_U, $F_{U^{-1}}$ are defined. If the operator U_{12} is compact, then $F_U(\mathfrak{K}^1_+)$ is compact in the w.o.t.; if the operator U_{21} is compact, then $F_{U^{-1}}(\mathfrak{K}^2_-)$ is compact in the w.o.t.*

Proof. As above (see Proposition 1) we prove that U^{-1} is a bistrict minus-operator. Therefore both of the transformations F_U, $F_{U^{-1}}$ are defined. If U_{12} is compact, then [7] F_U is continuous in the w.o.t., so (see Theorem 1) $F_U(\mathfrak{K}^1_+)$ is compact in the w.o.t. The other proof is similar. □

Let us recall (see, for example, [4], [8]) that for a focused plus-operator A the following inequality holds:
$\| P_-Ax \| \leq (1-\gamma) \| P_+x \|$ for some $\gamma = \gamma(A) : 0 < \gamma \leq 1$ and for all $x \in \mathfrak{P}^1_+$.

We shall need

Proposition 2. *Let U be an invertible bistrict plus-operator. Then if at least one of the operators U, U^* and U^{-1} is focused, then each of them is focused.*

Proof. Consider a (V_2, V_1)-unitary operator $T : (\mathfrak{H}, V_2) \to (\mathfrak{H}, V_1)$. Then TU is a focused operator in $(\mathfrak{H}, V_1)$, hence [9] U^*T^* is focused, as is U^*. Now we show that U^{-1} is focused too. It is enough to prove that $F_{U^{-1}}(\mathfrak{K}^2_-) \subseteq (1-\gamma)\mathfrak{K}^1_-$ for some $\gamma = \gamma(U) : 0 < \gamma \leq 1$. Otherwise there exists a sequence $\{M^n_-\}_{n=1}^{\infty}$: $M^n_- \in \mathfrak{M}^2_-$, $L^n_- = U^{-1}M^n_- = (P^1_- + K^n_-)\mathfrak{H}_-$, $\| K^n_- \| \to 1$, $n \to \infty$. We take a sequence $\{x^n_-\}_{n=1}^{\infty}$, such that $\| x^n_- \| = 1$, $\| K^n_- x^n_- \| \to 1$, $n \to \infty$, and put $x_n = x^n_- + K^n_- x^n_- \in L^n_-$, $y_n = K^n_- x^n_- + K^{n*}_- K^n_- x^n_-$. We have $\| x_n - y_n \|^2 = \| x^n_- - K^{n*}_- K^n_- x^n_- \|^2 = (x^n_- - K^{n*}_- K^n_- x^n_-, x^n_- - K^{n*}_- K^n_- x^n_-) = 1 - 2 \| K^n_- x^n_- \|^2 + \| K^{n*}_- K^n_- x^n_- \|^2 \leq 1 - \| K^n_- x^n_- \|^2 \to 0$, $n \to \infty$. But as U is focused, there exists a constant $c_U > 0$ such that the following inequalities hold: $[Uy_n, Uy_n] \geq c_u \| y_n \|^2 \geq c_U \| K^n_- x^n_- \|$. Hence for $n \geq n_0$ we obtain $[Uy_n, Uy_n] \geq d > 0$. On the other hand, as $x_n = U^{-1}z_n$ for some $z_n \in M^n_- \subset \mathfrak{P}^2_-$ we have: $[Uy_n, Uy_n] = [z_n, z_n] + 2\,\mathrm{Re}\,[z_n, U(y_n - x_n)] + [U(y_n - x_n), U(y_n - x_n)]$. Therefore $[Uy_n, Uy_n] < d$ for $n \geq n_1$. Taking $n \geq \max\{n_0, n_1\}$ we get a contradiction. □

Corollary. *Let U be an invertible bistrict focused plus-operator. Then there exist canonical decompositions of $(\mathfrak{H}, V_1)$ and $(\mathfrak{H}, V_2)$ such that the transformations F_U, $F_{U^{-1}}$ are defined and their images $F_U(\mathfrak{K}^1_+)$, $F_{U^{-1}}(\mathfrak{K}^2_-)$ are compact in the w.o.t.*

Proof. Let T be a (V_2, V_1)-unitary operator as above. Then the operator TU is focused in $(\mathfrak{H}, V_1)$. By virtue of [10], there exists a pair of subspaces $L_\pm \in \mathfrak{M}^{02}_\pm$ such that $TUL_\pm = L_\pm$. In the space $(\mathfrak{H}, V_1)$ we take a new canonical decomposition $\mathfrak{H} = \hat{\mathfrak{H}}^1_+ \oplus \hat{\mathfrak{H}}^1_-$ with $\hat{\mathfrak{H}}^1_- = L^1_-$ and in the space $(\mathfrak{H}, V_2)$ we take a canonical decomposition $\mathfrak{H} = \hat{\mathfrak{H}}^2_+ \oplus \hat{\mathfrak{H}}^2_-$ with $\hat{\mathfrak{H}}^2_- = TL_-$. With respect to these canonical decompositions the matrix of the operator U is of the form

$$\begin{bmatrix} U_{11} & 0 \\ U_{21} & U_{22} \end{bmatrix}.$$

Therefore the operator U_{12} is compact (in fact it is zero) and so, by Theorem 2, $F_U(\mathfrak{K}^1_+)$ is compact in the w.o.t. Taking $\hat{\mathfrak{H}}^1_+ = L_+$, $\hat{\mathfrak{H}}^2_+ = TL_+$, we get $U_{21} = 0$, so $F_{U^{-1}}(\mathfrak{K}^1_-)$ is compact in the w.o.t. □

2 The General Case of a Non-Invertible Operator U

In this case the transformations F_U, $F_{U^{-1}}$ may not be defined for injective strict plus-operators U in the space Π_κ. Therefore we must generalize our considerations.

Define two sets $\mathfrak{C}_U^- = \{K_- \in \mathfrak{K}_-^1 : U(P_-^1 + K_-)\mathfrak{H}_-^1 \subset \mathfrak{P}_-^2\}$, $\mathfrak{C}_U^+ = \{K_+ \in \mathfrak{K}_+^2 : (P_+^2 + K_+)\mathfrak{H}_+^2 \subset U\mathfrak{P}_+^1\}$. In the case of an invertible bistrict plus-operator U we have $\mathfrak{C}_U^- = F_{U^{-1}}(\mathfrak{K}_-^2)$, $\mathfrak{C}_U^+ = F_U(\mathfrak{K}_+^1)$. In the general case we shall suppose that the operator U satisfies suitable conditions for the sets $\mathfrak{C}_U^\pm$ be non-empty.

2.1 The case of Π_κ

Let us recall that for a Pontryagin space Π_κ with signature (ν_+, ν_-), $\min\{\nu_+, \nu_-\} < \infty$.

Theorem 3. *If $\nu_+ < \infty$ and a plus-operator U does not annihilate vectors in $\mathfrak{P}_+^1 \setminus \mathfrak{P}_0^1$ or if $\nu_- < \infty$ and U is a bistrict plus-operator, then the sets $\mathfrak{C}_U^\pm$ are non-empty and compact in the w.o.t.*

Proof. Let $\nu_+ < \infty$. Then since $Ux \neq 0$ for $x \in \mathfrak{P}_+^1 \setminus \mathfrak{P}_0^1$, the set $\mathfrak{C}_U^+$ is non-empty and [11] the operator U^* is a plus-operator. Let T be a (V_2, V_1)-unitary operator as above. Then TU is a plus-operator in $(\mathfrak{H}, V_1)$ and U^*T^* is a plus-operator in $(\mathfrak{H}, V_2)$. In view of the results of [7], the operators TU, U^*T^* have invariant subspaces $L_+ \in \mathfrak{M}_+^1$, $L_+^* \in \mathfrak{M}_+^2$, respectively. Putting $L_- = L_+^{*\perp}$ we get $TUL \subseteq L_-$ and $UL_- \subseteq T^{-1}L_- \subset \mathfrak{P}_-^2$. Therefore $\mathfrak{C}_U^- \neq \emptyset$.

Let us show that $\mathfrak{C}_u^-$ is closed with respect to the w.o.t. Let $\{K_-^\alpha\}$ be a directedness in $\mathfrak{C}_U^-$ which converges in the w.o.t. to an operator $K_- \in \mathfrak{K}_-^1$. We have $U(P_-^1 + K_-^\alpha)\mathfrak{H}_-^1 \subseteq \mathfrak{P}_-^2$, i.e., $\| (U_{11}K_-^\alpha + U_{12})x_- \| \leq \| (U_{21}K_-^\alpha + U_{22})x_- \|$ for all $x_- \in \mathfrak{H}_-^1$. Passing to limit we obtain $K_- \in \mathfrak{C}_U^-$. Therefore $\mathfrak{C}_U^-$ is compact in the w.o.t. (since it is a closed subset of a compact set).

Now we show that $\mathfrak{C}_U^+$ is closed in the w.o.t. Let $K_+^\alpha \in \mathfrak{C}_U^+$, $K_+^\alpha \to K_+$, $K_+ \in \mathfrak{K}_+^2$. We have $(P_+^2 + K_+^\alpha)\mathfrak{H}_+^2 = (U_{11} + U_{12}Q_+^\alpha)\mathfrak{H}_+^1$ for some $Q_+^\alpha \in \mathfrak{K}_+^1$. Without loss of generality consider $Q_+^\alpha \to Q_+ \in \mathfrak{K}_+^1$ (recall that $\dim \mathfrak{H}_+^1 = \dim \mathfrak{H}_+^2 = \nu_+ < \infty$). Therefore $\| (U_{11} + U_{12}Q_+^\alpha)x_+ \| \geq \| (U_{21} + U_{22}Q_+^\alpha)x_+ \|$, $x_+ \in \mathfrak{H}_+^1$. Thus

$$\begin{aligned} \| (U_{21} + U_{22}Q_+)x_+ \| &\leq \overline{\lim} \| (U_{21} + U_{22}Q_+^\alpha)x_+ \| \\ &\leq \overline{\lim} \| (U_{11} + U_{12}Q_+^\alpha)x_+ \| = \| (U_{11} + U_{12}Q_+)x_+ \| . \end{aligned}$$

Hence $(P_+^2 + K_+)\mathfrak{H}_+^2 = (U_{11} + U_{12}Q_+)\mathfrak{H}_+^1$, so $K_+ \in \mathfrak{C}_U^+$. Just as above we conclude that $\mathfrak{C}_U^+$ is compact in the w.o.t.

Now let $\nu_- < \infty$, and suppose that U is a bistrict plus-operator. Then $\mathfrak{C}_U^+ = F_U(\mathfrak{K}_+^1)$ is compact in the w.o.t. (Theorem 1). As above the operator U^*T^* has an invariant subspace $L_+^* \in \mathfrak{M}_+^2$. Therefore $\mathfrak{C}_U^-$ is non-empty, and just as above may be shown to be closed in the w.o.t. Therefore it is compact. □

2.2 The General Case of a Krein space

Proposition 3. *If U is a bistrict plus-operator, then $\mathfrak{C}_U^+ \neq \emptyset$.*

Proof. By the condition of the proposition the transformation F_U is defined, and $G_U^+ = F_U(\mathfrak{K}_+^1) \neq \emptyset$. By Theorem 3 [12], the blockmatrix TU (where T is a (V_2, V_1)-unitary operator as above) is of the form

$$TU = \begin{bmatrix} \hat{U}_{11} & \hat{U}_{11}\hat{K}_-^0 \\ \hat{K}_+^0\hat{U}_{11} & \hat{U}_{22} \end{bmatrix},$$

where $K_\pm^0 \in \mathfrak{K}_\pm^{01}$. Putting $\hat{L} = (P_-^1 - \hat{K}_-^0)\mathfrak{H}_-^1 \in \mathfrak{M}_-^1$, we have $TU\hat{L}_- \subseteq \mathfrak{H}_-^1 \subset \mathfrak{P}_-^1$. Therefore, $UL_- = T^{-1}TU\hat{L}_- \in \mathfrak{P}_-^2$, i.e., $\mathfrak{C}_U^- \neq \emptyset$. □

Theorem 4. *Let U be a bistrict plus-operator. Then $\mathfrak{C}_U^\pm \neq \emptyset$; if U_{12} is compact, then $\mathfrak{C}_U^+$ is compact in the w.o.t.; if U_{21} is compact, then $\mathfrak{C}_U^-$ is compact in the w.o.t.*

Proof. By Proposition 3, the sets $\mathfrak{C}_U^\pm$ are non-empty. If U_{12} is compact, then we get the compactness of $\mathfrak{C}_U^+$ as in Theorem 2. Let U_{21} be compact. Then, just as in Theorem 3 we can take the limits in both sides of the following inequality

$$\| (U_{11}K_- + U_{12})x_- \| \leq \| (U_{21}K_- + U_{22})x_- \|, \quad x_- \in \mathfrak{H}_-^1,$$

which is equivalent to $U(P_-^1 + K_-)\mathfrak{H}_-^1 \subset \mathfrak{P}_-^2$. Hence we obtain the closedness of $\mathfrak{C}_U^-$ in the w.o.t. and so too its compactness. □

Corollary 2. *Let U be a focused bistrict plus-operartor. Then the sets $\mathfrak{C}_U^\pm$ are non-empty and compact in the w.o.t. with respect to some canonical decomposition of the spaces $(\mathfrak{H}, V_1)$ and $(\mathfrak{H}, V_2)$.*

Proof. This may be proved just as in Corollary 1. □

Finally we touch upon the case of a possibly non-regular metric based on a V with $0 \in \sigma_c(V) \cup \rho(V)$. In this case we may obtain possibly non-complete subspaces $\mathfrak{H}_\pm$ in the decomposition (1).

Theorem 5. *Let $\dim \mathfrak{H}_+ < \infty$. If a plus-operator U does not annihilate vectors in $\mathfrak{P}_+^1 \setminus \mathfrak{P}_+^0$, then the sets $\mathfrak{C}_U^\pm$ are non-empty and $\mathfrak{C}_U^-$ is compact in the w.o.t.*

Proof. We shall show that U is bounded with respect to the norm $\| \quad \|$ (note that we do not need U to be bounded with respect to the original norm of $\mathfrak{H}$). Evidently U_{11} is bounded. Because of the condition $U(\mathfrak{P}_+^1 \setminus \mathfrak{P}_0^1) \subset \mathfrak{P}_+^2 \setminus \{0\}$, the operator U_{12} is bounded too, as is the operator $P_+U = U_{11} + U_{12}$. Therefore the boundedness of U is a consequence of Lemma 2.1 of [13]. As in Theorem 3 we get $\mathfrak{C}_U^\pm \neq \emptyset$ (keeping in mind that U^* maps the completion of $(\mathfrak{H}, V_2)$ into the completion of $(\mathfrak{H}, V_1)$ and that $L_- = L_+^{*\perp} \in \mathfrak{M}_-^2$ for all maximal non-negative subspaces L_+^* of the completion of $(\mathfrak{H}, V_2)$).

Now we get the conclusion of the theorem with the help of considerations which are similar to corresponding considerations in the proof of Theorem 3. □

References

1. Daletskii, Yu.L., Krein, M.G.: Stability of solutions of Differential equations in Banach Spaces. Transl. of Math. Monogr., vol. 43, A.M.S., Providence, R. I., 1974.
2. Massera, J.L., Shaffer, J.J.: Linear differential equations and function spaces. Acad. Press, New-York, London, 1966.
3. Zelenko, L.: The manifolds of bounded solutions of nonlinear ordinary differential systems, Journ. of Diff. Equat., to appear.
4. Azizov, T.Ya., Iohvidov, I.S.: Linear operators in spaces with an indefinite metric. John Wiley & Sons, Chichester, 1989.
5. Krein, M.G., Shmulyan, Yu.L.: On plus-operators in spaces with indefinite metric, Matem. Issled., Kishinev, 1 (1966), 131–161.
6. Krein, M.G., Shmulyan, Yu.L.: On fractional-linear transformations with operator coefficients, Matem. Issled., Kishinev, 2 (1967), 64–96.
7. Krein, M.G.: On one new application of fixed-point principle in the operator theory in spaces with indefinite metric, Doklady Acad. Nauk S.S.S.R., 154 (1964), 1023–1026.
8. Khatskevich, V.: Invariant subspaces and spectral properties of plus-operators with quasifocused powers, Functional Anal. Appl. vol. 18, N 1 (1984), 86–87.
9. Khatskevich, V.: On characteristic spectral properties of focused operators, Doklady Acad. Nauk. Arm. S.S.R., 79 (1984), 102–105.
10 Sobolev, A.V., Khatskevich, V.: On definite invariant subspaces and spectral structure of focused plus-operators, Functional Anal. Appl., vol. 15, N 1 (1981), 84–85.
11. Khatskevich, V.: On the symmetry of properties of the plus-operator and its conjugate operator, Funct. Analysis, Ulyanovsk, vol. 14 (1980), 177–186.
12. Khatskevich, V., Senderov, V.: Powers of plus-operators, Integral Equations Operator Theory, vol. 15 (1992), 784–795.
13. Khatskevich, V., Senderov, V., On normed J_ν-spaces and some classes of linear operators in such spaces, Matem. Issled., Kishinev, 8 (1973), 56–75.

Department of Mathematics and Computer Sciences
University of Haifa
Afula Research Institute
Mount Carmel, Haifa 31905
Israel

MSC: 47B50 47A53

Operator Theory:
Advances and Applications, Vol. 73

BOUNDARY VALUES OF BEREZIN SYMBOLS

To M.S. Livsic with great respect

Eric Nordgren and Peter Rosenthal

1. Introduction

A *functional Hilbert space* is a collection $\mathcal{H}$ of complex-valued functions on some set S such that $\mathcal{H}$ is a Hilbert space with respect to the usual vector operations on functions and which has the property that point evaluations are continuous (i.e., for each $z \in S$, the map $f \to f(z)$ is a continuous linear functional on $\mathcal{H}$).

A prototypical functional Hilbert space is the Hardy space $\mathcal{H}^2$, the space of all functions analytic on the open unit disk having Taylor coefficients that are square summable.

If $\mathcal{H}$ is a functional Hilbert space, the Riesz representation theorem ensures that for each $z \in S$ there is a unique element k_z of $\mathcal{H}$ such that $f(z) = (f, k_z)$ for all $f \in \mathcal{H}$. The collection $\{k_z : z \in S\}$ is called the *reproducing kernel* of $\mathcal{H}$. It is very easy to verify that the reproducing kernel of $\mathcal{H}^2$ is given by $k_z(w) = \frac{1}{1-\bar{z}w}$.

For A any bounded linear operator on a functional Hilbert space, Berezin [4,5] introduced the following numerical function associated with the operator. For $z \in S$, let $\hat{k}_z = \frac{1}{\|k_z\|}k_z$ be the normalized reproducing kernel of $\mathcal{H}$. For A a bounded linear operator on $\mathcal{H}$, the function $\tilde{A}$ defined on S by $\tilde{A}(z) = (A\hat{k}_z, \hat{k}_z)$ for $z \in S$ is the *Berezin symbol* of A. It is not hard to see that, on the most familiar functional Hilbert spaces, the Berezin symbol uniquely determines the operator (i.e., $\tilde{A}(z) = \tilde{B}(z)$ for all z implies $A = B$).

We say that a functional Hilbert space is *standard* if the underlying set S is a subset of a topological space and the boundary ∂S is non-empty and has the property that $\{\hat{k}_{z_n}\}$ converges weakly to 0 whenever $\{z_n\}$ is a sequence in S that converges to a point in ∂S. The common functional Hilbert spaces of analytic functions, including $\mathcal{H}^2$ and the Bergman space (and their analogues on the open unit ball of $\mathbf{C}^n$), are standard in this sense. (For $\mathcal{H}^2$, note that $(f, \hat{k}_z) = \sqrt{1-|z|^2}f(z)$ for $f \in \mathcal{H}^2$, and this obviously approaches 0 for $f \in \mathcal{H}^\infty$, and hence for all $f \in \mathcal{H}^2$, whenever $|z| \to 1^-$.)

For $\mathcal{H}$ a standard functional Hilbert space and A a compact operator on $\mathcal{H}$,

$\lim_{n\to\infty} \tilde{A}(z_n) = 0$ whenever $\{z_n\}$ converges to a point of ∂S, (since compact operators send weakly convergent sequences into strongly convergent ones). In this sense, the Berezin symbol of a compact operator on a standard functional Hilbert space vanishes on the boundary.

Berezin symbols have been extensively studied in the cases of Toeplitz [3,10] and Hankel [1,12] operators, mainly on Bergman spaces of functions on the unit ball of $\mathbf{C}^n$.

This note was stimulated by a question of C.A. Berger and L.A. Coburn: on the Hardy and Bergman spaces, must a bounded linear operator be compact if its Berezin symbol vanishes on the boundary?

We present several counterexamples to this question (section 2 below). However, we show that a stronger hypothesis yields an affirmative result. Namely, if the Berezin symbols of all unitary equivalents of an operator on a standard functional Hilbert space vanish on the boundary, then the operator is compact (Corollary 2.8). More generally, we show that the essential numerical range of an operator on a standard functional Hilbert space can be characterized as the set of all cluster values of all Berezin symbols of operators unitarily equivalent to the given operator (Theorem 2.7). The proof is elementary.

Similar considerations provide a proof that all Berezin symbols of an operator on a standard functional Hilbert space have continuous extensions to $S \cup \partial S$ if and only if the operator is a translate of a compact operator (Theorem 3.1).

2. Compactness criterion

There are several examples showing that the vanishing of the Berezin symbol on the boundary does not imply compactness.

EXAMPLE 2.1. *Let A be the diagonal operator on $\mathcal{H}^2$ with $\{(-1)^n\}$ on the main diagonal; i.e., $Ae_n = (-1)^n e_n$ for each n, where $e_n(z) = z^n$. Then A is not compact, but $\tilde{A}(z)$ vanishes on the boundary.*

PROOF. It is easily verified that

$$\begin{aligned}\tilde{A}(z) &= (1-|z|^2)\sum_{n=0}^{\infty}(-1)^n|z|^{2n} \\ &= \frac{1-|z|^2}{1+|z|^2}.\end{aligned}$$

It follows that $\lim_{z\to 1^-} \tilde{A}(z) = 0$. □

Before giving the next example, some preparation is necessary. For ϕ an analytic function mapping $\mathbf{D}$ into itself, the composition operator C_ϕ is defined by $(C_\phi f)(z) = f(\phi(z))$ for $f \in H^2$. Each C_ϕ is bounded — see [11] and [7] for surveys of the many known results about composition operators.

It is easy to compute the Berezin symbol of a composition operator. For any ϕ,

$$\begin{aligned} \langle C_\phi \hat{k}_z, \hat{k}_z \rangle &= \langle \frac{(1-|z|^2)^{1/2}}{1-\overline{z}\phi}, (1-|z|^2)^{1/2} k_z \rangle \\ &= \frac{1-|z|^2}{1-\overline{z}\phi(z)}. \end{aligned}$$

If ϕ has the form $\phi(z) = z\psi(z)$, where ψ is an analytic function mapping the disc into itself, then the Berezin symbol of C_ϕ has a particularly tractable expression:

$$\tilde{C}_\phi(z) = \frac{1-|z|^2}{1-|z|^2\psi(z)}.$$

Note that Example 2.1 above is the case where $\phi(z) = -z$.

EXAMPLE 2.2. *Let $\mathcal{R}$ be any open, connected, simply connected subset of the disk whose boundary has the following properties:*
(i) it is a simple closed Jordan curve;
(ii) its intersection with $\{z : |z| = 1\}$ contains a nontrivial arc;
(iii) it does not contain 1.
(For example, $\mathcal{R}$ could be $\{z : |z| < 1$ and $\operatorname{Re} z < 0\}$.) By the Riemann mapping theorem, there are conformal maps of the disk onto $\mathcal{R}$. Thus the following theorem gives a class of examples.

THEOREM 2.3. *If ψ is any conformal map of the unit disk onto a region $\mathcal{R}$ with the above properties and ϕ is defined by $\phi(z) = z\psi(z)$, then*

$$\lim_{|z|\to 1^-} \tilde{C}_\phi(z) = 0$$

but C_ϕ is not compact.

PROOF. Since the boundary of $\mathcal{R}$ does not contain 1, there is an $\varepsilon > 0$ such that $|1-\psi(z)| \geq \varepsilon$ for $|z| < 1$. Thus for $z \in \mathbf{D}$, $|1-|z|^2\psi(z)| \geq \varepsilon$, and it follows from the above formula for C_ϕ that $\lim_{|z|\to 1^-} \tilde{C}_\phi = 0$.

To see that such a C_ϕ is not compact, first note that (by Carathéodory's well known theorem) ψ extends to a continuous map of the closed disk onto the closure of $\mathcal{R}$. Now $\{e_n\}$ converges weakly to 0 in H^2, but

$$\begin{aligned} \|C_\phi e_n\|^2 &= \|\phi^n\|^2 \\ &= \int_{|z|=1} |\phi|^{2n} dm \\ &= \int_{|z|=1} |\psi|^{2n} dm \\ &\geq m\{z : |\psi(z)| = 1\}, \end{aligned}$$

where m is normalized Lebesgue measure on $\partial\mathbf{D}$.

Thus, by property (ii) above, $\{C_\phi e_n\}$ does not converge to 0, and C_ϕ is not compact. □

It should be mentioned that Shapiro [15] has found necessary and sufficient conditions that a composition operator be compact. The observation that compactness of C_ϕ implies that the boundary values of ϕ have modulus less than 1 a. e. goes back to Schwartz [13].

After the above examples were informally circulated, Sheldon Axler considered whether the result might hold for positive operators. He produced the following counter-example, which is included here with his kind permission.

EXAMPLE 2.4. *([2]) Let D be the diagonal operator on $\mathcal{H}^2$ defined by*

$$Dz^n = \begin{cases} z^n & \text{if } n = 2^k \text{ for } k = 0,1,2,\ldots \\ 0 & \text{otherwise} \end{cases}$$

Then D is a non-compact self-adjoint projection but $\lim_{|\lambda|\to 1^-} \tilde{D}(\lambda) = 0$.

PROOF. Axler [2] shows that the vanishing on the boundary follows from known results on functions belonging to the little Bloch space.

Here we present a direct proof. Fix any integer $N > 1$ and let $r = 2^N$. Then, for $|\lambda| < 1$,

$$\begin{aligned} |\tilde{D}(\lambda)| &= (1 - |\lambda^2|)\sum_{k=0}^{\infty} |\lambda^2|^{2^k} \\ &\leq (1 - |\lambda^2|)\sum_{k=0}^{N} |\lambda^2|^{2^k} + (1 - |\lambda^2|)\sum_{k=0}^{\infty} |\lambda^2|^{rk} \end{aligned}$$

Now $(1 - |\lambda^2|)\sum_{k=0}^{N} |\lambda^2|^{rk} = \frac{1-|\lambda^2|}{1-|\lambda^2|^r}$, and

$$\lim_{|\lambda|\to 1^-} \frac{1 - |\lambda^2|}{1 - |\lambda^2|^r} = \frac{1}{r} \qquad \text{(by L'Hôpital's Rule).}$$

Therefore $\limsup_{|\lambda|\to 1^-} |\tilde{D}(\lambda)| \leq \frac{1}{r}$, and, since this holds for all $r = 2^N$, it follows that $\lim_{|\lambda|\to 1^-} \tilde{D}(\lambda) = 0$. □

We need the following lemma to obtain the compactness criterion. As A.S. Markus has kindly pointed out, it goes back to Dixmier [8].

LEMMA 2.5. *Suppose $\{f_n\}$ is a weak null sequence of unit vectors and $\{\delta_n\}$ is a sequence of positive numbers. Then there exists a subsequence $\{f_{n_m}\}$ and an orthonormal sequence $\{h_m\}$ such that $\|f_{n_m} - h_m\| < \delta_m$ for every m.*

PROOF. Choose $n_1 = 1$ and $h_1 = f_1$. Proceeding inductively, suppose we have chosen $n_1, n_2, \ldots, n_m$ and $h_1, h_2, \ldots, h_m$, and put

$$g_{m,k} = \frac{f_k - \sum_{j=1}^{m} \langle f_k, h_j\rangle h_j}{\|f_k - \sum_{j=1}^{m} \langle f_k, h_j\rangle h_j\|}$$

for $k > n_m$. Because $\langle f_k, h\rangle \to 0$ for every vector h, it follows easily that $\|f_k - g_{m,k}\| \to 0$ as $k \to \infty$, and thus we may put $h_{m+1} = g_{m,k}$ and $n_{m+1} = k$ for sufficiently large k. □

We also record the following lemma for easy reference. Its simple proof is omitted.

LEMMA 2.6. *If A is a bounded operator on a Hilbert space $\mathcal{H}$, then the map $f \mapsto \langle Af, f\rangle$ is uniformly continuous on the unit sphere of $\mathcal{H}$.*

The essential numerical range of an operator was introduced in [14]. In [9] it was shown that, for every operator A, the essential numerical range of A consists of all complex numbers λ such that for some weak null sequence of unit vectors $\{f_n\}$, $\lim_{n\to\infty} \langle Af_n, f_n\rangle = \lambda$; or, equivalently, there exists an orthonormal sequence $\{g_n\}$ such that $\lim_{n\to\infty} \langle Ag_n, g_n\rangle = \lambda$. A corollary to a different characterization is that an operator is compact if and only if its essential numerical range consists of zero alone.

For a bounded function f on S and a point z_0 on the boundary of S, a complex number λ is a *cluster value of f at z_0* provided there exists a sequence $\{z_n\}$ in S converging to z_0 such that $\{f(z_n)\}$ converges to λ. For any operator A, and let $\mathcal{B}(A)$ be the set of Berezin symbols of all operators unitarily unitarily equivalent to A.

THEOREM 2.7. *The essential numerical range of an operator A on a standard functional Hilbert space coincides with the set of all cluster values of members of $\mathcal{B}(A)$. In fact, the set of cluster values of the members of $\mathcal{B}(A)$ at any point of the boundary of S includes the essential numerical range.*

Let $\{z_n\}$ be a sequence that converges to some point in ∂S and is such that $\{\tilde{A}(z_n)\}$ converges to λ. Since $\{\hat{k}_{z_n}\}$ converges weakly to 0, it follows from the above mentioned result of Fillmore, Stampfli and Williams [9] that λ is in the essential numerical range of A.

Conversely, suppose λ is in the essential numerical range of A. Then there exists an orthonormal sequence $\{g_n\}$ such that $\lim_{n\to\infty} \langle Ag_n, g_n\rangle = \lambda$. By taking a subsequence if necessary, we may suppose that $\{g_n\}$ spans a subspace with an infinite dimensional orthocomplement.

Let $\{z_n\}$ be a sequence in S that converges to a point in ∂S, and put $f_n = \hat{k}_{z_n}$. Apply Lemma 2.5 to produce an orthonormal sequence $\{h_m\}$ and a subsequence $\{f_{n_m}\}$ such that $\lim_{m\to\infty} \|h_m - f_{n_m}\| = 0$. On relabeling if necessary, we may suppose

$$\lim_{n\to\infty} \|h_n - \hat{k}_{z_n}\| = 0. \tag{1}$$

Also, we may suppose that the orthogonal complement of the sequence $\{h_n\}$ is infinite dimensional. Thus there exists a unitary operator U on H such that $Uh_n = g_n$ for all n. Put $B = U^*AU$.

Since $\langle Bh_n, h_n\rangle = \langle Ag_n, g_n\rangle$, it follows that $\lim_{n\to\infty}\langle Bh_n, h_n\rangle = \lambda$. By Lemma 2.6 it follows that $\lim_{n\to\infty}\langle B\hat{k}_{z_n}, \hat{k}_{z_n}\rangle = \lambda$. Thus λ is a cluster value of $\tilde{B}$.

COROLLARY 2.8. *An operator on a standard functional Hilbert space is compact if and only if all the Berezin symbols of all unitarily equivalent operators vanish on the boundary.*

PROOF. This is immediate from the theorem since an operator is compact if and only if its essential numerical range consists of zero alone. □

3. Continuous Berezin symbols

The operators A such that all members of $\mathcal{B}(A)$ vanish on the boundary of S are characterized in Corollary 2.8. What are the operators such that all members of $\mathcal{B}(A)$ are continuous on the closure of S?

THEOREM 3.1. *For any operator A, the members of $\mathcal{B}(A)$ all have continuous extensions to the closure of S if and only if A is a translate of a compact operator.*

PROOF. If $A = \lambda + K$, where K is compact, then it follows from Corollary 2.8 that $\lim_{|z|\to 1^-}\tilde{A}(z) = \lim_{|z|\to 1^-}(\lambda + \tilde{K}(z)) = \lambda$.

Conversely, suppose A is an operator that is not a translate of a compact operator. Then the essential numerical range of A contains distinct points λ and μ. Choose orthonormal sequences $\{f_n\}$ and $\{g_n\}$ such that $\lim_{n\to\infty}\langle Af_n, f_n\rangle = \lambda$ and $\lim_{n\to\infty}\langle Ag_n, g_n\rangle = \mu$.

The argument that was used to prove Lemma 2.5 can easily be modified to produce an orthonormal sequence $\{h_m\}$ and subsequences $\{f_{n_m}\}$ and $\{g_{n_m}\}$ such that $\lim_{m\to\infty}\|h_{2m-1}-f_{n_m}\| = 0$ and $\lim_{m\to\infty}\|h_{2m}-g_{n_m}\| = 0$. Thus by Lemma 2.6, $\lim_{n\to\infty}\langle Ah_{2n-1}, h_{2n-1}\rangle = \lambda$ and $\lim_{n\to\infty}\langle Ah_{2n}, h_{2n}\rangle = \mu$.

Lemma 2.5 also yields an orthonormal sequence $\{h'_n\}$ and a sequence $\{z_n\}$ in S that converges to a point in ∂S such that $\lim_{n\to\infty}\|h'_n - \hat{k}_{z_n}\| = 0$. Again, there is no harm in assuming that the orthocomplements of both the sequences $\{h_n\}$ and $\{h'_n\}$ are infinite dimensional, and thus there exists a unitary operator U such that $Uh'_n = h_n$ for every n. If $B = U^*AU$, then $\langle Bh'_n, h'_n\rangle = \langle Ah_n, h_n\rangle$, and consequently by Lemma 2.6, $\lim_{n\to\infty}\langle B\hat{k}_{z_{2n-1}}, \hat{k}_{z_{2n-1}}\rangle = \lambda$ and $\lim_{n\to\infty}\langle B\hat{k}_{z_{2n}}, \hat{k}_{z_{2n}}\rangle = \mu$. Thus $\tilde{B}$ is not continuous at 1. □

4. Two questions

Can anything be said about compactness in spaces without the property that the normalized kernel functions go weakly to 0 at the boundary? It would also be interesting to know how Schatten class operators are characterized in terms of their Berezin symbols.

REFERENCES

[1] Arazy, S. Fisher, S. Janson, J. Peetre, *An identity for reproducing kernels in a planar domain and the Hilbert-Schmidt Hankel operators,* J. Reine Angew. Math. **406** (1990), 179–199.

[2] S. Axler, *Berezin symbols and non-compact operators,* unpublished manuscript, August, 1988.

[3] C. A. Berger and L. A. Coburn, *A symbol calculus for Toeplitz operators,* Proc. Nat. Acad. Sci. U.S.A. **83** (1986), 3072-3073.

[4] F. A. Berezin, *Covariant and contravariant symbols for operators,* Math. USSR-Izv. **6** (1972), 1117–1151.

[5] F. A. Berezin, *Quantization,* Math. USSR-Izv. **8** (1974), 1109–1163.

[6] L. A. Coburn, *Toeplitz operators, quantum mechanics, and mean oscillation in the Bergman metric,* Proc. Symposia in Pure Math. **51**, Part 1 (1990), 97–104.

[7] C. C. Cowen, *Composition operators on Hilbert spaces of analytic functions: A status report,* in Operator Theory Operator Algebras and Applications, W. B. Arveson and R. G. Douglas, eds., Proc. Symposia Pure Math., vol. 51, Amer. Math. Soc., Providence, RI, 1990, pp. 131–145.

[8] J. Dixmier, *Etude sur les varietés et les opérateurs de Julia, avec quelques applications,* Bull. Soc. Math. France **77** (1949), 11–101.

[9] P. A. Fillmore, J. G. Stampfli and J. P. Williams, *On the essential numerical range, the essential spectrum, and a problem of Halmos,* Acta Sci. Math. (Szeged) 33 (1972), 179–192.

[10] V. Guillemin, *Toeplitz operators in n-dimensions,* Integral Equations Operator Theory **7** (1984),145–205.

[11] E. A. Nordgren, *Composition operators in Hilbert spaces,* in Hilbert space operators, Lecture Notes in Math. 693, Springer-Verlag, Berlin-Heidelberg-New York, 1978, pp. 37–63.

[12] J. Peetre, *The Berezin transform and Ha-plitz operators,* J. Oper. Theory **24** (1990), 165–189.

[13] H. J. Schwartz, *Composition operators on H^p,* Thesis, University of Toledo, Toledo, Ohio, 1969.

[14] J. G. Stampfli and J. P. Williams, *Growth conditions and the numerical range in a Banach algebra,* Tôhoku Math. J. 20 (1968), 417–424.

[15] J. H. Shapiro, *The essential norm of a composition operator,* Annals of Math.(2) 125 (1987),375–404.

[16] K. Zhu, *Operator theory in function spaces,* M. Dekker, New York, 1990.

Department of Mathematics
University of New Hampshire
Durham, New Hampshire

Department of Mathematics
University of Toronto
Toronto, Ontario

AMS Classification nos: 47B38, 47B35, 47B07.

Operator Theory:
Advances and Applications, Vol. 73

GENERALIZED HERMITE POLYNOMIALS AND THE BOSE-LIKE OSCILLATOR CALCULUS

MARVIN ROSENBLUM

Dedicated to Moshe Livšic

ABSTRACT. This paper studies a suitably normalized set of generalized Hermite polynomials and sets down a relevant Mehler formula, Rodrigues formula, and generalized translation operator. Weighted generalized Hermite polynomials are the eigenfunctions of a generalized Fourier transform which satisfies an F. and M. Riesz theorem on the absolute continuity of analytic measures. The Bose-like oscillator calculus, which generalizes the calculus associated with the quantum mechanical simple harmonic oscillator, is studied in terms of these polynomials.

1. INTRODUCTION

The generalized Hermite polynomials $\mathbb{H}_n^\mu(x), n \in \mathbb{N} = \{0,1,2,\dots\}$, were defined by Szego [29 , p380, Problem 25] as a set of real polynomials orthogonal with respect to the weight $|x|^{2\mu}e^{-x^2}$, $\mu > -\frac{1}{2}$, with the degree of $\mathbb{H}_n^\mu$ equal n. Thus $\int_{-\infty}^{\infty} \mathbb{H}_m^\mu(x)\mathbb{H}_n^\mu(x)e^{-x^2}|x|^{2\mu}\,dx = 0,\ m \neq n$. These polynomials can be exhibited in terms of certain confluent hypergeometric polynomials, or in terms of certain generalized Laguerre polynomials.

We refer to Erdélyi [15, Vol 1] for the definition and properties of the confluent hypergeometric function Φ , and generally of other special functions. The m-th confluent hypergeometric polynomial with parameter $\gamma + 1 > 0$ is given by

$$\Phi(-m,\gamma+1,x) := \sum_{k=0}^{m} (-1)^k \binom{m}{k} \frac{\Gamma(\gamma+1)}{\Gamma(k+\gamma+1)} x^k,$$

$$\text{which } = \frac{m!\Gamma(\gamma+1)}{\Gamma(m+\gamma+1)} L_m^\gamma(x).$$

Necessarily $\mathbb{H}_{2m}^\mu(x) = c_{2m}\Phi(-m,\mu+\frac{1}{2},x^2)$ and $\mathbb{H}_{2m+1}^\mu(x) = c_{2m+1}x\Phi(-m,\mu+\frac{3}{2},x^2)$, where $m \in \mathbb{N} = \{0,1,2\dots\}$, and the $c_.$ are real constants. (See Chihara [5,p43,p157].)

In his Ph.D. thesis Chihara [4] normalized these polynomials so that the coefficient of x^n in $\mathbb{H}_n^\mu$ is 2^n. Others studying these polynomials, in general with varying normalizations, are Dickinson and Warski [10] and Dutta, Chatterjea, More [13]. We shall set down a different normalization, one that is appropriate for our applications. We shall denote the Chihara polynomials by $\{\mathbb{H}_n^\mu\}_0^\infty$ and our class by $\{H_n^\mu\}_0^\infty$.

We study the generalized Hermite polynomials in section 2. In section 3 we define a relevant Fourier transform and heat equation, and in section 4 a relevant translation operator. Finally, in section 5, we study some basic aspects of the Bose-like oscillator calculus which intrinsically connect with the earlier sections. It will develop that the Bose-like oscillator calculus is a remarkable, fully structured generalization of the calculus associated with the quantum mechanical harmonic oscillator, that is, the Boson calculus. Our main result, Theorem 5.12, is a generalization of the von Neumann uniqueness theorem to the Bose-like oscillator calculus.

2. Generalized Hermite Polynomials

Let C be the set of complex numbers and let $C_o = C \setminus \{-\frac{1}{2}, -\frac{3}{2}, -\frac{5}{2}, \dots\}$.

2.1 Definition. *Suppose $\mu \in C_o$. The* **generalized Hermite polynomials** *$\{H_n^\mu\}_0^\infty$ are defined for n even by*

(2.1.1)

$$\begin{aligned} H_{2m}^\mu(x) &:= (-1)^m \frac{(2m)!}{m!} \Phi(-m, \mu + \tfrac{1}{2}, x^2) \\ &= (-1)^m \frac{(2m)!}{m!} \sum_{k=0}^{m} (-1)^k \binom{m}{k} \frac{\Gamma(\mu + \frac{1}{2})}{\Gamma(k + \mu + \frac{1}{2})} x^{2k}. \end{aligned}$$

They are defined for n odd by

(2.1.2)

$$\begin{aligned} H_{2m+1}^\mu(x) &:= (-1)^m \frac{(2m+1)!}{m!} \frac{x}{\mu + \frac{1}{2}} \Phi(-m, \mu + \frac{3}{2}, x^2) \\ &= (-1)^m \frac{(2m+1)!}{m!} \sum_{k=0}^{m} (-1)^k \binom{m}{k} \frac{\Gamma(\mu + \frac{1}{2})}{\Gamma(k + \mu + \frac{3}{2})} x^{2k+1}. \end{aligned}$$

2.2 On H_n^μ, γ_μ, and $\mathbf{e}_\mu$.

$$\begin{aligned} H_{2m}^\mu(x) &= \frac{\Gamma(\mu + \frac{1}{2})}{\Gamma(\frac{1}{2})} \frac{\Gamma(m + \frac{1}{2})}{\Gamma(m + \mu + \frac{1}{2})} \mathbb{H}_{2m}^\mu(x) \\ &= (-1)^m (2m)! \frac{\Gamma(\mu + \frac{1}{2})}{\Gamma(m + \mu + \frac{1}{2})} L_m^{\mu - \frac{1}{2}}(x^2). \end{aligned} \tag{2.2.1}$$

(2.2.2)

$$H^{\mu}_{2m+1}(x) = \frac{\Gamma(\mu+\frac{1}{2})}{\Gamma(\frac{1}{2})}\frac{\Gamma(m+\frac{3}{2})}{\Gamma(m+\mu+\frac{3}{2})}\mathbb{H}^{\mu}_{2m+1}(x)$$
$$= (-1)^m(2m+1)!\frac{\Gamma(\mu+\frac{1}{2})}{\Gamma(m+\mu+\frac{3}{2})}xL_m^{\mu+\frac{1}{2}}(x^2).$$

We list the first few generalized Hermite polynomials: $H^{\mu}_0(x) = 1$,

$$H^{\mu}_1(x) = (1+2\mu)^{-1}2x, \qquad H^{\mu}_2(x) = (1+2\mu)^{-1}4x^2 - 2$$
$$H^{\mu}_3(x) = (1+2\mu)^{-1}(3+2\mu)^{-1}24x^3 - (1+2\mu)^{-1}12x$$
$$H^{\mu}_4(x) = \left((1+2\mu)(3+2\mu)\right)^{-1}48x^4 - (1+2\mu)^{-1}48x^2 + 12.$$

This class of generalized Hermite polynomials has a rather nice generating function formula involving the confluent hypergeometric function Φ . If $\mu \in C_o$ we define

(2.2.3)
$$\mathbf{e}_\mu(x) := e^x\Phi(\mu, 2\mu+1, -2x), \qquad \text{which}$$
$$= \Gamma(\mu+\tfrac{1}{2})(2/x)^{\mu-\frac{1}{2}}\left(I_{\mu-\frac{1}{2}}(x) + I_{\mu+\frac{1}{2}}(x)\right)$$
$$= (2x)^{\frac{1}{2}-\mu}M_{-\frac{1}{2},\mu}(2x),$$

where I_μ is the modified Bessel function and $M_{..}$ is the Whittaker function. $\mathbf{e}_\mu$ plays the role of a generalized exponential function in what follows, and indeed $\mathbf{e}_0(x) = e^x$. $\mathbf{e}_\mu$ is an entire function, say,

(2.2.4)
$$\mathbf{e}_\mu(x) = \sum_{m=0}^{\infty}\frac{z^m}{\gamma_\mu(m)},$$

where the power series representation for the associated Bessel function yields

(2.2.5)
$$\gamma_\mu(2m) = \frac{2^{2m}m!\Gamma(m+\mu+\frac{1}{2})}{\Gamma(\mu+\frac{1}{2})} = (2m)!\frac{\Gamma(m+\mu+\frac{1}{2})}{\Gamma(\mu+\frac{1}{2})}\frac{\Gamma(\frac{1}{2})}{\Gamma(m+\frac{1}{2})}, \quad \text{and}$$

(2.2.6)
$$\gamma_\mu(2m+1) = \frac{2^{2m+1}m!\Gamma(m+\mu+\frac{3}{2})}{\Gamma(\mu+\frac{1}{2})} = (2m+1)!\frac{\Gamma(m+\mu+\frac{3}{2})}{\Gamma(\mu+\frac{1}{2})}\frac{\Gamma(\frac{1}{2})}{\Gamma(m+\frac{3}{2})}.$$

γ_μ plays the role of a generalized factorial. We list a few of the γ_μ : $\gamma_\mu(0) = 1$, $\gamma_\mu(1) = 1+2\mu$, $\gamma_\mu(2) = (1+2\mu)2$, $\gamma_\mu(3) = (1+2\mu)2(3+2\mu)$, $\gamma_\mu(4) = (1+$

$2\mu)2(3+2\mu)4$, and $\gamma_\mu(5) = (1+2\mu)2(3+2\mu)4(5+2\mu)$. We note the recursion relation for the γ_μ :

$$\gamma_\mu(n+1) = (n+1+2\mu\theta_{n+1})\gamma_\mu(n), \qquad n \in \mathrm{N}, \tag{2.2.7}$$

where θ_{n+1} is defined to be 0 if n + 1 is even and 1 if n + 1 is odd. It follows from (2.1.1) and (2.1.2) that for all $n \in \mathrm{N}$

$$H_n^\mu(x) = n! \sum_{k=0}^{[\frac{n}{2}]} \frac{(-1)^k (2x)^{n-2k}}{k!\gamma_\mu(n-2k)}. \tag{2.2.8}$$

We note that the coefficient of x^n in the expansion of H_n^μ is $2^n n!/\gamma_\mu(n)$. We set down for later reference integral expressions for the beta function $B(\cdot,\cdot)$.

2.3 LEMMA. *i) Suppose* $\mu > 0, \alpha > -\frac{1}{2}$, $n \in \mathrm{N}$, *and* $x \in C$.

$$B(n+\alpha+\tfrac{1}{2},\mu) = \int_{-1}^{1} t^{2n}|t|^{2\alpha}(1-t)^{\mu-1}(1+t)^\mu dt \tag{2.3.1}$$

$$B(n+\alpha+\tfrac{3}{2},\mu) = \int_{-1}^{1} t^{2n+1}|t|^{2\alpha}(1-t)^{\mu-1}(1+t)^\mu dt \tag{2.3.2}$$

$$\frac{\gamma_\alpha(n)}{\gamma_{\mu+\alpha}(n)} = \frac{1}{B(\alpha+\frac{1}{2},\mu)}\int_{-1}^{1} t^n|t|^{2\alpha}(1-t)^{\mu-1}(1+t)^\mu dt \tag{2.3.3}$$

$$\mathbf{e}_{\mu+\alpha}(x) = \frac{1}{B(\alpha+\frac{1}{2},\mu)}\int_{-1}^{1} \mathbf{e}_\alpha(xt)|t|^{2\alpha}(1-t)^{\mu-1}(1+t)^\mu dt \tag{2.3.4}$$

$$\mathbf{e}_\mu(x) = \frac{1}{B(\frac{1}{2},\mu)}\int_{-1}^{1} e^{xt}(1-t)^{\mu-1}(1+t)^\mu \, dt \tag{2.3.5}$$

(2.3.6) *ii) If now* $0 < \mu < \frac{1}{2}$,

$$e^x = \frac{1}{B(\frac{1}{2}-\mu,\mu)}\int_{-1}^{1} \mathbf{e}_{-\mu}(xt)|t|^{-2\mu}(1-t)^{\mu-1}(1+t)^\mu dt$$

(2.3.7)
iii) Suppose $\alpha > -\frac{1}{2}$, $\mu > -1$ *and* $\mu+\alpha > -\frac{1}{2}$. *Then* $\mathbf{e}_{\mu+\alpha}(x) = \mathbf{e}_\alpha(x) +$

$$\frac{\mu}{\mu+\alpha+\frac{1}{2}}\frac{1}{B(\alpha+\frac{1}{2},\mu+1)}\int_{-1}^{1} (\mathbf{e}_\alpha(xt)-\mathbf{e}_\alpha(x))\,|t|^{2\alpha}(1-t)^{\mu-1}(1+t)^\mu dt$$

Proof. Start with the usual integral representation for the beta function, $B(x,y) = \int_0^1 t^{x-1}(1-t)^{y-1}\,dt$. Use this to then derive (2.3.1) and (2.3.2). (2.3.3) and (2.3.4) follow from this and (2.2.5), (2.2.6), (2.2.4). Set $\alpha = -\mu$ in (2.3.4) to get (2.3.6) (2.3.7) is obtained from (2.3.4) by analytically continuing μ. We use the functional equation $(x+y)B(x,y+1) = yB(x,y)$ for the beta function. The rest follow easily.

Next we associate with the generalized exponential function $\mathbf{e}_\mu$ a generalized derivative operator $\mathfrak{D}_\mu$. These objects are special cases of functions and operators set down by C.F. Dunkl in his work on root systems associated with finite reflection groups. The papers [11] , [12] are particularly relevant here.

For the sake of simplicity we study the action of $\mathfrak{D}_\mu$ on entire functions.

2.4 Definition. *i) The linear operator $\mathfrak{D}_\mu$ is defined on all entire functions ϕ on C by*

$$\mathfrak{D}_\mu\phi(x) = \phi'(x) + \frac{\mu}{x}(\phi(x) - \phi(-x)),\ x \in C. \tag{2.4.1}$$

We use the notation $\mathfrak{D}_{\mu,x}$ when we wish to emphasize that $\mathfrak{D}_\mu$ is acting on functions of the variable x. Thus $\mathfrak{D}_{\mu,x}(\phi(x)) := (\mathfrak{D}_\mu\phi)(x)$.
ii) $\mathfrak{Q}$ is defined on all functions ϕ on C by

$$\mathfrak{Q}\phi(x) = x\phi(x). \tag{2.4.2}$$

2.5 Properties of $\mathfrak{D}_\mu$, $\mathbf{e}_\mu$, and H_n^μ. *Suppose $\mu \in C_o$, $n \in \mathbb{N}$, $x, z \in C$ and ϕ, ψ are entire functions.*

$$(\mathfrak{D}_\mu^2\phi)(x) = \phi''(x) + \frac{2\mu}{x}\phi'(x) - \frac{\mu}{x^2}(\phi(x) - \phi(-x)) \tag{2.5.1}$$

$$\mathfrak{D}_\mu^j : x^n \longmapsto \frac{\gamma_\mu(n)}{\gamma_\mu(n-j)}x^{n-j}, j = 0,1,\ldots,n; \quad \mathfrak{D}_\mu^j : 1 \longmapsto 0 \tag{2.5.2}$$

If ψ is an even entire function, then

$$\mathfrak{D}_\mu(\phi\psi) = \mathfrak{D}_\mu(\phi)\psi + \phi\mathfrak{D}_\mu(\psi) \tag{2.5.3}$$

$$H_n^\mu(x) = \frac{1}{B(\frac{1}{2},\mu)}\int_{-1}^{1} H_n(xt)(1-t)^{\mu-1}(1+t)^\mu dt \qquad \text{if} \qquad \mu > 0. \tag{2.5.4}$$

Exponential property of $\mathbf{e}_\mu$:

$$\mathfrak{D}_{\mu,x} : \mathbf{e}_\mu(\lambda x) \longmapsto \lambda\mathbf{e}_\mu(\lambda x) \tag{2.5.5}$$

Differential equation for $\mathbf{e}_\mu$ *:*

$$x\mathbf{e}_\mu''(x) + (1+2\mu)\mathbf{e}_\mu'(x) - (1+x)\mathbf{e}_\mu(x) = 0 \tag{2.5.6}$$

$$\mathbf{e}_\mu''(x) = \mathbf{e}_\mu(x) - \frac{2\mu}{2\mu+1}\mathbf{e}_{\mu+1}(x) \tag{2.5.7}$$

Generating function for the H_n^μ *:*

$$\exp(-z^2)\mathbf{e}_\mu(2xz) = \sum_{n=0}^{\infty} H_n^\mu(x)\frac{z^n}{n!}, \quad \mu \in C_o \tag{2.5.8}$$

Proof. (2.5.1) to (2.5.3) follow from the definition in (2.4.1), and also (2.2.7). To prove (2.5.4) substitute the expansion for $H_n = H_n^0$ provided by 2.1 in the right side of (2.5.4) and then employ (2.3.3) with $\alpha = 0$, so $\gamma_\alpha(n) = n!$.

We prove (2.5.5) using (2.5.2) . $\mathbf{e}_\mu(\lambda x) = \sum_{j=0}^{\infty}(\lambda x)^j/\gamma_\mu(j)$ is mapped by $\mathfrak{D}_\mu$ to $\sum_{j=1}^{\infty}\lambda^j x^{j-1}/\gamma_\mu(j-1) = \lambda\mathbf{e}_\mu(\lambda x)$.

For (2.5.6) refer to Slater [28,p94].

To prove (2.5.7) first assume $\mu > 0$ and check the result using (2.3.5) and the functional equation for the beta function. Then analytically continue μ to C_o.

The generating function formula for the classical Hermite polynomials is $\exp(-z^2+2xz) = \sum_{n=0}^{\infty} H_n(x)z^n/n!$. Use this result, (2.5.4), and (2.3.7) to prove (2.5.8) for $\mu > 0$. The result for $\mu \in C_o$ follows by analytically continuing μ.

2.6 Properties of H_μ. *Suppose* $n \in \mathbb{N}$, $\lambda, x, z \in C$, $|z| < 1$.

$$\mathfrak{D}_{\mu,x} : H_n^\mu(\lambda x) \longrightarrow 2\lambda n H_{n-1}^\mu(\lambda x) \tag{2.6.1}$$

$$(2\mathfrak{Q} - \mathfrak{D}_\mu)H_n^\mu = \frac{\gamma_\mu(n+1)}{(n+1)\gamma_\mu(n)}H_{n+1}^\mu = \left(1 + \frac{2\mu\theta_{n+1}}{n+1}\right)H_{n+1}^\mu \tag{2.6.2}$$

Three term recursion: Set $H_{-1}^\mu(x) = 0$. *Then*

$$2nH_{n-1}^\mu + \frac{\gamma_\mu(n+1)}{(n+1)\gamma_\mu(n)}H_{n+1}^\mu = 2\mathfrak{Q}H_n^\mu \tag{2.6.3}$$

$$\mathfrak{D}_{\mu,x}(e^{-\lambda^2x^2}H_n^\mu(\lambda x)) = -\lambda e^{-\lambda^2x^2}\frac{\gamma_\mu(n+1)}{(n+1)\gamma_\mu(n)}H_{n+1}^\mu(\lambda x) \tag{2.6.4}$$

Rodrigues formula:

$$(2.6.5)\qquad (-1)^n e^{\lambda^2x^2}\mathfrak{D}^n_{\mu,x}e^{-\lambda^2x^2} = \lambda^n\frac{\gamma_\mu(n)}{n!}H^\mu_n(\lambda x)$$

$$(2.6.6)\qquad H^\mu_n = \frac{n!}{\gamma_\mu(n)}(2\Omega - \mathfrak{D}_\mu)^n H^\mu_0$$

Inversion formula:

$$(2.6.7)\qquad \frac{(2x)^n}{\gamma_\mu(n)} = \sum_{k=0}^{[\frac{n}{2}]}\frac{H^\mu_{n-2k}(x)}{k!(n-2k)!}$$

Mehler formula:

(2.6.8)

$$\sum_{n=0}^{\infty}\frac{\gamma_\mu(n)}{2^n(n!)^2}H^\mu_n(x)H^\mu_n(y)z^n = \frac{1}{(1-z^2)^{\mu+\frac{1}{2}}}\exp\left(-(x^2+y^2)\frac{z^2}{1-z^2}\right)\mathrm{e}_\mu\left(2xy\frac{z}{1-z^2}\right)$$

$$(2.6.9)\qquad \sum_{n=0}^{\infty}\frac{\gamma_\mu(2n)}{(2n)!}H^\mu_{2n}(x)\frac{(-1)^n}{n!}\left(\frac{z}{2}\right)^{2n} = \frac{1}{(1-z^2)^{\mu+\frac{1}{2}}}\exp\left(\frac{-x^2z^2}{1-z^2}\right).$$

Proof. Apply $\mathfrak{D}_{\mu,x}$ to both sides of (2.5.8) with $'x'$ replaced by $'\lambda x'$. Then, using (2.5.5), $\sum_{n=0}^{\infty}\mathfrak{D}_{\mu,x}H^\mu_n(\lambda x)z^n/n! = \sum_{n=0}^{\infty}2\lambda nH^\mu_{n-1}(\lambda x)z^n/n!$, and thus (2.6.1) follows.

$2xH^\mu_n(x) - 2nH^\mu_{n-1}(x) = (1+2\mu\theta_{n+1}/(n+1))H^\mu_{n+1}(x)$ follows, upon separately considering the even and odd polynomials, from (2.2.8) and derives (2.6.3). From this and (2.6.1) we deduce (2.6.2).

Use (2.5.3) to infer that $\mathfrak{D}_{\mu,x}(\exp(-x^2)H^\mu_n(x)) = ((-2x+\mathfrak{D}_{\mu,x})H^\mu_n(x))e^{-x^2}$, which by (2.6.1) and (2.6.3) equals the right side of (2.6.4) with $\lambda = 1$. (2.6.4) for general λ follows easily by the easily derived chain rule formula $\mathfrak{D}_{\mu,x} : f(\lambda x) \longrightarrow \lambda(\mathfrak{D}_\mu f)(\lambda x)$.

(2.6.5) is proved by induction using (2.6.1). Note that $H^\mu_0(x) = 1$.

(2.6.6) is proved by induction using (2.6.2).

We prove (2.6.7): $\mathrm{e}_\mu(2xz) = e^{z^2}\sum_{n=0}^{\infty}H^\mu_n(x)z^n/n!$, so $\sum_{n=0}^{\infty}(2xz)^n/\gamma_\mu(n) = (\sum_{j=0}^{\infty}z^{2j}/j!)\sum_{n=0}^{\infty}H^\mu_n(x)z^n/n!$. Equate like powers of z to deduce (2.6.7).

Use (2.2.1), (2.2.2),(2.2.5), (2.2.6) and the bilateral generating function for the Laguerre polynomials Erdélyi [15 vol 2, p189] to prove that the left side of (2.6.8) equals

$$\sum_{n=0}^{\infty}(n!\Gamma(\mu+\tfrac{1}{2})/\Gamma(n+\mu+\tfrac{1}{2}))\,L_n^{\mu-\frac{1}{2}}(x^2)L_n^{\mu-\frac{1}{2}}(y^2)z^{2n}$$

$$+\sum_{n=0}^{\infty}(n!\Gamma(\mu+\tfrac{1}{2})/\Gamma(n+\mu+\tfrac{3}{2}))\,xyL_n^{\mu+\frac{1}{2}}(x^2)L_n^{\mu+\frac{1}{2}}(y^2)z^{2n+1}$$

$$=\left(\frac{\Gamma(\mu+\frac{1}{2})}{(1-z^2)}\right)\exp(-(x^2+y^2)\frac{z^2}{1-z^2})(xyz)^{-\mu+\frac{1}{2}}\Big(I_{\mu-\frac{1}{2}}(2xy\frac{z}{1-z^2})$$

$$+I_{\mu+\frac{1}{2}}((2xy\frac{z}{1-z^2})\Big),$$

which by (2.2.3) equals the right side of (2.6.8).

To derive (2.6.9) set $y=0$ in Mehler's formula (2.6.8) and note from Definition 2.1 that $H_{2n}^{\mu}(0)=(-1)^n\frac{(2n)!}{n!}$ and $H_{2n+1}^{\mu}(0)=0$ for all $n\in\mathbb{N}$.

2.7 Further properties of H_n^{μ}.

$$\exp(-y^2\mathfrak{D}_{\mu,x}^2)x^n=\frac{\gamma_\mu(n)}{n!}H_n^{\mu}(\frac{x}{2y})y^n,\quad 0\neq y\in C,\quad n\in\mathbb{N}. \tag{2.7.1}$$

$$\exp(\frac{1}{4}t\mathfrak{D}_{\mu,x}^2)(\exp(-\alpha x^2)\mathrm{e}_\mu(2zx))= \tag{2.7.2}$$

$$(1+\alpha t)^{-\mu-1/2}\exp(\frac{tz^2}{1+\alpha t})\exp(-\frac{\alpha x^2}{1+\alpha t})\mathrm{e}_\mu(\frac{2zx}{1+\alpha t}),$$
$$\Re\alpha>0,\ \Re t>0,\ x\in R,\ z\in C.$$

Proof. We prove (2.7.1). We deduce from (2.5.5) that $\exp(-y^2\mathfrak{D}_{\mu,x}^2)\mathrm{e}_\mu(\lambda x)=\exp(-\lambda^2y^2)\mathrm{e}_\mu(\lambda x)$, so

$$\exp(-y^2\mathfrak{D}_{\mu,x}^2)\sum_{n=0}^{\infty}\frac{\lambda^nx^n}{\gamma_\mu(n)}=\exp(-\lambda^2y^2)\mathrm{e}_\mu(2\lambda y\frac{x}{2y})=\sum_{n=0}^{\infty}H_n^{\mu}(\frac{x}{2y})\frac{y^n\lambda^n}{n!}.$$

(2.7.1) follows by equating the coefficients of λ^n.

Assume $|2vy|<1$ and use (2.7.1) and (2.6.8) to obtain

$$\exp(-y^2\mathfrak{D}_{\mu,x}^2)\exp(-v^2x^2)\mathrm{e}_\mu(2uvx)=\exp(-y^2\mathfrak{D}_{\mu,x}^2)\sum_{n=0}^{\infty}H_n^{\mu}(u)\frac{(vx)^n}{n!}=$$

$$\sum_{n=0}^{\infty} \frac{\gamma_\mu(n)}{n!} H_n^\mu(u) H_n^\mu(\frac{x}{2y}) \frac{y^n v^n}{n!} = \sum_{n=0}^{\infty} \frac{\gamma_\mu(n)}{(n!)^2 2^n} H_n^\mu(u) H_n^\mu(\frac{x}{2y})(2yv)^n =$$

$$\frac{1}{(1-4v^2y^2)^{\mu+\frac{1}{2}}} \exp\big((-u^2 - \frac{x^2}{4y^2})\frac{4v^2y^2}{1-4v^2y^2}\big) e_\mu(\frac{2uvx}{1-4v^2y^2})$$

Next set $t = -4y^2$, $z = uv$, and $\alpha = v^2$. Thus (2.7.2) holds at least if $|\alpha t| < 1$. (2.7.2) follows by analytic continuation for the parameters.

The identities (2.7.1) and (2.7.2) are generalizations of classical Hermite polynomial identities set down, using Boson calculus techniques, by J.D. Louck [20].

The generalized heat equation problem

$$\mathfrak{D}^2_{\mu,x}\psi(x,t) = \frac{\partial\psi(x,t)}{\partial t}, \text{ with } \psi(x,0) = \phi(x),\ t \geq 0,\ x \in \mathfrak{C}, \tag{2.7.3}$$

where ϕ is given, has the formal solution $\exp(t\mathfrak{D}^2_{\mu,x})\phi(x)$. This problem is related to the work [6] of Cholewinski and Haimo and reduces to it when the function ϕ is assumed to be even. From (2.7.1) we see that if $n \in \mathrm{N}$,

$$\exp(t\mathfrak{D}^2_{\mu,x})x^n = \frac{\gamma_\mu(n)}{n!} H_n^\mu(\frac{x}{2i\sqrt{t}})(i\sqrt{t})^n = \frac{\gamma_\mu(n)}{n!} \sum_{k=0}^{[\frac{n}{2}]} \frac{x^{n-2k}t^k}{k!\gamma_\mu(n-2k)} \tag{2.7.4}$$

The functions given by (2.7.4) are for even integers n the generalized heat polynomials of Cholewinski and Haimo [6] and [8,section 14].

We shall pursue the $L^2_\mu(R)$ theory for the generalized heat equation in section 3.

3. The generalized Fourier transform

3.1 Definition. *i) $L^2_\mu(R)$ is the Hilbert space of Lebesgue measurable functions f on R with*

$$\|f\|_\mu := \left(\int_{-\infty}^{\infty} |f(t)|^2 |t|^{2\mu}\, dt\right)^{\frac{1}{2}} < \infty.$$

ii) The **generalized Fourier transform operator** $\mathcal{F}_\mu$, $\mu > -\frac{1}{2}$ *is defined on the linear span $\mathfrak{S}$ of $\{\exp(-x^2)x^n : n \in \mathrm{N}\}$ in $L^2_\mu(R)$ by*

$$\mathcal{F}_\mu f(x) := \left(2^{\mu+\frac{1}{2}}\Gamma(\mu+\tfrac{1}{2})\right)^{-1} \int_{-\infty}^{\infty} e_\mu(-ixt)f(t)|t|^{2\mu}dt \tag{3.1.1}$$

This transform appears in the physics literature on Bose-like oscillators, [24, p294] and [22], and as a special case of a general transform in Dunkl [12].

We see from (2.2.3) that if x is real,

$$\mathbf{e}_\mu(-ix) = \Gamma(\mu+\tfrac{1}{2})2^{\mu-\frac{1}{2}}\frac{J_{\mu-\frac{1}{2}}(x) - iJ_{\mu+\frac{1}{2}}(x)}{x^{\mu-\frac{1}{2}}} =, \quad say, \quad \mathbf{c}_\mu(x) - i\mathbf{s}_\mu(x), \tag{3.1.2}$$

where $\mathbf{c}_\mu$ is real and even and $\mathbf{s}_\mu$ is real and odd. The integral in (3.1.1) is well-defined since

$$|\mathbf{e}_\mu(-ix)| \le C_\mu\big(|x|^{|\mu|} + 1\big) \text{ where } C_\mu \in R \text{ if } -\tfrac{1}{2} < \mu < 0, \tag{3.1.3}$$

$$\text{and } |\mathbf{e}_\mu(-ix)| \le 1 \text{ if } \mu \ge 0. \tag{3.1.4}$$

(3.1.3) is proved using the asymptotic formula for Bessel functions [30, Chapt 7] and (3.1.4) follows easily from (2.3.5). (2.3.5) implies that $\mathbf{e}_\mu(-ix)$ is a positive definite function of x. Thus the following result follows. In the limiting case $\mu = 0$ the result degenerates into the obvious $cos^2(x) + sin^2(x) \le 1$.

3.2 Remark. *Suppose $\mu > 0$. Then*

$$\big(J_{\mu-\frac{1}{2}}(x)\big)^2 + \big(J_{\mu+\frac{1}{2}}(x)\big)^2 \le \frac{1}{\Gamma(\mu+\frac{1}{2})^2}|x/2|^{2\mu-1} \tag{3.2.1}$$

for $x \in R$. Equality holds in (3.2.1) if and only if $x = 0$.

Proof. Use (3.1.4) and (3.1.2.).

3.3 Some Fourier transform integrals. *Suppose $\Re\lambda > 0$.*

$$\int_0^\infty \mathbf{c}_\mu(xt)\exp(-\tfrac{1}{2}\lambda t^2)t^{2\mu}\,dt = \frac{\Gamma(\mu+\frac{1}{2})}{2\lambda^{\mu+\frac{1}{2}}}\exp(-x^2/(4\lambda)) \tag{3.3.1}$$

$$\int_{-\infty}^\infty \mathbf{e}_\mu(-ixt)\exp(-\lambda t^2)|t|^{2\mu}\,dt = \frac{\Gamma(\mu+\frac{1}{2})}{\lambda^{\mu+\frac{1}{2}}}\exp(-x^2/(4\lambda)) \tag{3.3.2}$$

(3.3.3)

$$\int_{-\infty}^\infty \mathbf{e}_\mu(-ixt)t^n\exp(-\lambda t^2)|t|^{2\mu}\,dt = (-\frac{i}{2})^n\frac{\Gamma(\mu+\frac{1}{2})}{\lambda^{\frac{1}{2}n+\frac{1}{2}+\mu}}\frac{\gamma_\mu(n)}{n!}\exp(-\frac{x^2}{4\lambda})H_n^\mu(\frac{x}{2\lambda^{\frac{1}{2}}})$$

(3.3.4)

$$\int_{-\infty}^{\infty} \mathrm{e}_\mu(-ixt)\mathrm{e}_\mu(iyt)\exp(-\lambda t^2)|t|^{2\mu}\,dt = \frac{\Gamma(\mu+\frac{1}{2})}{\lambda^{\mu+\frac{1}{2}}}\exp(-\frac{x^2+y^2}{4\lambda})\mathrm{e}_\mu(xy/(2\lambda)),$$

$x, y \in \mathfrak{C}$

(3.3.5) $$\int_{-\infty}^{\infty} \mathrm{e}_\mu(-ixt)\exp(-\lambda^2t^2)H_n^\mu(\beta t)|t|^{2\mu}\,dt$$

$$= (-i)^n\Gamma(\mu+\tfrac{1}{2})\lambda^{-2\mu-1}\left((\beta/\lambda)^2-1\right)^{n/2}\exp(-x^2/(4\lambda^2))H_n^\mu\left(\frac{\beta x}{2\lambda(\beta^2-\lambda^2)^{\frac{1}{2}}}\right),$$

if $\beta^2 > \lambda^2 > 0$.

(3.3.6)

$$\int_{-\infty}^{\infty} \mathrm{e}_\mu(-ixt)\exp(-\tfrac{1}{2}t^2)H_n^\mu(t)|t|^{2\mu}\,dt = 2^{\mu+\frac{1}{2}}\Gamma(\mu+\tfrac{1}{2})(-i)^n\exp(-\tfrac{1}{2}x^2)H_n^\mu(x)$$

Proof. (3.3.1) is listed as a Hankel transform in Erdélyi [16 vol 2, p29]. (3.3.2) follows from (3.3.1) and (3.1.2). Apply $\mathfrak{D}_{\mu,x}^n$ to both sides of (3.3.2) and use the Rodrigues formula (2.6.5) to derive (3.3.3).(3.3.4) follows from (3.3.3) by use of (2.5.8) and (2.2.4).

(3.3.5) follows from (3.3.4) by use of the expansions $(\exp y^2)\mathrm{e}_\mu(i2\beta yt) = \sum_{n=0}^{\infty} H_n^\mu(\beta t)(iy)^n/n!$ and $\exp(-A^2y^2)\mathrm{e}_\mu(2By) = \sum_{n=0}^{\infty} H_n^\mu(B/A)(Ay)^n/n!$, which are a consequence of (2.5.8). (3.3.6) follows from (3.3.5) by setting $\lambda^2 = \frac{1}{2}$ and $\beta = 1$.

The Hilbert Space $L_\mu^2(R)$ has the inner product $\langle f, g\rangle_\mu := \int_{-\infty}^{\infty} f(t)g^*(t)|t|^{2\mu}\,dt$, where $f, g \in L_\mu^2(R)$ and g^* is the complex conjugate of g. Notice that $\|f\|_\mu = \langle f, f\rangle_\mu^{\frac{1}{2}}, f \in L_\mu^2(R)$.

3.4 Definition. *i) Define the* **generalized Hermite functions** ϕ_n^μ *on R by*

(3.4.1) $$\phi_n^\mu(x) := \left(\frac{\gamma_\mu(n)}{\Gamma(\mu+\frac{1}{2})}\right)^{\frac{1}{2}}\frac{1}{2^{n/2}n!}\exp(-\tfrac{1}{2}x^2)H_n^\mu(x), n \in \mathrm{N}.$$

ii) Define the operators $\boldsymbol{P}_\mu, \boldsymbol{Q}_\mu, \boldsymbol{H}_\mu$ *and* $\boldsymbol{A}_\mu$ *on the finite span* $\mathfrak{S}$ *of the generalized Hermite functions by*

(3.4.2)

$$\boldsymbol{P}_\mu\phi(x) := -i(\phi'(x) + \frac{\mu}{x}(\phi(x)-\phi(-x)))\,,\ \boldsymbol{Q}_\mu\phi(x) := \mathfrak{Q}\phi(x) = x\phi(x),\ \text{and}$$

(3.4.3) $$\boldsymbol{A}_\mu := 2^{-\frac{1}{2}}(\boldsymbol{Q}_\mu + i\boldsymbol{P}_\mu)\,,\ \boldsymbol{H}_\mu := \tfrac{1}{2}({\boldsymbol{Q}_\mu}^2 + {\boldsymbol{P}_\mu}^2).$$

3.5 PROPERTIES OF ϕ_n^μ AND $\mathcal{F}_\mu$.

(3.5.1) $\{\phi_n\}_{n\in\mathrm{N}}$ *is a complete orthonormal set in* $L^2_\mu(R)$.

$$\mathcal{F}_\mu\phi_n^\mu = (-i)^n\phi_n^\mu, n \in \mathrm{N}. \tag{3.5.2}$$

Mehler formula:

$$\sum_{n=0}^{\infty}\phi_n^\mu(x)\phi_n^\mu(y)z^n = \tag{3.5.3}$$

$$\frac{1}{\Gamma(\mu+\frac{1}{2})}\frac{1}{(1-z^2)^{\mu+\frac{1}{2}}}\exp(-\tfrac{1}{2}(x^2+y^2)\frac{1+z^2}{1-z^2})\mathrm{e}_\mu(2xy\frac{z}{1-z^2})$$

Proof. Suppose $u, v \in C$. Then

$$\sum_{j,k\in\mathrm{N}}\langle\phi_j^\mu,\phi_k^\mu\rangle_\mu(\gamma_\mu(j))^{-1/2}(\gamma_\mu(k))^{-1/2}(2^{\frac{1}{2}}u)^j(2^{\frac{1}{2}}v)^k$$

$$= (\Gamma(\mu+\tfrac{1}{2}))^{-1}\int_{-\infty}^{\infty}\sum_{j,k\in\mathrm{N}}H_j^\mu(t)H_k^\mu(t)\frac{u^jv^k}{j!k!}\exp(-t^2)|t|^{2\mu}\,dt,$$

which by (2.5.8)

$$= (\Gamma(\mu+\tfrac{1}{2}))^{-1}\int_{-\infty}^{\infty}\exp(-u^2)\mathrm{e}_\mu(2ut)\exp(-v^2)\mathrm{e}_\mu(2vt)\exp(-t^2)|t|^{2\mu}\,dt$$

which by (3.3.4) $= \mathrm{e}_\mu(2uv) = \sum_{j\in\mathrm{N}}2^ju^jv^j/\gamma_\mu(j)$.

Thus we see that $\{\phi_n^\mu\}_{n\in\mathrm{N}}$ is an orthonormal set in $L^2_\mu(R)$. It is complete by much the same argument used to prove that the classical $\mu = 0$ Hermite functions form a complete orthogonal set in $L^2_0(R)$, see, for example, Ahiezer and Glazman, [1], Chapt 1, paragraph 11.

(2.6.8) implies (3.5.4) and (3.3.6) yields (3.5.3).

3.6 THEOREM. *$\mathcal{F}_\mu$ is a unitary transformation on $L^2_\mu(R)$ with eigenvalues 1, -1, i, -i . $\{\phi_n^\mu\}_{n\in\mathrm{N}}$ is a complete orthonormal set of eigenvectors of $\mathcal{F}_\mu$. The inverse Fourier transform is given by :*

$$\mathcal{F}_\mu^*f(x) = \left(2^{\mu+\frac{1}{2}}\Gamma(\mu+\tfrac{1}{2})\right)^{-1}\int_{-\infty}^{\infty}\mathrm{e}_\mu(ixt)f(t)|t|^{2\mu}dt, \quad f\in\mathfrak{S}. \tag{3.6.1}$$

Proof. The first statement is a direct consequence of 3.5. From (3.5.3) we see that $\mathcal{F}_\mu^\star\phi_n^\mu = i^n\phi_n^\mu$, and it follows that $(\mathcal{F}_\mu^\star\psi)(t) = (\mathcal{F}_\mu\psi)(-t)$ for any $\psi\in\mathfrak{S}$. Then (3.6.1) follows from (3.1.1)

3.7 MORE ON ϕ_n^μ AND $\mathcal{F}_\mu$. *Define* $\phi_{-1}^\mu = 0$ *and assume* $n \in \mathbb{N}$.

(3.7.1) $$A_\mu \phi_n^\mu = \left(\frac{\gamma_\mu(n)}{\gamma_\mu(n-1)}\right)^{\frac{1}{2}} \phi_{n-1}^\mu$$

(3.7.2) $$A_\mu^\star \phi_n^\mu = \left(\frac{\gamma_\mu(n+1)}{\gamma_\mu(n)}\right)^{\frac{1}{2}} \phi_{n+1}^\mu$$

(3.7.3) $$\phi_n^\mu = (\gamma_\mu(n))^{-1/2} A^{\star n} \phi_0^\mu$$

(3.7.4)
$$H_\mu \phi_n^\mu = \tfrac{1}{2}(P_\mu^2 + Q_\mu^2)\phi_n^\mu = \tfrac{1}{2}(AA^\star + A^\star A)\phi_n^\mu = (n + \mu + \tfrac{1}{2})\phi_n^\mu \text{ on } \mathfrak{S}.$$

(3.7.5) $$J_\mu \phi_n^\mu = (-1)^n \phi_n^\mu$$

(3.7.6) $$i(P_\mu Q_\mu - Q_\mu P_\mu) = (AA^\star - A^\star A) = I_\mu + 2\mu J_\mu \text{ on } \mathfrak{S}.$$

(3.7.7) $$\mathcal{F}_\mu^\star Q_\mu \mathcal{F}_\mu = P_\mu \text{ on } \mathfrak{S}$$

Here I_μ *is the identity operator and* J_μ *is the unitary involution defined by* $J_\mu \phi(x) = \phi(-x)$, $\phi \in L_\mu^2(R)$, $x \in R$.

Proof. By use of (2.5.3) one obtains

$$\mathfrak{D}_{\mu,x}\left(\exp(-\tfrac{1}{2}x^2)\phi(x)\right) = \exp(-\tfrac{1}{2}x^2)\left(\mathfrak{D}_\mu - \mathfrak{Q}_\mu\right)\phi(x)$$

for all smooth ϕ. Thus (3.7.1) and (3.7.2) follow from (2.6.1) and (2.6.2). (3.7.3) is a consequence of (3.7.2).

P_μ, Q_μ, H_μ can be written in terms of A_μ, $A_\mu^\star$, so (3.7.4) and (3.7.6) can be derived from (3.7.1) and (3.7.2). (3.7.5) is true since ϕ_{2n} is even and ϕ_{2n+1} is odd. Now, the multiplication operator $\mathfrak{Q}_\mu$ clearly has a unique closed extension to a selfadjoint operator on the set $\{f \in L_\mu^2(R) : \mathfrak{Q}f \in L_\mu^2(R)\}$, and this operator we name Q_μ. Then $\mathcal{F}_\mu^\star Q_\mu \mathcal{F}_\mu$ is again a selfadjoint operator. One shows that (3.7.7) is true by using (3.5.2), (3.7.1) and (3.7.2).

A class of generalized harmonic and conjugate harmonic functions and a Hilbert transform operator associated with the generalized Fourier transform operator $\mathcal{F}_\mu$ was introduced and studied by Muckenhoupt and Stein in [21]. They sketch a proof of the following interesting generalization of the F. and M. Riesz theorem on the absolute continuity of analytic measures. If $\mu = 0$, then the result is the classical one.

THEOREM 3.8 F. AND M. RIESZ THEOREM ON ABSOLUTE CONTINUITY OF ANALYTIC MEASURES. *Assume that* $\mu \in (-\frac{1}{2}, \infty)$, $a \in R$ *and* ν *is a complex Borel measure on* $R_a := [a, \infty)$ *such that* ν *is finite if* $\mu \in [0, \infty)$ *and*

$\int_{R_a}(|t|^{-\mu}+1)|\nu|(dt)<\infty$ *if* $\mu\in(-\frac{1}{2},0)$. *Assume* ν *is an analytic measure, that is,*

$$\int_{R_a} e_\mu(ixt)\nu(dt)=0 \tag{3.8.1}$$

for all real x. Then ν *is absolutely continuous with respect to linear Lebesgue measure.*

Proof. See [21,p88] for the $a=0$ and $\mu>0$ case. In case $\mu=0$ then the theorem is the classical F. and M. Riesz theorem, [14,p45]. Our proof consists in showing that the theorem's hypotheses and (2.3.5), (2.3.6) imply that (3.8.1) holds with $\mu=0$.

Assume first that $-\frac{1}{2}<\mu<0$ and (3.8.1) holds. It follows from (2.3.6) that $\int_{R_a} e^{ixt}\nu(dt)=0$ for all real x, so ν is absolutely continuous by the classical result.

Assume next that $\mu>0$. Then an application of (2.3.7) yields $\int_{R_a}\mathrm{e}_{\mu_o}(ixt)\nu(dt)=0$ for some $\mu_o<\mu$, so repeated applications of (2.3.7) reduce the problem to the case when $\mu=0$.

We generalize the Gauss-Weierstrass operator semi-group of Hille and Phillips [18,p570] and continue to study the generalized heat equation (2.7.3) .

3.9 DEFINITION. *For* $t>0$ *and* $\mu>-\frac{1}{2}$ *the* $L^2_\mu(R)$ *operator* $\boldsymbol{T}_\mu(t)$ *is defined by*

$$(\boldsymbol{T}_\mu(t)\phi)(x):=\frac{1}{(4t)^{\mu+\frac{1}{2}}\Gamma(\mu+\frac{1}{2})}\int_R \exp(-\frac{x^2+y^2}{4t})\mathrm{e}_\mu(\frac{xy}{2t})\phi(y)|y|^{2\mu}\,dy, \tag{3.9.1}$$

$$x,y\in R.$$

3.10 THEOREM.

$$\boldsymbol{T}_\mu(t)\phi=\exp(-t\boldsymbol{P}^2_\mu)\phi \text{ for every } t>0 \text{ and } \phi\in L^2_\mu(R). \tag{3.10.1}$$

Proof. Assume $t>0$ and set $u(\cdot,t)=\exp(-t\boldsymbol{P}^2_\mu)\phi$, which by (3.7.5) equals $\mathcal{F}^*_\mu\exp(-t\boldsymbol{Q}^2_\mu)\mathcal{F}_\mu$. Then if $\phi\in L^2_\mu(R)$,

$$u(x,t)=$$

$$(2^{\mu+\frac{1}{2}}\Gamma(\mu+\tfrac{1}{2}))^{-2}\int_{-\infty}^{\infty}\mathrm{e}_\mu(ix\tau)\exp(-t\tau^2)|\tau|^{2\mu}(\int_{-\infty}^{\infty}\mathrm{e}_\mu(-iy\tau)\phi(y)|y|^{2\mu}dy)\,d\tau=$$

$$(2^{\mu+\frac{1}{2}}\Gamma(\mu+\tfrac{1}{2}))^{-2}\int_{-\infty}^{\infty}(\int_{-\infty}^{\infty}\mathrm{e}_\mu(-iy\tau)\mathrm{e}_\mu(ix\tau)\exp(-t\tau^2)|\tau|^{2\mu}d\tau)\phi(y)|y|^{2\mu}dy$$

which by (3.3.4) $=(\boldsymbol{T}_\mu(t)\phi)(x)$

3.11 THEOREM.

$$(3.11.1)\quad \boldsymbol{T}_\mu(t/4) : \big(\exp(-\alpha x^2)\mathrm{e}_\mu(2zx)\big) \longrightarrow$$
$$(1+\alpha t)^{-\mu-1/2}\exp(\frac{tz^2}{1+\alpha t})\exp(-\frac{\alpha x^2}{1+\alpha t})\mathrm{e}_\mu(\frac{2zx}{1+\alpha t}),$$
$$\Re t > 0,\ z \in C,\ \Re\alpha > 0.$$

$$(3.11.2)\quad \boldsymbol{T}_\mu(t) : \exp(-\alpha x^2) \longrightarrow (1+4\alpha t)^{-\mu-1/2}\exp(-\frac{\alpha x^2}{1+4\alpha t}),$$
$$\Re t > 0,\ \Re\alpha > 0.$$

Proof. (3.11.1) is a consequence of (2.7.2) and (3.11.2) is obtained by setting $z = 0$.

3.12 THEOREM. *Suppose* $\phi \in L^2_\mu(R)$. *Then* $\psi(x,t) := (\boldsymbol{T}_\mu(t)\phi)(x)$ *satisfies*

$$(3.12.1)\qquad \frac{\partial\psi}{\partial t} = -\boldsymbol{T}^2_\mu\psi = \mathfrak{D}^2_\mu\psi \in L^2_\mu(R)$$

$$(3.12.2)\qquad \text{and } \lim_{t\searrow 0}\|\psi(\cdot,t)-\phi\|_\mu = 0.$$

Proof. By (3.7.5) $\psi(\cdot,t) = \exp(-t\boldsymbol{P}^2_\mu)\phi = \mathcal{F}^*_\mu\exp(-t\boldsymbol{Q}^2_\mu)\mathcal{F}_\mu\phi$. Since $\boldsymbol{Q}^2_\mu\exp(-t\boldsymbol{Q}^2_\mu)\phi \in L^2_\mu(R)$ for every $\phi \in L^2_\mu(R)$, it follows that $\boldsymbol{P}^2_\mu\exp(-t\boldsymbol{P}^2_\mu)\phi \in L^2_\mu(R)$ for every $\phi \in L^2_\mu(R)$. Thus we see that (3.12.1) holds.

$$\|\psi(\cdot,t)-\phi\|^2_\mu = \|\mathcal{F}^\star\exp(-t\boldsymbol{Q}^2_\mu)\mathcal{F}\phi-\phi\|^2_\mu = \|\exp(-t\boldsymbol{Q}^2_\mu)\mathcal{F}\phi-\mathcal{F}\phi\|^2_\mu$$
$$= \int_{-\infty}^{\infty}|\exp(-ty^2)-1|^2|(\mathcal{F}\phi)(y)|^2|y|^{2\mu}dy \longrightarrow 0 \text{ as } y\searrow 0.$$

(3.12.2) follows.

3.13 COROLLARY. *Assume the hypotheses and notation of Theorem 3.12 , so* $\psi(t,x)$ *satisfies (3.12.1) and (3.12.2). i) Suppose that* ϕ *is also an even function. Then*

$$(3.13.1)\qquad \frac{\partial\psi}{\partial t} = \frac{\partial^2\psi}{\partial x^2} + \frac{2\mu}{x}\frac{\partial\psi}{\partial x}$$

ii) Suppose that ϕ *is also an odd function. Then*

$$(3.13.2)\qquad \frac{\partial\psi}{\partial t} = \frac{\partial^2\psi}{\partial x^2} + \frac{2\mu}{x}\frac{\partial\psi}{\partial x} - \frac{2\mu}{x^2}\psi$$

Proof. Use (2.5.1) and (3.12).

4. Generalized translation

4.1 Definition. *i) The* **generalized translation operator** $\mathfrak{T}_y$, $y \in R$ *is defined by*

$$\mathfrak{T}_y \phi := \mathbf{e}_\mu(y\mathfrak{D}_\mu)\phi = \sum_{n=0}^{\infty} \frac{y^n}{\gamma_\mu(n)} \mathfrak{D}_\mu^n \phi$$

for all entire functions ϕ on C for which the series converges pointwise.

ii) The linear operator $\mathcal{T}_y$, $y \in R$ is defined on $L^2_\mu(R)$ by

$$\mathcal{T}_y \phi := \mathbf{e}_\mu(iy\boldsymbol{P}_\mu)\phi. \tag{4.1.1}$$

We use the notation $\mathfrak{T}_{y,x}$ when we wish to emphasize the functional dependence on the variable x.

*iii) The μ–***binomial coefficients** *are defined by*

$$\binom{n}{k}_\mu := \gamma_\mu(n)/(\gamma_\mu(j)\gamma_\mu(n-j)), \quad k = 0, \ldots, n,\ n \in \mathbb{N}. \tag{4.1.2}$$

*iv) The μ–***binomial polynomials** $\{p_{n,\mu}(\cdot,\cdot)\}_{n\in\mathbb{N}}$ *are defined by*

$$p_{n,\mu}(x,y) := \mathfrak{T}_{y,x} x^n = \sum_{j=0}^{\infty} \frac{y^j}{\gamma_\mu(j)} \mathfrak{D}_{\mu,x}^j x^n. \tag{4.1.3}$$

Notice that $\mathcal{T}_y = \mathfrak{T}_y$ for almost all real y on the class of $L^2_\mu(R)$ entire functions of the form $\{p(x)\exp(-\lambda x^2) : p$ is a complex polynomial and $\lambda > 0\}$.

4.2 Properties of $\mathfrak{T}_y$. *μ-binomial expansion :*

$$p_{n,\mu}(x,y) = \sum_{j=0}^{n} \binom{n}{k}_\mu x^j y^{n-j}. \tag{4.2.1}$$

The first few μ–binomial polynomials are 1, $x+y$, $x^2 + \frac{2}{1+2\mu}xy + y^2$, *and* $x^3 + \frac{3+2\mu}{1+2\mu}(x^2y + xy^2) + y^3$, $x^4 + 4\frac{1}{1+2\mu}(x^3y + xy^3) + 2\frac{3+2\mu}{1+2\mu}x^2y^2 + y^4$. *Thus it is clear that $\mathfrak{T}_y$ in general does not take nonnegative functions into nonnegative functions. Consider, for example,* $(x-1)^2$.

$$\mathfrak{T}_{y,x}\mathbf{e}_\mu(\lambda x) = \mathbf{e}_\mu(\lambda y)\mathbf{e}_\mu(\lambda x), \quad \lambda \in C \tag{4.2.2}$$

Generating function :

$$\text{(4.2.3)} \qquad \mathbf{e}_\mu(\lambda x)\mathbf{e}_\mu(\lambda y) = \sum_{n=0}^{\infty} \frac{p_{n,\mu}(x,y)}{\gamma_\mu(n)} \lambda^n, \quad \text{if} \quad \lambda \in C.$$

(4.2.4)

$$\mathcal{T}_y : \exp(-\lambda x^2) \longmapsto \exp(-\lambda(x^2+y^2))\mathbf{e}_\mu(-2\lambda xy) \quad \text{if} \quad \lambda > 0$$

(4.2.5)

$$\mathcal{T}_y : x\exp(-\lambda x^2) \longmapsto (x+y)\exp(-\lambda(x^2+y^2))\mathbf{e}_\mu(-2\lambda xy) \quad \text{if} \quad \lambda > 0$$

Proof. Use (2.5.2) so

$$p_{n,\mu}(x,y) = \sum_{j=0}^{\infty} \frac{y^j}{\gamma_\mu(j)} \mathfrak{D}^j_{\mu,x} x^j = \sum_{j=0}^{n} \gamma_\mu(n)/(\gamma_\mu(j)\gamma_\mu(n-j))x^j y^{n-j},$$

which is equivalent to (4.2.1).

$$\mathfrak{T}_{y,x}\mathbf{e}_\mu(\lambda x) = \mathbf{e}_\mu(y\mathfrak{D}_{\mu,x})\mathbf{e}_\mu(\lambda x) =$$

$$\sum_{n=0}^{\infty} \frac{y^n}{\gamma_\mu(n)} \mathfrak{D}^n_{\mu,x}\mathbf{e}_\mu(\lambda x) = \sum_{n=0}^{\infty} \frac{\lambda^n y^n}{\gamma_\mu(n)} \mathbf{e}_\mu(\lambda x) = \mathbf{e}_\mu(\lambda y)\mathbf{e}_\mu(\lambda x)$$

This proves (4.2.2).
(4.2.3) is a consequence of (4.2.2) and (4.1.3) . (4.2.4) follows from (2.6.5),(2.5.8) since:

Set $\lambda = \alpha^2$. Then $\mathfrak{T}_y : \exp(-\alpha^2 x^2) \longmapsto \mathbf{e}_\mu(y\mathfrak{D}_{\mu,x})\exp(-\alpha^2 x^2)$

$$= \exp(-\alpha^2 x^2) \sum_{n=0}^{\infty} \frac{y^n}{\gamma_\mu(n)} \mathfrak{D}^n_{\mu,x} \exp(-\alpha^2 x^2) = \sum_{n=0}^{\infty} (-1)^n \frac{\alpha^n y^n}{n!} H^\mu_n(\alpha x)$$

$$= \exp(-\alpha^2 x^2)\exp(-\alpha^2 y^2)\mathbf{e}_\mu(-2\alpha^2 xy)$$

(4.2.5) follows when one applies $\mathfrak{D}_{\mu,x}$ to both sides of (4.2.4).

In the rest of section 4 we assume $\mu > 0$. Define the probability measure α_μ and the function $\widetilde{\omega}$ by

$$\text{(4.2.6)} \qquad \alpha_\mu(dt) := \frac{1}{B(\frac{1}{2},\mu)}(1-t)^{\mu-1}(1+t)^\mu\,dt, \quad t \in (-1,1),$$

$$\text{(4.2.7)} \qquad \widetilde{\omega}(t) := (x^2+2xyt+y^2)^{\frac{1}{2}}, \quad t \in [-1,1] \text{ and } x,y \in R.$$

4.3 LEMMA. *Suppose ϕ is an $L^\infty(R)$ function. Then*

$$\mathfrak{T}_{y,x}(\phi(x)) = \frac{1}{2}\int_{-1}^{1}(1+\frac{x+y}{\widetilde{\omega}(t)})\phi(\widetilde{\omega}(t))\,\alpha_\mu(dt)+ \frac{1}{2}\int_{-1}^{1}(1-\frac{x+y}{\widetilde{\omega}(t)})\phi(-\widetilde{\omega}(t))\,\alpha_\mu(dt) \tag{4.3.1}$$

Proof. It follows from (4.2.4),(4.2.5) and (2.3.5) that

$$(\mathfrak{T}_y\phi)(x) = \int_{-1}^{1}\psi(\widetilde{\omega}(t))\,\alpha_\mu(dt), \qquad (\mathfrak{T}_y\psi)(x) = (x+y)\int_{-1}^{1}\frac{1}{\widetilde{\omega}(t)}\psi(\widetilde{\omega}(t))\,\alpha_\mu(dt)$$

if ϕ is an even and ψ is an odd $L^\infty(R)$ function. Thus (4.3.1) follows.

4.4 COROLLARY. *Suppose $\mu > 0$ and ϕ is an entire function. Then*

$$(\mathfrak{T}_y\phi)(x) = (\mathfrak{T}_x\phi)(y) \text{ for all real x and } y\,.$$
$$\mathfrak{T}_{y,x}(x\phi(x)) = (x+y)\mathfrak{T}_{y,x}(\phi(x)) \quad \text{if } \phi \text{ is even}\,.$$
$$p_{2n+1,\mu}(x,y) = (x+y)p_{2n,\mu}(x,y),\; n \in \mathrm{N}.$$

Proof. These follow from (4.3.1).

4.5 NOTATION. *i) Suppose $x, y, \xi \in R$. Define*

$$\begin{aligned} \Psi(x,y,\xi) &:= \frac{1}{16}((x+y)^2-\xi^2)(\xi^2-(x-y)^2) \\ \Xi &:= \{(x,y,\xi)\in R^3 : \Psi(x,y,\xi) > 0\} \\ \Xi(x,y) &:= \{\xi \in R : \Psi(x,y,\xi) > 0\} \end{aligned} \tag{4.5.1}$$

Thus

$$\Xi(x,y) = (-|x+y|, -|x-y|) \cup (|x-y|, |x+y|) \text{ if } xy > 0$$
$$\Xi(x,y) = (-|x-y|), -|x+y|) \cup (|x+y|, |x-y|) \text{ if } xy < 0.$$

ii) We note that Ψ is the homogeneous symmetric polynomial that is relevant in Heron's formula for the area of a triangle, and $\Psi(x,y,\xi) = \Psi(|x|,|y|,|\xi|)$. It appears in the expressions for the generalized translation operator that form a basis for the Bessel calculus studied by Cholewinski [8] and by others. See [2,p35-36] and [8] for references but note that operator they study acts only on even functions or on functions on a half-line. The translation operator we will set down acts more generally on functions on R. iii) Suppose $x, y, \xi \in R$. Define

$\Delta(x,y,\xi)$ to be the area of the triangle formed, if possible, with sides of length $|x|, |y|, |\xi|$, and 0 otherwise. Then Heron's formula, see [9,p12] states that

$$\begin{aligned} &\Delta(x,y,\xi) = (\Psi(x,y,\xi))^{\frac{1}{2}} \quad \textit{if } (x,y,\xi) \in \Xi \text{ and} \\ &\Delta(x,y,\xi) = 0 \quad \textit{if } (x,y,\xi) \notin \Xi \end{aligned} \tag{4.5.2}$$

iv) Define the measure $\beta_{\mu,x,y}$ on $\Xi(x,y)$ by

$$\beta_{\mu,x,y}(d\xi) := \frac{1}{B(\frac{1}{2},\mu)}\left(\frac{\Delta(x,y,\xi)}{|xy|}\right)^{2\mu} d\xi. \tag{4.5.3}$$

4.6 LEMMA. *Assume $x,y \in R\setminus\{0\}$, $\phi,\psi \in L^\infty(R)$ with ϕ an even and ψ an odd function. Then*

$$\int_{\Xi(x,y)} \frac{sgn(xy\xi)}{x+y-\xi}\phi(\xi)\,\beta_{\mu,x,y}(d\xi). = \int_{-1}^{1} \phi(\widetilde{\omega}(t))\alpha_\mu(dt) \tag{4.6.1}$$

$$\int_{\Xi(x,y)} \frac{sgn(xy\xi)}{x+y-\xi}\psi(\xi)\,\beta_{\mu,x,y}(d\xi) = (x+y)\int_{-1}^{1} \frac{\psi(\widetilde{\omega}(t))}{\widetilde{\omega}(t)}\alpha_\mu(dt) \tag{4.6.2}$$

.

Proof. Set $\xi = \widetilde{\omega}(t) = (x^2+y^2+2xyt)^{\frac{1}{2}}$, so $t = \frac{\xi^2-x^2-y^2}{2xy}$ Thus

$$1+t = (\xi^2-(x-y)^2)/(2xy), \quad 1-t = ((x+y)^2-\xi^2)/(2xy), \text{ and}$$

$$(1-t^2)^\mu = \left(\frac{2\Delta(x,y,\xi)}{|xy|}\right)^{2\mu}.$$

First assume $xy > 0$. Then $B(\frac{1}{2},\mu)$ times the right side of (4.6.1)

$$= \left(\int_{-|x+y|}^{-|x-y|} + \int_{|x-y|}^{|x+y|}\right)\frac{sgn(xy\xi)}{x+y-\xi}\phi(\xi)\,\beta_{\mu,x,y}(d\xi) \quad =$$

$$\int_{|x-y|}^{|x+y|} \frac{2\xi}{(x+y)^2-\xi^2}\left(\frac{2\Delta(x,y,\xi)}{|xy|}\right)^{2\mu}\phi(\xi)d\xi = \int_{-1}^{1}\phi(\widetilde{\omega}(t))(1-t)^{\mu-1}(1+t)^\mu\,dt.$$

This implies (4.6.1) in case $xy > 0$. The case $xy < 0$ follows similarly. From (4.6.1) we deduce that the right term in (4.6.2) equals

$$(x+y)\int_{\Xi(x,y)} \frac{sgn(xy\xi)}{(x+y-\xi)\xi}\psi(\xi)\,\beta_{\mu,x,y}(d\xi)$$

$$= \int_{\Xi(x,y)} sgn(xy\xi)\left(\frac{1}{x+y-\xi}+\frac{1}{\xi}\right)\psi(\xi)\,\beta_{\mu,x,y}(d\xi) = \text{ the left side of (4.6.2).}$$

4.7 THEOREM. *Suppose $\phi \in L^\infty(R)$, $\mu > 0$, and $x, y \in R \setminus \{0\}$.*

(4.7.1)
$$(\mathfrak{T}_y\phi)(x) == \int_{\Xi(x,y)} \frac{sgn(xy\xi)}{x+y-\xi}\phi(\xi)\,\beta_{\mu,x,y}(d\xi)$$

Proof. (4.7.1) holds for even and odd ϕ by Lemma 4.6 and Theorem 4.3. Thus it holds for all ϕ.

4.8 THEOREM. *Suppose $\phi \in L^\infty(R)$ and $\mu > 0$.*

(4.8.1)
$$(\mathfrak{T}_y\phi)(x) = (\mathfrak{T}_x\phi)(y) \text{ for almost all real x and y}$$

(4.8.2)
$$\|\mathfrak{T}_y\phi\|_\mu \leq \|\phi\|_\mu, \phi \in L^2_\mu(R), y \in R.$$

Proof. (4.8.1) is clearly implied by (4.7.1), and (4.8.2) by (3.1.4) and (4.1.1).

5. THE BOSE-LIKE OSCILLATOR

Suppose that $\mathfrak{H}$ is a complex Hilbert space. We will be examining certain equations of motion and commutation relations that relate several unbounded operators. In order to avoid the pitfalls associated with formal computation involving unbounded operators [26,p270-274] we shall postulate the existence of a suitably tailored linear invariant set of analytic vectors.

DEFINITION 5.1. *i)Suppose $\boldsymbol{P}, \boldsymbol{Q}$ and $\boldsymbol{H}$ are possibly unbounded selfadjoint operators on $\mathfrak{H}$. $\boldsymbol{P}$ and $\boldsymbol{Q}$ are dominated* by $\boldsymbol{H}$ if

$\mathfrak{S}$ is a linear invariant set of analytic vectors for $\boldsymbol{P}, \boldsymbol{Q}$, and $\boldsymbol{H}$,

and $\mathfrak{S} := \bigcup\{\boldsymbol{E}((-n,n))\mathfrak{H} : n \in \mathrm{N}\}$, where $\boldsymbol{E}$ is the spectral measure of $\boldsymbol{H}$.

ii) We next specify $\boldsymbol{H}$. Let $V(\cdot)$ be a continuously differentiable real function on the real line with derivative $V'(\cdot)$ and specify the associated *Hamiltonian operator* $\boldsymbol{H}$ by $\boldsymbol{H} = \frac{1}{2}\boldsymbol{P}^2 + V(\boldsymbol{Q})$. Then the *equations of motion associated with the Hamiltonian* are

(EM)
$$i[\boldsymbol{P},\boldsymbol{H}] := i(\boldsymbol{PH}-\boldsymbol{HP}) = V'(\boldsymbol{Q}) \quad \text{and} \quad i[\boldsymbol{Q},\boldsymbol{H}] := i(\boldsymbol{QH}-\boldsymbol{HQ}) = -\boldsymbol{P}$$

on $\mathfrak{S}$. In case $V(\boldsymbol{Q}) = \frac{1}{2}\boldsymbol{Q}^2$ the equations of motion are that of the *quantum mechanical harmonic oscillator.*

iii) Suppose $\boldsymbol{P}$ and $\boldsymbol{Q}$ are selfadjoint operators that are dominated by the selfadjoint operator $\boldsymbol{H}$, where $\boldsymbol{H} := \frac{1}{2}(\boldsymbol{P}^2 + \boldsymbol{Q}^2)$ on $\mathfrak{S}$, where $\mathfrak{S}$ is as in 5.1i).

Then $(\mathfrak{H}, \boldsymbol{P}, \boldsymbol{Q}, \boldsymbol{H})$ is a **Bose-like (quantum mechanical simple harmonic) oscillator** , *or a* **para-Bose oscillator** *if the equations of motion are*

$$i[\boldsymbol{P}, \boldsymbol{H}] = \boldsymbol{Q} \quad \text{and} \quad i[\boldsymbol{Q}, \boldsymbol{H}] = -\boldsymbol{P}. \tag{5.1.1}$$

iv) The Bose-like oscillator is **irreducible** *if whenever* $\boldsymbol{B}$ *is an everywhere defined bounded operator on* $\mathfrak{H}$ *to* $\mathfrak{H}$ *such that*

$$\exp(i\lambda \boldsymbol{P})\boldsymbol{B} = \boldsymbol{B}\exp(i\lambda \boldsymbol{P}) \quad \textit{and} \quad \exp(i\lambda \boldsymbol{Q})\boldsymbol{B} = \boldsymbol{B}\exp(i\lambda \boldsymbol{Q}) \tag{5.1.2}$$

for all real λ , *then there exists* $c \in C$ *such that* $\boldsymbol{B} = c\boldsymbol{I}$.

REMARK 5.2. $\boldsymbol{Q}$ *is the* **position** , $\boldsymbol{P}$ *is the* **momentum** , $\boldsymbol{A} := 2^{-\frac{1}{2}}(\boldsymbol{Q} + i\boldsymbol{P})$ *is the* **annihilation** *and* $\boldsymbol{A}^\star = 2^{-\frac{1}{2}}(\boldsymbol{Q} - i\boldsymbol{P})$ *is the* **creation** *operator.* $\boldsymbol{H} = \frac{1}{2}(\boldsymbol{A}^\star \boldsymbol{A} + \boldsymbol{A}\boldsymbol{A}^\star)$.

REMARK 5.3. *i)By a quantum mechanical system consisting of a single particle moving in one dimension* we mean a triple of self adjoint operators $\boldsymbol{P}$ and $\boldsymbol{Q}$ dominated by $\boldsymbol{H} = \frac{1}{2}\boldsymbol{P}^2 + V(\boldsymbol{Q})$ that satisfy the commutation relation

$$i[\boldsymbol{P}, \boldsymbol{Q}] := i(\boldsymbol{P}\boldsymbol{Q} - \boldsymbol{Q}\boldsymbol{P}) = \boldsymbol{I} \text{ on } \mathfrak{S} \tag{$\mathbb{CR}$}$$

If $\mathbb{CR}$ holds, then the equations of motion $\mathbb{EM}$ holds on $\mathfrak{S}$.

Proof. The following computations are valid on $\mathfrak{S}$:

$$i[\boldsymbol{P}, \boldsymbol{H}] = i[\boldsymbol{P}, V(\boldsymbol{Q})] = V'(\boldsymbol{Q}), \text{ and}$$

$$2i[\boldsymbol{Q}, \boldsymbol{H}] = i[\boldsymbol{Q}, \boldsymbol{P}^2] = i[\boldsymbol{Q}, \boldsymbol{P}]\boldsymbol{P} + i\boldsymbol{P}[\boldsymbol{Q}, \boldsymbol{P}] = -2\boldsymbol{P},$$

which imply $\mathbb{EM}$.

In 1950 E. P. Wigner [31] posed the question 'Do the equations of motion determine the quantum mechanical commutation relations? 'He considered formal operator equations of motion of the form $\mathbb{EM}$. In the case where the equations of motion are that of a Bose-like oscillator, Wigner noted that there exists a one parameter family of inequivalent operator representations for position $\boldsymbol{Q}$ and momentum $\boldsymbol{P}$. There is now an extensive physical literature on these representations, see [24] and [22]. We shall give a self-contained treatment of the theory , and indicate their relationship to the generalized Fourier transform and generalized Hermite functions.

Our goal is to study a generalization of the Boson calculus. The *Boson calculus* is the collection of operators, functions, and analysis associated with the quantum mechanical harmonic oscillator. It is often studied using Lie group theory, by noting that $\mathbb{CR}$ gives rise to the Heisenberg group through the Weyl commutation relations. An alternate perspective to the Boson calculus appears in Glimm and Jaffe [17,Chapt1] and Biedenharn and Louck [3,Chapt5]. Our perspective uses operator equations and operator theory directly to implement the analysis, without emphasizing the Lie group aspects of the algebraic structure of the operator equations.

The generalization, the *Bose-like oscillator calculus,* is set down in Ohnucki and Kamefuchi [24,Chapt23] and in [22]. For a related calculus see Cholewinski [7].

From now on we consider a Bose-like oscillator and fix the notation of definition 5.1. We do not assume that $\mathbb{CR}$ holds.

5.4 Alternate formulations of the equations of motion. *The following statements are equivalent to (5.1.1):*

$$i[\boldsymbol{P},\boldsymbol{Q}^2]=2\boldsymbol{Q} \quad \text{and} \quad i[\boldsymbol{P}^2,\boldsymbol{Q}]=2\boldsymbol{P} \text{ on } \mathfrak{S}; \tag{5.4.1}$$

$$[\boldsymbol{A},\boldsymbol{H}]=\boldsymbol{A} \quad or \quad [\boldsymbol{A}^\star,\boldsymbol{H}] = -\boldsymbol{A}^\star \text{ on } \mathfrak{S}; \tag{5.4.2}$$

$$[\boldsymbol{A},\boldsymbol{A}^{\star 2}]=2\boldsymbol{A}^\star \quad or \quad [\boldsymbol{A}^\star,\boldsymbol{A}^2] = -2\boldsymbol{A} \text{ on } \mathfrak{S}. \tag{5.4.3}$$

Proof. (5.4.1)-(5.4.3) follow from (5.1.1) and 5.1, 5.2.

5.5 Remark. *Suppose $n\in\mathbb{N}$. When considered on $\mathfrak{S}$:*

$$[\boldsymbol{A},\boldsymbol{A}^\star]=i[\boldsymbol{P},\boldsymbol{Q}] \text{ commutes with } \boldsymbol{A}^2,\boldsymbol{A}^{\star 2}, \text{ and } \boldsymbol{P}^2,\boldsymbol{Q}^2,\boldsymbol{H}; \tag{5.5.1}$$

$$[\boldsymbol{A},\boldsymbol{A}^{\star(2n)}]=2n\boldsymbol{A}^{\star(2n-1)}, \tag{5.5.2}$$

$$[\boldsymbol{A},\boldsymbol{A}^{\star(2n+1)}]=\boldsymbol{A}^{\star(2n)}(2n+[\boldsymbol{A},\boldsymbol{A}^\star]); \tag{5.5.3}$$

$$i[\boldsymbol{P},\boldsymbol{Q}^{2n}]=2n\boldsymbol{Q}^{2n-1}, \qquad i[\boldsymbol{P},\boldsymbol{Q}^{2n+1}]=\boldsymbol{Q}^{2n}(2n+i[\boldsymbol{P},\boldsymbol{Q}]); \tag{5.5.4}$$

$$i[\boldsymbol{P}^{2n},\boldsymbol{Q}]=2n\boldsymbol{P}^{2n-1}, \qquad i[\boldsymbol{P}^{2n+1},\boldsymbol{Q}]=\boldsymbol{P}^{2n}(2n+i[\boldsymbol{P},\boldsymbol{Q}]) \tag{5.5.5}$$

Proof. (5.5.1) follows from 5.4 and the Jacobi identity

$$[[\boldsymbol{X},\boldsymbol{Y}],\boldsymbol{Z}]+[[\boldsymbol{Y},\boldsymbol{Z}],\boldsymbol{X}]+[[\boldsymbol{Z},\boldsymbol{X}],\boldsymbol{Y}]=0,$$

which holds for suitably defined operators $\boldsymbol{X},\boldsymbol{Y},\boldsymbol{Z}$ selected from $\boldsymbol{A}^2,\boldsymbol{A}^{\star 2},\boldsymbol{H}$ and $\boldsymbol{A},\boldsymbol{A}^\star$.

(5.5.2) is true if $n=1$ by 5.4. Assume that it is true for n. Then $(2n+2)\boldsymbol{A}^{\star(2n+1)} = (\boldsymbol{A}\boldsymbol{A}^{\star 2n}-\boldsymbol{A}^{\star 2n}\boldsymbol{A})\boldsymbol{A}^{\star 2}+2\boldsymbol{A}^{\star(2n+1)}$ which equals $\boldsymbol{A}\boldsymbol{A}^{\star(2n+2)}-\boldsymbol{A}^{\star 2n}(\boldsymbol{A}\boldsymbol{A}^{\star 2}-2\boldsymbol{A}^\star)$. By (5.4.3) this $=[\boldsymbol{A},\boldsymbol{A}^{\star(2n+2}]$. This proves (5.5.2).

(5.5.3) is clearly true if for $n=0$. Assume that it is true for n. Then $[\boldsymbol{A},\boldsymbol{A}^{\star(2n+3)}]=\left(\boldsymbol{A}\boldsymbol{A}^{\star(2n+1)}-\boldsymbol{A}^{\star(2n+1)}\boldsymbol{A}\right)\boldsymbol{A}^{\star 2}+\boldsymbol{A}^{\star(2n+1)}\left(\boldsymbol{A}\boldsymbol{A}^{\star 2}-\boldsymbol{A}^{\star 2}\boldsymbol{A}\right)$. But this equals $\boldsymbol{A}^{\star(2n+2)}(2n+[\boldsymbol{A},\boldsymbol{A}^\star])+\boldsymbol{A}^{\star(2n+1)}(2\boldsymbol{A}^\star)$, so (5.5.3) follows.

(5.5.4) and (5.5.5) are proved similarly.

5.6 LEMMA. *The following operator identities hold for all real* λ, ν.

(5.6.1) $$\exp(i\lambda \boldsymbol{H})\boldsymbol{Q}\exp(-i\lambda \boldsymbol{H}) = \boldsymbol{Q}cos\lambda + \boldsymbol{P}sin\lambda \text{ on } \mathfrak{S}$$

(5.6.2) $$\exp(i\lambda \boldsymbol{H})\boldsymbol{P}\exp(-i\lambda \boldsymbol{H}) = -\boldsymbol{Q}sin\lambda + \boldsymbol{P}cos\lambda \text{ on } \mathfrak{S}$$

(5.6.3) $$\exp(i\lambda \boldsymbol{H})\boldsymbol{A}\exp(-i\lambda \boldsymbol{H}) = \exp(-i\lambda)\boldsymbol{A} \text{ on } \mathfrak{S}$$

(5.6.4) $$\exp(i\lambda \boldsymbol{H})\exp(i\nu \boldsymbol{Q})\exp(-i\lambda \boldsymbol{H}) = \exp\big(i\nu(\boldsymbol{Q}cos\lambda + \boldsymbol{P}sin\lambda)\big) \text{ on } \mathfrak{H}$$

(5.6.5) $$\exp(i\lambda \boldsymbol{H})\exp(i\nu \boldsymbol{P})\exp(-i\lambda \boldsymbol{H}) = \exp\big(i\nu(-\boldsymbol{Q}sin\lambda + \boldsymbol{P}cos\lambda)\big) \text{ on } \mathfrak{H}$$

Proof. Let $F(\lambda)$ and $G(\lambda)$ equal the left sides of (5.6.1) and (5.6.2). Then

$$F'(\lambda) = -i\exp(i\lambda \boldsymbol{H})[\boldsymbol{Q},\boldsymbol{H}]\exp(-i\lambda \boldsymbol{H}) = G(\lambda) \text{ and}$$
$$G'(\lambda) = -i\exp(i\lambda \boldsymbol{H})[\boldsymbol{P},\boldsymbol{H}]\exp(-i\lambda \boldsymbol{H}) = -F(\lambda).$$

Also $F(0) = \boldsymbol{Q}$ and $G(0) = \boldsymbol{P}$. One obtains (5.6.1) and (5.6.2) by solving the system of differential equations for F and G on $\mathfrak{S}$. (5.6.3) follows from (5.6.1) and (5.6.2).

Since $\mathfrak{S}$ is a dense set of analytic vectors for $\boldsymbol{P},\boldsymbol{Q},\boldsymbol{H}$ one deduces from (5.6.1) and (5.6.2) that (5.6.4) and (5.6.5) hold when acting on a fixed vector in $\mathfrak{S}$ provided $|\nu|$ is small. Both sides are groups of unitary operators so (5.6.4) and (5.6.5) hold generally.

From now on we assume that the Bose-like oscillator is irreducible.

5.7 STRUCTURE THEOREM FOR THE BOSE-LIKE OSCILLATOR, I. *Suppose that* $(\mathfrak{H},\boldsymbol{P},\boldsymbol{Q},\boldsymbol{H})$ *is an irreducible Bose-like oscillator. Then there exists a real number* $\mu \in (-\frac{1}{2},\infty)$ *and* $\phi_0 \in \mathfrak{H}$ *as follows :*

$\boldsymbol{H}$ *has pure point spectra. The spectrum* $\sigma(\boldsymbol{H})$ *of* $\boldsymbol{H}$ *is given by*

(5.7.1) $$\sigma(\boldsymbol{H}) = \{\mu+\tfrac{1}{2},\mu+\tfrac{3}{2},\mu+\tfrac{5}{2},\dots\}.$$

$\mu+\frac{1}{2}$ *is the smallest eigenvalue of* $\boldsymbol{H}$. *Select* $\phi_0 \in \mathfrak{H}$ *so* $\boldsymbol{H}\phi_0 = (\mu+\frac{1}{2})\phi_0$, $\|\phi_0\| = 1$.

(5.7.2) *Define* $\phi_n := (\gamma_\mu(n))^{-1/2}\boldsymbol{A}^{\star n}\phi_0$, $n \in \mathrm{N}$ *and* $\phi_{-1} = 0$.

(5.7.3) $\{\phi_n\}_{n\in\mathrm{N}}$ *is a complete orthonormal set in* $\mathfrak{H}$.

(5.7.4) - (5.7.9) hold for all $n \in \mathbb{N}$.

(5.7.4)
$$A^m A^{\star n}\phi_0 = \frac{\gamma_\mu(n)}{\gamma_\mu(n-m)} A^{\star(n-m)}\phi_0 \ \textit{if}\ n \geq m \geq 0$$
$$= 0 \ \textit{if}\ m > n$$
$$A^\star A\phi_n = (n + 2\mu\theta_n)\phi_n$$
$$AA^\star\phi_n = (n + 1 + 2\mu\theta_{n+1})\phi_n$$

(5.7.5)
$$H\phi_n = \tfrac{1}{2}(AA^\star + A^\star A)\phi_n = (n + \mu + \tfrac{1}{2})\phi_n.$$

The commutation relation for the Bose-like oscillator:

(5.7.6) $i(PQ - QP) = I + 2\mu J$ on $\mathfrak{S}$, where $J := \exp\big(-\pi i(H - (\mu + \tfrac{1}{2})I)\big)$.

(5.7.7)
$$JP = -PJ, \ \text{and}\ JQ = -QJ \ \text{ on } \mathfrak{S}$$

(5.7.8)
$$J\phi_n = (-1)^n\phi_n, \ n \in \mathbb{N}.$$

(5.7.9)
$$J = J^\star = J^{-1}, \ \text{and}\ \exp\big(-2\pi i(H - \mu - \tfrac{1}{2})\big) = I.$$

Proof. Consider the operator $J_0 := \exp(-i\pi H)$ and assume that $\nu \in R$. Then from (5.6.4) and (5.6.5) we get $J_0^\star \exp(i\nu Q)J_0 = \exp(-i\nu Q)$ and $J_0^\star \exp(i\nu P)J_0 = \exp(-i\nu P)$. Thus $J_o Q = -QJ_o$ and $J_o P = -PJ_o$ on $\mathfrak{S}$. Also $\exp(i\nu Q)J_0^2 = J_0^2 \exp(i\nu Q)$ and $\exp(i\nu P)J_0^2 = J_0^2 \exp(i\nu P)$. It follows from definition 5.1 iv) that $J_0^2 = cI$ for some complex number c. But since J_0^2 is a unitary operator necessarily $J_0^2 = \exp(-2\pi i\alpha)I$ for some real number α. Next set $J = \exp(\pi i\alpha)J_0 = \exp(-i\pi(H - \alpha I))$. Clearly (5.7.7) is true. Since $J^2 = \exp(-2\pi i(H - \alpha I)) = I$, we deduce that (5.7.9) holds provided $\mu + \frac{1}{2} := \alpha$. When applied to (5.7.9) the spectral mapping theorem implies that $\sigma(H) \subseteq \mu + \frac{1}{2} + \mathbb{Z}$, where $\mathbb{Z}$ is the set of integers. Now, H is a non-negative operator, so we infer that (5.7.1) is valid if '$=$' is replaced by '$\subseteq$'. Clearly $\mu + \frac{1}{2} \geq 0$.

Select ϕ_0 and ϕ_n as in the statement of the theorem. From (5.4.2) we see that $AH\phi_0 = (H + I)A\phi_0$, so $(H - (\mu - \frac{1}{2})I)A\phi_0 = 0$. But $\mu + \frac{1}{2}$ is the smallest eigenvalue of H, so $A\phi_0 = 0$. Also $[A, A^\star]\phi_0 = AA^\star\phi_0 = 2H\phi_0 = (2\mu + 1)\phi_0$. Notice that $H\phi_0 = (\mu + \frac{1}{2})\phi_0$ implies that $J\phi_0 = \phi_0$.

From (5.5.2) and (5.5.3) we have $AA^{\star n}\phi_0 - A^{\star n}A\phi_0 = (n + 2\mu\theta_n)A^{\star(n-1)}\phi_0$ for all $n \in \mathbb{N} \setminus \{0\}$, so by (2.2.7) $AA^{\star n}\phi_0 = \frac{\gamma_\mu(n)}{\gamma_\mu(n-1)}A^{\star(n-1)}\phi_0$. This in turn implies the first statements in (5.7.4), and the rest of (5.7.4) follows from this and (2.2.7). (5.7.5) is a consequence of (5.7.4).

Assume $m \leq n$, $m, n \in \mathrm{N}$. Then, using (5.7.4), the inner product $\langle \phi_n, \phi_m \rangle = \left(\gamma_\mu(n)\gamma_\mu(m)\right)^{-1/2}\langle \boldsymbol{A}^{\star m}\boldsymbol{A}^{\star n}\phi_0, \phi_0 \rangle = \delta_{m,n}$. Thus $\{\phi_n\}$ is an orthonormal set in $\mathfrak{H}$. Let $\mathfrak{M}$ be the linear span of $\boldsymbol{A}^{\star n}\phi_0, n \in \mathrm{N}$. Then (5.7.4) implies that $\boldsymbol{A}$ and $\boldsymbol{A}^\star$ map $\mathfrak{M}$ into itself. Thus $\boldsymbol{Q}$ and $\boldsymbol{P}$ map $\mathfrak{M}$ into itself, and since the oscillator is irreducible, necessarily the closure of $\mathfrak{M}$ equals $\mathfrak{H}$. Hence (5.7.3) is proved.

Finally $i(\boldsymbol{PQ} - \boldsymbol{QP})\phi_n = (\boldsymbol{AA}^\star - \boldsymbol{A}^\star\boldsymbol{A})\phi_n = (1 + 2\mu\theta_{n+1} - 2\mu\theta_n)\phi_n = (\boldsymbol{I} + 2\mu\boldsymbol{J})\phi_n$ for all $n \in \mathrm{N}$, proving the commutator identity in (5.7.6)

The non-negativity of $\boldsymbol{H}$ implied that $\mu + \frac{1}{2} \geq 0$. Suppose that $\mu + \frac{1}{2} = 0$. Then by (5.7.4) $\boldsymbol{A}\phi_0 = \boldsymbol{A}^\star\phi_0 = 0$. It follows that $\boldsymbol{P}$ and $\boldsymbol{Q}$ both commute with the projection on ϕ_0, contradicting the irreducibility assumption of Definition 5.1iv) . It follows that necessarily $\mu \in (-\frac{1}{2}, \infty)$.

5.8 Lemma. *Suppose $n \in \mathrm{N} \setminus \{0\}$.*

(5.8.1) $$2^{-\frac{1}{2}}\boldsymbol{A}^\star\phi_0 = \boldsymbol{Q}\phi_0 = -i\boldsymbol{P}\phi_0 \quad \text{and} \quad i[\boldsymbol{P},\boldsymbol{Q}]\phi_0 = (1+2\mu)\phi_0$$

(5.8.2)
$$i[\boldsymbol{P},\boldsymbol{Q}^n]\phi_0 = \frac{\gamma_\mu(n)}{\gamma_\mu(n-1)}\boldsymbol{Q}^{n-1}\phi_0 \quad , \quad i[\boldsymbol{P}^n,\boldsymbol{Q}]\phi_0 = \frac{\gamma_\mu(n)}{\gamma_\mu(n-1)}\boldsymbol{P}^{n-1}\phi_0$$

(5.8.3) $$[\boldsymbol{A},\boldsymbol{A}^{\star n}]\phi_0 = \frac{\gamma_\mu(n)}{\gamma_\mu(n-1)}\boldsymbol{A}^{\star(n-1)}\phi_0$$

Proof. (5.8.1) is true since $(\boldsymbol{Q}+i\boldsymbol{P})\phi_0 = 2^{\frac{1}{2}}\boldsymbol{A}\phi_0 = 0$, and $\boldsymbol{Q} - i\boldsymbol{P} = 2^{\frac{1}{2}}\boldsymbol{A}^\star$. From (5.5.4) and (5.7.6) we obtain $i(\boldsymbol{PQ}^n - \boldsymbol{Q}^n\boldsymbol{P})\phi_0 = (n\boldsymbol{Q}^{n-1} + 2\mu\theta_n\boldsymbol{Q}^{n-1}\boldsymbol{J})\phi_0$. By (5.7.8) this yields the first equation in (5.8.2). The other equations in (5.8.1), (5.8.2) and (5.8.3) have similar proofs.

5.9 Theorem. *Suppose $p(\cdot)$ is a complex polynomial and $\mathfrak{D}_\mu$ is the generalized differentiation operator of (2.4.1). Then*

$$i[\boldsymbol{P}, p(\boldsymbol{Q})]\phi_0 = (\mathfrak{D}_\mu p)(\boldsymbol{Q})\phi_0, \quad i[p(\boldsymbol{P}),\boldsymbol{Q}]\phi_0 = (\mathfrak{D}_\mu p)(\boldsymbol{P})\phi_0$$

$$[\boldsymbol{A}, p(\boldsymbol{A}^\star)]\phi_0 = (\mathfrak{D}_\mu p)(\boldsymbol{A}^\star)\phi_0$$

In case $p(x) = H_n^\mu(\lambda x)$ one has $(\mathfrak{D}_\mu p)(x) = 2\lambda n H_{n-1}^\mu(\lambda x)$, $\lambda \in C$.

Proof. By (2.5.2) $p(x) = \sum c_n x^n$ is mapped by $\mathfrak{D}_\mu$ to $\sum \frac{\gamma_\mu(n)}{\gamma_\mu(n-1)} c_n x^{n-1}$, so the theorem follows easily from lemma 5.8. The last statement is proved in (2.6.1).

5.10 THEOREM. *Suppose $n \in \mathbb{N}$. Then the following formulas hold:*
Rodrigues formula:

$$(5.10.1) \qquad \boldsymbol{P}^n\phi_0 = i^n \frac{\gamma_\mu(n)}{2^{n/2}n!} H_n^\mu(2^{-\frac{1}{2}}\boldsymbol{Q})\phi_0 = i^n \frac{\gamma_\mu(n)}{2^n n!} H_n^\mu(2^{-\frac{1}{2}}\boldsymbol{A}^\star)\phi_0$$

Dual Rodrigues formula:

$$(5.10.2) \qquad \boldsymbol{Q}^n\phi_0 = (-i)^n \frac{\gamma_\mu(n)}{2^{n/2}n!} H_n^\mu(2^{-\frac{1}{2}}\boldsymbol{P})\phi_0 = (-i)^n \frac{\gamma_\mu(n)}{2^n n!} H_n^\mu(i2^{-\frac{1}{2}}\boldsymbol{A}^\star)\phi_0$$

$$(5.10.3) \qquad \boldsymbol{A}^{\star n}\phi_0 = \frac{\gamma_\mu(n)}{2^{n/2}n!} H_n^\mu(\boldsymbol{Q})\phi_0$$

$$(5.10.4) \qquad \phi_n = \frac{(\gamma_\mu(n))^{\frac{1}{2}}}{2^{n/2}n!} H_n^\mu(\boldsymbol{Q})\phi_0 = (-i)^n \frac{(\gamma_\mu(n))^{\frac{1}{2}}}{2^{n/2}n!} H_n^\mu(\boldsymbol{P})\phi_0$$

Proof. We prove (5.10.1) using Theorem 5.9 and the three term recursion relation (2.6.3) written in the form

$$2^{-\frac{1}{2}} \frac{\gamma_\mu(n+1)}{(n+1)\gamma_\mu(n)} H_{n+1}^\mu(2^{-\frac{1}{2}}x) = xH_n^\mu(2^{-\frac{1}{2}}x) - 2^{\frac{1}{2}} n H_{n-1}^\mu(2^{-\frac{1}{2}}x). \quad \text{Then}$$

$i[\boldsymbol{P}, H_n^\mu(2^{-\frac{1}{2}}\boldsymbol{Q})]\phi_0 = 2^{\frac{1}{2}} n H_{n-1}^\mu(2^{-\frac{1}{2}}\boldsymbol{Q})\phi_0$, so $i\boldsymbol{P}H_n^\mu(2^{-\frac{1}{2}}\boldsymbol{Q})\phi_0 + \boldsymbol{Q}H_n^\mu(2^{-\frac{1}{2}}\boldsymbol{Q})\phi_0 = 2^{\frac{1}{2}} n H_{n-1}^\mu(2^{-\frac{1}{2}}\boldsymbol{Q})\phi_0$. The induction proof of (5.10.1) proceeds from this equation.

The remaining equations have similar proofs using 5.9 and (2.6.2).

5.11 STRUCTURE THEOREM FOR THE BOSE-LIKE OSCILLATOR, II. *Suppose that $(\mathfrak{H}, \boldsymbol{P}, \boldsymbol{Q}, \boldsymbol{H})$ is an irreducible Bose-like oscillator, and μ and ϕ_n are as in Theorem 5.7. Define the unitary operator $\mathcal{F}$ on $\mathfrak{H}$ to $\mathfrak{H}$ by*

$$(5.11.1) \qquad \mathcal{F} = \exp(-\tfrac{1}{2}\pi i(\boldsymbol{H} - (\mu + \tfrac{1}{2})\boldsymbol{I})). \quad \textit{Then}$$

$$(5.11.2) \qquad \mathcal{F}^2 = \boldsymbol{J} \quad \text{and} \quad \mathcal{F}^\star = \boldsymbol{J}\mathcal{F} = \mathcal{F}\boldsymbol{J};$$

$$(5.11.3) \qquad \boldsymbol{P} = \mathcal{F}^\star\boldsymbol{Q}\mathcal{F} \text{ on } \mathfrak{S};$$

$$(5.11.4) \qquad \mathcal{F}\phi_n = (-i)^n\phi_n,\ n \in \mathbb{N}.$$

Proof. Compare the definitions of $\mathcal{F}$ and $\boldsymbol{J}$ in (5.11.1) and (5.7.6) to derive (5.11.2). (5.11.3) comes from (5.6.1) with $\lambda = \pi/2$. The eigenvectors of $\boldsymbol{H}$ are necessarily the eigenvectors of $\mathcal{F}$ and thus (5.11.4) is true.

We show, finally, that given any abstract irreducible Bose-like oscillator, there exists a number $\mu \in (-\frac{1}{2}, \infty)$ such that the abstract Bose-like oscillator is unitarily equivalent to the concrete irreducible Bose-like oscillator on $L^2_\mu(R)$ specified in 3.4 to 3.7. Thus one has a generalization of the von Neumann uniqueness theorem [26,p275]. Formal aspects of the physical theory are detailed in Ohnuki and Kamefuchi [24], Chapter 23, entitled *The wave-mechanical representation for a Bose-like oscillator* .

5.12 REPRESENTATION THEOREM FOR THE BOSE-LIKE OSCILLATOR. *Suppose* $(\mathfrak{H}, \boldsymbol{P}, \boldsymbol{Q}, \boldsymbol{H})$ *is an irreducible Bose-like oscillator, and maintain the notation of Theorems 5.7 and 5.11. Define the unitary mapping* $\boldsymbol{U}$ *of* $\mathfrak{H}$ *onto the Hilbert space* $L^2_\mu(R)$ *by* $\boldsymbol{U} : \phi_n \longmapsto \phi^\mu_n, n \in \mathrm{N}$. *Then* $\boldsymbol{U}$ *maps* $\boldsymbol{P}, \boldsymbol{Q}, \boldsymbol{H}, \boldsymbol{J}$, *and* $\mathcal{F}$ *onto* $\boldsymbol{P}_\mu, \boldsymbol{Q}_\mu, \boldsymbol{H}_\mu, \boldsymbol{J}_\mu$, *and* $\mathcal{F}_\mu$ *respectively* .

Proof. (5.7.3) and (3.5.2) assure us that $\boldsymbol{U}$ maps a complete orthonormal set onto a complete orthonormal set and thus $\boldsymbol{U}$ is a unitary mapping. In addition, $\boldsymbol{U}$ maps $\boldsymbol{A}$ of 5.2 onto $\boldsymbol{A}_\mu$ of (3.4.3) because of the action of these operators on the orthonormal sets, see (5.7.2) and (3.7). The other assertions then follow.

REFERENCES

1. Ahiezer and Glazman, *Theory of Linear Operators in Hilbert Space , Vol 1*, Frederick Ungar, New York, 1961.
2. R. Askey, *Orthogonal Polynomials and Special Functions*, Regional Conference Series in Applied Mathematics, Society for Industrial and Applied Mathematics, Philadelphia, Pennsylvania, 1975.
3. L. C. Biedenharn and J. D. Louck, *Angular Momentum in Quantum Physics*, Encyclopedia of Mathematics and its Applications, Vol 9, Addison-Wesley, Reading, Massachusetts, 1981.
4. T. S. Chihara, *Generalized Hermite Polynomials*, Thesis, Purdue, 1955.
5. T. S. Chihara, *An Introduction to Orthogonal Polynomials*, Gordon and Breach, New York, London, Paris, 1984.
6. F. M. Cholewinski and D. T. Haimo, *Classical analysis and the generalized heat equation*, SIAM Review **10** (1968), 67-80.
7. F. M. Cholewinski, *Generalized Foch spaces and associated operators*, SIAM J. Math. Analysis **15** (1984), 177-202.
8. F.M. Cholewinski, *The Finite Calculus Associated with Bessel Functions*, Contemporary Mathematics Vol. 75, American Mathematics Society, Providence, Rhode Island, 1988.
9. H.S.M. Coxeter, *Introduction to Geometry*, John Wiley, N. Y., London, Sydney, Toronto, 1969.
10. D.J.Dickinson and S.A. Warsi, *On a generalized Hermite polynomial and a problem of Carlitz*, Boll. Un. Mat. Ital. **(3) 18** (1963), 256-259.
11. C. F. Dunkl, *Integral kernels with reflection group invariance*, Canadian J. Math. **43** (1991), 1213-1227.
12. C. F. Dunkl, *Hankel transforms associated to finite reflection groups*, Contemporary Math. **to appear**.
13. M. Dutta, S.K. Chatterjea and K. L. More, *On a class of generalized Hermite polynomials*, Bull. of the Inst. of Math. Acad. Sinica **3** (1975), 377-381.

14. H. Dym and H. P. McKean, *Gaussian Processes, Function Theory, and the Inverse Spectral Problem*, Vol 31 , Probability and Mathematical Statistics, Academic Press, New York, San Francisco, London, 1976.
15. A. Erdélyi, *Higher Transcendental Functions, Vol 1, 2 ,3*, McGraw-Hill, New York, 1980.
16. A. Erdélyi, *Tables of Integral Transforms, Vol 1, 2*, McGraw-Hill, New York, 1954.
17. J. Glimm and A. Jaffe, *Quantum Physics*, Springer-Verlag, New York, 1987.
18. E. Hille and R. S. Phillips, *Functional Analysis and Semi-Groups*, Amer. Math. Soc. Colloquium Publ. Vol. 31, American Mathematics Society, Providence, Rhode Island, 1957.
19. N. N. Lebedev, *Special Functions and their Applications*, Translated by R. A. Silverman, Dover, New York, 1972.
20. J. D. Louck, *Extension of the Kibble-Slepian formula to Hermite polynomials using Boson operator methods*, Advances in Applied Math. **2** (1981), 239-249.
21. B. Muckenhoupt and E. M. Stein, *Classical expansions and their relation to conjugate harmonic functions*, Trans. Amer. Math. Soc. **118** (1965), 17-92.
22. N. Mukunda, E.C.G. Sudershan, J.K. Sharma, and C.L. Mehta, *Representations and properties of para-Bose oscillator operators. I. Energy position and momentum eigenstates*, J. Math. Phys. **21** (1980), 2386-2394.
23. E. Nelson, *Analytic vectors*, Annals of Math. **70** (1959), 572-615.
24. Y. Ohnuki and S.Kamefuchi, *Quantum Field Theory and Parastatistics*, University of Tokyo Press, Springer-Verlag Berlin Heidelberg New York, 1982.
25. E. D. Rainville, *Special Functions*, Chelsea, Bronx, New York, 1971.
26. M. Reed and B. Simon, *Methods of Modern Mathematical Physics*, I, Functional Analysis, Academic Press, San Diego New York Berkeley, 1980.
27. M. Reed and B. Simon, *Methods of Modern Mathematical Physics*, II Fourier Analysis, Self-Adjointness, Academic Press, New York San Francisco London, 1975.
28. L. J. Slater, *Confluent Hypergeometric Functions*, Cambridge University Press, London and New York, 1960.
29. G. Szego, *Orthogonal Polynomials*, Amer. Math. Soc. Colloquium Publ. Vol. 23, American Mathematics Society, New York, 1939.
30. G. N. Watson, *A Treatise on the Theory of Bessel Functions,2nd Edition*, Cambridge University Press, Cambridge, Great Britain, 1966.
31. E. P. Wigner, *Do the equations of motion determine the quantum mechanical commutation relations?*, Phys. Rev. **77** (1950), 711-712.

Department of Mathematics
University of Virginia
Math/Astro Building
Charlottesville, Virginia 22903-3199
U. S. A.
mr1t@virginia.edu

MSC 1991 Primary 33C45, 81Q05
Secondary 44A15

Operator Theory:
Advances and Applications, Vol. 73

A General Theory of Sufficient Collections of Norms with a Prescribed Semigroup of Contractions

Nahum Zobin[1] and Veronica Zobina[2]

To Professor Moshe Livsic

The paper contains a complete exposition and proofs of our results on the theory of sufficient collections of norms. Consider a uniformly bounded irreducible semigroup G of operators in a finite dimensional real linear space V. We consider G-contractive norms on V, i.e., such norms that every operator from G is contractive with respect to each of them. A collection of G-contractive norms $\{ \ \|\cdot\|_\alpha \ \}_{\alpha\in A}$ is called sufficient if for any linear operator $T : V \to V$ the following implication holds: if T is contractive in every norm $\|\cdot\|_\alpha$ $(\alpha \in A)$ then T is contractive in every G-contractive norm. We describe all sufficient collections, construct two canonical sufficient collections and study their extremal properties.

Introduction

This paper is devoted to an interpolation theory of norms with a prescribed semigroup of contractions. The core of the interpolation may be described as follows: if you need to estimate the norm of a complicated linear operator A acting on a complicated Banach space E you try to find a collection $\{E_s, \ s \in S\}$ of «simpler» Banach spaces such that

(i) the operator A acts on the spaces E_s $(s \in S)$ and it is not very difficult to estimate the norms $\| A : E_s \to E_s \|$;

(ii) the space E is a strict interpolation space for the collection $\{E_s, \ s \in S\}$, i.e., for any linear operator T acting simultaneously on all E_s, $s \in S$, it is true that T acts also on E and satisfies the estimate

$$\| T : E \to E \| \leq \max\{\| T : E_s \to E_s \|, \ s \in S\}.$$

The latter inequality gives the needed estimate.

---

1) The research was supported in part by a grant from the Ministry of Absorption and the Rashi Foundation.

2) The research was supported in part by a grant from the Ministry of Science and the «Maagara» — a special project for absorption of new immigrants, at the Department of Mathematics, Technion.

This method is successful if we are fortunate enough to find such a collection $\{E_s,\ s \in S\}$ for our problem.

One general idea for this purpose has been developed since 1978 in a series of papers [8, 9, 13–17]. Given a Banach space E, we can usually find a more or less rich set of contractions (or isometries) of this space. We may consider the semigroup G generated by this set of contractions. If we are lucky and the initial operator lies in this semigroup G, then we have already obtained an estimate of its norm, and no interpolation is needed. If we are not so lucky and $A \notin G$ then we must do something more. We may consider all those norms for which the operators from G are contractive and try to find simple norms of this class, satisfying (ii), and then estimate the operator A in these norms. We propose some constructions of such collections of norms (which satisfy (ii)).

This paper continues investigations started in [8–11, 13, 14, 17].

Our constructions were applied to the case when G is a Coxeter group, we have calculated the canonical collections explicitly and obtained some final results on their extremal properties [11, 12, 16]; these results give a broad generalization and elaboration of a finite dimensional version of the well-known Calderon-Mityagin Theorem, describing interpolation spaces for the couple (l_1, l_∞); a complete exposition is contained in [12]. Similar ideas were applied in [7] to a description of extreme mixed norms, here we dealt with a special semigroup of operators. Our further investigations [6] showed that it is necessary to develop a theory for a general semigroup G. Preliminary results were announced in [10, 11, 17]. This paper is devoted to a complete exposition and proofs.

1 Formulation of the Problem

Let us remind the problem and the main notions. Let V and V' be a pair of real finite dimensional spaces in a duality. The duality is given by a nondegenerate bilinear form $\langle\cdot,\cdot\rangle$ on the space $V \times V'$. Let G be a semigroup of linear operators (containing 1 and -1) acting on V.

Definition 1. A closed convex set $U \subset V$ is called G-**symmetric** if for any point $x \in U$ the G-orbit of x also belongs to U.

We let $\mathbf{Co_G x}$ denote the closed convex hull of the G-orbit of x.

Definition 2. A collection of G-symmetric sets $\{U_\alpha\}_{\alpha \in A}$ is called **sufficient** if for any linear operator L the inclusions $LU_\alpha \subset U_\alpha$ $(\forall \alpha \in A)$ imply the inclusion $LU \subset U$ for any G-symmetric set U.

The problem is to describe sufficient collections, to construct certain canonical sufficient collections and to investigate them.

Example. Let $V = \mathbb{R}^n$, $G = B_n$ be the group of permutations and sign changes of coordinates of the canonical basis in V. Then the collection consisting of the following two sets

$$\{x = (x_1, \dots, x_n) : \sum |x_i| \leq 1\} \text{ and } \{x = (x_1, \dots, x_n) : \max |x_i| \leq 1\}$$

is a sufficient collection for this group. This is a formulation of the famous Calderon-Mityagin theorem [2, 5] in terms of sufficient collections: any space with a B_n-invariant norm is a strict interpolation space between the spaces l_1^n and l_∞^n.

Our approach to the construction and investigation of sufficient collections is based on a systematic exploitation of the canonical duality between the space End V of linear operators on V and the tensor product space $V \otimes V'$. This duality makes it possible to reformulate the initial problem into a dual one and to make use of the connections between the dual problems.

In the previous papers [8, 9, 13, 14, 16] we have completely investigated the case when G is a group of linear operators (-1 does not necessarily belong to G). The case when G is a semigroup was partially studied in [10, 17].

2 Notions

The sufficient collections are described in terms of a geometry of certain sets: $\mathfrak{A}$, $K(\mathfrak{A})$, $\operatorname{Extr} K(\mathfrak{A})$, U^0, $S(U)$. We remind their definitions:

1. $\mathfrak{A} = \{a \otimes f \in V \otimes V' : \sup_{g \in G} \langle ga, f \rangle \leq 1\}$.

   In [17] it was established that the set $\mathfrak{A}$ is compact if and only if the semigroup G acts irreducibly. Henceforth we assume that the semigroup G is uniformly bounded and acts irreducibly.

2. $K(\mathfrak{A}) = \operatorname{conv} \mathfrak{A}$ is the closed convex hull of the set $\mathfrak{A}$.
3. $\operatorname{Extr} K(\mathfrak{A})$ is the set of extreme points of $K(\mathfrak{A})$.

   Since $\mathfrak{A}$ is a compact set then $K(\mathfrak{A})$ is also compact. Then according to the Krein-Milman theorem (see, e.g., [1]) $K(\mathfrak{A})$ is the closed convex hull of its extreme points. According to the Milman theorem (see, e.g., [1]) $\operatorname{Extr} K(\mathfrak{A}) \subset \mathfrak{A}$.

4. $U^0 = \{f \in V' : \langle x, f \rangle \leq 1, \ \forall x \in U\}$ — is the polar set of U
5. $S(U) = \{x \otimes f \in V \otimes V' : x \in U, f \in U^0\}$.

3 Formulations of Results

3.1 Description of sufficient collections

In [17] we have obtained the following description of all sufficient collections.

Theorem 1. *A collection $\{U_\alpha\}_{\alpha\in A}$ of G-symmetric sets is sufficient if and only if the following equality holds:*

$$\mathbf{K}(\mathfrak{A}) = \mathbf{conv} \bigcup_{\alpha\in \mathbf{A}} \mathbf{S}(\mathbf{U}_\alpha)$$

Collections consisting of sets of the form $Co_G x$ are called **simple collections.** Collections consisting of sets of the form $(Co_{G^*} f)^0$ are called **dual simple collections.** In [17] two important sets — $\mathfrak{N}$ and $\mathfrak{N}'$ — were defined, and two canonical collections: the canonical simple collection $\{Co_G a\}_{a\in\mathfrak{N}}$ and the canonical dual simple collection $\{(Co_{G^*} f)^\circ\}_{f\in\mathfrak{N}'}$ were constructed. In [10] we studied the problem of minimality of the canonical sufficient collections, i.e. the following question: do there exist sufficient collections which are smaller than the canonical ones? This problem for general semigroups turns out to be more difficult, compared to the analogous problem considered earlier for a general group G [13, 14]. In this paper we investigate the problem of existence of the smallest sufficient collection.

For the sake of completeness we recall some notions and the main results of the paper [10] in the next items 3.2–3.5.

3.2 Hausdorff topology

The set of all compact convex subsets of the space V can be equipped with the following **Hausdorff topology** [3]. If B is a fixed neighborhood of the origin of the space V then the Hausdorff distance $d_H(U_1, U_2)$ between two compact sets U_1 and U_2 is defined by the formula

$$d_H(U_1, U_2) = \inf\{\lambda > 0 : U_1 \subset U_2 + \lambda B,\ U_2 \subset U_1 + \lambda B\}.$$

It is clear that the Hausdorff topology does not depend on the choice of B. Let $\lim_H$, $\overline{}^H$, $\overset{H}{\rightarrow}$, denote the limit, the closure, and the convergence in the Hausdorff topology, correspondingly.

Let U be a compact convex set. Let $p^U(f)$ denote the support function of the set U:

$$p^U(f) = \sup_{x\in U} \langle x, f\rangle$$

Recall some properties of the support function:

(1) $p^U(.)$ is a convex homogeneous function, bounded on the unit ball of the space V';

(2) $p^{U_1+U_2}(f) = p^{U_1}(f) + p^{U_2}(f)$;

(3) $U_1 \subseteq U_2$ implies $p^{U_1}(f) \le p^{U_2}(f)$ for any $f \in V'$;

(4) the mapping $U \to p^U$ establishes a $1-1$ correspondence between the set of compact convex subsets of V and the set of convex homogeneous continuous functions, bounded on the unit ball of the space V'.

The Hausdorff distance $d_H(U_1, U_2)$ between two convex compact sets U_1 and U_2 is the supremum norm (on the unit ball of the space V') of the difference of support functions p^{U_1} and p^{U_2} of the sets U_1 and U_2.

Really, it follows from the property (3) that $d_H(U_1, U_2) \leq \epsilon$ if and only if $\forall f \in V'$

$$p^{U_1}(f) \leq p^{U_2}(f) + \epsilon \parallel f \parallel \text{ and } p^{U_2}(f) \leq p^{U_1}(f) + \epsilon \parallel f \parallel$$

Therefore $d_H(U_1, U_2) \leq \epsilon$ if and only if

$$\left| \frac{p^{U_1}(f) - p^{U_2}(f)}{\parallel f \parallel} \right| \leq \epsilon \qquad (\forall f \in V')$$

$$d_H(U_1, U_2) = \sup_f \frac{|p^{U_1}(f) - p^{U_2}(f)|}{\parallel f \parallel} = \sup_{\|f\| \leq 1} \left| p^{U_1}(f) - p^{U_2}(f) \right|$$

The Blaschke theorem [3] asserts that the set of convex closed subsets of a bounded convex closed set is compact in the Hausdorff topology. This easily follows from the Ascoli Theorem.

3.3 Equivalence of sufficient collections

Let $\{U_\alpha\}_{\alpha \in A}$ be a sufficient collection of bounded G-symmetric sets

Remark 1. It is obvious that the collection $\{\lambda_\alpha U_\alpha\}$ ($\lambda_\alpha \in \mathbb{R} \backslash \{0\}$) is also sufficient.

Remark 2. Let $\{W_\beta\}_{\beta \in B}$ be a collection of bounded G-symmetric sets such that any set U_α is a Hausdorff limit of sets from the collection $\{W_\beta\}_{\beta \in B}$. It is obvious that the collection $\{W_\beta\}_{\beta \in B}$ is also sufficient.

These Remarks justify the following

Definition 3. Sufficient collections $\{U_\alpha\}_{\alpha \in A}$ and $\{W_\beta\}_{\beta \in B}$ of G-symmetric bounded sets are called **equivalent** if

$$\overline{\{\lambda U_\alpha\}}^H_{\lambda \in \mathbb{R}, \alpha \in A} = \overline{\{\lambda W_\beta\}}^H_{\lambda \in \mathbb{R}, \beta \in B}$$

If

$$\overline{\{\lambda U_\alpha\}}^H_{\lambda \in \mathbb{R}, \alpha \in A} \subset \overline{\{\lambda W_\beta\}}^H_{\lambda \in \mathbb{R}, \beta \in B}$$

then the collection $\{U_\alpha\}_{\alpha \in A}$ is said to be **smaller** than the collection $\{W_\beta\}_{\beta \in B}$.

3.4 Majorization in $K(\mathfrak{A})$

We remind that we assume the semigroup G to be uniformly bounded and irreducible.

Consider an action of the semigroup $G \times G^*$

$$g \times h^* : \sum a_i \otimes f_i \to \sum g a_i \otimes h^* f_i$$

in the space $V \otimes V'$.

Definition 4. A point $a_1 \otimes f_1$ majorizes $a_2 \otimes f_2$, if

$$a_2 \otimes f_2 \in Co_{G\times G} * (a_1 \otimes f_1) \quad (\text{denote } a_1 \otimes f_1 > a_2 \otimes f_2)$$

Such a majorization for the group B_n (of permutations and sign changes of coordinates) is called the Schur majorization, see, e.g., [4].

Theorem 2. *For any extreme point $a \otimes f \in \operatorname{Extr} K(\mathfrak{A})$ there exists a maximal element $\hat{a} \otimes \hat{f} \in \operatorname{Extr} K(\mathfrak{A})$ which majorizes the element $a \otimes f$.*

3.5 Definition of canonical collections

Let $M(\mathfrak{A})$ denote the set of maximal elements of $\operatorname{Extr} K(\mathfrak{A})$. Take

$$\mathfrak{N} = \{a \in V : \exists f_a \in V',\ a \otimes f_a \in M(\mathfrak{A})\}$$
$$\mathfrak{N}' = \{f \in V' : \exists a_f \in V,\ a_f \otimes f \in M(\mathfrak{A})\}$$

Definition 5. The collection $\{Co_G a\}_{a\in\mathfrak{N}}$ is called the **canonical simple collection.** The collection $\{(Co_{G^*} f)^0\}_{f\in\mathfrak{N}'}$ is called the **canonical dual simple collection.**

Theorem 3. *The canonical collections are sufficient.*

Theorem 4. *The canonical simple collection is the smallest simple sufficient collection. The canonical dual simple collection is the smallest dual simple collection.*

3.6 Existence of the smallest sufficient collection

As we have already mentioned the subject of this paper is the problem of existence of the smallest sufficient collection (smallest up to the equivalence). The main result is the following

Theorem 5. *The smallest sufficient collection exists if and only if the canonical collections coincide.*

In [9] we have obtained that the set consisting of the balls of the spaces l_1^n and l_∞^n form the coinciding canonical collections — the simple and the dual simple ones — for the group $G = B_n$ (permutations and sign changes of coordinates). The main theorem asserts that this sufficient collection is the smallest one. In particular, l_∞ is not a strictly interpolation space for any finite collection of symmetric spaces, which does not contain l_∞.

4 Proofs of Results

4.1 Proof of Theorem 1

The sufficiency of a collection $\{U_\alpha\}_{\alpha\in A}$ means by the definition that for any linear operator L the inclusions $LU_\alpha \subset U_\alpha$ ($\forall\alpha$) imply $LU \subset U$ for any G-symmetric set U. Denote

$$T(U) = \{L \in \operatorname{End} V : LU \subset U\}$$

the operator unit ball, then the above condition can be rewritten in the following way: a collection $\{U_\alpha\}_{\alpha\in A}$ is sufficient if and only if

$[L \in T(U_\alpha), \forall\alpha \in A]$ implies $[L \in T(U)$ for any G-symmetric set $U]$,

or $L \in \bigcap_\alpha T(U_\alpha)$ implies $L \in T(U)$ for any G-symmetric set U,

or $\bigcap_\alpha T(U_\alpha) \subset T(U)$ for any G-symmetric set U.

Take polars of both sides of this inclusion: $(\bigcap_\alpha T(U_\alpha))^0 \supset (T(U))^0$, i.e.

$$\operatorname{conv}\bigcup_\alpha T(U_\alpha)^0 \supset (T(U))^0 \tag{$*$}$$

So, what is $T(U)$? It is easy to see that $T(U) = (S(U))^0$. Indeed

$$\begin{aligned}(S(U))^0 &= \{L \in \operatorname{End} V : \langle L, x\otimes f\rangle \le 1 \text{ for } x\otimes f \in S(U)\}\\ &= \{L \in \operatorname{End} V : \langle Lx, f\rangle \le 1 \text{ for } x \in U, f \in U^0\}\\ &= \{L \in \operatorname{End} V : LU \subset U\} = T(U).\end{aligned}$$

Then $(T(U))^0 = (S(U))^{00} = \operatorname{conv} S(U)$ (the Bipolar theorem, see, e.g., [1]). Then we obtain the following form of $(*)$:

$$\operatorname{conv}\bigcup_\alpha \operatorname{conv} S(U_\alpha) \supset \operatorname{conv} S(U)$$

for any G-symmetric set U. Hence

$$\operatorname{conv}\bigcup_\alpha S(U_\alpha) \supset \operatorname{conv}\bigcup S(U) \tag{$**$}$$

the last union is taken over all G-symmetric sets U.

It is easy to see that $S(U_\alpha) \subset K(\mathfrak{A})$, $\forall\alpha$. Indeed, if $x\otimes f \in S(U_\alpha)$, then $x \in U_\alpha, f \in U_\alpha^0$, hence $\langle x, f\rangle \le 1$. And since $gx \in U_\alpha$ for any operator $g \in G$ then also $\langle gx, f\rangle \le 1$, and $\sup_{g\in G}\langle gx, f\rangle \le 1$. It means that $x \otimes f \in \mathfrak{A}$, hence $S(\mathcal{U}_\alpha) \subset \mathfrak{A}$ $(\forall\alpha)$. Then $\operatorname{conv}\bigcup_\alpha S(\mathcal{U}_\alpha) \subset K(\mathfrak{A})$. Thus it follows from the inclusion (**), that

$$K(\mathfrak{A}) \supset \operatorname{conv}\bigcup_\alpha S(\mathcal{U}_\alpha) \supset \operatorname{conv}\bigcup S(U) \qquad (***)$$

the last union is taken over all G-symmetric sets U.

On the other hand, $\operatorname{conv}\bigcup S(U) = K(\mathfrak{A})$. Indeed, first we show that $\mathfrak{A} \subset \bigcup S(U)$, where the union is taken over all G-symmetric sets U. If $x \otimes f \in \mathfrak{A}$ then $\langle gx, f\rangle \le 1$ $(\forall g \in G)$, i.e., $f \in (Co_G x)^0$, hence $x \otimes f \in S(Co_G x)$, hence $\mathfrak{A} \subset \bigcup S(U)$. Thus, combining the last inclusion and the inclusion (* * *) we obtain

$$K(\mathfrak{A}) \supset \operatorname{conv}\bigcup_\alpha S(\mathcal{U}_\alpha) \supset \operatorname{conv}\bigcup S(U) \supset K(\mathfrak{A})$$

Hence — all these inclusions must be equalities and we obtain the statement of the theorem: $K(\mathfrak{A}) = \operatorname{conv}\bigcup_\alpha S(\mathcal{U}_\alpha)$. □

4.2 Proof of Theorem 2

Proof of Theorem 2 is based on two lemmas.

Lemma 1. *Let $x_i \to x$, then $Co_G x_i \overset{H}{\to} Co_G x$.*

(The semigroup G is considered to be uniformly bounded.)

Proof. For any $\varepsilon > 0$ there exists a number N, such that for $i > N$ the inclusions $x - x_i \in \varepsilon B$ (B is a unit ball in V) hold. Then

$$x = x_i + \varepsilon b \qquad (b \in B)$$

Consider any element of $Co_G x$

$$\sum_{g\in G} \lambda_g gx = \sum_{g\in G} \lambda_g gx_i + \varepsilon \sum_{g\in G} \lambda_g gb, \quad \lambda_g \ge 0, \quad \sum_{g\in G} \lambda_g = 1.$$

Since G is uniformly bounded, then

$$\| \varepsilon \sum_{g\in G} \lambda_g gb \| \le \varepsilon C \| b \| \sum_{g\in G} \lambda_g = \varepsilon C.$$

Hence,

$$Co_G x \subset Co_G x_i + \varepsilon CB$$

Similarly we obtain

$$Co_G x_i \subset Co_G x + \varepsilon CB$$

This means that

$$Co_G x_i \overset{H}{\to} Co_G x$$ □

Lemma 2. *Let U be a G-symmetric set, $a_1 \otimes f_1 \in S(U)$ and $a_1 \otimes f_1 > a_2 \otimes f_2$. Then $a_2 \otimes f_2 \in \operatorname{conv} S(U)$.*

Proof. Let $a_1 \otimes f_1 > a_2 \otimes f_2$, then by the definition

$$a_2 \otimes f_2 \in Co_{G \times G^*}(a_1 \otimes f_1).$$

This means that

$$a_2 \otimes f_2 = \sum_{g,h} \lambda_{g,h} g a_1 \otimes h^* f_1 \quad \Big(\sum \lambda_{g,h} = 1,\ \lambda_{g,h} \geq 0\Big)$$

Also, $g a_1 \otimes h^* f_1 \in S(U)$ ($\forall g \in G$, $\forall h^* \in G^*$), because of the G-symmetricity of U. Hence, $a_2 \otimes f_2 \in \operatorname{conv} S(U)$. □

Now we are able to prove Theorem 2.

(i) Consider any chain of extreme points $\{a_\alpha \otimes f_\alpha\}_{\alpha \in A}$

$$a_\alpha \otimes f_\alpha \in \operatorname{Extr} K(\mathfrak{A})$$

such that $a_\alpha \otimes f_\alpha > a \otimes f$.

First we show that this chain has a majorant in $\operatorname{Extr} K(\mathfrak{A})$. Since this chain is a linearly ordered set of elements then the sets $Co_{G \times G^*}(a_\alpha \otimes f_\alpha)$ are included one into another and all are included into $K(\mathfrak{A})$ because $K(\mathfrak{A})$ is convex, closed and $G \times G^*$-invariant. The sets $Co_{G \times G^*}(a_\alpha \otimes f_\alpha)$ converge to some set in the Hausdorff topology. Really, their support functions $p^\alpha(\cdot)$ increase and stay smaller than the support function $p(\cdot)$ of the set $K(\mathfrak{A})$; therefore the net $p^\alpha(\cdot)$ converges pointwise to a function $\hat{p}(\cdot) \leq p(\cdot)$, $p^\alpha(\cdot) \to \hat{p}(\cdot)$. Now we show that $p^\alpha(\cdot)$ converge to $\hat{p}(\cdot)$ uniformly on the ball of the space $(V \otimes V')'$. Take $\varepsilon > 0$ and find a finite $\frac{\varepsilon}{3}$-net with respect to the norm $p(\cdot)$ on the ball of the space $(V \otimes V')'$: $\{m_i\}_{i=1}^N$. Choose α_0 such that for $\alpha > \alpha_0$ the inequalities

$$\hat{p}(m_i) - p^\alpha(m_i) < \frac{\varepsilon}{3} \quad i = 1, \ldots, N$$

hold. Then for any point m of the ball of the space $(V \otimes V')'$ we can write the following

$$\begin{aligned} \hat{p}(m) - p^\alpha(m) &= \hat{p}(m) - \hat{p}(m_i) + p^\alpha(m_i) - p^\alpha(m) + \hat{p}(m_i) - p^\alpha(m_i) \\ &\leq \hat{p}(m - m_i) + p^\alpha(m_i - m) + \frac{\varepsilon}{3} < \varepsilon \end{aligned}$$

therefore, $p^\alpha(\cdot) \to \hat{p}(\cdot)$ uniformly on the ball of the space $(V \otimes V')'$. Hence $\hat{p}(\cdot)$ is continuous. $\hat{p}(\cdot)$ is obviously convex and homogeneous. Therefore $\hat{p}(\cdot)$ is the support function of the set $(\lim_\alpha {}_H Co_{G \times G^*}(a_\alpha \otimes f_\alpha))$.

Now we need to show that this limit set can be represented in the form of $Co_{G\times G^*}\hat{a}\otimes\hat{f}$.

Choose a subnet $a_{\alpha_i}\otimes f_{\alpha_i}$ converging to some element $\hat{a}\otimes\hat{f}$, $\hat{a}\otimes\hat{f}\in K(\mathfrak{A})$ ($K(\mathfrak{A})$ is compact). Applying Lemma 1, we obtain

$$Co_{G\times G^*}a_{\alpha_i}\otimes f_{\alpha_i}\xrightarrow{H}Co_{G\times G^*}\hat{a}\otimes\hat{f}$$

Hence

$$\lim_H Co_{G\times G^*}a_\alpha\otimes f_\alpha = Co_{G\times G^*}\hat{a}\otimes\hat{f}.$$

As $\lim_H Co_{G\times G^*}a_\alpha\otimes f_\alpha\supset Co_{G\times G^*}a_\beta\otimes f_\beta$ $(\forall\beta)$ we get that

$$a_\alpha\otimes f_\alpha<\hat{a}\otimes\hat{f}.$$

Thus a majorant is found.

(ii) If $\hat{a}\otimes\hat{f}\in\operatorname{Extr}K(\mathfrak{A})$, then $\hat{a}\otimes\hat{f}$ is just the majorant we were looking for. If $\hat{a}\otimes\hat{f}\notin\operatorname{Extr}K(\mathfrak{A})$ we can represent

$$\hat{a}\otimes\hat{f}=\lambda m_1+(1-\lambda)m_2,\quad m_1,m_2\in K(\mathfrak{A}),\quad 0<\lambda<1$$

We can take $m_1\in\operatorname{Extr}K(\mathfrak{A})$, i.e., $m_1=a'\otimes f'$. Let us show that m_1 is the needed majorant. For this purpose we need to prove that

$$m_1>a_\alpha\otimes f_\alpha\quad(\forall\alpha)$$

Note that $a_\alpha\otimes f_\alpha\in\overline{orb}_{G\times G^*}\hat{a}\otimes\hat{f}$, since $a_\alpha\otimes f_\alpha\in\operatorname{Extr}K(\mathfrak{A})$ and $a_\alpha\otimes f_\alpha\in Co_{G\times G^*}\hat{a}\otimes\hat{f}\subset K(\mathfrak{A})$. Then for any α_0 there exist nets of operators $\{g_\beta\}$ and $\{h^*_\beta\}$ such that

$$(g_\beta\times h^*_\beta)(\hat{a}\otimes\hat{f})\to a_{\alpha_0}\otimes f_{\alpha_0}.$$

But

$$\begin{aligned}(g_\beta\times h^*_\beta)(\hat{a}\otimes\hat{f})&=(g_\beta\times h^*_\beta)(\lambda m_1+(1-\lambda)m_2)\\&=\lambda(g_\beta\times h^*_\beta)m_1+(1-\lambda)(g_\beta\times h^*_\beta)m_2\to a_{\alpha_0}\otimes f_{\alpha_0}\end{aligned}$$

Both entries of the sum belong to $K(\mathfrak{A})$ and $K(\mathfrak{A})$ is compact. Therefore passing to subnets we may assume that

$$\lambda(g_\beta\times h^*_\beta)m_1\to\lambda c_1 \text{ and } (1-\lambda)(g_\beta\times h^*_\beta)m_2\to(1-\lambda)c_2,$$

i.e., $a_{\alpha_0}\otimes f_{\alpha_0}=\lambda c_1+(1-\lambda)c_2$. Obviously, $c_1,c_2\in K(\mathfrak{A})$. Since $a_{\alpha_0}\otimes f_{\alpha_0}\in\operatorname{Extr}K(\mathfrak{A})$, then $c_1=c_2=a_{\alpha_0}\otimes f_{\alpha_0}$. Hence $a_{\alpha_0}\otimes f_{\alpha_0}\in\overline{orb}_{G\times G^*}m_1$. This means that $a_{\alpha_0}\otimes f_{\alpha_0}<m_1$. As m_1 is an extreme point we obtain that m_1 is the majorant, we are looking for. Applying the Zorn lemma we complete the proof. □

4.3 Proof of Theorem 3

Let us prove the sufficiency of the collection $\{Co_G a\}_{a\in\mathfrak{N}}$. Due to Theorem 1 we need to prove the equality

$$\operatorname{conv}\bigcup_{a\in\mathfrak{N}}(Co_G a) = K(\mathfrak{A})$$

The inclusion $\operatorname{conv}\bigcup_{a\in\mathfrak{N}} S(Co_G a) \subset K(\mathfrak{A})$ was already proved (see the Proof of Theorem 1). Now we prove the converse inclusion: $K(\mathfrak{A}) \subset \operatorname{conv}\bigcup_{a\in\mathfrak{N}} S(Co_G a)$. It is enough to prove that $\operatorname{Extr} K(\mathfrak{A}) \subset \overline{\bigcup_{a\in\mathfrak{N}} S(Co_G a)}$.

Let $a\otimes f \in \operatorname{Extr} K(\mathfrak{A})$. If $a\otimes f$ is a maximal element in $\operatorname{Extr} K(\mathfrak{A})$ (i.e. $a\otimes f \in M(\mathfrak{A})$) then $a \in \mathfrak{N}$ and $a\otimes f \in S(Co_G a)$. So $a\otimes f \in S(Co_G a) \subset \bigcup_{x\in\mathfrak{N}} S(Co_G x)$.

If $a\otimes f \notin M(\mathfrak{A})$ then there exists a maximal element $a_1\otimes f_1 \in M(\mathfrak{A})$ such that $a_1\otimes f_1 > a\otimes f$, i.e. $a\otimes f \in Co_{G\times G^*} a_1\otimes f_1$. But $a_1\otimes f_1 \in S(Co_G a_1)$, hence

$$Co_{G\times G^*} a_1\otimes f_1 \subset \operatorname{conv} S(Co_G a_1)$$

and

$$a\otimes f \in \operatorname{conv}\bigcup_{x\in\mathfrak{N}} S(Co_G x).$$

So in any case the inclusion $a\otimes f \in \operatorname{Extr} K(\mathfrak{A})$ implies the inclusion $a\otimes f \in \operatorname{conv}\bigcup_{x\in\mathfrak{N}} S(Co_G x)$. Therefore $K(\mathfrak{A}) \subset \operatorname{conv}\bigcup_{a\in\mathfrak{N}} S(Co_G a)$.

The sufficiency of the collection $\{(Co_{G^*} f)^0\}_{f\in\mathfrak{N}'}$ can be proved analogously. □

4.4 Proof of Theorem 4

Let us prove that the simple canonical collection $\{Co_G a\}_{a\in\mathfrak{N}}$ is the smallest simple sufficient collection. Let $\{Co_G c\}_{c\in Q}$ be any other simple sufficient collection. We shall show that the simple canonical collection is smaller than $\{Co_G c\}_{c\in Q}$, i.e., we shall prove the inclusion

$$\{\overline{\lambda Co_G a}\}^H_{a\in\mathfrak{N},\lambda\in\mathbb{R}} \subset \{\overline{\lambda Co_G c}\}^H_{c\in Q,\lambda\in\mathbb{R}}$$

It is clear that it is enough to prove that

$$\{Co_G a\}_{a\in\mathfrak{N}} \subset \{\overline{\lambda Co_G c}\}^H_{c\in Q,\lambda\in\mathbb{R}}.$$

Since the collection $\{Co_G c\}_{c\in Q}$ is sufficient then by Theorem 1

$$K(\mathfrak{A}) = \operatorname{conv}\bigcup_{c\in Q} S(Co_G c) = \operatorname{conv}\bigcup_{c\in Q}\operatorname{conv} S(Co_G c).$$

According to the Krein-Milman theorem [1] the closed convex set $\operatorname{conv} S(Co_G c)$ is the convex hull of its extreme points, therefore

$$\operatorname{conv} \bigcup_{c \in Q} \operatorname{conv} S(Co_G c) = \operatorname{conv} \bigcup_{c \in Q} \operatorname{conv} \operatorname{Extr}(\operatorname{conv} S(Co_G c))$$

$$= \operatorname{conv} \bigcup_{c \in Q} \operatorname{Extr}(\operatorname{conv} S(Co_G c)).$$

So, $K(\mathfrak{A}) = \operatorname{conv} \bigcup_{c \in Q} \operatorname{Extr}(\operatorname{conv} S(Co_G c))$.

According to the Milman theorem [1] extreme points of the closed convex hull of a compact set lie in this set: $\operatorname{Extr} K(\mathfrak{A} \subset \overline{\bigcup \operatorname{Extr}(\operatorname{conv} S(Co_G c))}$. By the same reason the extreme points of the sets $\operatorname{conv} S(Co_G c)$ lie in the sets $\overline{S(Co_G c)}$ and they are of the form $b \otimes \varphi$, where $\lambda b \in \overline{orb}_G c$. Then $\operatorname{Extr} K(\mathfrak{A}) \subset \overline{\{gc \otimes \varphi\}}_{c \in Q, g \in G, \varphi \in V'}$.

Let $a \in \mathfrak{N}$. Then according to the definition of $\mathfrak{N}$ there exists f such that $a \otimes f \in M(\mathfrak{A})$ (i.e., $a \otimes f$ is a maximal element of $\operatorname{Extr} K(\mathfrak{A})$). Hence,

$$a \otimes f \in \overline{\{gc \otimes \varphi\}}_{c \in Q, g \in G, \varphi \in V'}$$

i.e.,

$$a \otimes f = \lim_{\iota} g_\iota c_\iota \otimes \varphi_\iota, \quad g_\iota \in G.$$

We can assume, that $a = \lim_{\iota} g_\iota c_\iota$, $f = \lim \varphi_\iota$. Since the semigroup G is uniformly bounded there exists a convergent subsequence $g_{\iota_n} \to g$ (g does not necessarily belong to G). We can assume $c_\iota \to c$, then $a = \lim_{\iota} g_\iota c_\iota = gc$. Thus we obtain $a \otimes f = \lim_{\iota} g_\iota c_\iota \otimes f$. It is clear that $gc \in Co_G c$. Then $Co_G a = Co_G gc \subset Co_G c$, hence it follows that $Co_{G \times G^*} a \otimes f \subset Co_{G \times G^*} c \otimes f$. Therefore $a \otimes f < c \otimes f$. It means that these two elements $a \otimes f$ and $c \otimes f$ are comparable. But the maximality of $a \otimes f$ implies that $a \otimes f > c \otimes f$. Hence

$$c \otimes f \in Co_{G \times G^*}(a \otimes f) \Rightarrow c \otimes f = \sum_{\iota} \lambda_\iota g_\iota a \otimes h_\iota^* f, \quad \sum_{\iota} \lambda_\iota = 1, \quad 0 \leq \lambda_\iota \leq 1.$$

Take $m \in V$ such that $\langle f, m \rangle = \| f \| \cdot \| m \| = 1$ and apply the operators from the equality above to m:

$$c \langle f, m \rangle = \sum_{\iota} \lambda_\iota g_\iota a \langle h_\iota^* f, m \rangle.$$

We see that $c = \sum_{\iota} \lambda_\iota g_\iota a \langle h_\iota^* f, m \rangle$, and

$$\sum_{\iota} |\lambda_\iota \langle h_\iota^* f, m \rangle| \leq \sum_{\iota} \lambda_\iota \| h_\iota^* \| \cdot \| f \| \cdot \| m \| \leq \sum_{\iota} \lambda_\iota = 1.$$

Hence, $c \in Co_G a$ and $Co_G c \subset Co_G a$. Recalling the converse inclusion, obtained earlier we conclude that $Co_G a = Co_G c$. Hence,

$$\{Co_G a\}_{a \in \mathfrak{N}} \subset \overline{\{Co_G c\}}^H_{c \in Q}.$$

The statement about the dual simple canonical collection can be proved analogously. □

4.5 Proof of Theorem 5

Proof of Theorem 5 is based on some lemmas.

Lemma 3. *(i) If $f \in \mathfrak{N}'$ and $(Co_{G^*} f)^0 = Co_G b$ then $Co_G b = Co_G a_f$, $a_f \otimes f \in M(\mathfrak{A})$, $a_f \in \mathfrak{N}$.*

(ii) If $a \in \mathfrak{N}$ and $Co_G a = (Co_{G^*}\varphi)^0$ then $(Co_{G^*}\varphi)^0 = (Co_{G^*} f_a)^0$, $a \otimes f_a \in M(\mathfrak{A})$, $f_a \in \mathfrak{N}'$.

Proof. 1. Let $f \in \mathfrak{N}'$ and $(Co_{G^*} f)^0 = Co_G b$, then $\operatorname{Extr}(Co_{G^*} f)^0 = \operatorname{Extr}(Co_G b)$. Since $f \in \mathfrak{N}'$ then by the definition of $\mathfrak{N}'$ there exists an element $a_f \in V$ such that $a_f \otimes f \in M(\mathfrak{A})$. But $M(\mathfrak{A}) \subset \operatorname{Extr} K(\mathfrak{A})$, hence $a_f \otimes f \in \operatorname{Extr} K(\mathfrak{A})$. It is obvious that $a_f \in \mathfrak{N}$ and $a_f \in \operatorname{Extr}(Co_{G^*} f)^0 = \operatorname{Extr}(Co_G b)$. So, $Co_G a_f \subset Co_G b$, and hence $Co_{G \times G^*} a_f \otimes f \subset Co_{G \times G^*} b \otimes f$. By Definition 4 $a_f \otimes f < b \otimes f$. This relation means that these two elements are comparable.

If we prove that $a_f \otimes f > b \otimes f$, then it will mean that

$$Co_{G \times G^*} a_f \otimes f = Co_{G \times G^*} b \otimes f$$

This implies the equality $Co_G a_f = Co_g b$.

2. Really, we know already that $Co_G a_f \subset Co_G b$. Suppose the opposite: $Co_G a_f$ does not coincide with $Co_G b$ and $b \notin Co_G a_f$, it is equivalent to the existence of a functional $\varphi \in V'$ such that $\sup_{g \in G} \langle g a_f, \varphi \rangle \le 1$, but $\langle b, \varphi \rangle > 1$.

Consider elements of the space $V \otimes V'$ as operators from V' to V': for ψ, $\chi \in V'$, $x \in V$ $(x \otimes \psi)\chi = \langle x, \chi \rangle \psi$. Apply operators from $Co_{G \times G^*} b \otimes f = Co_{G \times G^*} a_f \otimes f$ to φ. We obtain

$$\operatorname{Conv}\{(gb \otimes h^* f)\varphi : g, h \in G\} = \operatorname{Conv}\{(g a_f \otimes h^* f)\varphi : g, h \in G\}$$

or

$$\operatorname{Conv}\{h^* f \langle gb, \varphi \rangle : g, h \in G\} = \operatorname{Conv}\{h^* f \langle g a_f, \varphi \rangle : g, h \in G\}$$

Since $\langle g a_f, \varphi \rangle \le 1$ for any $g \in G$, and $\langle b, \varphi \rangle > 1$ then the left part of the above equality is

$$\operatorname{Conv}\{h^* f \langle gb, \varphi \rangle : g, h \in G\} = \nu Co_{G^*} f, \quad \nu > 1$$

and the right part is

$$\operatorname{Conv}\{h^* f \langle g a_f, \varphi \rangle : g, h \in G\} = Co_{G^*} f.$$

Thus, $Co_{G^*} = \nu Co_{G^*}f$, $\nu > 1$ a contradiction. Hence $Co_G a_f = Co_G b$.

3. Thus, all what is remaining to be proved is the relation

$$a_f \otimes f > b \otimes f \qquad (*)$$

If $b \otimes f \in \operatorname{Extr} K(\mathfrak{A})$ then $(*)$ follows immediately from the maximality of the element $a_f \otimes f$ in $\operatorname{Extr} K(\mathfrak{A})$.

If $b \otimes f \notin \operatorname{Extr} K(\mathfrak{A})$ then $b \otimes f$ can be represented as a convex combination of the extreme points of the set $K(\mathfrak{A})$ (since $b \otimes f \in K(\mathfrak{A})$):

$$b \otimes f = \sum_n \lambda_n m_n, \text{ where } m_n \in \operatorname{Extr} K(\mathfrak{A}), \text{ and } \sum_n \lambda_n = 1.$$

Since $a_f \otimes f \in \operatorname{Extr} K(\mathfrak{A})$ and $Co_{G\times G^*} b \otimes f \subset K(\mathfrak{A})$ and $a_f \otimes f \in Co_{G\times G^*} b \otimes f$, then $a_f \otimes f \in \operatorname{Extr} Co_{G\times G^*} b \otimes f \subset \overline{orb}\, b \otimes f$. It means that there exist sequences of operators $g_i \in G$, $h_i^* \in G^*$ such that

$$a_f \otimes f = \lim(g_i \otimes h_i^*)\,(b \otimes f) = \lim(g_i \otimes h_i^*) \sum_n \lambda_n m_n$$

$$= \sum_n \lambda_n \lim_i (g_i \otimes h_i^*) m_n.$$

As $m_n \in \operatorname{Extr} K(\mathfrak{A}) \subset K(\mathfrak{A})$ and $K(\mathfrak{A})$ is compact one may assume, passing to a subsequence that

$$(g_i \otimes h_i^*) m_n \to c_n \in K(\mathfrak{A}) \qquad (**)$$

Hence, $a_f \otimes f = \sum_n \lambda_n c_n$. But $a_f \otimes f$ is an extreme point of $K(\mathfrak{A})$. This means that the last equality is possible only if $c_n = a_f \otimes f$, $\forall n$. Moreover it is known that $c_n \in \overline{orb}_{G\times G^*} m_n$ $(\forall n)$ (see $(**)$). Hence, $a_f \otimes f < m_n$ $(\forall n)$. So $a_f \otimes f$ and m_n are comparable extreme points of $K(\mathfrak{A})$. By the maximility of $a_f \otimes f$ in $\operatorname{Extr} K(\mathfrak{A})$ we have $a_f \otimes f > m_n$ $(\forall n)$. It follows that $Co_{G\times G^*} a_f \otimes f \supset Co_{G\times G^*} m_n$ $(\forall n)$. Therefore $m_n \in Co_{G\times G^*} a_f \otimes f$ and $b \otimes f = \sum_n \lambda_n m_n \in Co_{G\times G^*} a_f \otimes f$, hence $b \otimes f < a_f \otimes f$.

The proof of the statement (ii) is analogous. □

Lemma 4. *If the canonical collections are equivalent, then they coincide.*

Proof. Let the canonical collections $\{Co_G a\}_{a\in\mathfrak{N}}$ and $\{(Co_{G^*} f)^0\}_{f\in\mathfrak{N}'}$ be equivalent, i.e.,

$$\overline{\{\lambda Co_G a\}}^H_{a\in\mathfrak{N},\lambda\in\mathbb{R}} = \overline{\{\lambda (Co_{G^*} f)^0\}}^H_{f\in\mathfrak{N}',\lambda\in\mathbb{R}}$$

We shall prove that

$$\{Co_g a\}_{a\in\mathfrak{N}} = \{(Co_{G^*} f)^0\}_{f\in\mathfrak{N}'}$$

Let $f \in \mathfrak{N}'$ and $(Co_{G^*} f)^0 = \lim_H Co_G a_i$ $(a_i \in \mathfrak{N})$ (one can assume that all coefficients λ are included into sets). Then we assert that for some $b : (Co_{G^*} f)^0 = Co_G b$.

Really, if $a_i \in \mathfrak{N}$ then, by Definition 1 of the set $\mathfrak{N}$, for every a_i there exists $f_i \in V'$ such that $a_i \otimes f_i \in M(\mathfrak{A})$. But $M(\mathfrak{A}) \subset \operatorname{Extr} K(\mathfrak{A}) \subset K(\mathfrak{A})$. $\mathfrak{A}$ is compact and hence $K(\mathfrak{A})$ is compact. It follows that there exists a subsequence $a_{i_n} \otimes f_{i_n} \to b \otimes \varphi$. One may assume that $a_{i_n} \to b$. By the Lemma 1 we see that $Co_G a_{i_n} \to Co_G b$ and then $Co_G b = (Co_{G^*} f)^0$. Applying Lemma 3 (i), we obtain that there exists an element $a \in \mathfrak{N}$ such that $Co_G b = Co_G a$, and hence $(Co_{G^*} f)^0 = Co_G a$, $a \in \mathfrak{N}$. This means that $\{(Co_{G^*} f)^0\}_{f \in \mathfrak{N}'} \subset \{Co_G a\}_{a \in \mathfrak{N}}$.

The inverse inclusion can be proved by the analogous reasoning using Lemma 3 (ii), and Lemma 7, which we adduce below.

Definition 6. A collection of sets $\{\hat{U}_\beta\}$ is called the Hausdorff completion of the collection $\{U_\alpha\}$ of compact convex sets if

$$\{\hat{U}_\beta\} = \{\overline{\lambda U_\alpha}\}^H_{\lambda,\alpha}$$

It is obvious that the collections $\{U_\alpha\}$ and $\{\hat{U}_\beta\}$ are equivalent (by Definition 3). As it was noted above, if the collection $\{U_\alpha\}$ is sufficient then its completion $\{\hat{U}_\beta\}$ is sufficient too.

Lemma 5. *Let $\{U_\alpha\}$ be a collection of closed bounded G-symmetric sets. Then the following equality holds:*

$$\overline{\bigcup_\alpha S(U_\alpha)} = \bigcup_\beta S(\hat{U}_\beta).$$

To prove this lemma we need the following two lemmas.

Lemma 6. *Let A and A_i $(i = 1, 2, \ldots)$ be bounded closed convex neighborhoods of the origin and $A_i \xrightarrow{H} A$. Then there exists a number $\nu > 0$ such that for any i $A_i \supset \nu B$ (B is the unit ball in V).*

Proof. Since $A_i \xrightarrow{H} A$ then for any $\varepsilon > 0$ there exists N such that for $i > N$ the following inclusions hold:

$$A_i \subset A + \varepsilon B \text{ and } A \subset A_i + \varepsilon B \quad i = 1, 2, \ldots$$

Assume that the required ν does not exist. Take $\nu_i = \sup\{\nu : A_i \supset \nu B\}$, then $\inf \nu_i = 0$. Passing to a subsequence one can assume, that $\nu_i \to 0$. Take $\hat{\nu}_i = 3\nu_i$. It follows from the definition of ν_i that $\nu_i B \bigcap \partial A_i \neq \emptyset$. Let $x_i \in \nu_i B \bigcap \partial A_i$. Then $\hat{x}_i = 2x_i \in \frac{2}{3}\hat{\nu}_i B \setminus A_i$. Let $\{x \in V : \langle x, f_i \rangle = \langle x_i, f_i \rangle\}$ be a support hyperplane to A_i at the point x_i (and therefore it is a support hyperplane to $\nu_i B$ also), and $\| f \| = 1$. Then

$$\sup_{x \in A_i} \langle x, f_i \rangle \leq \frac{1}{2} \langle \hat{x}_i, f_i \rangle$$

Take $\nu_0 \neq 0$ such that $A \supset \nu_0 B$ (A is a neighborhood of the origin). Then we have (for $i > N$)

$$\frac{\nu_0}{\nu_i} x_i \in \nu_0 B \subset A \subset A_i + \varepsilon_i B \quad (\varepsilon_i \to 0), \text{ i.e.}$$

$$\frac{\nu_0}{\nu_i} x_i = y_i + \varepsilon_i a_i \quad (y_i \in A_i,\ a_i \in B).$$

Thus we first obtain

$$\frac{\nu_0}{\nu_i} \langle x_i, f_i \rangle = \nu_0 \sup_{x \in B} \langle x, f_i \rangle = \nu_0 \| f_i \| = \nu_0 \neq 0.$$

And we also see that

$$\frac{\nu_0}{\nu_i} \langle x_i, f_i \rangle = \langle y_i, f_i \rangle + \varepsilon_i \langle a_i, f_i \rangle \leq \sup_{x \in A_i} \langle x, f_i \rangle + \varepsilon_i \sup_{x \in B} \langle x, f_i \rangle$$

$$\leq \frac{1}{2} \langle \hat{x}_i, f_i \rangle + \varepsilon_i \| f_i \| \leq \frac{1}{2} \sup_{x \in \frac{2}{3} \hat{\nu}_i B} \langle x, f_i \rangle + \varepsilon_i$$

$$= \frac{1}{3} \hat{\nu}_i \sup_{x \in B} \langle x, f_i \rangle + \varepsilon_i = \nu_i \| f_i \| + \varepsilon_i = \nu_i + \varepsilon_i \to 0 \text{ if } \varepsilon_i \to 0.$$

This contradiction proves the lemma. □

Lemma 7. *Let A and A_i $(i = 1, 2, \ldots)$ be bounded closed convex neighborhoods of the origin and $A_i \xrightarrow{H} A$. Then $A_i^0 \xrightarrow{H} A^0$.*

Proof. 1. Take $f \in A^0$, i.e. $\langle x, f \rangle \leq 1$ for any $x \in A$. Since $A_i \xrightarrow{H} A$ then for any ε there exists N such that $i > N$

$$A \subset A_i + \varepsilon B \text{ and } A_i \subset A + \varepsilon B.$$

The inclusion $A_i \subset A + \varepsilon B$ implies that for any $y \in A_i$ there exist $x \in A$ and $a \in B$ such that $y = x + \varepsilon a$ and

$$\langle y, f \rangle = \langle x + \varepsilon a, f \rangle = \langle x, f \rangle + \varepsilon \langle a, f \rangle \leq 1 + \varepsilon \sup_{x \in B} \langle a, f \rangle = 1 + \varepsilon \| f \| .$$

Hence,

$$\langle y, \frac{1}{1 + \varepsilon \| f \|} f \rangle \leq 1$$

i.e.,

$$\frac{1}{1+\varepsilon \parallel f \parallel} f \in A_i^0$$

Then one can write the following representation for f:

$$f = \frac{1}{1+\varepsilon \parallel f \parallel} f + \frac{\varepsilon \parallel f \parallel}{1+\varepsilon \parallel f \parallel} f$$

where the first term belongs to A_i^0, and the second — to $\frac{\varepsilon\|f\|^2}{1+\varepsilon\|f\|} B^0$. Moreover, if $A \supset \nu B$, then $A^0 \subset \frac{B^0}{\nu}$ and hence $\parallel f \parallel \leq 1/\nu$, then $\frac{\varepsilon\|f\|^2}{1+\varepsilon\|f\|} \leq \varepsilon \parallel f \parallel^2 \leq \varepsilon/\nu^2$. Therefore $A^0 \subset A_i^0 + (\varepsilon/\nu^2)B$.

2. Acting in the same manner we show that $A_i^0 \subset A^0 + (\varepsilon/\nu^2)B$. Let $f \in A_i^0$, i.e. $\langle x, f\rangle \leq 1$ ($\forall x \in A_i$). Since $A \subset A_i + \varepsilon B$ then for any $y \in A$ we obtain as in p. 1

$$\langle y, f\rangle = \langle x + \varepsilon a, f\rangle = \langle x, f\rangle + \varepsilon\langle a, f\rangle \leq 1 + \varepsilon \parallel f \parallel,$$

where $x \in A_i$, $a \in B$. Hence $\langle y, \frac{f}{1+\varepsilon\|f\|}\rangle \leq 1$, i.e. $\frac{f}{1+\varepsilon\|f\|} \in A^0$. Writing the same presentation for f (as in p. 1) and recalling that by Lemma 6 there exists ν such that $A_i \supset \nu B$ (and hence $f \in A_i^0 \subset \frac{B^0}{\nu}$, i.e. $\parallel f \parallel < \frac{1}{\nu}$) we obtain $\frac{\varepsilon\|f\|^2}{1+\varepsilon\|f\|} \leq \varepsilon \parallel f \parallel^2 \leq \varepsilon/\nu^2$. Therefore $A_i^0 \subset A^0 + (\varepsilon^2/\nu^2)B^0$. Thus $A_i^0 \xrightarrow{H} A^0$ and the lemma is proved.

Proof of Lemma 5. So we need to prove that

$$\overline{\bigcup_\alpha S(U_\alpha)} = \bigcup_\beta S(\hat{U}_\beta).$$

First we prove the inclusion $\overline{\bigcup_\alpha S(U_\alpha)} \supset \bigcup_\beta S(\hat{U}_\beta)$. Take $a \otimes f \in S(\hat{U}_\beta)$. This means that $\mu a \in \hat{U}_\beta$ and $\frac{1}{\mu} f \in \hat{U}_\beta^0$. Let us contract (or dilate) the sets $\hat{U}_\beta$ so that $a \in \hat{U}_\beta$, then $f \in \hat{U}_\beta^0$. Since $\hat{U}_\beta = \lim_H U_\alpha$ (we assume that the coefficients λ are included in the sets U_α) then for any $\varepsilon > 0$ and α sufficiently large we get inclusions: $\hat{U}_\beta \subset U_\alpha + \varepsilon B$, $U_\alpha \subset \hat{U}_\beta + \varepsilon B$. Since U_α are bounded neighborhoods of the origin then by Lemma 7 $\hat{U}_\beta^0 = \lim_H U_\alpha^0$. Hence there exist sequences $\{a_\alpha\}$ and $\{f_\alpha\}$ such that $a_\alpha \in U_\alpha$, $a_\alpha \to a$, and $f_\alpha \in U_\alpha^0$, $f_\alpha \to f$. This means that $a_\alpha \otimes f_\alpha \in S(U_\alpha)$ and $a_\alpha \otimes f_\alpha \to a \otimes f$, and therefore $a \otimes f \in \overline{\bigcup_\alpha S(U_\alpha)}$.

Now we show the converse inclusion: $\overline{\bigcup_\alpha S(U_\alpha)} \subset \bigcup_\beta S(\hat{U}_\beta)$.

Let $a \otimes f \in \overline{\bigcup_\alpha S(U_\alpha)}$, i.e. $a \otimes f = \lim_i a_i \otimes f_i$, $a_i \otimes f_i \in S(U_i)$. This means that $\lambda_i a_i \in U_i$ and $\frac{1}{\lambda_i} f_i \in U_i^0$ for some λ_i. One may assume that the coefficients λ_i and $\frac{1}{\lambda_i}$ are included in a_i and f_i.

We show that the sequence of sets $\{U_i\}$ is uniformly bounded.

Really as $f_i \in U_i^0$ then $Co_{G^*} f_i \subset U_i^0$ (since U_i^0 is G^*-symmetric) and therefore $(Co_{G^*} f_i)^0 \supset U_i^{00} = U_i$.

Further, if $f_i \to f$, then $Co_{G^*} f_i \overset{H}{\to} Co_{G^*} f$ (see Lemma 1) and $(Co_{G^*} f_i)^0 \overset{H}{\to} (Co_{G^*} f)^0$ (see Lemma 7). Then for any $\varepsilon > 0$ and sufficiently large i we have $(Co_{G^*} f_i)^0 \subset (Co_{G^*} f)^0 + \varepsilon B$ (B is the unit ball in V') and $U_i \subset (Co_{G^*} f_i)^0 \subset (Co_{G^*} f)^0 + \varepsilon B$. So U_i are uniformly bounded.

Then by the Blaschke theorem ([3], p. 9) there exists a subsequence $\{U_{i_k}\}$ such that $U_{i_k} \overset{H}{\to} \hat{U}_{\beta_0}$ and hence $U_{i_k}^0 \overset{H}{\to} \hat{U}_{\beta_0}^0$, i.e. $\lim a_{i_k} = \lim a_i = a \in \hat{U}_\beta$ and

$$\lim f_{i_k} = \lim f_i = f \in \hat{U}_{\beta_0}^0$$

Hence, $a \otimes f \in S(\hat{U}_{\beta_0})$ and the required inclusion is proved. This completes the proof of Lemma 5. □

Lemma 8. *Let $a \otimes f \in S(U)$, then there exists λ such that*

$$\lambda Co_G a \subset U \subset \lambda (Co_{G^*} f)^0.$$

Proof. As $a \otimes f \in S(U)$ then $\lambda a \in U$ and $\frac{1}{\lambda} f \in U^0$ for some λ. From the G-symmetricity of the set U it follows that

$$\lambda Co_G a \subset U \text{ and } \frac{1}{\lambda} Co_{G^*} f \subset U^0.$$

Hence $\lambda Co_G a \subset U \subset \lambda (Co_{G^*} f)^0$. □

Proof of Theorem 5. Suppose canonical collections do not coincide then by Lemma 4 they are not equivalent, then there exist at least two unequivalent minimal sufficient collections. So the smallest sufficient collection does not exist.

Conversely, let the canonical collections coincide.

$$\{Co_G a\}_{a \in \mathfrak{N}} = \{(Co_{G^*} f)^0\}_{f \in \mathfrak{N}'}.$$

We prove that this is exactly the smallest sufficient collection.

Let $\{U_\alpha\}$ be any other sufficient collection. Replace it by its completion $\{\hat{U}_\beta\}$. We shall show that the canonical collections are included into the collection $\{\hat{U}_\beta\}$. And this will prove the theorem.

Since $\{U_\alpha\}$ is a sufficient collection then by Theorem 1

$$K(\mathfrak{A}) = \operatorname{conv} \bigcup_\alpha S(U_\alpha).$$

Hence $\operatorname{Extr} K(\mathfrak{A}) \subset \overline{\bigcup_\alpha S(U_\alpha)}$. By Lemma 5, $\overline{\bigcup_\alpha S(U_\alpha)} = \bigcup_\beta S(\hat{U}_\beta)$, hence $\operatorname{Extr} K(\mathfrak{A}) \subset \bigcup_\beta S(\hat{U}_\beta)$.

If $a \otimes f \in M(\mathfrak{A})$ ($M(\mathfrak{A}) \subset \operatorname{Extr} K(\mathfrak{A})$!) then $a \otimes f \in S(U_{\beta_0})$ for some β_0 and then by Lemma 7 $\lambda Co_G a \subset U_{\beta_0} \subset \lambda(Co_{G^*} f)^0$. Since the canonical collections coincide we have $Co_G a = (Co_{G^*} f)^0$ and the collection $\{\hat{U}_\beta\}$ contains the canonical collection (see Lemma 3). □

References

1. Bourbaki, N.: Espaces Vectoriels Topologiques, Herman, Paris, 1968.
2. Calderon, A.P.: Spaces between L^1 and L^∞ and the theorem of Marcinkiewicz, Studia Math., 26(1966), 229–273.
3. Leichtweiss, K.: Konvexe Mengen, VEB Deutscher Verlag der Wissenschaften, Berlin, (1980).
4. Marshall, A.W., Olkin, I.: Inequalities: Theory of Majorization and its Applications, Mathematics in Science and Engineering, 143(1979).
5. Mityagin, B.S.: An interpolation theorem for modular spaces, Matem. Sbornik, v. 66, 4(1965), 473–482 (Russian).
6. Veselova, L.V., Zobin, N.M.: A general theory of interpolation and duality, in «Constructive theory of functions and funct. analysis» (edited by A.N. Sherstnev), Kazan University Press, (1990), 14–29 (Russian).
7. Zobin, N.M.: On extreme mixed norms, Uspekhi Mat. Nauk, 39(1984), 157–158 (Russian).
8. Zobin, N.M., Zobina, V.G.: Interpolation in spaces possessing the prescribed symmetries, Funct. Anal. Pril., v. 12, 4(1978), 85–86 (Russian).
9. Zobin, N.M., Zobina, V.G.: Interpolation in the spaces with prescribed symmetries. Finite dimension, J. Izvestiya vusov, Matematika, 4(1981), 20–28 (Russian).
10. Zobin, N.M., Zobina, V.G.: Minimal sufficient collections for semigroups of operators, J. Izvestiya vusov, Matematika, 11(1989), 31–35 (Russian).
11. Zobin, N.M., Zobina, V.G.: Duality in operator spaces and problems of interpolation of operators, Pitman Research Notes in Math., Longman Sci. & Tech., London, v. 257 (1992), 123–144.
12. Zobin, N., Zobina, V.: Coxeter groups and interpolation of operators, Preprint Max-Planck-Inst. für Math. – Bonn, MPI/93-39 (1993), 1–44.
13. Zobina, V.G.: Interpolation in spaces with prescribed symmetries and the uniqueness of sufficient collections, Soobshenya AN Gruzin. SSR, v. 93, 2(1979), 301–303 (Russian).

14. Zobina, V.G.: Duality in interpolation of operators, Soobshęnya AN Gruzin. SSR, V. 95, 1(1979), 45–48 (Russian).
15. Zobina, V.G.: Interpolation in spaces with prescribed symmetries. Group $B_2 \otimes B_2$, preprint VINITI, 2.08.78, 2828-78(1978), (Russian).
16. Zobina, V.G.: Interpolation of operators in spaces with a Coxeter group as the group of symmetries, preprint VINITI 16.02.82, 3318-82(1982), (Russian).
17. Zobina, V.G.: Sufficient collections for semigroups of operators, J. Izvestiya vusov, Matematika, 1(1989), 33–35 (Russian).

N. Zobin
Research Institute of Afula &
Department of Mathematics and CS
University of Haifa
Haifa, 31905
Israel

V. Zobina
Department of Mathematics
Technion – I.I.T.
Haifa, 32000
Israel

1991 Mathematics Subject Classification
Primary 46M35, Secondary 52A21

Titles previously published in the series

OPERATOR THEORY: ADVANCES AND APPLICATIONS

BIRKHÄUSER VERLAG

1. **H. Bart, I. Gohberg, M.A. Kaashoek:** Minimal Factorization of Matrix and Operator Functions, 1979, (3-7643-1139-8)
2. **C. Apostol, R.G. Douglas, B.Sz.-Nagy, D. Voiculescu, Gr. Arsene** (Eds.): Topics in Modern Operator Theory, 1981, (3-7643-1244-0)
3. **K. Clancey, I. Gohberg:** Factorization of Matrix Functions and Singular Integral Operators, 1981, (3-7643-1297-1)
4. **I. Gohberg** (Ed.): Toeplitz Centennial, 1982, (3-7643-1333-1)
5. **H.G. Kaper, C.G. Lekkerkerker, J. Hejtmanek:** Spectral Methods in Linear Transport Theory, 1982, (3-7643-1372-2)
6. **C. Apostol, R.G. Douglas, B. Sz-Nagy, D. Voiculescu, Gr. Arsene** (Eds.): Invariant Subspaces and Other Topics, 1982, (3-7643-1360-9)
7. **M.G. Krein:** Topics in Differential and Integral Equations and Operator Theory, 1983, (3-7643-1517-2)
8. **I. Gohberg, P. Lancaster, L. Rodman:** Matrices and Indefinite Scalar Products, 1983, (3-7643-1527-X)
9. **H. Baumgärtel, M. Wollenberg**: Mathematical Scattering Theory, 1983, (3-7643-1519-9)
10. **D. Xia:** Spectral Theory of Hyponormal Operators, 1983, (3-7643-1541-5)
11. **C. Apostol, C.M. Pearcy, B. Sz.-Nagy, D. Voiculescu, Gr. Arsene** (Eds.): Dilation Theory, Toeplitz Operators and Other Topics, 1983, (3-7643-1516-4)
12. **H. Dym, I. Gohberg** (Eds.): Topics in Operator Theory Systems and Networks, 1984, (3-7643-1550-4)
13. **G. Heinig, K. Rost:** Algebraic Methods for Toeplitz-like Matrices and Operators, 1984, (3-7643-1643-8)
14. **H. Helson, B. Sz.-Nagy, F.-H. Vasilescu, D.Voiculescu, Gr. Arsene** (Eds.): Spectral Theory of Linear Operators and Related Topics, 1984, (3-7643-1642-X)
15. **H. Baumgärtel:** Analytic Perturbation Theory for Matrices and Operators, 1984, (3-7643-1664-0)
16. **H. König:** Eigenvalue Distribution of Compact Operators, 1986, (3-7643-1755-8)
17. **R.G. Douglas, C.M. Pearcy, B. Sz.-Nagy, F.-H. Vasilescu, D. Voiculescu, Gr. Arsene** (Eds.): Advances in Invariant Subspaces and Other Results of Operator Theory, 1986, (3-7643-1763-9)

18 **I. Gohberg** (Ed) I Schur Methods in Operator Theory and Signal Processing, 1986, (3-7643-1776-0)

19 **H. Bart, I. Gohberg, M.A. Kaashoek** (Eds) Operator Theory and Systems, 1986, (3-7643-1783-3)

20 **D. Amir:** Isometric characterization of Inner Product Spaces, 1986, (3-7643-1774-4)

21 **I. Gohberg, M.A. Kaashoek** (Eds) Constructive Methods of Wiener-Hopf Factorization, 1986, (3-7643-1826-0)

22 **V.A. Marchenko:** Sturm-Liouville Operators and Applications, 1986, (3-7643-1794-9)

23 **W. Greenberg, C. van der Mee, V. Protopopescu:** Boundary Value Problems in Abstract Kinetic Theory, 1987, (3-7643-1765-5)

24 **H. Helson, B. Sz.-Nagy, F.-H. Vasilescu, D. Voiculescu, Gr. Arsene** (Eds) Operators in Indefinite Metric Spaces, Scattering Theory and Other Topics, 1987, (3-7643-1843-0)

25 **G.S. Litvinchuk, I.M. Spitkovskii:** Factorization of Measurable Matrix Functions, 1987, (3-7643-1883-X)

26 **N.Y. Krupnik:** Banach Algebras with Symbol and Singular Integral Operators, 1987, (3-7643-1836-8)

27 **A. Bultheel:** Laurent Series and their Pade Approximation, 1987, (3-7643-1940-2)

28 **H. Helson, C.M. Pearcy, F.-H. Vasilescu, D. Voiculescu, Gr. Arsene** (Eds) Special Classes of Linear Operators and Other Topics, 1988, (3-7643-1970-4)

29 **I. Gohberg** (Ed) Topics in Operator Theory and Interpolation, 1988, (3-7634-1960-7)

30 **Yu.I. Lyubich:** Introduction to the Theory of Banach Representations of Groups, 1988, (3-7643-2207-1)

31 **E.M. Polishchuk:** Continual Means and Boundary Value Problems in Function Spaces, 1988, (3-7643-2217-9)

32 **I. Gohberg** (Ed) Topics in Operator Theory Constantin Apostol Memorial Issue, 1988, (3-7643-2232-2)

33 **I. Gohberg** (Ed) Topics in Interplation Theory of Rational Matrix-Valued Functions, 1988, (3-7643-2233-0)

34 **I. Gohberg** (Ed) Orthogonal Matrix-Valued Polynomials and Applications, 1988, (3-7643-2242-X)

35 **I. Gohberg, J.W. Helton, L. Rodman** (Eds) Contributions to Operator Theory and its Applications, 1988, (3-7643-2221-7)

36 **G.R. Belitskii, Yu.I. Lyubich:** Matrix Norms and their Applications, 1988, (3-7643-2220-9)

37 **K. Schmüdgen:** Unbounded Operator Algebras and Representation Theory, 1990, (3-7643-2321-3)

38 **L. Rodman:** An Introduction to Operator Polynomials, 1989, (3-7643-2324-8)

39 **M. Martin, M. Putinar:** Lectures on Hyponormal Operators, 1989, (3-7643-2329-9)

40 **H. Dym, S. Goldberg, P. Lancaster, M.A. Kaashoek** (Eds) The Gohberg Anniversary Collection, Volume I, 1989, (3-7643-2307-8)

41 **H. Dym, S. Goldberg, P. Lancaster, M.A. Kaashoek** (Eds.): The Gohberg Anniversary Collection, Volume II, 1989, (3-7643-2308-6)

42 **N.K. Nikolskii** (Ed.): Toeplitz Operators and Spectral Function Theory, 1989, (3-7643-2344-2)

43 **H. Helson, B. Sz.-Nagy, F.-H. Vasilescu, Gr. Arsene** (Eds.): Linear Operators in Function Spaces, 1990, (3-7643-2343-4)

44 **C. Foias, A. Frazho:** The Commutant Lifting Approach to Interpolation Problems, 1990, (3-7643-2461-9)

45 **J.A. Ball, I. Gohberg, L. Rodman:** Interpolation of Rational Matrix Functions, 1990, (3-7643-2476-7)

46 **P. Exner, H. Neidhardt** (Eds.): Order, Disorder and Chaos in Quantum Systems, 1990, (3-7643-2492-9)

47 **I. Gohberg** (Ed.): Extension and Interpolation of Linear Operators and Matrix Functions, 1990, (3-7643-2530-5)

48 **L. de Branges, I. Gohberg, J. Rovnyak** (Eds.): Topics in Operator Theory. Ernst D. Hellinger Memorial Volume, 1990, (3-7643-2532-1)

49 **I. Gohberg, S. Goldberg, M.A. Kaashoek:** Classes of Linear Operators, Volume I, 1990, (3-7643-2531-3)

50 **H. Bart, I. Gohberg, M.A. Kaashoek** (Eds.): Topics in Matrix and Operator Theory, 1991, (3-7643-2570-4)

51 **W. Greenberg, J. Polewczak** (Eds.): Modern Mathematical Methods in Transport Theory, 1991, (3-7643-2571-2)

52 **S. Prössdorf, B. Silbermann:** Numerical Analysis for Integral and Related Operator Equations, 1991, (3-7643-2620-4)

53 **I. Gohberg, N. Krupnik:** One-Dimensional Linear Singular Integral Equations, Volume I, Introduction, 1992, (3-7643-2584-4)

54 **I. Gohberg, N. Krupnik:** One-Dimensional Linear Singular Integral Equations, Volume II, General Theory and Applications, 1992, (3-7643-2796-0)

55 **R.R. Akhmerov, M.I. Kamenskii, A.S. Potapov, A.E. Rodkina, B.N. Sadovskii:** Measures of Noncompactness and Condensing Operators, 1992, (3-7643-2716-2)

56 **I. Gohberg** (Ed.): Time-Variant Systems and Interpolation, 1992, (3-7643-2738-3)

57 **M. Demuth, B. Gramsch, B.W. Schulze** (Eds.): Operator Calculus and Spectral Theory, 1992, (3-7643-2792-8)

58 **I. Gohberg** (Ed.): Continuous and Discrete Fourier Transforms, Extension Problems and Wiener-Hopf Equations, 1992, (3-7643-2809-6)

59 **T. Ando, I. Gohberg** (Eds.): Operator Theory and Complex Analysis, 1992, (3-7643-2824-X)

60 **P.A. Kuchment:** Floquet Theory for Partial Differential Equations, 1993, (3-7643-2901-7)

61 **A. Gheondea, D. Timotin, F.-H. Vasilescu** (Eds.): Operator Extensions, Interpolation of Functions and Related Topics, 1993, (3-7643-2902-5)

62 **T. Furuta, I. Gohberg, T. Nakazi** (Eds) Contributions to Operator Theory and its Applications The Tsuyoshi Ando Anniversary Volume, 1993, (3-7643-2928-9)

63 **I. Gohberg, S. Goldberg, M.A. Kaashoek:** Classes of Linear Operators, Volume 2, 1993, (3-7643-2944-0)

64 **I. Gohberg** (Ed) New Aspects in Interpolation and Completion Theories, 1993, (3-7643-2948-3)

65 **M.M. Djrbashian:** Harmonic Analysis and Boundary Value Problems in the Complex Domain, 1993, (3-7643-2855-X)

66 **V. Khatskevich, D. Shoiykhet:** Differentiable Operators and Nonlinear Equations, 1993, (3-7643-2929-7)

67 **N.V. Govorov †:** Riemann's Boundary Problem with Infinite Index, 1994, (3-7643-2999-8)

68 **A. Halanay, V. Ionescu:** Time-Varying Discrete Linear Systems Input-Output Operators Riccati Equations Disturbance Attenuation, 1994, (3-7643-5012-1)

69 **A. Ashyralyev, P.E. Sobolevskii:** Well-Posedness of Parabolic Difference Equations, 1994, (3-7643-5024-5)

70 **M. Demuth, P. Exner, G. Neidhardt, V. Zagrebnov** (Eds) Mathematical Results in Quantum Mechanics International Conference in Blossin (Germany), May 17-21, 1993, 1994, (3-7643-5025-3)

71 **E.L. Basor, I. Gohberg** (Eds) Toeplitz Operators and Related Topics The Harold Widom Anniversary Volume Workshop on Toeplitz and Wiener-Hopf Operators, Santa Cruz, California, September 20–22, 1992, 1994 (3-7643-5068-7)

72 **I. Gohberg, L.A. Sakhnovich** (Eds) Matrix and Operator Valued Functions The Vladimir Petrovich Potapov Memorial Volume, (3-7643-5091-1)

73 **A. Feintuch, I. Gohberg** (Eds) Nonselfadjoint Operators and Related Topics Workshop on Operator Theory and Its Applications, Beersheva, February 24–28, 1994, (3-7643-5097-0)

Monographs in Mathematics

The foundations of this outstanding book series were laid in 1944. Until the end of the 1970s, a total of 77 volumes appeared, including works of such distinguished mathematicians as Carathéodory, Nevanlinna and Shafarevich, to name a few. The series came to its name and present appearance in the 1980s. According to its well-established tradition, only monographs of excellent quality will be published in this collection. Comprehensive, in-depth treatments of areas of current interest are presented to a readership ranging from graduate students to professional mathematicians. Concrete examples and applications both within and beyond the immediate domain of mathematics illustrate the import and consequences of the theory under discussion.

Published in the series since 1983

Volume 78 **H. Triebel, Theory of Function Spaces I**
1983, 284 pages, hardcover, ISBN 3-7643-1381-1.

Volume 79 **G.M. Henkin /J. Leiterer, Theory of Functions on Complex Manifolds**
1984, 228 pages, hardcover, ISBN 3-7643-1477-X.

Volume 80 **E. Giusti, Minimal Surfaces and Functions of Bounded Variation**
1984, 240 pages, hardcover, ISBN 3-7643-3153-4.

Volume 81 **R.J. Zimmer, Ergodic Theory and Semisimple Groups**
1984, 210 pages, hardcover, ISBN 3-7643-3184-4.

Volume 82 **V.I. Arnold / S.M. Gusein-Zade / A.N. Varchenko, Singularities of Differentiable Maps – Vol. I**
1985, 392 pages, hardcover, ISBN 3-7643-3187-9.

Volume 83 **V.I. Arnold / S.M. Gusein-Zade / A.N. Varchenko, Singularities of Differentiable Maps – Vol. II**
1988, 500 pages, hardcover, ISBN 3-7643-3185-2.

Volume 84 **H. Triebel, Theory of Function Spaces II**
1992, 380 pages, hardcover, ISBN 3-7643-2639-5.

Volume 85 **K.R. Parthasarathy, An Introduction to Quantum Stochastic Calculus**
1992, 300 pages, hardcover, ISBN 3-7643-2697-2.

Volume 86 **M. Nagasawa, Schrödinger Equations and Diffusion Theory**
1993, 332 pages, hardcover, ISBN 3-7643-2875-4.

Volume 87 **J. Prüss, Evolutionary Integral Equations and Applications**
1993, 392 pages, hardcover, ISBN 3-7643-2876-2.

Volume 88 **R.W. Bruggeman, Families of Automorphic Forms**
1994, 328 pages, hardcover, ISBN 3-7643-5046-6.